Wildfire Risks and Management

Wildfire Risks and Management

Edited by Harry Jones

Syrawood Publishing House
New York

Published by Syrawood Publishing House,
750 Third Avenue, 9th Floor,
New York, NY 10017, USA
www.syrawoodpublishinghouse.com

Wildfire Risks and Management
Edited by Harry Jones

International Standard Book Number: 978-1-68286-417-3 (Hardback)

Cataloging-in-publication Data

Wildfire risks and management / edited by Harry Jones.
p. cm.
Includes bibliographical references and index.
ISBN 978-1-68286-417-3
1. Wildfire risk. 2. Wildfire risk--Management. 3. Wildfires--Prevention and control. 4. Wildfires--Climatic factors.
I. Jones, Harry.
SD421 .W55 2017
634.9618--dc23

Printed in China.

TABLE OF CONTENTS

PREFACE

Wildfire risks include human activity and natural developments around forest areas that cause fire hazards. Wildfire risks and management deals with the constant threat of wildfires and its risk for wildlife, the environment and the community at large. Strategies for minimizing such damage include methods of early detection, suppression, fire retardation and controlled wildfires. The various advancements in wildfire risks assessment and management are glanced at and their applications as well as ramifications are looked at in detail. This book is a compilation of chapters that discuss the most vital concepts and emerging trends in this field. The extensive content of this book provides the readers with a thorough understanding of the subject. This book will provide guidance to specialists and students researching in grassland management, sustainable forestry, wildlife management, wildfire management, resource management and risk management.

This book is the end result of constructive efforts and intensive research done by experts in this field. The aim of this book is to enlighten the readers with recent information in this area of research. The information provided in this profound book would serve as a valuable reference to students and researchers in this field.

At the end, I would like to thank all the authors for devoting their precious time and providing their valuable contributions to this book. I would also like to express my gratitude to my fellow colleagues who encouraged me throughout the process.

Editor

Evidence of Climate-Induced Range Contractions in Bull Trout *Salvelinus confluentus* in a Rocky Mountain Watershed, U.S.A.

Lisa A. Eby[1], Olga Helmy[1], Lisa M. Holsinger[2], Michael K. Young[3]*

1 University of Montana, Wildlife Biology Program and Department of Ecosystem and Conservation Sciences, College of Forestry and Conservation, Missoula, Montana, United States of America, **2** U.S. Department of Agriculture Forest Service, Rocky Mountain Research Station, Fire Sciences Laboratory, Missoula, Montana, United States of America, **3** U.S. Department of Agriculture Forest Service, Rocky Mountain Research Station, Forestry Sciences Laboratory, Missoula, Montana, United States of America

Abstract

Many freshwater fish species are considered vulnerable to stream temperature warming associated with climate change because they are ectothermic, yet there are surprisingly few studies documenting changes in distributions. Streams and rivers in the U.S. Rocky Mountains have been warming for several decades. At the same time these systems have been experiencing an increase in the severity and frequency of wildfires, which often results in habitat changes including increased water temperatures. We resampled 74 sites across a Rocky Mountain watershed 17 to 20 years after initial samples to determine whether there were trends in bull trout occurrence associated with temperature, wildfire, or other habitat variables. We found that site abandonment probabilities (0.36) were significantly higher than colonization probabilities (0.13), which indicated a reduction in the number of occupied sites. Site abandonment probabilities were greater at low elevations with warm temperatures. Other covariates, such as the presence of wildfire, nonnative brook trout, proximity to areas with many adults, and various stream habitat descriptors, were not associated with changes in probability of occupancy. Higher abandonment probabilities at low elevation for bull trout provide initial evidence validating the predictions made by bioclimatic models that bull trout populations will retreat to higher, cooler thermal refuges as water temperatures increase. The geographic breadth of these declines across the region is unknown but the approach of revisiting historical sites using an occupancy framework provides a useful template for additional assessments.

Editor: Marcus Sheaves, James Cook University, Australia

Funding: This project was primarily funded by the Joint Fire Sciences Project (JFSP-09-1002-9). MT NSF EPSCoR (EPS-0701906) and the Longwell Foundation also provided fellowship support to O.H. The funders had no role in study design, data collection and analysis, decision to publish, or preparation of the manuscript.

Competing Interests: The authors have declared that no competing interests exist.

* E-mail: mkyoung@fs.fed.us

Introduction

Freshwater ecosystems host a disproportionately large amount of the Earth's biodiversity, including many fish species of economic and cultural value [1], yet account for an outsize share of globally imperiled species [2], [3]. Aquatic organisms in freshwater ecosystems are expected to be particularly sensitive to climate shifts because most are ectothermic and have a relatively narrow thermal range for growth and survival [4], [5]. Bioclimatic models accounting for climate change predict an array of phenological changes and range shifts in freshwater aquatic species [6], [7]. Alteration in the timing of life history events has been relatively widely observed [8], [9]. In contrast, confirmation of predictions that stenothermic cold-water fishes should be undergoing distributional shifts to cooler, high-elevation refuges has been elusive [10], particularly in North America [11].

The northern Rocky Mountains, U.S.A. is undergoing climate-mediated shifts e.g., reduced annual snowpack, earlier annual peak snowmelt, and winter precipitation switching from snow to rain, that are contributing to changes in hydrologic and thermal regimes [12], [13], [14]. Summer water temperatures have increased up to 0.3°C/decade [15] and summer base flows are declining [16]. The most stenothermic coldwater fish in the northern Rocky Mountains is the bull trout (*Salvelinus confluentus*) [17], which is listed as threatened under the U.S. Endangered Species Act. Juvenile bull trout rear in cold stream reaches across the upper elevations of river networks, with the upstream extent of individual populations limited by channel size and gradient [18]. Regional temperature increases associated with climate change have led to dire predictions about the persistence of this species in the U.S. [17], [19], but there is little empirical evidence of climate-related shifts. An additional complexity is attributing changes in occupancy directly to climate change [20]. Wildfire is a frequent natural disturbance that can lead to the decades-long elevation of summer stream temperatures because of the loss of shade from riparian vegetation [21], [22]. A recent increase in fire severity and size in the western U.S. has been linked to climate change [23]. Similarly, climate projections favor headwater invasions by less thermally restricted nonnative species such as brook trout (*S. fontinalis*) [24] that can reduce bull trout occupancy [25].

In this study, we repeated a late 20th-century inventory of bull trout occupancy within a river network that encompasses a broad temperature and elevation gradient. Our objective was to compare site-scale abandonment and colonization probabilities to determine whether they differed and if they were associated with landscape features such as wildfire occurrence and severity, habitat attributes including gradient, width, large wood, temperature, and elevation, and biotic variables such as proximity to strongholds of migratory adults and brook trout presence. If the range of bull trout contracted in response to climate change, we expected site-level abandonment probabilities to be greatest at the warmest sites and to exceed those for colonization.

Materials and Methods

Study area

The East Fork Bitterroot River basin is a 1,055-km^2 watershed in west-central Montana, U.S.A. (Figure 1). The basin is mainly a forested landscape with lower elevations dominated by ponderosa pine (*Pinus ponderosa*) and Douglas-fir (*Pseudotsuga menziesii*) and higher elevations by lodgepole pine (*Pinus contorta* var. *latifolia*), subalpine fir *(Abies lasiocarpa)*, and Engelmann spruce (*Picea engelmannii*). The watershed is a temperate, snowmelt-dominated system with a range of elevations from 1,220 to 2,887 m. In 2000, wildfires burned 52.0% (29.2% at moderate to high severity) of the basin and 3.8% (2.5% at moderate to high severity) in 2007. Maximum summer stream temperatures in reaches where moderate- to high-severity fires burned in riparian stands remain elevated 1.4 to 2.2°C above those from reaches adjacent to unburned stands [22]. Over a comparable interval (1994–2007) maximum summer stream temperatures at some unburned sites also increased 1.9–2.6°C [22], which is higher than the July/August 0.24°C/decade increase described for the Greater Yellowstone area [26] and 0.22°C/decade increase across the U.S. Northwest [15]. Average daily maximum summer water temperature have been increasing in recent years in the main-stem East Fork Bitterroot River, as have summer air temperatures at the weather station nearest our study area (Sula, MT; Figure 1).

The East Fork Bitterroot River is a core conservation area for bull trout [27]. This watershed consists primarily of public land administered by the U.S. Forest Service (USFS) and has no known barriers to fish movement within our study area [27]. Bull trout in this basin exhibit partial migration, with resident and migratory individuals in most spawning tributaries. Resident individuals spend their entire lives within their natal stream or tributary, moving only short distances (e.g., <2 km), whereas migratory individuals spawn in headwater tributaries but migrate to the river to forage [28], [29]. Other native fish in the basin include westslope cutthroat trout *Oncorhynchus clarkii lewisi*, slimy sculpin (*Cottus cognatus*), longnose dace (*Rhinichthys cataractae*), mountain whitefish (*Prosopium williamsoni*), and longnose suckers (*Catostomus catostomus*). Non-native brown trout (*Salmo trutta*), brook trout, and rainbow trout (*O. mykiss*) are present throughout the main stem and in several tributaries in the basin.

A number of factors influence bull trout habitat occupancy in its U.S. range. As noted earlier, site occupancy of juvenile or resident bull trout is strongly correlated with maximum temperatures [21], [30]. However large individuals with migratory life histories are not restricted to cold thermal environments and can move through much warmer waters (e.g., 21°C 7-day average daily maximum temperature; [31]) before reaching spawning areas. Bull trout are also associated with relatively large patches of connected, complex habitat [17], [32]. In the Bitterroot River basin, the probability of bull trout presence in stream reaches was positively correlated with large wood, stream width, and relative abundance of main-stem bull trout at a tributary mouth, and negatively correlated with stream gradient and the presence of brook trout [33]. Bull trout in this basin tend to occupy streams to their headwaters (until stream width $<$ 2 m; [33]), thus increases in occupancy are only likely at downstream locations or in previously unoccupied streams.

Data collection

First- through 4th-order streams were sampled between 1992 and 1995 to determine bull trout occupancy patterns in the Bitterroot River basin [33]. In this sampling, three 500-m study sites were equally spaced over the estimated length of suitable habitat in each tributary. Between 2009 and 2011, we revisited 74 sites on streams sampled previously [33] within the connected portions of the East Fork Bitterroot River basin to examine whether bull trout occupancy had changed. We relocated the sites and replicated the sampling methods of the earlier study. Fish were collected with a single-pass survey using a backpack electrofishing unit during the summer low-flow period. Care was taken to electroshock slowly and inclusively through all areas of cover. Our resampling was confined to 1st- through 3rd-order streams that were small enough to effectively sample with these techniques. As in the previous study, we divided each site into five sequential 100-m sections for sampling. All fish were identified to species, counted, and measured (total length). As in the earlier study, we avoided basing bull trout occupancy on the ephemeral presence of a large, migratory adult; presence in a section was defined by the capture of $\geq$ 2 bull trout, at least one of which was less than 250 mm (and thus likely to be a juvenile or small resident adult of local origin) [21]. All sampling was performed in accordance with guidelines specified under scientific collection permits issued to Lisa Eby by Montana Fish, Wildlife and Parks and under the protocols approved in the University of Montana Institutional Animal Care and Use Committee AUP 031-09.

We used field measures and GIS spatial data layers to assess covariates potentially related to bull trout occupancy. We counted large wood in the first 100-m section of every 500-m site. We estimated bankfull width (m) for the section by measuring it at three representative locations. Elevation (m) was noted in the field from the GPS unit (Garmin 60CSx) and validated from the 30-m cell size National Elevation Dataset [34]. Gradient at each site was derived from this dataset using TauDEM software [35]. We used estimates of bull trout abundance in the main-stem East Fork Bitterroot River (http://fwp.mt.gov/fishing/mFish/newSearch.html; accessed 8/2/13) to assess the proximity of sites to main-stem locations where adults were commonly captured. We categorized sites into three groups: (0) bull trout common in the main-stem East Fork Bitterroot River at the tributary mouth; (1) bull trout common in the main-stem within 2.5 km of the tributary mouth; and (2) bull trout common in the main-stem $>$ 2.5 km from the tributary mouth. Because we did not have a single year with temperature data at every site and annual variation in temperature is large, we used a locally calibrated stream temperature model that allowed standardized representation of relative temperatures among sites (Text S1).

We obtained fire severity GIS layers from the Bitterroot National Forest (Hamilton, MT). Burn severity is used to describe the amount of fire-related change including overstory vegetation mortality, soil heating, and fuel consumption [36], [37]. Our sites within moderate- and high-severity riparian burns had the majority of the riparian area (and watershed) burned, thus fire severity and proportion of site burned were positively related. We grouped burn severity into two categories and at each site we indicated whether the riparian area experienced no-to-low severity burns or medium-to-high severity burns.

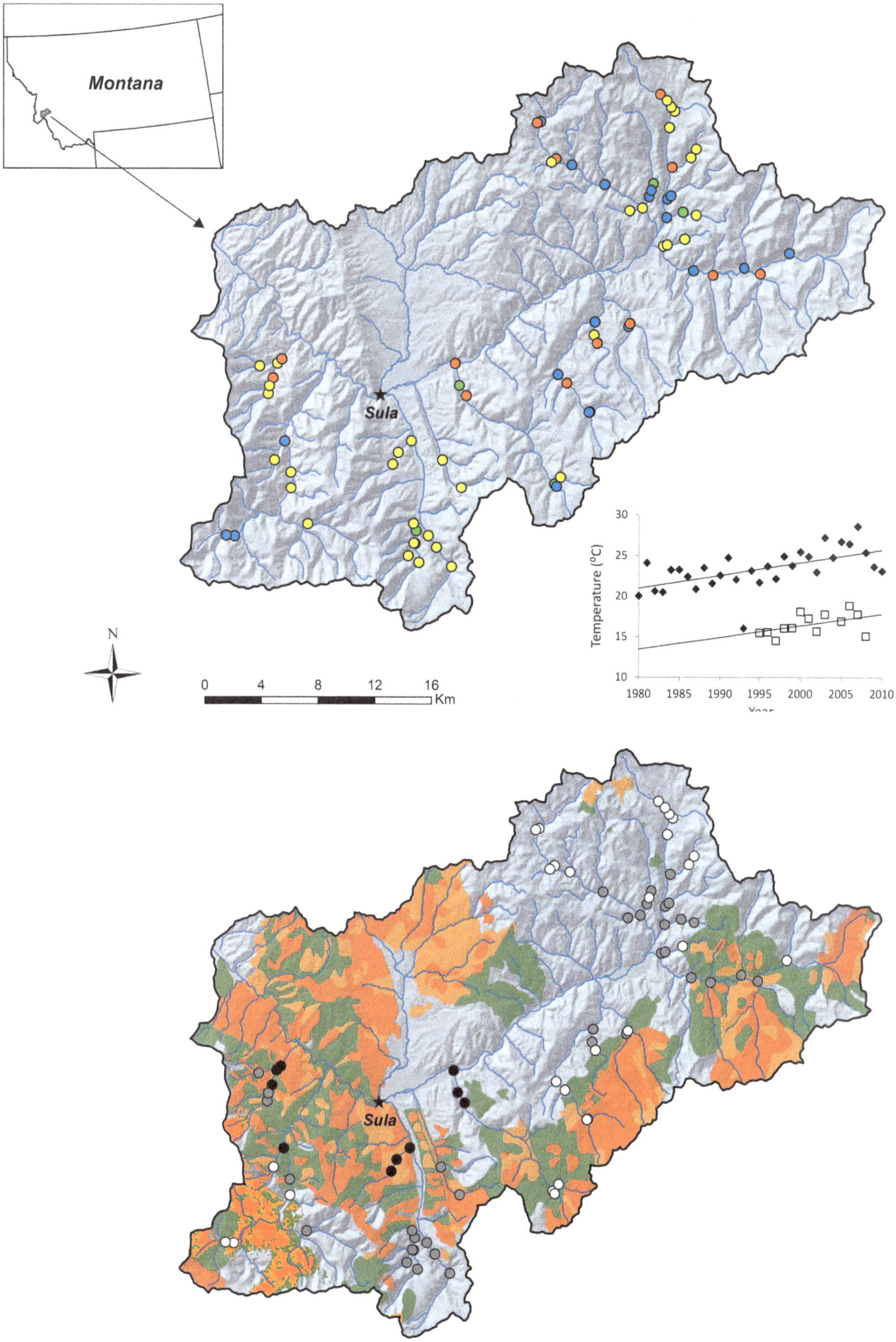
Montana
Sula
N
0 4 8 12 16
Km
Temperature (°C)
30
25
20
15
10
1980 1985 1990 1995 2000 2005 2010
Sula

Figure 1. Study area. Sampling locations (500-m sites that were initially visited in 1992–1995 and resampled in 2009–2011) in the East Fork Bitterroot River watershed. Top panel: patterns in occupancy (yellow, not occupied in either period; blue, occupied in both periods; red, occupied in first but not second period; green, occupied in second but not first period). These reflect observed patterns not corrected for probability of detection. Water and air temperature patterns within the East Fork Bitterroot River basin are inset. Black diamond symbols are average daily summertime (July and August air temperature recorded over the study period at the closest weather station at Sula, MT ($y = 0.1567x - 289.27$, $R^2 = 0.34$, $p = 0.0006$). Hollow squares are average daily maximum water temperatures over the summer season (July 15 to September 30) from the East Fork Bitterroot River main stem 28.6 km upstream of the confluence with the West Fork Bitterroot River ($y = 0.1441x - 271.96$, $r^2 = 0.22$, $p = 0.10$). Bottom panel: sampling locations shaded to indicate estimates of abandonment probability (white: 0.21–0.32, grey: 0.32–0.47, and black: 0.47–0.62). Burn severity for fires in the watershed is indicated by low severity in green, moderate severity in orange, and high severity in red. Grey is outside of the fire perimeters.

Data analyses

We used program PRESENCE 4.1 [38] to estimate detection probability, occupancy, and abandonment and colonization probabilities. We constructed a survey history based on the five, 100-m sections for each site across the 1992–1995 and 2009–2011 surveys to estimate these parameters. These separate survey intervals were regarded as seasons. Estimates for probability of detection were modeled as a function of standardized values for season, large wood, fire, width, and gradient. Because large wood and the occurrence of medium- to high-severity fire at the site change over time, we treated these as survey-specific covariates. Given that both survey events had imperfect detectability and different crews (but the same field protocol), we compared detection probability between the earlier and more recent surveys. In addition, we explored all possible combinations of covariates in competing models to examine which covariates best described the probability of detection across sites (based on maximum likelihood estimators and Akaike Information Criterion (AIC) [39].

We used a multi-season model to estimate site abandonment and colonization probabilities of bull trout. The model uses initial occupancy estimates for the first sampling period and derives estimates for the abandonment and colonization probabilities that determine whether a species occurs at a site during the second sampling period [40]. To examine changes in occupancy in our data set, we first determined whether estimates of bull trout abandonment or colonization probabilities were significantly different from zero and from one another. Any changes were regarded as apparent because occupancy was only estimated twice, not at repeated intervals. Finally, we tested whether large wood, gradient, width, brook trout presence, proximity to where bull trout were common in the main-stem, occurrence of medium- or high-severity fire, or relative temperature co-varied with these probabilities. In addition, we examined how elevation, an occasional surrogate for water temperature [21], degree of connectivity, channel gradient, and human land use, individually co-varied with colonization and abandonment probabilities. Models were fit using maximum likelihood estimators and ranked based on AIC scores. We considered all models within 2 AIC units of the top model, but disregarded uninformative parameters i.e., covariates for which approximate 85% confidence intervals overlapped zero [41]. We also performed stepwise variable removal, based on the minimum absolute value of β/SE, stopping when variable exclusion led to a decrease in AIC score for the model [41].

Results

Presence of bull trout in previously occupied sites declined (from 33 sites in 24 streams to 20 sites in 22 streams), and absence from previously unoccupied sites decreased (from 41 sites in 26 streams to 36 sites in 26 streams). Brook trout occupancy declined between the earlier (12 sites in 6 streams) and later (7 sites in 5 streams) surveys. In addition, thirty-one of the 74 sites had adjacent riparian burns during the period between surveys, 12 of which were from medium- to high-severity fires.

Significant correlations among covariates used to explain probability of detection were weak or absent (Table S1), so all were considered in models of detection probability. There were three top models (within 2 AIC units of each other; Table 1). Our naïve (no covariates) probability of detection estimate was 0.54 (SE, 0.03). Naïve detection probabilities did not differ between seasons (season 1: 0.54, SE 0.04; season 2: 0.54, SE 0.06) and season did not increase the model AIC score. Therefore, the top model for estimating probability of detection included only large wood and width as covariates. Based on this model, site-specific probability of detection varied from 0.09 to 0.99 (Table S2).

The overall (no site or survey covariates) abandonment probability (0.36, SE 0.07) was almost 3-fold greater than that for colonization (0.13, SE 0.07). Most covariates in models for estimating colonization or abandonment probabilities were uninformative because of small effect sizes or large standard errors. The three top models with informative covariates for abandonment probabilities included either no covariate or the single covariates of elevation or temperature (Table 2). We model-averaged the top three models with informative parameters. Estimated abandonment probabilities increased approximately three-fold from cooler to warmer sites (Figure 2) and high- to low-elevation sites (Figure 3). Elevation and temperature were negatively correlated (−0.59). No informative covariates were retained in the top model for estimating colonization.

Discussion

By revisiting historically sampled sites within a river network that encompasses a broad temperature and elevation gradient, we demonstrated that site abandonment probabilities of bull trout were highest at warmer, low-elevation sites over the last two decades. This coincided with increases in summer stream temperatures in the East Fork Bitterroot River basin. Neither colonization nor abandonment probabilities were related to variables reflecting habitat, biotic interactions, or recent disturbance, and probabilities of abandonment were three-fold greater than those of colonization. Collectively, these results represent the first empirical evidence supporting predicted declines in the distribution of bull trout as a consequence of climate change [17], [26], [42]. We acknowledge that the observed relation between climate-related warming and reduction in bull trout occupancy is correlative and that other covariates we did not consider may have influenced this outcome, but the effects of temperature on bull trout distributions are consistent across its historical range [17]. A complete mechanistic understanding of the effects of warming temperatures on bull trout has not been realized, but the restricted scope for growth of juvenile life history stages at warm temperatures is clear [4], [43]. Given that much of the historical range of bull trout is undergoing relatively rapid warming as a consequence of climate change [14], [26], testing the generality of

Table 1. Models within two Akaike Information Criterion (AIC) units of the top model for estimating probability of detection of bull trout with probability of occupancy, colonization, and abandonment held constant.

Model[a,b]	AIC	DeltaAIC	AIC Weight	Model likelihood	Parameters	-2Loglikelihood
p($W_{4.22}$,$LW_{4.47}$)[c]	479.22	0.00	0.5467	1.0000	6	467.22
p($S_{0.55}$ $W_{4.22}$,$LW_{4.47}$)	480.88	1.66	0.2384	0.436	7	468.04
p($W_{4.22}$,$LW_{4.47}$,$G_{0.35}$)	481.09	1.87	0.2146	0.393	7	467.09

[a]Variables subscripted with β/SE absolute values; variables with values < 1.4 are regarded as uninformative (Arnold 2010).
[b]Abbreviations: W, width; LW, large wood; S, season; G, gradient.
[c]Only this model lacks uninformative variables.

our observations will be straightforward if comparable long-term data sets become available.

Only those covariates most closely related to the decreased thermal suitability of bull trout habitat—water temperature and elevation—appeared in the top models for estimating abandonment probabilities. This may seem surprising, given that occupancy models for bull trout have included an array of habitat and biotic variables [18], [33]. We did not, however, model where bull trout are currently found, but examined what influenced recent changes in that distribution. Consequently, it might be expected that elevation (represented in a linear function across a broad range) would not explain the distribution of bull trout in the Bitterroot River basin [33], but was a top predictor of locations abandoned by bull trout. We attribute most of the explanatory power of elevation in our model to its relation to water temperature, but acknowledge that it can, in part, represent effects of other variables, such as the presence of nonnative species, the effects of fire, or proximity to population strongholds. In this study, however, none of these were informative contributors to models of changes in bull trout occupancy. At the sites we examined, brook trout occupancy was low and appeared to decline during the study. Elsewhere, replacement of bull trout by brook trout appears to be associated with particular valley morphologies [18], [25], that may not be prevalent in the study area. We regard it as unlikely that warming temperatures also reduced brook trout occupancy [24] because this species prefers warmer temperatures than does bull trout [4]. More plausible is that brook trout were declining in response to fire effects [44] but this species was too poorly represented in the data to evaluate this trend.

Stand-replacing fires tend to cause warmer stream temperatures [22], [45], thus it might be expected that reductions in bull trout occupancy would be associated with fire directly, or indirectly via fire's relation to elevation. Although low-elevation sites adjacent to burned areas were some of our warmest sites, there was no significant correlation ($r = 0.45$) between relative temperature and fire because wildfires burned across the entire watershed. Moreover, changes in water temperature may be ameliorated by other fire-related changes in habitat, such as increased autochthonous productivity, macroinvertebrate community shifts, or channel alteration [46], and previous observations of bull trout in the study area did not reveal population declines following fire [44]. Nevertheless, because stand-replacing fire in riparian zones leads to chronic increases in summer stream temperatures it has the potential to contribute to local, site-specific changes in occupancy by bull trout. In addition, anticipated increases in fire extent or frequency attributable to climate change [23], [47] may lead to more profound shifts, or outright extirpations, of populations across the landscape, where populations are isolated or landscapes are prone to large, post-fire debris flows [21].

Connectivity has long been thought to influence the persistence of salmonid populations because these fishes can be highly mobile and frequently form metapopulations [48]. In our analyses, rank distance to the main-stem river sections where bull trout are

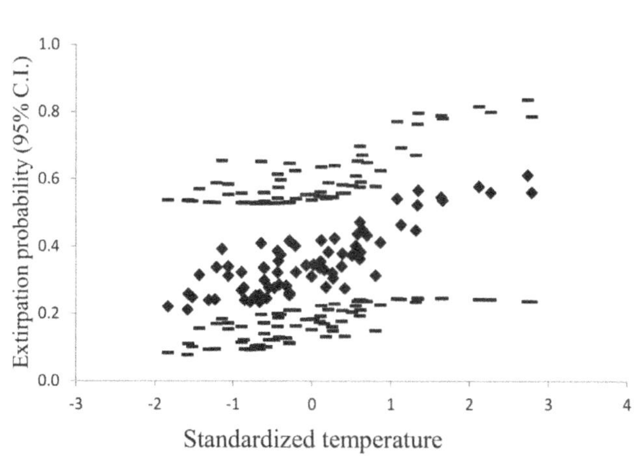

Figure 2. Effect of temperature on abandonment probabilities. Model-averaged abandonment probabilities (filled diamonds) from the top three informative models (Table 2) with their upper and lower 95% confidence intervals (dashes) versus standardized relative temperature across sites in the East Fork Bitterroot River basin.

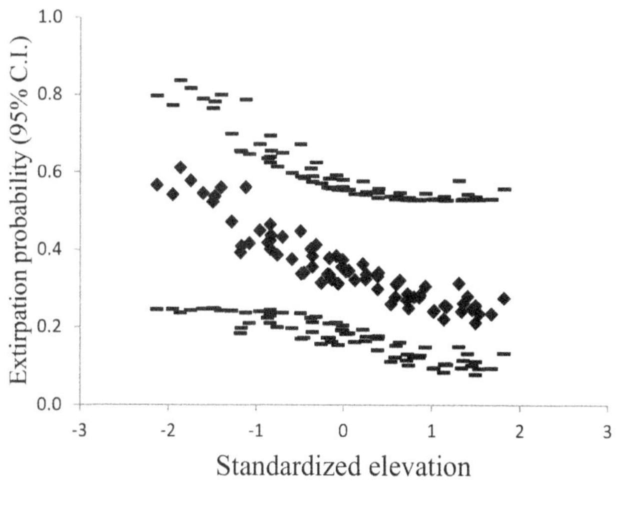

Figure 3. Effect of elevation on abandonment probabilities. Model-averaged abandonment probabilities (filled diamonds) from the top three informative models (Table 2) with their upper and lower 95% confidence intervals (dashes) versus standardized elevation.

Table 2. Models within two AIC units of the top model for using changes in occupancy (ψ) to estimate colonization (γ) and abandonment (ε) probabilities of bull trout.

Model[a,b,c]	AIC	DeltaAIC	AIC weight	Model likelihood	Parameters	-2Loglikelihood
$\psi, \gamma, \varepsilon(E_{1.59})^d$	478.38	0	0.10	1	7	464.38
$\psi,\gamma,\varepsilon(B_{0.5})$	478.62	0.24	0.088	0.88	7	464.62
$\psi,\gamma,\varepsilon(T_{1.43})^d$	478.94	0.56	0.075	0.76	7	464.94
$\psi,\gamma,\varepsilon^d$	479.22	0.84	0.065	0.66	6	467.22
$\psi,\gamma,\varepsilon(B_{0.47}, D_{0.73})$	480.11	1.73	0.042	0.42	8	464.11
$\psi,\gamma,\varepsilon(B_{0.46}, T_{0.72})$	480.12	1.74	0.042	0.42	8	464.12
$\psi,\gamma,\varepsilon(B_{0.66}, F_{0.73})$	480.12	1.74	0.042	0.42	8	464.12
$\psi,\gamma,\varepsilon(D_{1.04})$	480.17	1.79	0.040	0.41	7	466.17
$\psi,\gamma,\varepsilon(B_{0.24},G_{0.62})$	480.24	1.86	0.039	0.40	8	464.24
$\psi,\gamma,\varepsilon(T_{1.43}, LW_{0.68})$	480.45	2.0	0.035	0.36	8	464.45

[a]All models contained the probability of detection function p(W, LW).
[b]Variables subscripted with β/SE absolute values.
[c]Abbreviations: E, elevation; T, temperature; LW, large wood; W, stream width; G, gradient; B, brook trout presence at site; D, distance to where bull trout are common in the main-stem East Fork Bitterroot River; F, occurrence of moderate- to high-severity fire.
[d]Models without uninformative variables. Model-averaged parameter estimates for the untransformed coefficients: E, −0.73 (0.46 SE); T, 0.63 (0.44 SE).

common did not account for the probability of abandonment or colonization at tributary sites. It may have been that the distances involved (a few tens of kilometers at most) do not represent meaningful levels of isolation for bull trout, migratory forms of which often traverse much longer distances [31], [49]. Genetic evidence indicates that despite declines in the abundance of migratory bull trout in the East Fork Bitterroot River basin over the last few decades, dispersal by bull trout among tributaries and between tributaries and the main-stem East Fork Bitterroot River remains common [50]. If climate change or anthropogenic habitat alteration increases the energetic and demographic costs of migration beyond some threshold, the influence of connectivity may become more evident [51].

An ongoing paradox is that demographic shifts among freshwater species have been difficult to detect despite that these taxa may be among the most sensitive to climate change [11]. Although a few studies have suggested that declines in freshwater fish abundance could be related to climate change [52], [53], [54], beyond the present study only one other [55] has reported changes in their distributions. This could be partly attributable to a paucity of adequately georeferenced historical data sets, particularly those that permit detectability estimation. Nevertheless, the enormity of current and historical fish monitoring efforts by state, tribal, and federal agencies throughout North America suggests that many such data sets exist. For areas lacking historical temperature data, recent advances in modeling dendritic ecological networks [56] can facilitate accurate hindcasting and prediction of stream temperatures for basins well represented by recent temperature records [21].

Other obstacles to detecting the effects of climate change on cold-water fishes reflect their habitat and biology. Although rising water temperature appears to be a consistent trend in many portions of the historical range of bull trout and other western North American salmonids [15], [26], uncertainty about the response of particular watersheds [57] or certain species [55] remains high. Fish abundance is exceptionally temporally variable and sometimes requires decades of sampling before statistically significant trends emerge [58], [59]. In addition, such trends may be superimposed on long-term variation in abundance dictated by climate cycles such as the Pacific Decadal Oscillation [60]. Nevertheless, correlations between bull trout abundance and broad-scale climate cycles (or this species' abundance and that of other salmonids) are weak [61], [62], [63]. Analyses that rely on occupancy, rather than abundance, may be less vulnerable to temporal fluctuations.

In summary, we found that in this core, connected conservation area for bull trout, patterns in site abandonment were consistent with the predicted effects of stream temperature warming. Extending the current time series of observation of bull trout in the East Fork Bitterroot River basin, as well as similar studies in other basins, is essential to evaluating the generality of this trend. Monitoring designs that focus on stenothermic species and on locations most likely to undergo rapid change offer the greatest power for detecting responses of aquatic species to climate change [11]. For bull trout, these include low-elevation reaches that are warming rapidly and high-elevation reaches undergoing flow reductions from declining snowpacks [64]. In the absence of such targeted designs, however, revisiting historically sampled sites across a range of elevations within a stream network to examine changes in occupancy constitutes a practical alternative [10], [65], [66].

Supporting Information

Table S1 Correlation coefficients. Pearson correlation coefficients for standardized variables in analyses including elevation (E), large wood (LW), bank-full width (W), gradient at site (G), relative temperature (T), the presence of medium to high severity burns at the site (F), the presence of brook trout (B), and the distance from the tributary confluence to where bull trout are common in the main-stem (D). An asterisk indicates a significant correlation ($P \leq 0.05$).

Table S2 Detection probabilities. Detection probabilities associated with each site and each survey considering large wood during each survey and stream width at each site. Probability of detection without covariates was 0.54 (SE, 0.03).

Acknowledgments

We thank T. Smith, W. Glenn, L. Hall, J. Leppi, and T. Moua for help with fieldwork and T. McMahon and C. Sestrich for sharing their previous dataset. D. Isaak, S. Wenger, R. Al-Chokhachy, and A. Ruesch provided helpful comments on drafts of this manuscript.

Author Contributions

Conceived and designed the experiments: LAE OH LMH MKY. Performed the experiments: OH. Analyzed the data: LAE OH LMH. Wrote the paper: LAE MKY. Manuscript comments: OH LMH.

References

1. Dudgeon D, Arthington AH, Gessner MO, Kawabata ZI, Knowler DJ, et al. (2006) Freshwater biodiversity: importance, threats, status and conservation challenges. Biol Rev Camb Philos Soc 81: 163–182.
2. Williams JE, Williams RN, Thurow RF, Elwell L, Philipp DP, et al. (2011) Native fish conservation areas: a vision for large-scale conservation of native fish communities. Fisheries 36: 267–277.
3. Burkhead NM (2012) Extinction rates in North American freshwater fishes, 1900–2010. BioScience 62: 798–808.
4. McMahon T, Zale A, Barrows F, Selong J, Danehy R (2007) Temperature and competition between bull trout and brook trout: a test of the elevation refuge hypothesis. Trans Am Fish Soc 136: 1313–1326.
5. Elliott JM, Elliott JA (2010) Temperature requirements of Atlantic salmon *Salmo salar*, brown trout *Salmo trutta* and Arctic charr *Salvelinus alpinus*: predicting the effects of climate change. J Fish Biol 77: 1793–1817.
6. Ficke AD, Myrick CA, Hansen LJ (2007) Potential impacts of global climate change on freshwater fisheries. Rev Fish Biol Fish 17: 581–613.
7. Beer WN, Anderson JJ (2013) Sensitivity of salmonid freshwater life history in western US streams to future climate conditions. Glob Chang Biol 8: 2547–2556.
8. Crozier LG, Scheuerell MD, Zabel RW (2011) Using time series analysis to characterize evolutionary and plastic responses to environmental change: a case study of a shift toward earlier migration date in sockeye salmon. Am Nat 178: 755–773.
9. Warren DR, Robinson JM, Josephson DC, Sheldon DR, Kraft CE (2012) Elevated summer temperatures delay spawning and reduce redd construction for resident brook trout (*Salvelinus fontinalis*). Glob Chang Biol 18: 1804–1811.
10. Comte L, Buisson L, Daufresne M, Grenouillet G (2012) Climate-induced changes in the distribution of freshwater fish: observed and predicted trends. Freshw Biol 58: 625–639.
11. Isaak DJ, Rieman BE (2013) Stream isotherm shifts from climate change and implications for distributions of ectothermic organisms. Glob Chang Biol 19: 742–751.
12. Mote PW, Hamlet AF, Clark MP, Lettenmaier DP (2005) Declining mountain snowpack in western North America. Bull Am Meteorol Soc 86: 39–49.
13. Knowles N, Dettinger MD, Cayan DR (2006) Trends in snowfall versus rainfall in the western United States. J Clim 19: 4545–4559.
14. Pederson GT, Graumlich LJ, Fagre DB, Kipfer T, Muhlfeld CC (2010) A century of climate and ecosystem change in western Montana: what do temperature trends portend? Clim Change 98: 133–154.
15. Isaak DJ, Wollrab S, Horan D, Chandler G (2012b) Climate change effects on stream and river temperatures across the northwest US from 1980–2009 and implications for salmonid fishes. Clim Change 113: 499–524.
16. Leppi JC, DeLuca TH, Harrar SW, Running SW (2012) Impacts of climate change on August stream discharge in the Central-Rocky Mountains. Clim Change 112: 997–1014.
17. Rieman BE, Isaak D, Adams S, Horan D, Nagel D, et al. (2007) Anticipated climate warming effects on bull trout habitats and populations across the interior Columbia River basin. Trans Am Fish Soc 136: 1552–1565.
18. Wenger SJ, Isaak DJ, Dunham JB, Fausch KD, Luce CH, et al. (2011a) Role of climate and invasive species in structuring trout distributions in the interior Columbia River Basin, USA. Can J Fish Aquat Sci 68: 988–1008.
19. Wenger SJ, Som NA, Dauwalter DC, Isaak DJ, Neville HM, et al. (2013) Probabilistic accounting of uncertainty in forecasts of species distributions under climate change. Glob Chang Biol 10.1111/gcb.12294.
20. Parmesan C, Burrows MT, Duarte CM, Poloczanska ES, Richardson AJ, et al. (2013) Beyond climate change attribution in conservation and ecological research. Ecol Lett 16: 58–71.
21. Isaak DJ, Luce CH, Rieman BE, Nagel DE, Peterson EE, et al. (2010) Effects of climate change and wildfire on stream temperatures and salmonid thermal habitat in a mountain river network. Ecol Appl 20: 1350–1371.
22. Mahlum SK, Eby LA, Young MK, Clancy CG, Jakober M (2011) Effects of wildfire on stream temperatures in the Bitterroot River Basin, Montana. Int J Wildl Fire 20: 240–247.
23. Westerling AL, Hidalgo HG, Cayan DR, Swetnam TW (2006) Warming and earlier spring increase western U.S. forest wildfire activity. Science 313: 940–943.
24. Wenger SJ, Isaak DJ, Luce CH, Neville HM, Fausch KD, et al. (2011b) Flow regime, temperature, and biotic interactions drive differential declines of trout species under climate change. Proc Natl Acad Sci U S A 108: 14175–14180.
25. Rieman BE, Peterson JT, Myers DL (2006) Have brook trout (*Salvelinus fontinalis*) displaced bull trout (*Salvelinus confluentus*) along longitudinal gradients in central Idaho streams? Can J Fish Aquat Sci 63: 63–78.
26. Isaak DJ, Muhlfeld CC, Todd AS, Al-Chokhachy R, Roberts J, et al. (2012a) The past as prelude to the future for understanding 21st-century climate effects on Rocky Mountain trout. Fisheries 37: 542–556.
27. Montana Bull Trout Restoration Team (2000) Restoration plan for bull trout in the Clark Fork River basin and Kootenai River basin Montana. Montana Department of Fish, Wildlife and Parks, Helena. 114 pp.
28. Nelson ML, McMahon TE, Thurow RF (2002) Decline of the migratory form in bull charr, *Salvelinus confluentus*, and implications for conservation. Env Biol Fish 64: 321–332.
29. Al-Chokhachy R, Budy P (2008) Demographic characteristics, population structure, and vital rates of a fluvial population of bull trout in Oregon. Trans Am Fish Soc 137: 1709–1722.
30. Dunham J, Rieman B, Chandler G (2003) Influences of temperature and environmental variables on the distribution of bull trout within streams at the southern margin of its range. N Am J Fish Manage 23: 894–904.
31. Howell PJ, Dunham JB, Sankovich PM (2010) Relationships between water temperatures and upstream migration, cold water refuge use, and spawning of adult bull trout from the Lostine River, Oregon, USA. Ecol Freshw Fish 19: 96–106.
32. Dunham JB, Rieman BE (1999) Metapopulation structure of bull trout: Influences of physical, biotic, and geometrical landscape characteristics. Ecol Appl 9: 642–655.
33. Rich CF, McMahon TE, Rieman BE, Thompson WL (2003) Local-habitat, watershed, and biotic features associated with bull trout occurrence in Montana streams. Trans Am Fish Soc 132: 1053–1064.
34. U.S. Geological Survey (2006) National elevation dataset (http://ned.usgs.gov/).
35. Tarboton DG (2008) Terrain analysis using digital elevation models (TauDEM). (http://hydrology.neng.usu.edu/taudem/).
36. Keane RE, Agee JK, Fule P, Keeley JE, Key C, et al. (2008) Ecological effects of large fires on US landscapes: benefit or catastrophe? Int J Wildl Fire 17: 696–712.
37. Keeley JE (2009) Fire intensity, fire severity, and burn severity: a brief review and suggested usage. Int J Wildl Fire 18: 116–126.
38. Hines JE (2006) PRESENCE2 (v4.1). Software to estimate patch occupancy and related parameters. USGS-PWRC. http://www.mbr-pwrc.usgs.gov/software/presence.html.
39. Burnham KP, Anderson DA (2002) Model selection and multimodel inference: a practical information–theoretic approach. Springer-Verlag, New York.
40. Donovan TM, Hines J (2007) Exercises in occupancy modeling and estimation. <http://www.uvm.edu/envnr/vtcfwru/spreadsheets/occupancy.htm>
41. Arnold T (2010) Uninformative parameters and model selection using Akaike's Information Criterion. J Wildl Manage 74: 1175–1178.
42. Ruesch AS, Torgersen CE, Lawler JJ, Olden JD, Peterson EE, et al. (2012) Projected climate-induced habitat loss for salmonids in the John Day River network, Oregon, U.S.A. Con Biol 26: 873–882.
43. Selong JH, McMahon TE, Zale AV, Barrows FT (2001) Effect of temperature on growth and survival of bull trout, with application of an improved method for determining thermal tolerance in fishes. Trans Am Fish Soc 130: 1026–1037.
44. Sestrich CM, McMahon TE, Young MK (2011) Influence of fire on native and nonnative salmonid populations and habitat is a western Montana basin. Trans Am Fish Soc 140: 136–146.
45. Dunham JB, Rosenberger AE, Luce CH, Rieman BE (2007) Influence of wildfire and channel reorganization on spatial and temporal variation in stream temperature and the distribution of fish and amphibians. Ecosystems 10: 335–346.
46. Gresswell RE (1999) Fire and aquatic ecosystems in forested biomes in North America. Trans Am Fish Soc 128: 193–221.
47. McKenzie D, Gedalof Z, Peterson DL, Mote P (2004) Climatic change, wildfire, and conservation. Con Biol 18: 890–902.
48. Rieman BE, Dunham JB (2000) Metapopulations and salmonids: a synthesis of life history patterns and empirical observations. Ecol Freshw Fish 9: 51–64.
49. Bjornn TC, Mallet J (1964) Movements of planted and wild trout in an Idaho river system. Trans Am Fish Soc 93: 70–76.
50. Nyce LG, Eby LA, Clancy CG, Painter S, Leary RF (2013) Genetic population structure of bull trout in the East Fork Bitterroot River drainage, Montana. N Am J Fish Manage 33: 432–445.

51. Starcevich SJ, Howell PJ, Jacobs SE, Sankovich PM (2012) Seasonal movement and distribution of fluvial adult bull trout in selected watersheds in the mid-Columbia River and Snake River basins. PLoS ONE 7: e37257.
52. Hari RE, Livingstone DM, Siber R, Burkhart-Holm P, Guettinger H (2006) Consequences of climatic change for water temperature and brown trout populations in alpine rivers and streams. Glob Chang Biol 12: 10–26.
53. Winfield IJ, Hateley J, Fletcher JM, James JB, Bean CW, et al. (2010) Population trends of Arctic charr (*Salvelinus alpinus*) in the UK: assessing the evidence for a widespread decline in response to climate change. Hydrobiologia 650: 55–65.
54. Almodóvar A, Nicola GG, Ayllón D, Elvira B (2011) Global warming threatens the persistence of Mediterranean brown trout. Glob Chang Biol 18: 1549–1560.
55. Comte L, Grenouillet G (2013) Do stream fish track climate change? Assessing distribution shifts in recent decades. Ecography 39: 1–11.
56. Peterson EE, Ver Hoef JM, Isaak DJ, Falke JA, Fortin MJ, et al. (2013) Modelling dendritic ecological networks in space: an integrated network perspective. Ecol Lett 16: 707–719.
57. Arismendi I, Johnson SL, Dunham JB, Haggerty R (2013) Descriptors of natural thermal regimes in streams and their responsiveness to change in the Pacific Northwest of North America. Freshw Biol 58: 880–894.
58. Maxwell D, Jennings S (2005) Power of monitoring programmes to detect decline and recovery of rare and vulnerable fish. J Appl Ecol 42: 25–37,
59. Al-Chokhachy R, Budy P, Conner M (2009) Detecting declines in the abundance of a bull trout (*Salvelinus confluentus*) population: understanding the accuracy, precision, and costs of our efforts. Can J Fish Aquat Sci 66: 649–658.
60. Mantua NJ, Hare SR, Zhang Y, Wallace JM, Francis RC (1997) A Pacific interdecadal climate oscillation with impacts on salmon production. Bull Am Meteorol Soc 78: 1069–1079.
61. High B, Meyer KA, Schill DJ, Mamer ERJ (2008) Distribution, abundance, and population trends of bull trout in Idaho. N Am J Fish Manage 28: 1687–1701.
62. Copeland T, Meyer KA (2011) Interspecies synchrony in salmonid densities associated with large-scale bioclimatic conditions in central Idaho. Trans Am Fish Soc 140: 928–942.
63. Meyer KA, Garton EO, Schill DJ (2014) Bull trout trends in abundance and probabilities of persistence in Idaho. N Amer J Fish Manage 34: 202–214.
64. Luce CH, Abatzoglou JT, Holden ZA (2013) The missing mountain water: slower westerlies decrease orographic enhancement in the Pacific Northwest USA. Science. DOI:10.1126/science.1242335
65. Altwegg R, Wheeler M, Erni B (2008) Climate and the range dynamics of species with imperfect detection. Biol Lett 4: 581–584.
66. Tingley MW, Beissinger SR (2009) Detecting range shifts from historical species occurrences: new perspectives on old data. Trends Ecol Evol 24: 625–633.

2

Four Decades of Forest Persistence, Clearance and Logging on Borneo

David L. A. Gaveau[1]*, Sean Sloan[2], Elis Molidena[1], Husna Yaen[1], Doug Sheil[1,3], Nicola K. Abram[4,5], Marc Ancrenaz[5,6,7,8], Robert Nasi[1], Marcela Quinones[9], Niels Wielaard[9], Erik Meijaard[1,5,10]

1 Center for International Forestry Research, Bogor, Indonesia, **2** Centre for Tropical Environmental and Sustainability Science, School of Marine and Tropical Biology, James Cook University, Cairns, QLD, Australia, **3** Department of Ecology and Natural Resource Management (INA), Norwegian University of Life Science (NMBU), Ås, Norway, **4** Durrell Institute of Conservation and Ecology, University of Kent, Canterbury, Kent, United Kingdom, **5** Borneo Futures project, People and Nature Consulting International, Ciputat, Jakarta, Indonesia, **6** Sabah Wildlife Department, Kota Kinabalu, Sabah, Malaysia, **7** HUTAN, Kinabatangan Orang-utan Conservation Programme, Kota Kinabalu, Sabah, Malaysia, **8** North England Zoological Society, Chester Zoo, Chester, United Kingdom, **9** SarVision, Wageningen, The Netherlands, **10** School of Biological Sciences, University of Queensland, Brisbane, QLD, Australia

Abstract

The native forests of Borneo have been impacted by selective logging, fire, and conversion to plantations at unprecedented scales since industrial-scale extractive industries began in the early 1970s. There is no island-wide documentation of forest clearance or logging since the 1970s. This creates an information gap for conservation planning, especially with regard to selectively logged forests that maintain high conservation potential. Analysing LANDSAT images, we estimate that 75.7% (558,060 km^2) of Borneo's area (737,188 km^2) was forested around 1973. Based upon a forest cover map for 2010 derived using ALOS-PALSAR and visually reviewing LANDSAT images, we estimate that the 1973 forest area had declined by 168,493 km^2 (30.2%) in 2010. The highest losses were recorded in Sabah and Kalimantan with 39.5% and 30.7% of their total forest area in 1973 becoming non-forest in 2010, and the lowest in Brunei and Sarawak (8.4%, and 23.1%). We estimate that the combined area planted in industrial oil palm and timber plantations in 2010 was 75,480 km^2, representing 10% of Borneo. We mapped 271,819 km of primary logging roads that were created between 1973 and 2010. The greatest density of logging roads was found in Sarawak, at 0.89 km km^{-2}, and the lowest density in Brunei, at 0.18 km km^{-2}. Analyzing MODIS-based tree cover maps, we estimate that logging operated within 700 m of primary logging roads. Using this distance, we estimate that 266,257 km^2 of 1973 forest cover has been logged. With 389,566 km^2 (52.8%) of the island remaining forested, of which 209,649 km^2 remains intact. There is still hope for biodiversity conservation in Borneo. Protecting logged forests from fire and conversion to plantations is an urgent priority for reducing rates of deforestation in Borneo.

Editor: Kamal Bawa, University of Massachusetts, United States of America

Funding: This work was funded by the Arcus foundation and the CGIAR Research Program on Forests, Trees, and Agroforestry. The funders had no role in study design, data collection and analysis, decision to publish, or preparation of the manuscript.

Competing Interests: The authors have declared that no competing interests exist.

* Email: d.gaveau@cgiar.org

Introduction

Conservationists have historically prioritized the protection of 'pristine', 'old-growth' tropical forests over human-modified ones [1]. Pristine tropical forests are becoming increasingly rare, however, particularly in the lowlands of South-East Asia [2] due to widespread timber extraction ("logging"), conversion to other land uses and increased vulnerability to fire. The conservation value of selectively-logged forests has been increasingly highlighted as requiring recognition [3–5]. Selective timber extraction prevails in tropical forests meaning that only a few stems (typically 4–10) are removed from each hectare leaving a diverse forest still standing [5–10]. These modified habitats retain appreciable biodiversity [3,5,9] and serve as effective buffers and corridors for wildlife moving between intact forest fragments [11].

Borneo's forests include old-growth lowland, hill and montane dipterocarp forests, freshwater and peat swamp forests, heath forests (*kerangas*), and mangrove forests (including areas dominated by the palm *Nypa fruticans Wurmb* locally termed *nipah*) [12]. These forests possess some of the richest biological communities on the planet and should therefore be preserved [12]. But, much has already been logged [13] – between 1980 and 2000 more round wood was harvested from Borneo than from Africa and the Amazon combined [14] – or destroyed by fire [15,16], or converted to plantations. Forest conversion encompasses clearing forest to establish industrial oil palm (*Elaeis guineensis*) [17], and to a lesser extent acacia (*Acacia* spp) and rubber tree (*Hevea brasiliensis*) plantations [17,18]. There is no island-wide satellite-based documentation of forest clearance, conversion or logging and the island remains under studied in this respect compared to other major tropical regions, although a recent study filled this gap for the northern part of Borneo [19]. Confusion reigns over the actual extent of deforestation, remaining intact and logged-over forests hampering proper conservation planning. For example, Indonesia's pledge to maintain at least 45% of forest in the Indonesian part of Borneo (Kalimantan) was criticized by environmental groups reporting that Kalimantan retains only 30% forest cover

[20], appreciably less than the 55% reported by Indonesia's Ministry of Forestry [21].

Building on a detailed, LANDSAT-based, spatial inventory of forest cover, clearance and logging over the 1973–2010 period, the era of industrial-scale forest exploitation on the island [13], we address the following questions: (i) what was the extent of forest cover in the early 1970s; (ii) how much has been selectively logged or cleared since; and (iii) how are remaining intact and logged forests distributed across zones designated for protection, timber production, or conversion to plantations?

Methods

Overall approach

We mapped forest extent, and deforestation for the period 1973–2010 as well as total logged area since 1973 over the whole island of Borneo (737,188 km^2) at medium spatial resolution (pixel size = 60 m×60 m). To assess deforestation (forest clearance) we developed a baseline forest cover map for the year 1973 using LANDSAT MSS imagery (pixel size = 60 m×60 m). Combining our 1973 baseline map with a 2010 forest cover map (pixel size = 50 m×50 m) developed using ALOS PALSAR radar satellite imagery [22] we produced a map of deforestation between 1973 and 2010 (pixel size = 60 m×60 m). To estimate the area of forest converted to plantations since 1973 and to correct for the tendency of the ALOS PALSAR classification to confound mature plantations for forest [22], we mapped all industrial oil-palm and timber plantations as of 2010 using LANDSAT imagery. Finally, using LANDSAT imagery and a MODIS-based percent tree cover map, we mapped the total logged-over forest area from 1973 to 2010 using logging roads as a proxy. The following sections elaborate the datasets (Table 1) and associated methods.

Mapping of natural forest cover in 1973

To map 1973 forest cover we submitted 61 LANDSAT MSS images acquired during the 1970s and spanning the whole Borneo to a supervised tree-based classification algorithm [23] that iteratively classified 'Forest', 'Non-forest', 'Cloud', 'Cloud shadow', and 'Water' classes. All LANDSAT MSS imagery was viewed as a band 4-3-2 (or 3-4-2) false-color composite (Figure 1A).

We obtained 37 of the 61 images from the USGS Global Land Survey (GLS) 1970s collection of ortho-rectified LANDSAT MSS Imagery [24] and completed with 24 additional non-GLS LANDSAT MSS images (obtained from the USGS LANDSAT MSS archive) wherever available to reduce the area contaminated by clouds. The list of LANDSAT MSS images (n = 61) used to map forest cover in 1973 is provided in Table S1 in File S1. Persistently cloudy areas in the 1970s that were observed as 'Forest' in 2010 were reclassified as 'Forest' in 1973 accordingly. Areas classified as 'Cloud' in 1973 and 'Non-forest' in 2010 accounted for <5% of Borneo's area. LANDSAT images not sourced from the GLS collection (non-GLS) were geometrically registered to the reference GLS imagery using a second or third order polynomial co-registration technique (RMS<1 pixel), but they were not ortho-rectified because precise data required for ortho-rectification (sensor angle, viewing angle, distance to object) for each individual Landsat image were often not available. Therefore a moderate horizontal displacement of high-elevation features (e.g., mountain peaks) from their actual coordinates might persist. The 61 images were acquired between 1972 and 1980, but because >80% of Borneo's area was imaged between 1972 and 1973, a weighted average by area takes 1973 as the nominal map year (see Figure S1a in File S1).

Measures were taken to ensure high accuracy of the finished 1973 forest cover map, in particular manually editing complex areas such as those obscured by haze or in steep topography, where the supervised classification algorithm often produced classification errors, but where a trained remote sensing analyst could still visually interpret the imagery. Two analysts only were involved as the use of multiple interpreters can compromise the consistency of the results [25].

Mapping deforestation: 1973–2010

To map deforestation between 1973 and 2010 we compared our baseline map with the 2010 forest cover map developed by SARVISION using ortho-rectified ALOS PALSAR radar imagery [22]. The SARVISION 2010 map distinguishes several forest types by altitude, soil type (peat and mineral), and condition. We collapsed select classes to constitute a general 2010 'Forest' class comparable to our 1973 'Forest' class. Classes within the SARVISION 2010 'Forest' class included tropical lowland forest, tropical mountain forest, freshwater and peat swamp closed and pole forest, riverine forest, mangrove forest and *nipah* mangrove forest. All other classes (e.g., towns, agricultural areas, open-cast mines, water ways) were classified as 'Non-forest'.

We took all possible measures to check for other discrepancies between the ALOS PALSAR-based 2010 forest classification and our interpretation of what constitutes 'Forest' and 'Non-forest' on the LANDSAT imagery. In particular, numerous visual checks ensured that small-scale deforestation patches were adequately captured in the final deforestation map by visually reviewing the entire LANDSAT database from 1972–2010 (n = 268). The list of all LANDSAT images (MSS, TM, ETM+) used in this study is provided in Table S1 in File S1. All LANDSAT TM and ETM+

Table 1. Summary of Spatial Data.

Description	Source Data	Spatial resolution	Classification Method	Years of observation
1973 Forest cover	LANDSAT MSS	60 m×60 m	Supervised classification followed by visual interpretation	1973
2010 Forest cover	ALOS-PALSAR [22]	50 m×50 m	Unsupervised classification	2010
Forest clearance (1973–2010)	LANDSAT MSS and ALOS-PALSAR	60 m×60 m	map comparison and visual interpretation	1973–2010
Total logged area since 1973	LANDSAT MSS, TM, ETM+ imagery; MODIS [35,36]	Vector data resampled to 60 m×60 m	Visual interpretation and buffering	1973, 1990, 2000, 2000, 2010
Industrial oil-palm (IOPP) and timber plantations (ITP)	LANDSAT MSS, TM, ETM+,	Vector data resampled to 60 m×60 m	Visual interpretation	1973, 1990, 1995, 2000, 2005, 2010

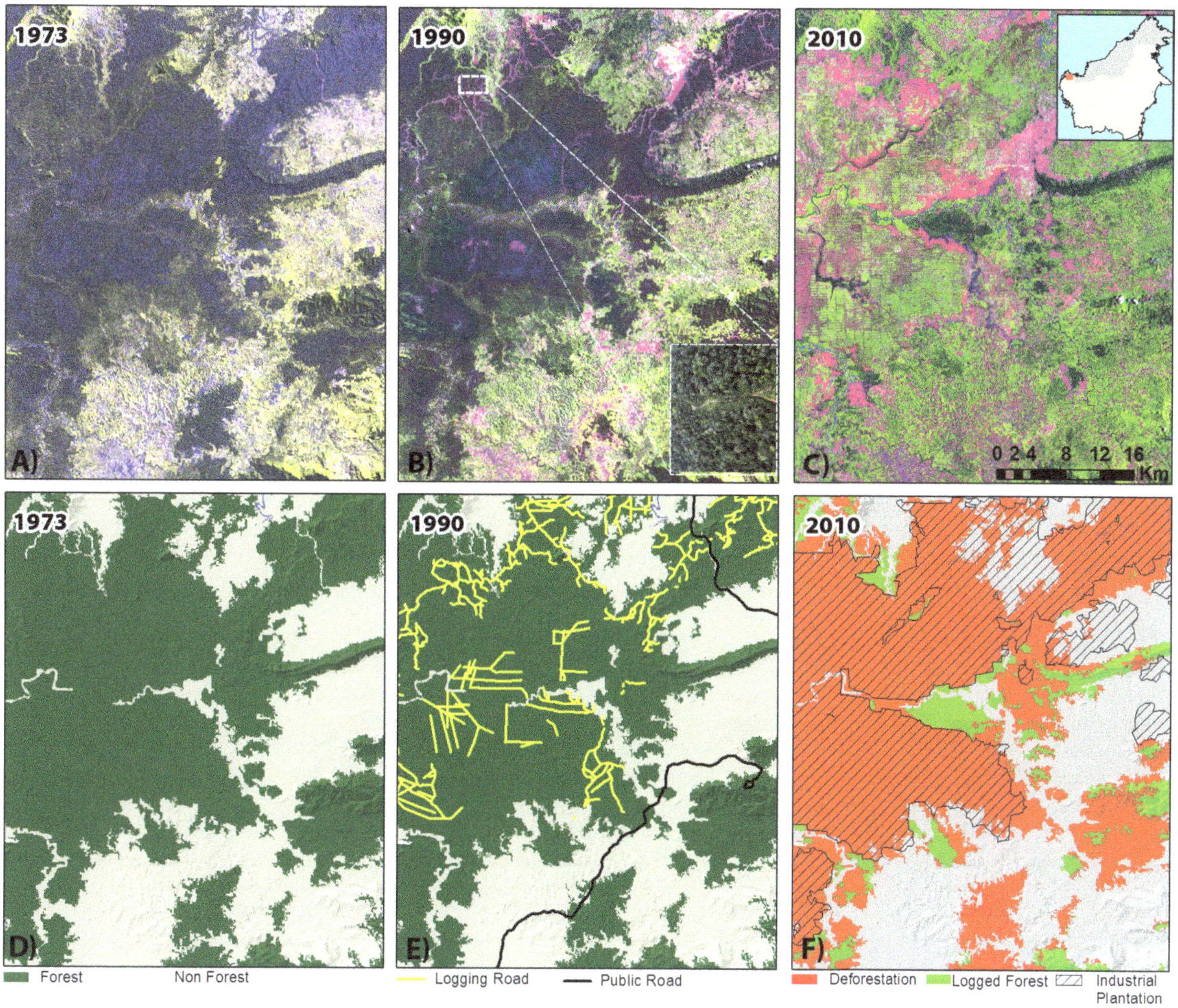

Figure 1. A close-up view of an area in West Kalimantan province, Borneo (see top-right inset for location). In this example, an intact forest in year 1973 was logged in 1990 and became converted to an industrial oil Palm Plantation in 2010. On the 1973 LANDSAT MSS imagery (false color composite: 3-4-2) forest appears dark green (panel A). The resulting forest non-forest supervised classification is shown in panel D. On the 1990 LANDSAT TM imagery (false color composite: 5-4-3) primary logging roads are seen carving through the forest (panel B). The logging roads (yellow lines) and public roads (black lines) are shown in panel E. An aerial view of a primary logging road (Photo by R. Butler [53]) is shown the inset of panel B. On the 2010 LANDSAT TM imagery), the forest has become converted to an industrial oil palm plantation (grid-like patterns). The boundary of the plantations was digitized by visual interpretation (Panel F).

imagery was viewed as a band 4-5-3 (or 5-4-3) false-color composite (Figure 1B&C).

All industrial plantations that existed in year 2010 were digitized via visual interpretation of 268 LANDSAT scenes from the 1970s, 1990, 2000 and 2010 (see Table S1 in File S1). The inspection of multiple images prior to 2010 ensured that those plantations that were mature in a given year and therefore difficult to detect in that year were fully accounted for (cf. [17]). Plantations were identified as large geometrically-shaped areas with distinctive, grid-like or contour-like patterns and homogeneous spectral signatures characteristic of monocultures or recently-cleared lands (Figure 1C&F). In instances where the 2010 'Forest' map from SARVISION overlapped with our LANDSAT-derived 2010 'Industrial Plantation' layer described below, the area of overlap was re-classified as 'Industrial plantation' accordingly. Areas that were 'Forest' in the 1973 map and 'Non-forest' in 2010 were classified as 'Forest clearance'.

Mapping logged-over forest: 1973-2010

We estimated the approximate extent of forests impacted by industrial-scale mechanized logging between 1973 and 2010 by first mapping all primary logging roads (large enough to be detected on MSS or TM images), next determining a distance from logging roads beyond which forest extraction typically extends, and then 'buffering' roads by this distance. A review of a satellite-based methods of estimating the extent of selectively-logged tropical forests finds that studies which similarly mapped and buffered logging roads [25–30] did so effectively, particularly at large spatial scales at which more complex approaches are challenging, inappropriate, or impossible [31].

We digitized the extent of primary logging roads by visually analyzing our 268 LANDSAT images acquired over 1972–2010 (see Table S1 in File S1). Wide logging roads were readily detectable in the LANDSAT imagery (Figure 1B&E). We were capable to detect logging roads under most areas of persistent haze, by zooming in closely and applying a local contrast enhancement, and by digitizing logging roads underneath haze by mouse-click. The expansion of the road network overtime was observed for *c.*1973, 1990, 2000, and 2010. Imagery acquired a year or two before and after these nominal years served to reduce cloud contamination. We also inspected imagery from ca. 1995 and 2005 to better detect disused logging roads less visible due to rapid forest regrowth.

Similarly to our approach for mapping industrial plantations LANDSAT 5&7 (TM and ETM+) images were viewed as band 4-5-3 (or 5-4-3) false color composites enhanced to optimize road detection. Likewise, LANDSAT MSS images were viewed as band 4-3-2 (or 3-4-2) (Figure 1A, B&C). We used ancillary public-road maps from the Indonesian Ministry of Public Works [32] and the Sabah-based NGO HUTAN for Sarawak and Sabah to help distinguish unpaved public roads from logging roads.

We defined our buffer as the maximum distance from primary logging roads at which median measures of percent tree cover measured by the MODIS Vegetation Continuous Field product [33] exhibited depressed values indicative of the effects of logging on canopy cover. Forests within this distance were considered logged. The MODIS dataset defines percent tree cover continuously per pixel across Borneo over twelve consecutive months from March 2000 at 230-metre pixel resolution. To ensure that observations of the relationship between tree cover and distance from roads reflected only the effects of contemporaneous logging, we confined our observations to areas within 2.5 km of logging roads detected in 2000 that were still forested in 2010 and that were at least 5 km from logging roads mapped in 1970, in 1990, or in 2000 where in the latter case they were deforested by 2010. We analyzed an area of 38,940.6 km^2 surrounding 16,336.4 km of selected logging roads (see Figure S2; Table S2 in File S1) and observed the relationship separately for Kalimantan, Sarawak, and Sabah based on the assumption that logging practices may differ between regions.

Finding the relationship between percent tree cover and distance from roads to be similar between Kalimantan, Sarawak and Sabah, we nominated a single distance threshold and buffered all logging roads by this distance to yield an initial estimate of the total area logged since 1973. This initial estimate was considered conservative, however, given its linear geometry and the fact that imagery detects only relatively marked instances of canopy damage [34]. To address this caveat, we delimited the 2010 forest areas enclosed by the buffered logging roads, and reclassified these 'enclosed' forests as logged wherever they met the following criteria: (i) area <100 ha; or ii) felling within the secondary-forest (*hutan sekunder*) class mapped by the Indonesian Ministry of Forestry (MoF) in 2009 and 2000 via visual interpretation of LANDSAT imagery [35,36]. The MoF secondary-forest class encompasses forests where logging concessions were granted, evidence of logging was observed (e.g. roads), and/or forest perturbation is known to exist, and it is taken generally as the official, broad estimate of logged-over forest. The combined extent of buffered roads and enclosed forests meeting the criteria represents a better approximation of the variable local shape/extent of logged areas. We therefore adopted this combined extent as the 'total area logged since 1973'. The intersection of this area with 2010 'Forest' in turn defines the extent of 'Logged forest' as of year 2010.

Map validation

We conducted validations of all our maps to assess accuracy. For the 1973 forest map, we validated the accuracy by comparing it to photos taken from the declassified KH7 satellite-borne camera purchased from [24], having a 1 m^2 resolution able to

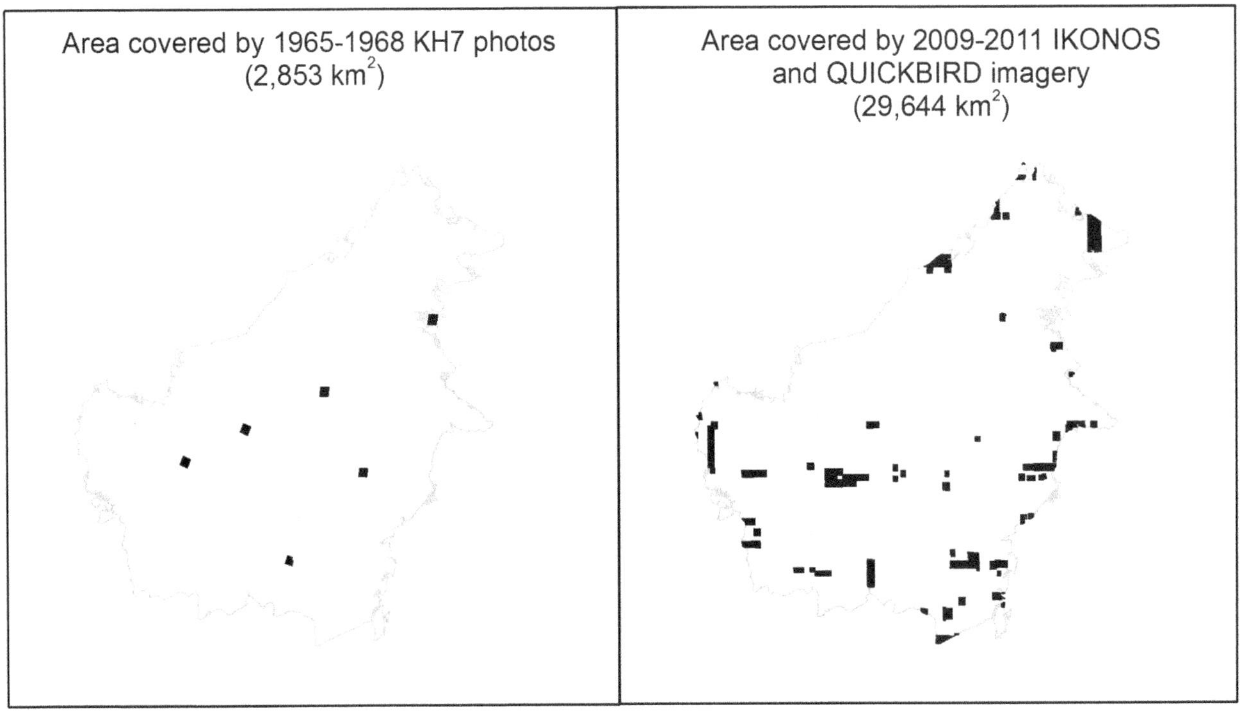

Figure 2. Map showing the areas where validation of the forest and plantation maps was performed. Area covered by 1965–68 KH7 imagery (Black) used to validate the 1973 forest map in Left Panel. Area covered by 2010 IKONOS and QUICKBIRD imagery (Black) used to validate the 2010 forest and plantation maps in Right Panel.

Table 2. Accuracy results of 1973 'Forest', and 'Non-forest' classification.

Class	User Accuracy	Producer Accuracy
Forest (n = 210)	92.3%	90.6%
Non-forest (n = 113)	82.3%	85.3%

Overall accuracy = 89.0%.

discriminate individual tree crowns. The KH7 images were acquired during 1965–1968, i.e., 5–8 years prior to our LANDSAT-MSS 1973 forest cover map. The KH7 images were geometrically registered to our orthographically rectified Global Land Survey (GLS) images using a second or third order polynomial co-registration technique. These photos cover 2,853 km^2 across six different sites encompassing a variety of topographic conditions (Figure 2). We randomly sampled 322 validation points within this area, with points being at least 1,000 m from each other to minimize spatial autocorrelation [37]. For each point, a 60 m×60 m area was visually interpreted as either 'Forest' or 'Non-forest' in the KH7 photos at 1:1,000 scale (see Figure S3 in File S1). A confusion matrix [38,39] determined the frequency of class agreement between our reference photos and forest map, as determined by overall accuracy and user's and producer's accuracy [40].

We checked the accuracy of the combined 2010 forest and industrial plantation map using Google Earth's collection of high-resolution QUICKBIRD and IKONOS images (0.6–1 m^2 resolution) accessed using the "Open Layer plug-in" of the open-source software QUANTUM GIS [41]. The total area of these images acquired during 2009–2011 in Borneo spanned 29,644 km^2 (Figure 2), within which we randomly sampled 1,921 points separated by more than 1000 m. As for the 1973 validation, we visually interpreted each 60 m×60 m area centered on a point to be 'Forest', 'Industrial oil palm plantation (IOPP)', Industrial timber plantation (ITP)' or 'Non-forest' in the high-resolution imagery. Several QUICKBIRD validation points are shown in Figure S4 (in File S1). We determined the frequency of class agreement between these reference points and our 2010 combined forest and plantation map using a confusion matrix.

Results

Validation of Maps

The overall accuracy for the 1973 'Forest' and 'Non-forest' classification is 89%. The per-class accuracies of 'Forest' and 'Non-forest' were similarly high, with user's accuracy ranging from 82–92%, and producer's accuracy ranging from 85–91% (Table 2; Table S3 in File S1). The overall accuracy for the 2010 map of 'Forest', 'Non-forest', 'Industrial oil palm plantation (IOPP)', and 'Industrial timber plantation (ITP)' is 93% (Table 3, Table S4 in File S1). The per-class accuracies of those four land cover classes were similarly high, with user's accuracies ranging from 83–97% and producer's accuracies from 75–94%. The lowest accuracy (75%) is for the ITP class, as there were few validation points for this class (n = 37).

Forest Clearance

We estimate that in 1973, 75.7% of Borneo remained under natural forest. That is 558,060 km^2 of mainly intact (i.e., unlogged) old-growth forest (Figure 3A; Table 4). By 2010, this forest area had been reduced by 168,493 km^2, representing a 30.2% forest loss over the previous four decades (Figure 3B). More than 97% (164,644 km^2) of this deforestation occurred in Borneo's coastal lowlands (<500 m asl). The Sultanate of Brunei and Sarawak have the lowest rates of deforestation with 8.4% and 23.1%, respectively. Sabah and Kalimantan have the highest rates (39.5% and 30.7%) (Table 4). Of the 168,493 km^2 total forested area lost since 1973, 51% (86,339 km^2) had been logged between 1973 and 2005, and 33% (56,080 km^2) had been converted to industrial plantations (IOPP and ITP). In 2010, the area planted in IOPPs and ITPs was 64,943 km^2 and 10,537 km^2, respectively, representing 10% of Borneo (Figure 3D, Table S5 in File S1).

Logging roads and logged forests

We calculated that 271,819 km of primary logging roads were opened between 1973 and 2010 (Figure 3C), equating to an overall density of 0.48 km road per km^2 of forest in 1973 (Table 5). The highest density is found in Sarawak (0.89 km km^{-2}) and the lowest density in Brunei (0.18 km km^{-2}).

Percent tree cover exhibited depressed values indicative of the effects of logging on canopy cover at least 460 m nominally from primary logging roads in Borneo (Figure 4). This nominal distance is measured from pixel center to pixel center and may be extended to 700 m practically upon accounting for the breadth of pixels and variation in where a logging road may fall therein. Taking 700 m as the upper distance thresholds and then applying our empirical 700 m buffer distance to all logging roads (Figure 3C) and

Table 3. Accuracy results of 2010 'Forest', 'Non-forest', 'Industrial oil palm plantation (IOPP)', and 'Industrial timber plantation (ITP)' classification.

Class	User Accuracy	Producer Accuracy
Forest(n = 458)	91.4%	90.3%
Non-forest (n = 1208)	95.4%	94.6%
IOPP (n = 218)	83.0%	94.7%
ITP (n = 37)	97.2%	75%

Overall accuracy = 93.1%.

reclassifying any enclosed forests as 'Logged forest', we estimate the total forest area logged since 1973 at 266,257 km^2. Of these 266,257 km^2 some 67.5% (179,917 km^2) remain standing as forest in 2010, 32.5% (86,340 km^2) were cleared.

Remaining intact forest in 2010

An estimated 389,566 km^2 (52.8%) of the island remained forested (either intact or logged) in 2010. Intact forests represent 53.8% (209,649 km^2) of the total remaining forest area or 28.4% of the whole Borneo (Table 4). Brunei has the highest proportion of intact forest area at 56.9%, compared to 32.6% in Kalimantan, 19.1% in Sabah and 14.6% in Sarawak.

Much of the remaining intact forests will be logged and converted under the current forest-use designations. Some 42% (88,150 km^2) of intact forests fall within the 'production forest' land-use class and will be logged (Table S5 in File S1). The actual area of intact forest in the land use designated for production is greater than the area of intact forests in protected areas (Table S5 in File S1). A further 16% (33,548 km^2) of these intact forests will be converted based on their intended land-use. Given the present distribution of intact forest across land use designations, the future extent of intact forest for all Borneo may decline to only 87,953 km^2, or 11.9% of Borneo, assuming that all production forests are ultimately logged, area designated for conversion is

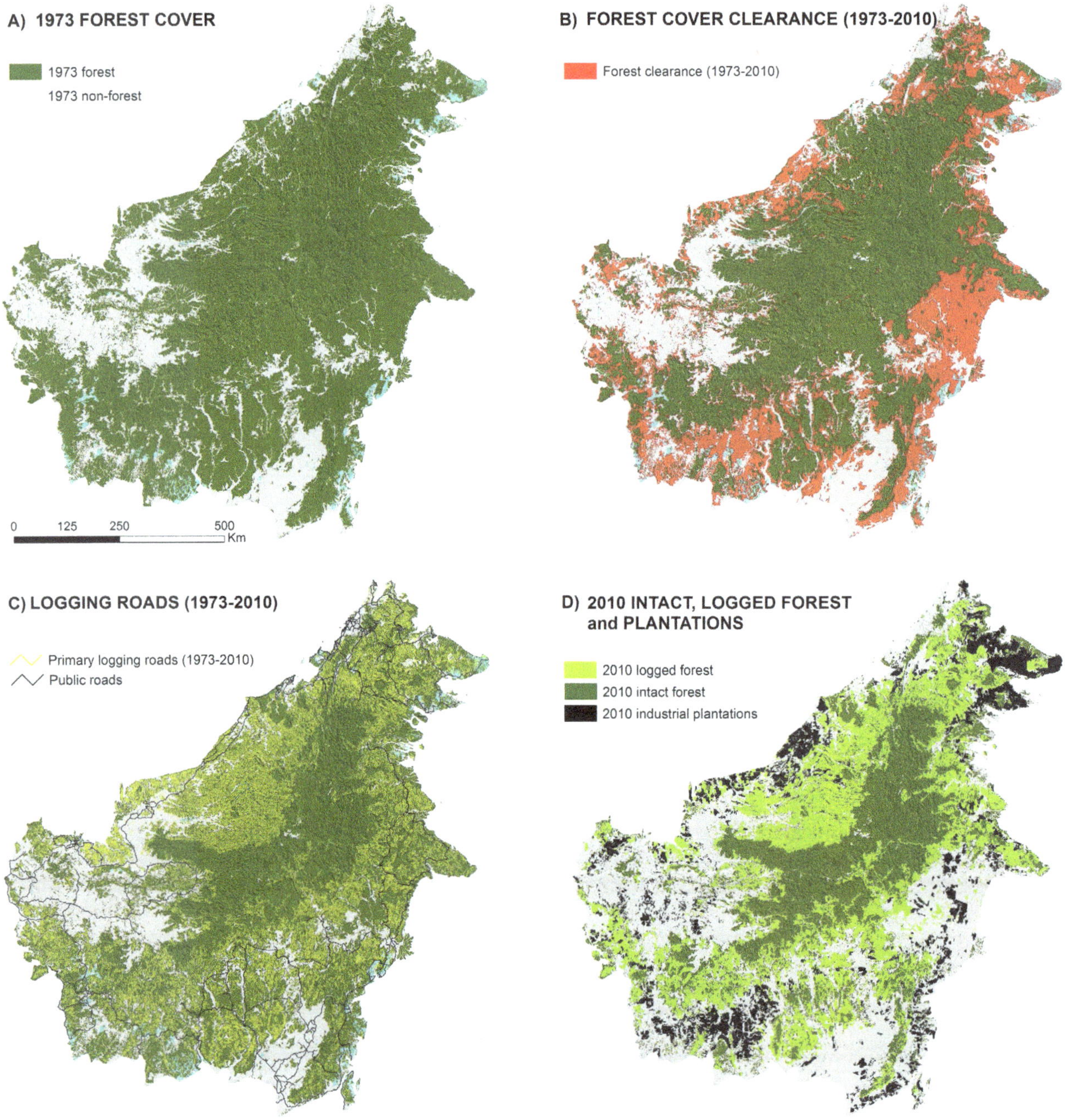

Figure 3. Four decades of forest persistence, clearance and logging on Borneo. Forest (dark green) and non-forest (white) in year 1973, and residual clouds (cyan) in Panel A. Areas of forest loss during 1973–2010 (red) in Panel B. Primary logging roads from 1973–2010 (yellow lines) in Panel C. Remaining intact forest (dark green), remaining logged forest (light green), and industrial oil palm and timber plantations (Black) in year 2010 in Panel D.

Table 4. Intact, logged and cleared forest area by country and elevation, 1973 and 2010.

	Borneo	Brunei	Kalimantan	Sabah	Sarawak
Total land area (km^2)					
0–500 m	601,276	5,584	446,995	54,371	94,326
501–1000 m	98,404	185	62,705	14,379	21,135
1001–1500 m	32,836	26	21,154	4,217	7,439
1501–2000 m	4,504	1	2,549	865	1,089
>2000 m	168	0	19	133	16
All elevations	*737,188*	*5,796*	*533,422*	*73965*	*124,005*
Forest area in 1973 (km^2)					
0–500 m	424,753	4,286	318,050	39,721	62,697
501–1000 m	96,159	184	61,811	13,220	20,944
1001–1500 m	32,504	26	21,118	3,950	7,411
1501–2000 m	4,481	0	2,544	852	1,085
>2000 m	163	0	19	129	16
All elevations	*558,060*	*4,496*	*403,541*	*57,871*	*92,152*
Intact Forest area in 2010 (km^2)					
0–500 m	114,017	3,100	95,123	7,344	8,450
501–1000 m	65,036	178	55,560	4,279	5,019
1001–1500 m	26,552	26	20,844	1,828	3,855
1501–2000 m	3,894	0	2,504	568	821
>2000 m	151	0	17	119	16
All elevations	*209,649*	*3,303*	*174,048*	*14,138*	*18,161*
Logged Forest area in 2010 (km^2)					
0–500 m	146,092	808	99,609	11,634	34,042
501–1000 m	27,531	7	5,633	7,015	14,876
1001–1500 m	5,710	0	269	1,941	3,500
1501–2000 m	573	0	40	270	263
>2000 m	11	0	2	9	0
All elevations	*179,917*	*815*	*105,553*	*20,868*	*52,681*
Forest area loss 1973–2010 (km^2)					
0–500 m	164,644	378	123,318	20,743	20,205
501–1000 m	3,593	0	618	1,926	1,048
1001–1500 m	241	0	5	181	56
1501–2000 m	14	0	0	14	0
>2000 m	1	0	0	1	0
All elevations	*168,493 (30.2%)*	*378 (8.4%)*	*123,941 (30.7%)*	*22,865 (39.5%)*	*21,309 (23.1%)*

Table 5. Length and density of primary logging roads by country.

	1973 forest cover (km^2)	Logging road length (km)	Logging road density (km/km^2)
Brunei	4,496	818	0.18
Kalimantan	403,541	151,101	0.37
Sabah	57,871	37,660	0.65
Sarawak	92,152	82,239	0.89
Borneo	**558,060**	**271,819**	**0.48**

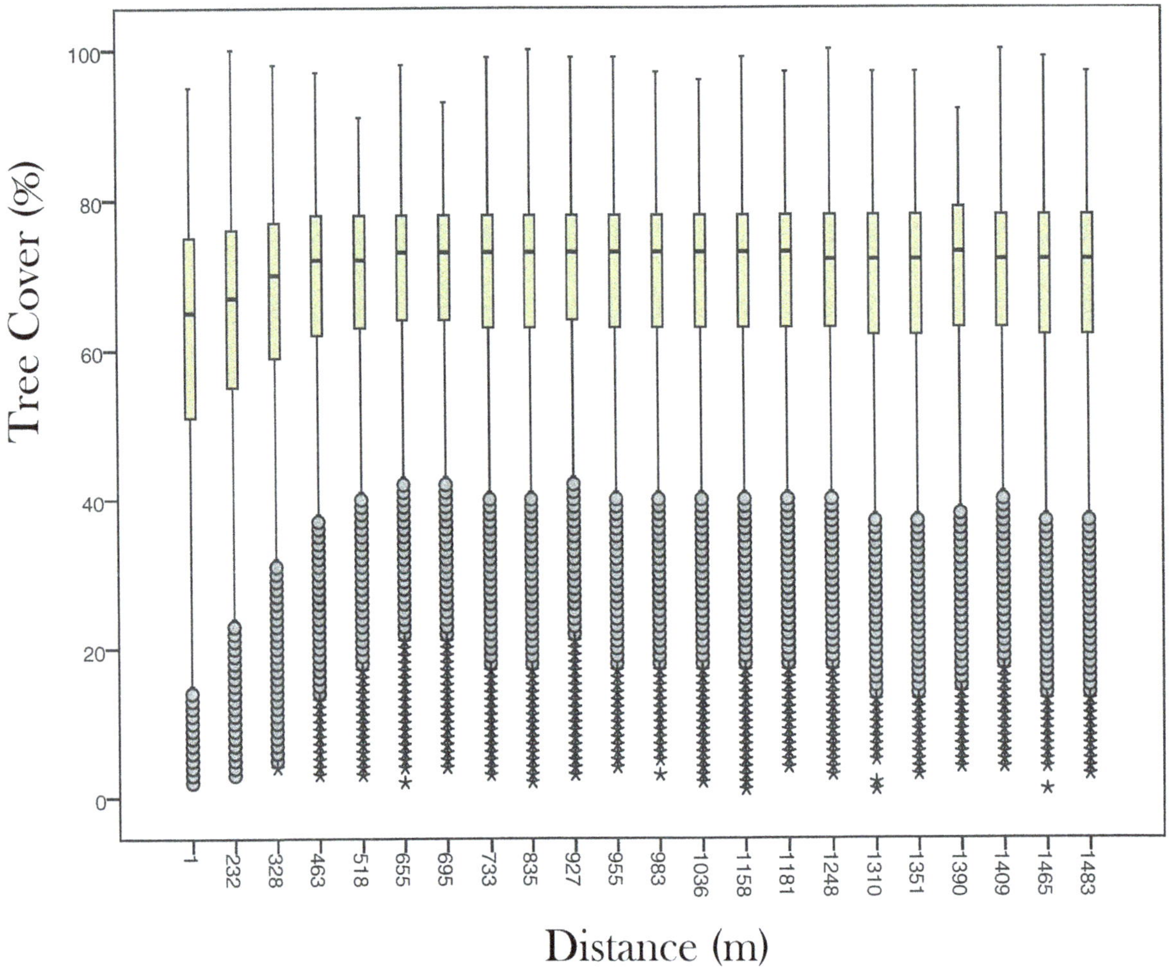

Figure 4. Relationship between distance to selected logging roads established in year 2000 and percent tree cover (2000/2001). The black points at approximately 70% on the y-axis are median values for each distance class. The boxes centered on each median are bounded by the 75th and 25th percentile values of each distance class. Grey circles flag 'outliers' having values lower/higher than the 25th/75th percentile by 1.5–3 times the inter-quartile range. Asterisks flag extreme 'outliers' having values >3 times lower/higher than the 25th/75th percentile value.

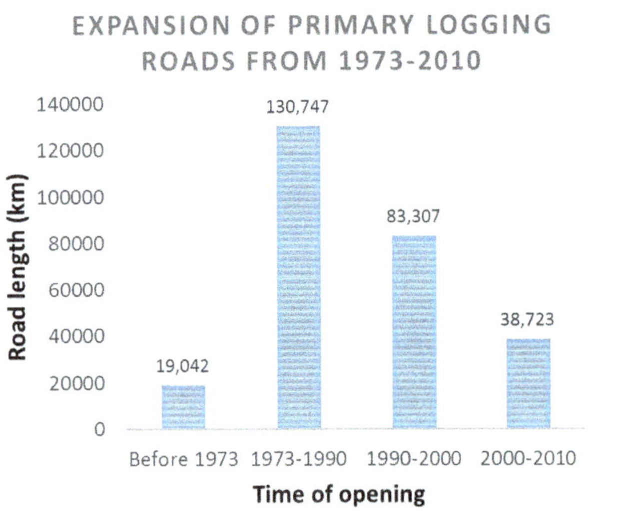

Figure 5. Expansion of the primary logging road network in Borneo from 1973 until 2010.

ultimately converted or degraded, assuming protected intact forests remain unperturbed and that the extents of these land-use decisions remain unchanged. For the lowland forests (<500 m asl), the most threatened, the corresponding figures are 33,773 km^2 and 5.6% of Borneo's lowlands.

We observed an appreciable area of logged forest within the protection designation, which prohibits logging. Of the 179,917 km^2 of current logged forest in Borneo (Table 4), 10.3% falls within the protected areas (Table S5 in File S1) wherein it constitutes 17% of all forest cover. Some, but not all, of this logged forest is evidence of illegal logging. In other cases, it evidences legally logged forests that were later re-designated for protection, e.g., the Sebangau National Park, declared in 2004 [42]. While our data do not allow distinguishing illegal from legal logging *per se*, it is reasonable to assume that some logged forests within the protection designation were exploited illegally.

Spatial and Temporal Trends in Logging, 1973–2010

Temporal and spatial trends in logging describe an initial 'boom' in the exploitation of easily accessible lowland forests, followed by a slower expansion into remaining, more marginal forests abutting highland protected forests. Over half of all logging

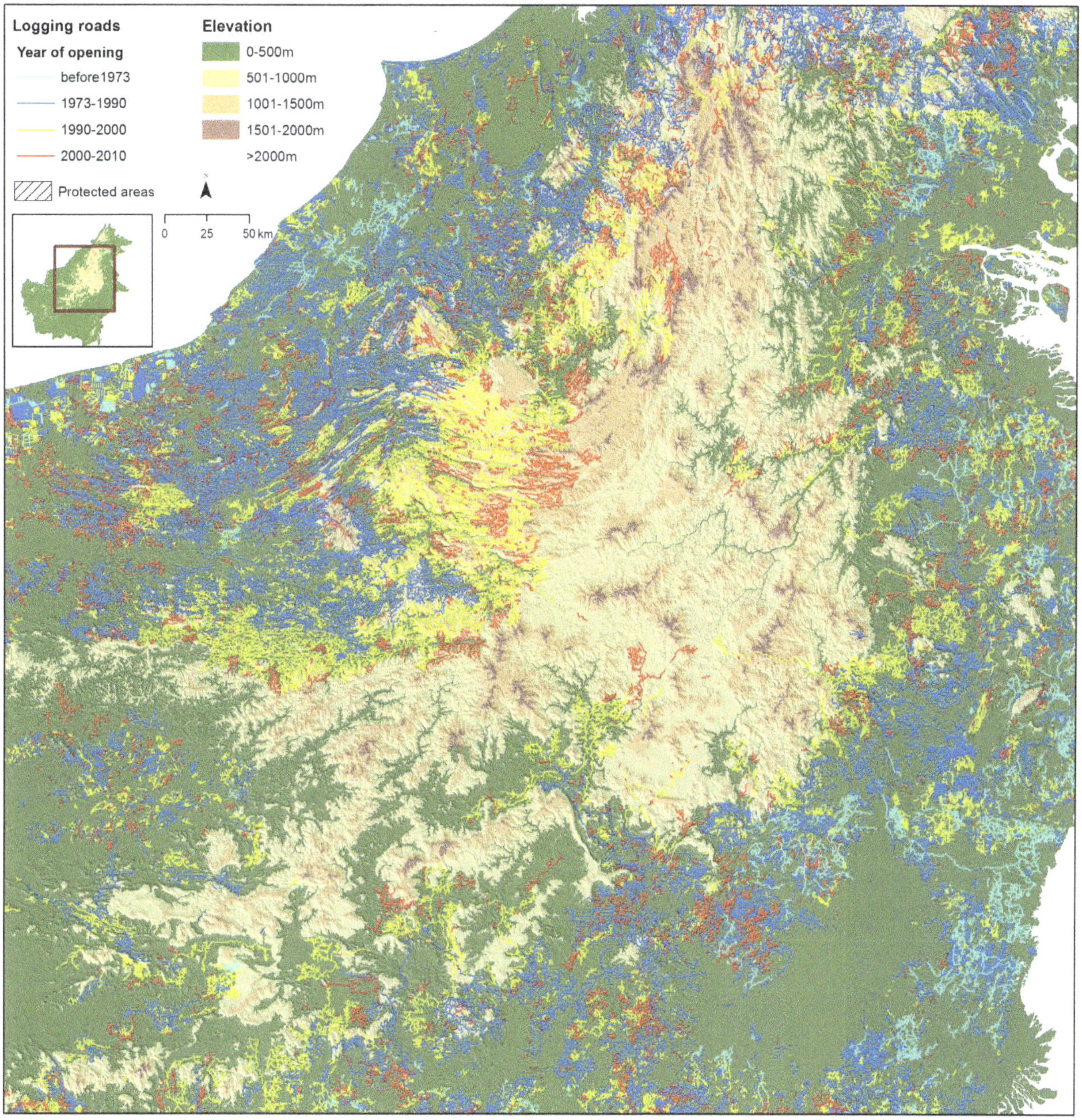

Figure 6. The heart of Borneo and the spatial progression of logging since 1973 depicting a logging 'frontier' moving steadily upward and inland, from the lowland coasts to the highlands. In many areas, logging roads are surrounding and abutting against the edges of highland protected forests – the last contiguous bastions of intact forest.

roads by length were established prior to 1990 (Figure 5), when the logging industry was relatively young and intact forests were easily accessible.

Since 2000 the rate of logging roads expansion has approximately halved (Figure 5), reflecting a growing scarcity of unlogged, accessible forests. The spatial progression of logging since 1973 therefore depicts a logging 'frontier' moving steadily upward and inland, from the lowland coasts to the highlands, in search of remaining unlogged forests (Figure 6). In many areas, logged forests have expanded to surround and abut against the edges of highland protected forests – the last contiguous bastions of intact forest.

Discussion

Methodological considerations

Our study has generated the first full island-wide maps of forest cover and forest cover change in Borneo since industrial logging began in early 1970s, with high classification accuracies, appropriate for analyzing fine scale changes in forest cover [43]. We note that >90% of the validation area (the black areas in Figure 2) is distributed in the lowlands (between 0–500 m asl), but the lack of validation area in mountains of the island did not overestimate our validation because forest cover loss occurred in the lowlands (0–500 m). Nevertheless, our accuracy assessments were conducted on just a 0.4–4% spatial subset, and therefore may vary across the study area.

We caution that our estimates of remaining logged forest area are likely an under-estimation because they did not capture most narrow, ephemeral, sub-canopy logging trails branching off the main logging roads or smaller-scale community-level logging activities. For example, the small trails created by illegal logging such as those found in Gunung Palung national park in 1999 [14] remained undetected by our method. In turn, this means that our remaining intact forest area may be over-estimated. A direct verification of this over-estimation is not possible because of a lack of contemporary, fine-scale annual satellite imagery in which to observe logging trails. We observe a general agreement between our empirically-estimated buffer distance of 700 m – the distance from logging roads within which timber extraction extends – and similar observations in the literature. It compares favorably with estimates of 350–1000 m in Sabah and Sarawak [19], <500 m in Papua New Guinea [44], and 1 km in Central Africa [27] and Sumatra [25]. Finally, we acknowledge the limitations of buffering primary logging roads with a fixed distance to estimate the extent of logged forests because this buffer distance varies with accessibility such as topography and soil type. Modeling the buffer distance across landscapes using spatial attributes is an area of research where we would hope to make further refinements.

Our results merit comparison with those from an earlier study of forest degradation in Malaysian Borneo. Bryan et al. [19] evaluated the impact of logging in Northern Borneo (Brunei, Sabah and Sarawak) and estimated that 45,391 km^2 of intact forest and 68,504 km^2 of degraded forest remained in 2009, representing 22% and 34% of northern Borneo, respectively. Our numbers for 2010 are of the same order of magnitude: 35,602 km^2 (17.4%) of intact forest and 74,364 km^2 (36.4%) of logged forest. Our methods were similar but not identical. As with our analysis, Bryan et al. [19] used a combination of supervised classification and visual interpretation to identify and map logging roads using LANDSAT, and used a buffer distance of only 350 m to estimate the area of forest degradation. Here, based on our observations (see Figure 4), we used a distance of 700 m to define the area impacted by timber harvesting, which likely provides the principle explanation for the differences in estimated intact and logged area between the two studies.

Relevance for conservation

We show that Borneo has lost forest cover nearly twice as fast as the rest of the world's humid tropical forests [45], in part to create and expand industrial plantations, and in particular oil palm. About 10% of Borneo has become covered in industrial scale monoculture plantations. In addition, Borneo has been intensively logged. The concentration of logging roads found on Borneo is high compared to international standards. Laporte et al. [27], using the same road mapping technique, mapped only 51,916 km (0.03 km of roads per km^2) of primary logging roads between 1973–2003 in Central Africa (Cameroon, Central African Republic, DRC, Equatorial Guinea, Gabon, Republic of Congo).This road density is 16 times less than our Borneo-wide estimated density. The remarkably high density of logging roads for Sarawak is corroborated by independent observations [19]. Little intact forest remains in that region. But, with 389,566 km^2 (52.8%) of the island remaining forested, of which 209,649 km^2 remains intact there is still much hope for biodiversity conservation in Borneo.

In all regions of Borneo it is conversion to plantation development that drives forest loss, and not logging *per se* [17]. A logged forest retains high conservation value [4]. In Indonesian Borneo, the law stipulates that forests in timber concessions are permanent, but the government of Indonesia has bypassed this law. For example, the government re-zoned over 25% of these concessions for plantation development and thus permitted their conversion [42]. In contrast, rates of forest clearance have been lower in Sarawak despite having the highest density of logging roads. Sarawak had developed the most effective techniques to penetrate the rugged forest interior to harvest natural timber, but it did not systematically convert forests to plantations because of the lack of skilled labor [46]. Regional climatic discrepancies between Kalimantan, Sarawak and Sabah may also partially explain the differences we observe. Vulnerability to drought and fire may drive more deforestation in Kalimantan and Sabah than in Sarawak's year-round wetter forests [15,16,47].

Protecting logged forests from fires and conversion to agricultural plantations is therefore an urgent priority for reducing rates of deforestation in Borneo. This could be achieved by reclassifying timber concessions in natural forests as protected areas under the IUCN Protected Area Category VI, as many other countries have done [42]. Land use economics, however, drive the rapid conversion of over-logged forests to oil palm plantations where short-term revenues are much higher [48,49]. Therefore, the law needs to stipulate more strongly that remaining forests need to be kept and conversion of natural forest minimized. There are signs that this is happening, such as the recent re-zoning of a former logging concession into Sebangau National park. Meanwhile oil palm plantations continue to expand, and a much improved understanding is needed on how these potentially conflicting targets can be reconciled and trade-offs minimized [50]. One important factor in this is a more complete valuation of the long-term economic and health values of natural forests managed for timber as well as other services. Especially, the role forests play in sequestering atmospheric carbon and buffer water run-off and thus prevent soil erosion and flooding are important [51], as well as the more traditional potential economic value of timber and other forest products [52].

Author Contributions

Conceived and designed the experiments: DG SS. Performed the experiments: DG SS. Analyzed the data: DG SS E. Molidena HH. Contributed reagents/materials/analysis tools: NW NA MA MQ. Wrote the paper: DG SS DS RN E. Meijaard.

References

1. Gibson L, Lee TM, Koh LP, Brook BW, Gardner TA, et al. (2011) Primary forests are irreplaceable for sustaining tropical biodiversity. Nature 478: 378–381.
2. Giam X, Clements GR, Aziz SA, Chong KY, Miettinen J (2011) Rethinking the 'back to wilderness' concept for Sundaland's forests. Biological Conservation.
3. Berry NJ, Phillips OL, Lewis SL, Hill JK, Edwards DP, et al. (2010) The high value of logged tropical forests: lessons from northern Borneo. Biodiversity and Conservation 19: 985–997.
4. Meijaard E, Sheil D (2007) A logged forest in Borneo is better than none at all. Nature 446: 974.

5. Putz FE, Zuidema PA, Synnott T, Peña-Claros M, Pinard MA, et al. (2012) Sustaining conservation values in selectively logged tropical forests: the attained and the attainable. Conservation Letters.
6. Billand A, Nasi R (2008) Production dans les forêts de conservation, conservation dans les forêts de production: vers des forêts tropicales durables, à partir du cas de l'Afrique centrale.
7. Clark C, Poulsen J, Malonga R, Elkan P Jr (2009) Logging concessions can extend the conservation estate for Central African tropical forests. Conservation Biology 23: 1281–1293.
8. Edwards DP, Laurance WF (2013) Biodiversity despite selective logging. Science (New York, NY) 339: 646.
9. Meijaard E (2005) Life after logging: reconciling wildlife conservation and production forestry in Indonesian Borneo. Center for International Forestry Research. Available: http://www.cifor.org/publications/pdf_files/books/BMeijaard0501E0.pdf. Accessed 10 June 2013.
10. Wilson KA, Meijaard E, Drummond S, Grantham HS, Boitani L, et al. (2010) Conserving biodiversity in production landscapes. Ecological Applications 20: 1721–1732.
11. Meijaard E, Sheil D (2008) The persistence and conservation of Borneo's mammals in lowland rain forests managed for timber: observations, overviews and opportunities. Ecological Research 23: 21–34.
12. McKinnon K (1996) The ecology of Kalimantan: [Hong Kong]: Periplus Editions.
13. Brookfield H, Byron Y (1990) Deforestation and timber extraction in Borneo and the Malay Peninsula: the record since 1965. Global Environmental Change 1: 42–56.
14. Curran LM, Trigg SN, McDonald AK, Astiani D, Hardiono YM, et al. (2004) Lowland forest loss in protected areas of Indonesian Borneo. Science 303: 1000–1003.
15. Goldammer JG, Seibert B (1990) The impact of droughts and forest fires on tropical lowland rain forest of East Kalimantan. Fire in the tropical biota: Springer. pp. 11–31.
16. Siegert F, Ruecker G, Hinrichs A, Hoffmann A (2001) Increased damage from fires in logged forests during droughts caused by El Nino. Nature 414: 437–440.
17. Carlson KM, Curran LM, Asner GP, Pittman AM, Trigg SN, et al. (2012) Carbon emissions from forest conversion by Kalimantan oil palm plantations. Nature Climate Change.
18. Casson A (2000) The hesitant boom: Indonesia's oil palm sub-sector in an era of economic crisis and political change: Center for International Forestry Research.
19. Bryan JE, Shearman PL, Asner GP, Knapp DE, Aoro G, et al. (2013) Extreme Differences in Forest Degradation in Borneo: Comparing Practices in Sarawak, Sabah, and Brunei. PloS one 8: e69679.
20. Satriastanti FE (2012) SBY Sees Kalimantan as the 'Lungs of the World'. The Jakarta Post January 20, 2012. Available: http://www.thejakartaglobe.com/home/sby-sees-kalimantan-as-the-lungs-of-the-world/492525. Accessed 25 June 2014.
21. Ministry of Forestry (2011) Forestry Statistics of Indonesia. Jakarta, Indonesia: Direktorat Jenderal Planologi Kehutanan, Kementerian Kehutanan.
22. Hoekman DH, Vissers MA, Wielaard N (2010) PALSAR wide-area mapping of Borneo: methodology and map validation. Selected Topics in Applied Earth Observations and Remote Sensing, IEEE Journal of 3: 605–617.
23. Rulequest (2007) Data mining tools SEE5. Available: http://www.rulequest.com/see5-info.html. Accessed 25 June 2014.
24. USGS (2013) Global Land Survey. U.S. Geological Survey. Available: http://earthexplorer.usgs.gov. Accessed 25 June 2014.
25. Margono BA, Turubanova S, Zhuravleva I, Potapov P, Tyukavina A, et al. (2012) Mapping and monitoring deforestation and forest degradation in Sumatra (Indonesia) using Landsat time series data sets from 1990 to 2010. Environmental Research Letters 7: 034010.
26. Souza C Jr, Barreto P (2000) An alternative approach for detecting and monitoring selectively logged forests in the Amazon. International Journal of Remote Sensing 21: 173–179.
27. Laporte NT, Stabach JA, Grosch R, Lin TS, Goetz SJ (2007) Expansion of industrial logging in Central Africa. Science 316: 1451–1451.
28. Potapov P, Yaroshenko A, Turubanova S, Dubinin M, Laestadius L, et al. (2008) Mapping the world's intact forest landscapes by remote sensing. Ecology and Society 13: 51.
29. Monteiro A, Souza C Jr, Barreto P (2003) Detection of logging in Amazonian transition forests using spectral mixture models. International Journal of Remote Sensing 24: 151–159.
30. Matricardi E, Skole D, Chomentowski M, Cochrane M (2001) Multi-temporal detection of selective logging in the Amazon using remote sensing. BSRSI Research Advances: 2003–2001.
31. Souza C Jr (2013) Degradation monitoring methods. GOFC-GOLD.
32. Ministry of Public Works (2013). Available: http//www.pu.go.id. Accessed 25 June 2014.
33. Townshead JRG, Carroll M, Dimiceli C, Sohlberg R, Hansen M, et al. (2011) Vegetation Continuous Fields MOD44B.5, 2010 Percent Tree Cover, Collection 5 Version 1 Collection 5 Version 1 ed. Maryland, USA: University of Maryland, College Park.
34. Peres CA, Barlow J, Laurance WF (2006) Detecting anthropogenic disturbance in tropical forests. Trends in Ecology & Evolution 21: 227–229.
35. Ministry of Forestry (2011) Landcover Map 2009/2010. Jakarta: Direktorat Inventarsi Dan Pemantauan Sumber Daya Hutan Dan Direktorat Jenderal Planologi Kehutanan Kementrian Kehuatanan (Directorate of Forest Resource Inventory and Monitoring and Directorate General of Forest Planning). Available: http://nfms.dephut.go.id/ipsdh/. Accessed 25 June 2014.
36. Ministry of Forestry (2000) Landcover Map 2000. Jakarta: Direktorat Inventarsi Dan Pemantauan Sumber Daya Hutan Dan Direktorat Jenderal Planologi Kehutanan Kementrian Kehuatanan (Directorate of Forest Resource Inventory and Monitoring and Directorate General of Forest Planning). Available: http://nfms.dephut.go.id/ipsdh/. Accessed 25 June 2014.
37. Koening WD (1999) Spatial autocorrelation of ecological phenomena. Trends in Ecology & Evolution 14: 22–26.
38. Foody GM (1992) On the compensation for chance agreement in image classification accuracy assessment. Photogrammetric Eng Remote Sensing 58: 1459–1460.
39. Aronoff S (1982) The map accuracy report:a user's view. Photogrammetric Eng Remote Sensing 48: 1309–1312.
40. Fielding AH, Bell JF (1997) A review of methods for the assessment of prediction errors in conservation presence/absence models. Environmental conservation 24: 38–49.
41. Quantum GIS (2013). Available: http://www.qgis.org/en/site/forusers/download.html. Accessed2 April 2013.
42. Gaveau DL, Kshatriya M, Sheil D, Sloan S, Molidena E, et al. (2013) Reconciling Forest Conservation and Logging in Indonesian Borneo. PloS one 8: e69887.
43. Thomlinson JR, Bolstad PV, Cohen WB (1999) Coordinating methodologies for scaling land-cover classifications from site-specific to global: steps towards validating global map products. Remote Sensing of the Environment 70: 16–28.
44. Shearman PL, Ash J, Mackey B, Bryan JE, Lokes B (2009) Forest conversion and degradation in Papua New Guinea 1972–2002. Biotropica 41: 379–390.
45. Achard F, Eva HD, Stibig HJ, Mayaux P, Gallego J, et al. (2002) Determination of deforestation rates of the world's humid tropical forests. Science 297: 999–1002.
46. Ngidang D (2002) Contradictions in land development schemes: the case of joint ventures in Sarawak, Malaysia. Asia Pacific Viewpoint 43: 157–180.
47. Walsh R, Newbery D (1999) The ecoclimatology of Danum, Sabah, in the context of the world's rainforest regions, with particular reference to dry periods and their impact. Philosophical Transactions of the Royal Society of London Series B: Biological Sciences 354: 1869–1883.
48. Fisher B, Edwards DP, Larsen TH, Ansell FA, Hsu WW, et al. (2011) Cost-effective conservation: calculating biodiversity and logging trade-offs in Southeast Asia. Conservation Letters 4: 443–450.
49. Venter O, Meijaard E, Possingham H, Dennis R, Sheil D, et al. (2009) Carbon payments as safeguard for threatened tropical mammals. Conservation letters 2: 123–129.
50. Abram NK, Meijaard E, Ancrenaz M, Runting RK, Wells JA, et al. (2014) Spatially explicit perceptions of ecosystem services and land cover change in forested regions of Borneo. Ecosystem Services. In press.
51. Wells JA, Meijaard E, Abram N, Wich S (2013) Forests, floods, people and wildlife on Borneo. A review of flooding and analysis of local perceptions of flooding frequencies and trends, and the role of forests and deforestation in flood regimes, with a view to informing government decision-making on flood monitioring, forest management and biodiversity conservation. Bangkok, Thailand: UNEP.
52. Meijaard E, Abram NK, Wells JA, Pellier A-S, Ancrenaz M, et al. (2013) People's Perceptions about the Importance of Forests on Borneo. PLOS ONE 8: e73008.
53. Butler R (2013) Tropical rainforest conservation. Available: http://rainforests.mongabay.com/0814.htm. Accessed 1 December 2013.

3

Land Management Practices Associated with House Loss in Wildfires

Philip Gibbons[1]*, Linda van Bommel[1], A. Malcolm Gill[1], Geoffrey J. Cary[1], Don A. Driscoll[1], Ross A. Bradstock[2], Emma Knight[3], Max A. Moritz[4], Scott L. Stephens[4], David B. Lindenmayer[1]

1 The Fenner School of Environment and Society, The Australian National University, Canberra, Australian Capital Territory, Australia, **2** Centre for Environmental Risk Management of Bushfires, University of Wollongong, Wollongong, New South Wales, Australia, **3** Centre for Mathematics and its Applications, The Australian National University, Canberra, Australian Capital Territory, Australia, **4** Ecosystem Sciences Division, Department of Environmental Science, Policy and Management, University of California, Berkeley, California, United States of America

Abstract

Losses to life and property from unplanned fires (wildfires) are forecast to increase because of population growth in peri-urban areas and climate change. In response, there have been moves to increase fuel reduction—clearing, prescribed burning, biomass removal and grazing—to afford greater protection to peri-urban communities in fire-prone regions. But how effective are these measures? Severe wildfires in southern Australia in 2009 presented a rare opportunity to address this question empirically. We predicted that modifying several fuels could theoretically reduce house loss by 76%–97%, which would translate to considerably fewer wildfire-related deaths. However, maximum levels of fuel reduction are unlikely to be feasible at every house for logistical and environmental reasons. Significant fuel variables in a logistic regression model we selected to predict house loss were (in order of decreasing effect): (1) the cover of trees and shrubs within 40 m of houses, (2) whether trees and shrubs within 40 m of houses was predominantly remnant or planted, (3) the upwind distance from houses to groups of trees or shrubs, (4) the upwind distance from houses to public forested land (irrespective of whether it was managed for nature conservation or logging), (5) the upwind distance from houses to prescribed burning within 5 years, and (6) the number of buildings or structures within 40 m of houses. All fuel treatments were more effective if undertaken closer to houses. For example, 15% fewer houses were destroyed if prescribed burning occurred at the observed minimum distance from houses (0.5 km) rather than the observed mean distance from houses (8.5 km). Our results imply that a shift in emphasis away from broad-scale fuel-reduction to intensive fuel treatments close to property will more effectively mitigate impacts from wildfires on peri-urban communities.

Editor: Rohan H. Clarke, Monash University, Australia

Funding: The authors received financial support from the Environmental Decisions Hub of the Australian Government's National Environmental Research Program and the Australian Research Council Centre of Excellence for Environmental Decisions. The funders had no role in study design, data collection and analysis, decision to publish, or preparation of the manuscript.

Competing Interests: The authors have declared that no competing interests exist.

* E-mail: Philip.Gibbons@anu.edu.au

Introduction

Peri-urban communities in fire-prone regions around the world are at increasing risk from unplanned fires (wildfires) because of population growth [1,2,3] and climate change [4,5,6,7]. The potential consequences of these factors were illustrated by recent major wildfires in California (26 deaths, 3361 houses lost) [8], Russia (54 deaths, circa. 2000 houses lost) [9] and Australia (173 deaths, 2133 houses lost) [10]. The behaviour of wildfires is primarily determined by weather, terrain and fuel [11]. Fuel in vegetation is often the easiest of these to manipulate [12]. Thus, there have been moves to increase the area of fuel reduction in many fire-prone regions [10,13,14].

Common fuel-reduction treatments employed in fire-prone landscapes are clearing, prescribed burning, grazing and mechanical removal of biomass (e.g., thinning) [6,12,13,15]. These treatments are often undertaken at broad-scales and distant from peri-urban communities. For example, in the United States of America, 89% of all fuel-reduction treatments undertaken on federal lands were >2.5km from the wildland urban interface [13]. Fuel treatments can be expensive [13] and can have undesirable health [16] and environmental [17] impacts (although not in all cases [18]). Yet, evidence that these treatments mitigate impacts on peri-urban communities from wildfires remains extremely limited [1].

Houses are a critical asset to protect during wildfires because most wildfire fatalities occur among people evacuating late from, sheltering in, or defending them [19]. Houses are destroyed during wildfires when exposed to flames in adjacent fuel, radiant heat from nearby fuel (≤40m) [20], or airborne embers and firebrands originating in nearby and distant fuel (typically<10 km) [21,22]. However, the relative importance of these different fuels—and therefore the relative effectiveness of different fuel treatments in protecting houses—have not been examined empirically. This is because wildfires are a difficult phenomenon to study [1,23]. Large, destructive fires cannot be lit experimentally, house loss during wildfires is often aggregated, preventing replication of landscape-scale variables, and adequate pre- and post-fire data are not always available. Thus, there are few wildfires that lend themselves to empirical research on the effects of the full range of fuel treatments on house loss.

Wildfires in south-eastern Australia in 2009 destroyed a large population of houses in landscapes with a mix of housing densities, terrains and fuel types, and occurred in landscapes where there were adequate pre- and post-fire data. These wildfires therefore provided a rare opportunity to robustly quantify and compare the effectiveness of different fuel types and different fuel treatments on house loss during wildfire.

Results

To quantify the relative effects of different fuels on house loss we sampled 499 houses and at each house recorded 24 potential explanatory variables representing the three principal drivers of fire behaviour (i.e. weather, terrain and fuel) [11]. We sampled extremes in several of these variables not achieved in previous studies. For example, the Forest Fire Danger Index (FFDI) [24] at sampled houses ranged from 5 to 189, slope ranged from 0.3 to 22.6°, the percent of cleared land upwind from houses ranged from 0 to 100% and the percent of land prescribe-burnt within 5 years upwind from houses ranged from 0 to 36.4%.

A model to predict house loss

A logistic regression model we selected to predict house loss contained eight significant explanatory variables (Table 1). This model indicated that a greater proportion of houses were lost where: there was a higher % cover of trees and shrubs within 40 m; the vegetation within 40 m was dominated by remnant native (rather than planted) vegetation; there were more buildings within 40 m; groups of trees or shrubs were closer in the upwind direction; forest burnt within 5 years in the upwind direction was distant rather than close; and houses were closer to public land (had less private land) in the upwind direction (Figure 1). In the best alternative logistic model we identified, variables representing the amount of land that is not State Forest and the amount of land that is not National Park replaced the amount of private land upwind from houses (together the former are negatively correlated with the latter). This alternative logistic model indicated that, other things being equal, houses were at similar risk when they occurred close to either National Park or State Forest. None of the interactions we tested (see Materials and Methods) were significant in the selected model.

The selected logistic regression model included several variables in addition to fuel that also affect fire behaviour. Other things being equal, weather had a strong effect, with a greater proportion of houses lost at higher levels of temperature and wind speed and lower levels of relative humidity (measured using FFDI). We included an "autocovariate" [25] in the selected logistic regression model to account for spatial autocorrelation between houses within 1 km of each other (see Materials and Methods). No variables representing terrain were significant in the selected model.

A Hosmer–Lemeshow test for the selected logistic model indicated that observed house loss was not significantly different from predicted house loss ($P = 0.487$). The area under the Receiver Operating Characteristic Curve (AUC) indicated that the fitted logistic model correctly discriminated between burnt and unburnt houses 80% of the time.

Predictions from the fitted logistic model indicated that reducing fuel could substantially reduce the number of houses destroyed during severe wildfires. With variables representing fuel held at observed minimum loads (i.e., 10% cover of planted trees and shrubs within 40 m from houses, 100 m to the nearest trees and shrubs in the upwind direction, 500 m to forest burnt ≤5 years ago in the upwind direction and no buildings within 40 m); and other variables fixed at their means (i.e., FFDI, the distance to public land and the covariate representing spatial autocorrelation), we predicted that 4.6% (±1.9% s.e.m) of all houses would be destroyed. Thus, under otherwise average conditions observed during these wildfires, minimizing key fuels at every house could potentially reduce the percent of houses destroyed from the observed value of 35.0% to a predicted mean value of 4.6% (±1.9% s.e.m). This equates to a reduction in the number of houses lost of 76%–97% (95% confidence interval). However, this level of fuel management is unlikely to be realised at all houses for reasons we outline in the Discussion.

Table 1. The selected logistic regression model used to predict the proportion of houses lost during the sampled wildfires.

Variable	Coefficient ± s.e.m	$P(>\|z\|)$
Intercept	−5.687±1.073	0.000
Tree and shrub cover (%) within 40 meters (m)	0.022±0.006	0.000
Log_{10} (FFDI)	1.062±0.3076	0.000
Log_{10} (amount of land not burnt within 5 years (m))	0.565±0.216	0.001
Vegetation type within 40 m (planted)	-	-
Vegetation type within 40 m (remnant)	0.726±0.246	0.003
Log_{10} (amount of private land (m)+1)	−0.479±0.199	0.016
Log_{10} (distance to nearest trees and shrubs (m)+1)	−0.574±0.191	0.003
Log_{10} (buildings within 40 m)	0.963±0.483	0.046
Autocovariate (spatial autocorrelation)	4.800±1.110	0.000

Significant explanatory variables, their coefficients and *P*-values in the logistic model selected to predict the (logit or log-odds) proportion of houses destroyed during wildfire. Vegetation type is a categorical variable with 'planted' being the reference level. The autocovariate represents spatial autocorrelation between neighbouring houses.

The relative effects of fuel variables on house loss

We used mean predictions from the selected logistic model to examine the relative effects of different fuels on house loss (Figure 1). Predictions for each fuel variable were made with the other significant explanatory variables held at their means, but with FFDI held at 100. This is the value for FFDI above which 64% of houses have been destroyed by wildfires in Australia [26] and the value for FFDI that invokes the highest level of public warning in fire-prone regions of Australia. We predicted that reducing remnant native vegetation around houses (within 40 m) from 90% cover (the observed maximum) to 5% cover reduced the likelihood of house loss by 43%. That is, every 10% reduction in remnant native vegetation cover around houses reduced the likelihood of house loss by approximately 5%. Thirty eight percent fewer houses were destroyed if surrounded (within 40 m) by predominantly planted vegetation rather than predominantly remnant native vegetation. Twenty six percent fewer houses were lost if further (100 m) from the nearest group of trees or shrubs in the upwind direction, compared with houses adjacent (0 m) to groups of trees or shrubs in the upwind direction. Compared with houses located 10 m from public forest, 14% fewer houses were lost if 200 m from public forest, and 26% fewer houses were lost if 2 km from public forest (the average distance between houses and public forest). On average, 15% fewer houses were lost if prescribed burning within 5 years was undertaken 0.5 km upwind from houses (the nearest distance between houses and prescribed

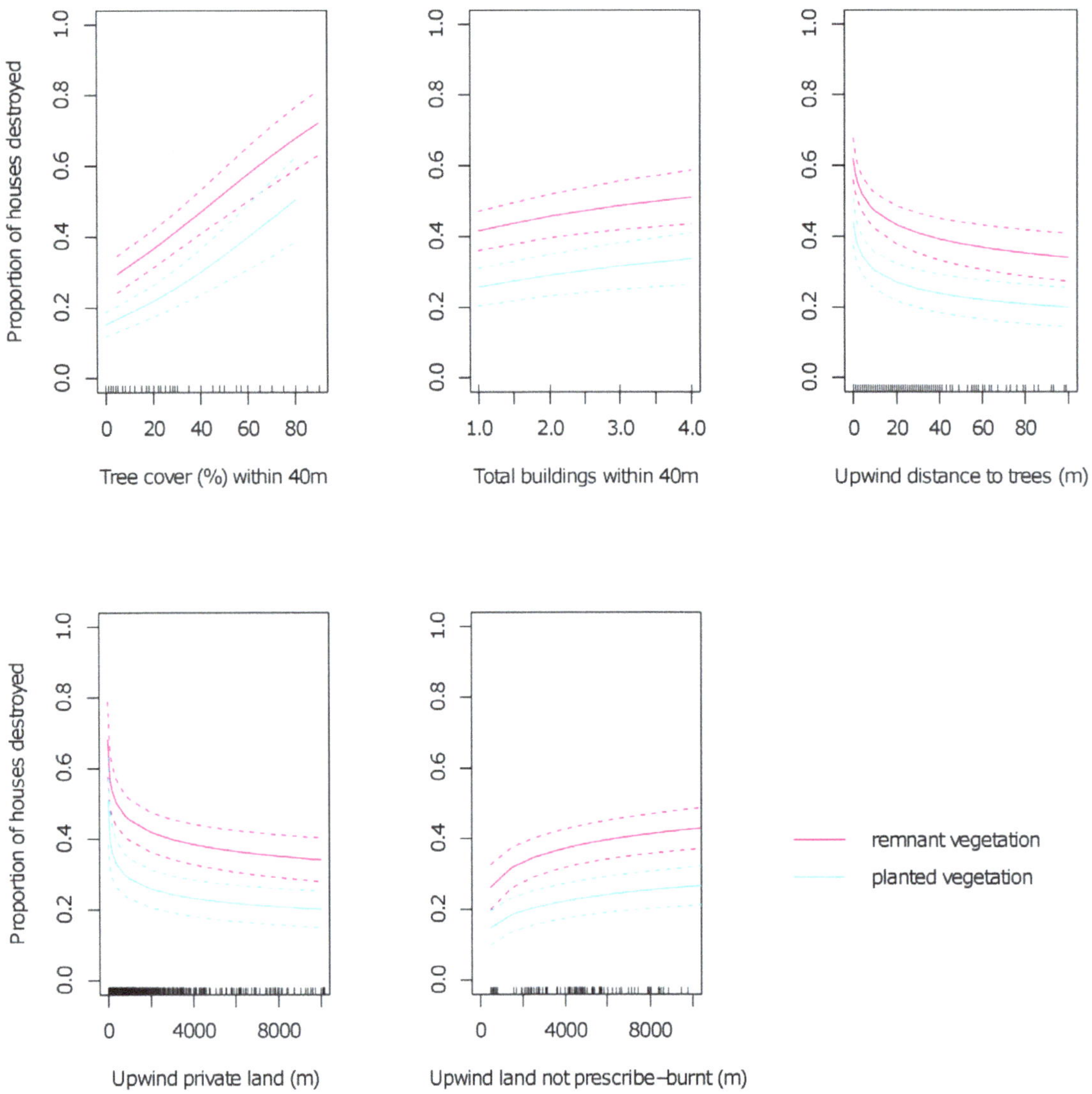

Figure 1. Individual effects (mean ± s.e.m.) of fuel variables in the logistic model used to predict the proportion of houses lost during wildfire. Each prediction was made with the other significant explanatory variables held at their means and FFDI fixed at 100, which is the value above which 64% of houses have been destroyed in wildfires in Australia [26]. Magneta (pink) lines are predictions for vegetation within 40 m of houses that is predominantly remnant native vegetation and cyan lines are predictions for vegetation within 40 m of houses that is predominantly planted vegetation.

burning), rather than 8.5 km upwind from houses (the average distance between houses and prescribed burning). We predicted a 3% increase in the number of destroyed houses with every additional building or shed located within 40 m.

Discussion

We predicted that modifying key fuels could substantially reduce house loss during wildfires that burn in extreme fire weather conditions. Many deaths occur among people sheltering in houses during wildfires (69% of lives lost during the 2009 wildfires examined here were in houses [10]), so managing key fuels could also save considerable lives.

The relative effects of fuel treatments on house loss

We found that modifying fuel closer to houses was a more effective way to reduce house loss than modifying fuel distant from houses. In severe fire weather (FFDI = 100), we predicted that reducing trees and shrubs from 90% cover to 5% cover within 40 m of houses could potentially reduce house loss by an average of 43%, making this the single most effective fuel treatment that we measured. We predicted that conversion from predominantly remnant to predominantly planted vegetation within 40 m from houses could reduce house loss by 38%. Increasing the upwind distance from houses to groups of trees and shrubs from zero to 100 m would reduce the number of houses lost by an average of 26%. The distance between houses and public forest had a similar effect. Twenty six percent fewer houses were lost 2 km from public land (the mean observed distance) compared with houses adjacent to public land. We predicted that an average of 15% fewer houses were destroyed when prescribed burning was undertaken 0.5 km from houses (the minimum distance we observed), compared with 8.5 km from houses (the mean distance we observed). One less building within 40 m from a house reduced the likelihood of destruction by an average of 3%, making this the least influential fuel variable in the selected logistic regression model.

Our finding that fuel management close to houses was more effective than fuel management further from houses can be explained by the behaviour of embers and radiant heat—the principal causes of house loss during wildfires [20,22,27]. The density of airborne embers [21] and the amount of radiant heat [20] are greatest closer to the fuel source, which is consistent with our results that fuel and fuel treatments closer to houses were more strongly associated with house loss. The reduction of fuel close to houses also increases 'defensible space', or the area around houses in which suppression is most likely to be successful [23].

Prescribed burning and house loss

Prescribed burning is a widely employed fuel treatment in many regions [12,13] and many commentators identified this as the key fuel treatment contributing to house loss in our study area. Although there was relatively limited prescribed burning in many parts of our study area, stratifying by this variable enabled us to sample houses with between 0% and 36.4% of the landscape burnt within 5 years in the upwind direction to the 2009 wildfire boundary. Within this range of variation, we found the effect of prescribed burning within 5 years was greatest closer to houses (Figure 1). This pattern is consistent with our results across all fuel variables (Figure 1). It is also noteworthy that prescribed burning was not a significant explanatory variable in any of the feasible logistic regression models when it was measured as the percentage of the landscape treated in the upwind direction from houses to the nearest 2009 wildfire boundary (rather than upwind distance from houses to the closest prescribed burning). This suggests that the proximity to houses of prescribed burning is more important than the total percentage of the landscape that is prescribe-burnt. These results are consistent with previous research indicating the effects of prescribed burning can diminish within a short period of time (2–6 years) [12,28,29] and in severe fire weather conditions [8,12,30], which are the conditions when most houses are destroyed [26]. Our results therefore indicated that prescribed burning—when executed at the scale observed in this study—was most effective when undertaken close to houses and at least every 5 years.

It is argued [10], based on relationships between prescribed burning and changes to the incidence and extent of wildfires [31,32], that prescribed burning can make control and suppression of wildfires before they reach houses more effective if executed in larger units and over a larger percentage of the landscape than observed in this study. However, it remains untested whether this strategy is effective in the extreme weather conditions. It is also important to note that the extent of prescribed burning that is feasible in many landscapes, including our study area, is restricted because of the number of days per year in which weather conditions are suitable and/or the proximity of public land to infrastructure [12].

Conclusions

Devastating wildfires provide a window into conditions that may become more common in the future [1,2,3,4,5,6] and therefore represent important learning opportunities for decision-makers. The typical response to destructive wildfires is to increase the total area of land that is fuel-reduced [10,13]. Our results instead indicate that a shift in emphasis from broad-scale fuel-reduction treatments to intensive fuel treatments close to houses will more effectively mitigate impacts from wildfires on houses. This result is consistent with observations that the density of airborne embers and amount of radiant heat (the principal causes of house loss during wildfires) are greatest closer to the fuel source. This suggests that the actions of private landholders, who manage fuel close to houses, are extremely important when reducing risks to houses posed by fuel. Our results are based on data collected at wildfires in south-eastern Australia. While it has been speculated that these conclusions apply to other regions around the world [13], the broader applicability of our results can only be confirmed with sampling across a broader range of fuel types and climates.

Although our results indicated that risks posed to peri-urban communities by severe wildfires can be reduced by effectively managing fuel, these risks cannot be eliminated by managing fuel alone. Fuel treatments can be expensive [13] and can have undesirable health [16] and environmental [17] impacts (but not in all cases [18]). Therefore, intensive fuel-reduction is not always an appropriate strategy to reduce risk posed by wildfire. Weather strongly influenced the effect of fuel variables (Table 1), hence other measures not accounted for here (e.g., architectural solutions, education of residents, suppression effort, safer places, early evacuation) [22,33,34,35] must remain part of a strategy to mitigate increasing risks to communities from wildfires.

Overall our results clearly imply that fuel close to housing plays a key role in house loss during wildfire, so fuel management should be considered as part of a strategy to mitigate increasing risks to peri-urban communities from wildfires. Future impacts from wildfires will be reduced, and the negative effects of fuel treatments avoided, if new peri-urban developments in fire prone regions are restricted to areas where there is adequate separation between high fuel loads and houses.

Materials and Methods

Study area and stratification

Houses were sampled within the boundaries of three wildfires that ignited in the State of Victoria, south-eastern Australia on 7 February, 2009. These fires were known as: the Kilmore East fire, which burnt 125,383ha, destroyed 1242 houses and claimed 119 lives; the Murrindindi fire, which burnt 43,159ha, destroyed 538 houses and claimed 40 lives [10]; and the Churchill fire which burnt 25,861ha, destroyed 145 houses and claimed 11 lives (Figure S1). The wildfire boundaries were as mapped in the FIRE_SEV09 GIS shape file provided by the Victorian Department of Sustainability and Environment (DSE). We stratified the study area by the three principal drivers of wildfire behaviour: weather, terrain and fuel [11]. Weather (measured using FFDI), ranged from 5 to 189. Slope of the terrain at each house ranged from 0.3° to 22.6°. Fuel, measured as % burnt upwind from houses within 5 years to the 2009 wildfire boundary and as the % of the landscape cleared upwind from houses to the 2009 wildfire boundary, ranged from 0% to 36% and 0% to 100% respectively.

Response variable

Our response was a binary variable representing house loss (intact or destroyed). To sample houses we allocated 499 points randomly to the study area in a Geographical Information System (GIS) in numbers proportional to the area of each stratum. We then selected the nearest house to each point using fine-scale (35 cm to 50 cm pixel resolution) orthorectified aerial imagery in the visible spectrum taken between 1 and 37 months prior to the fires. Our sampling of houses was blind in the sense that we did not know which had been destroyed. Several variables were measured within a circle with a 40 m radius from the centroid of each house (Table S1). To increase the likelihood of independence between responses we did not sample houses when these circles overlapped, instead choosing another random point until a non-overlapping house with 40 m circle was located. We recorded damage to each sampled house (intact or destroyed) based on a visual inspection of the house using fine-scale (15 cm pixel resolution) orthorectified imagery in the visible spectrum taken 17–23 days after the fires. We judged a house as destroyed if at least part of the roof had visibly collapsed or incinerated and judged a house as intact if the roof remained. In all cases this distinction was clear, which is consistent with on-ground observations by Wilson and Ferguson [33] who rejected using a continuous or ordinal variable to categorise house damage because virtually all houses affected by wildfire in their study had either been destroyed or sustained only superficial damage.

Potential explanatory variables

We recorded 24 potential explanatory variables at each house reflecting the three drivers of fire behaviour (i.e. weather, terrain and fuel) [11].

Weather conditions were measured with the Forest Fire Danger Index (FFDI) because house loss in Australian fires exhibits a higher correlation with this index rather than any of its individual components (i.e., wind, temperature, relative humidity and drought factor) [26]. FFDI was calculated using the formula [24]

$$FFDI = 2.0 \times \exp(-0.450 + 0.987\ln(DF) - 0.0345RH + 0.338T + 0.0234V)$$

where, DF is drought factor, RH is relative humidity (%), T is air temperature (°C) and V is wind speed (km hr^{-1}). Weather variables used to calculate FFDI (and to calculate wind direction for some fuel variables in Table S1) were derived from half-hourly data recorded at the closest weather station to each house [36]. We were advised by the Australian Bureau of Meteorology that these were the most reliable weather data available for our purpose. The estimated time that fire impacted on each house was estimated from fire progression maps provided by DSE for the Kilmore East and Murrindindi fires and a fire progression map prepared by the Victorian Country Fire Authority for the Churchill fire [10].

Terrain was measured as slope, topographic position and aspect (Table S1).

Fuel was measured (a) within 40 m of the centroid of each house, which is the approximate maximum distance that radiant heat is likely to ignite a wooden structure [20], (b) as a percentage of the landscape along a single transect in the upwind direction from each house to the nearest 2009 wildfire boundary, which ensured there was little overlap between measurements taken for different houses at this scale, and (c) as the distance from each house to the fuel variable in the upwind direction (Table S1). We could not measure the distance from each house to several fuel variables (public land, previous burning within 10 years, logging within 30 years, National Park, State Forest) because they did not always occur between sampled houses and the 2009 wildfire boundary in the upwind direction. If we excluded houses that did not have all of these fuel variables in the upwind direction then this would bias the sample (only 13 of the 499 sampled houses had all of the measured fuel variables in the upwind direction to the 2009 wildfire boundary). We therefore measured the amount of land upwind from houses that did NOT contain these fuel variables. For example, instead of recording the distance from houses to land burnt within 5 years in the upwind direction, we recorded the amount of land not burnt within 5 years in the upwind direction between each house and the 2009 wildfire boundary. This enabled us to include all randomly sampled houses in the analysis. The % cover of trees and shrubs within 40 m from the centroid of each house was estimated visually on the pre-fire aerial imagery by one person (P.G.). To test the accuracy of this method, we randomly selected 30 houses and compared our visual % cover estimates with estimates for the same houses derived by digitising tree and shrub cover in a GIS. Visual % cover estimates were highly correlated with % cover estimates derived from digitising trees and shrubs ($r = 0.95$, Pearson correlation coefficient). The mean (± s.e.m.) % cover of trees and shrubs derived from visual estimates (33.2±4.4) was not significantly different from estimates derived from digitising trees and shrubs (32.1±4.0) ($P = 0.462$, 2-tailed t-test).

Exploratory data analysis

Summary statistics for the continuous potential explanatory variables (see Table S1 for definitions) (mean, range) were: FFDI (48, 5–189), slope (8.5°, 0.3–22.6°), aspect (186°, 25–329°), number of buildings (2, 1–9), % cover of trees and shrubs (30%, 0–90%), distance to nearest tree or shrub (2.6 m, 0–108 m), upwind distance to nearest trees or shrubs (26 m, 0–686 m), upwind distance to nearest block of trees (272 m, 0–3021 m), upwind distance to mapped cleared land (773 m, 0–25121 m), amount of private land (2145 m, 0–15280 m), % cleared (32%, 0–100%), amount of land not burnt for ≤5 years (8848 m, 14–40041 m), % of land burnt ≤5 years ago (2.8%, 0–36.4%), amount of land not burnt for >5–10 years (10985 m, 14–35157 m), % of land burnt >5–10 years ago (0.4%, 0–37.7%), amount unlogged (9107 m, 14–36168 m), % logged (1.7%, 0–32.9%), amount not National Park (7457 m, 14–35157 m) and amount not State Forest (5501 m, 13–24703 m). The % of houses with the different measured fuel variables in the upwind direction within the 2009 wildfire boundary were: trees or shrubs (97%), block of trees (97%), land burnt within 5 years (25%), land burnt within >5–10 years (12%), logged within 30 years (26%), mapped cleared land (94%), public land (75%), National Park (41%) and State Forest (72%). The majority (86%) of the area burnt within 10 years was from prescribed fire, with the remainder burnt from wildfire.

We constructed a correlation matrix using the Pearson correlation coefficient (r) for all pairs of potential explanatory variables to determine those highly correlated ($r \geq 0.7$). One pair of potential explanatory variables was highly correlated: distance from houses to the 2009 wildfire boundary and amount of land not burnt for >5 to 10 years ($r = 0.83$). Only one of each of these variables was included in any model. Several other pairs of variables (amount of land not burnt for ≤5 years, amount unlogged, amount of land that is not National Park, amount of land that is private and amount of land that is not State Forest) were reasonably highly correlated ($r = 0.59$–0.68), but were included in all models. The following variables had highly skewed distributions and were therefore transformed by $\log_{10}(x)$ (or $\log_{10}(x+1)$ for variables with zeros) to give them a more symmetrical distribution prior to statistical modelling: upwind distance to nearest trees or shrubs, upwind distance to nearest block of trees, upwind distance to mapped cleared land, amount of land upwind from houses not burnt for ≤5 years and >5–10 years, amount of land upwind from houses that is privately owned, buildings within 40 m of houses and the amount of land upwind from houses that is not National Park or State Forest.

Statistical analysis

We initially examined relationships between the response and uncorrelated potential explanatory variables using mixed-effects modelling to account for the hierarchical structure of our data (i.e., houses were sampled within three separate fires which ignited at different times of the day and in different regions). That is, there was potential for the effect on houses of the same fire to be more alike than the effect on houses in a different fire. However, initial analyses using Generalized Linear Mixed Modelling (GLMM), implemented using the MGCV library in R [37], indicated that the variable representing the three different fires had a variance component approaching zero (<0.001). That is, the response did not change between fires. Remaining analyses were therefore undertaken using the more parsimonious Generalized Linear Modelling (GLM).

We used GLM with a logit link implemented using the MASS library in R [37] to identify fuel variables that were the best predictors of house loss. We accounted for the influence of weather and terrain by fitting FFDI slope, aspect and topographic position as co-variates during model selection. We included several interaction terms during model selection. To test whether effects of fuel variables varied with weather conditions we included interactions between FFDI and some fuel variables (% of landscape not burnt in the upwind direction, amount of land not burnt in the upwind direction, nearest upwind distance to trees or shrubs, % cover of trees and shrubs ≤40 m from houses). To test whether the effect of slope on the proportion of houses destroyed was influenced by aspect we fit an interaction term between slope and aspect. To test whether there was an interaction between defensible space created by clearing close to houses and broader-scale fuel reduction, we included interaction terms between the % cover of trees and shrubs ≤40 m from houses and the amount of previous burning in the upwind direction from houses.

We chose a model of 'best fit' using stepwise selection [38]. We first fitted a full model (with all terms) and then dropped terms sequentially if they did not lower Akaike's Information Criterion (AIC). Following Venebles and Ripley [38] (pp. 175–176), we then dropped any variables from this model if they were not significant ($P \leq 0.05$) using the traditional analysis of deviance, thus obtaining a more parsimonious result. All predictions were made from this single model of 'best fit'.

Because many houses in our study area occurred in a clustered spatial arrangement around towns (Figure S1), it follows that there is potential for spatial autocorrelation in our data. That is, if one house is destroyed then neighbouring houses are more likely to be destroyed, which would violate the assumption of independence in our fitted GLM. We tested whether residuals from the fitted GLM were spatially autocorrelated using Moran's I. This test was undertaken using the Ape package in R [37], which is based on the method described by Gittleman and Kot [39]. Moran's I, calculated using the residuals from the selected logistic model, was significantly different ($p < 0.001$) to the expected value of Moran's I if the residuals were distributed randomly, leading us to conclude that there is strong evidence for spatial autocorrelation in our data.

To account for this spatial autocorrelation we added an "autocovariate" [25] to the fitted GLM, which is a covariate representing spatial autocorrelation, following the methodology for non-normally distributed data reported in the appendix to Dormann et al. [40]. The autocovariate was scaled from zero (there was no relationship in the response between neighbouring houses) to 1 (the response was identical between neighbouring houses). The autocovariate was calculated using a matrix of neighbours within 1 km of each house. A correlogram of Moran's I indicated that most spatial autocorrelation in the residuals between neighbouring houses occurred within this distance. We added the autocovariate to the fitted GLM and then, again using the method of Gittleman and Kot [39], we confirmed that Moran's I calculated using the residuals in this new model was no longer significantly different to expected if distributed randomly ($p = 0.845$).

We reported goodness of fit for our selected logistic model using the Hosmer-Lemeshow test calculated using the Design library in the R statistical software [37]. We calculated AUC [41] from observed and predicted values for this model using the package Rocr in the R statistical software [37] to determine the probability that true positives rank above false positives. AUC has a value between 0.5 (a discriminating ability no better than chance) to 1 (perfect discriminating ability).

Supporting Information

Figure S1 Houses sampled in (A) the Kilmore East Murrindindi wildfires and (B) the Churchill wildfire. Sampled houses that were intact (clear houses) and destroyed (solid red houses) after the wildfires are illustrated.

Acknowledgments

Several data sets were supplied by the Victorian Department of Sustainability and Environment (DSE) and the Australian Government's Bureau of Meteorology. We acknowledge several officers of the Department of Sustainability and Environment for their cooperation and advice when seeking and using spatial data. Owen Price (University of Wollongong) supplied some data and provided helpful advice. Tony Bannister (Bureau of Meteorology) provided meteorological data and provided advice on its use. Jeff Wood, Ross Cunningham and Wade Blanchard (The Australian National University) provided advice on experimental design and statistical analysis. We thank Rohan Clarke and two anonymous referees for helpful comments on an earlier version of the manuscript.

Author Contributions

Conceived and designed the experiments: PG EK AMG DD GC RB DL LvB MM SS. Performed the experiments: PG LvB. Analyzed the data: PG LvB. Wrote the paper: PG EK AMG DD GC RB DL LvB MM SS.

References

1. Mell WE, Manzello SL, Maranghides A, Butry D, Rehm RG (2010) The wildland-urban interface fire problem - current approaches and research needs. International Journal of Wildland Fire 19: 238–251.
2. Hughes R, Mercer D (2009) Planning to reduce risk: the wildfire management overlay in Victoria, Australia. Geographical Research 47: 124–141.
3. Radeloff VC, Hammer RB, Stewart SI, Fried JS, Holcomb SS, et al. (2005) The wildland-urban interface in the United States. Ecological Applications 15: 799–805.
4. Piñol J, Terradas J, Lloret F (1998) Climate warming, wildfire hazard, and wildfire occurrence in coastal eastern Spain. Climatic Change 38: 345–357.
5. Pausas JG (2004) Changes in fire and climate in the eastern Iberian Peninsula (Mediterranean basin). Climatic Change 63: 337–350.
6. Westerling AL, Hidalgo HG, Cayan DR, Swetnam TW (2006) Warming and earlier spring increase western U.S. forest wildfire activity. Science 313: 940–943.
7. Liu Y, Stanturf J, Goodrick S. Trends in global wildfire potential in a changing climate. Forest Ecology and Management 259: 685–697.
8. Keeley JE, Fotheringham CJ, Moritz MA (2004) Lessons from the October 2003 wildfires in Southern California. Journal of Forestry 102: 26–31.
9. Vasquez T (2011) The Russian inferno of 2010. Weatherwise 62: 20–25.
10. Teague B, McLeod R, Pascoe S (2010) 2009 Victorian Bushfires Royal Commission. Final Report. Melbourne: Parliament of Victoria.
11. Pyne SJ, Andrews PL, Laven RD (1996) Introduction to Wildland Fire. Second Edition. New York: John Wiley & Sons.

12. Fernandes PM, Botelho HS (2003) A review of prescribed burning effectiveness in fire hazard reduction. International Journal of Wildland Fire 12: 117–128.
13. Schoennagel T, Nelson CR, Theobald DM, Carnwath GC, Chapman TB (2009) Implementation of National Fire Plan treatments near the wildland-urban interface in the western United States. Proceedings of the National Academy of Sciences of the United States of America 106: 10706–10711.
14. Agee JK, Skinner CN (2005) Basic principles of forest fuel reduction treatments. Forest Ecology and Management 211: 83–96.
15. Bradstock RA, Gill AM (2001) Living with fire and biodiversity at the urban edge: in search of a sustainable solution to the human protection problem in southern Australia. Journal of Mediterranean Ecology 2: 179–195.
16. Wotowa G, Trainer M (2000) The influence of Canadian forest fires on pollutant concentrations in the United States. Science 288: 324–328.
17. Driscoll DA, Lindenmayer DB, Bennett AF, Bode M, Bradstock RA, et al. (2010) Resolving conflicts in fire management using decision theory: asset-protection versus biodiversity conservation. Conservation Letters 3: 215–223.
18. Stephens SL, McIver JD, Boerner REJ, Fettig CJ, Fontaine JBH, B.R., et al. (in press) Effects of forest fuel reduction treatments in the United States. BioScience.
19. Haynes K, Handmer J, McAneney J, Tibbits A, Coates L (2010) Australian bushfire fatalities 1900–2008: exploring trends in relation to the 'Prepare, stay and defend or leave early' policy. Environmental Science & Policy 13: 185–194.
20. Cohen JD (2000) Preventing disaster: Home ignitability in the wildland-urban interface. Journal of Forestry 98: 15–21.
21. Koo E, Pagni PJ, Weise DR, Woycheese JP (2010) Firebrands and spotting ignition in large-scale fires. International Journal of Wildland Fire 19: 818–843.
22. Maranghides A, Mell WE (2011) A case study of a community affected by the Witch and Guejito wildland fires. Fire Technology 47: 379–420.
23. Gill AM, Stephens SL (2009) Scientific and social challenges for the management of fire-prone wildland-urban interfaces. Environmental Research Letters 4: 1–10.
24. Noble IR, Bary BAV, Gill AM (1980) McArthur's fire-danger meters expressed as equations. Australian Journal of Ecology 5: 201–203.
25. Augustin NH, Mugglestone MA, Buckland ST (1996) An Autologistic Model for the Spatial Distribution of Wildlife. Journal of Applied Ecology 33: 339–347.
26. Blanchi R, Lucas C, Leonard J, Finkele K (2010) Meteorological conditions and wildfire-related house loss in Australia. International Journal of Wildland Fire 19: 914–926.
27. Chen K, McAneney J (2004) Quantifying bushfire penetration into urban areas in Australia. Geophysical Research Letters 31: L12212.
28. Cary GJ, Flannigan MD, Keane RE, Bradstock RA, Davies ID, et al. (2009) Relative importance of fuel management, ignition management and weather for area burned: evidence from five landscape-fire-succession models. International Journal of Wildland Fire 18: 147–156.
29. Bradstock RA, Hammill KA, Collins L, Price O (2010) Effects of weather, fuel and terrain on fire severity in topographically diverse landscapes of south-eastern Australia. Landscape Ecology 25: 607–619.
30. Moritz MA, Keeley JE, Johnson EA, Schaffner AA (2004) Testing a basic assumption of shrubland management fire: how important is fuel age? Frontiers in Ecology and the Environment 2: 67–72.
31. Boer MM, Sadler RJ, Wittkuhn RS, McCaw L, Grierson PF (2009) Long-term impacts of prescribed burning on regional extent and incidence of wildfires-Evidence from 50 years of active fire management in SW Australian forests. Forest Ecology and Management 259: 132–142.
32. Finney MA, Seli RC, McHugh CW, Ager AA, Bahro B, et al. (2007) Simulation of long-term landscape-level fuel treatment effects on large wildfires. International Journal of Wildland Fire 16: 712–727.
33. Wilson AAG, Ferguson IS (1986) Predicting the probability of house survival during bushfires. Journal of Environmental Management 23: 259–270.
34. Abt R, Kelly D, Kuypers M (1987) The Florida Palm Coast fire: an analysis of fire incidence and residence characteristics. Fire Technology 23: 186–197.
35. Stephens SL, et al. (2009) Urban wildland fires: how California and other regions of the US can learn from Australia. Environmental Research Letters 4: 014010.
36. Bureau of Meteorology (2009) Meteorological Aspects of the 7 February 2009 Victorian Fires. An Overview. Canberra: Australian Government, Bureau of Meteorology.
37. R Development Core Team (2010) R: A Language and Environment for Statistical Computing. Vienna, Austria: R Foundation for Statistical Computing.
38. Venables WN, Ripley BD (2002) Modern Applied Statistics with S. New York: Springer.
39. Gittleman JL, Kot M (1990) Adaptation: statistics and a null model for estimating phylogenetic effects. Systematic Zoology 39: 227–241.
40. Dormann CF, McPherson JM, Araújo MB, Bivand R, Bolliger J, et al. (2007) Methods to account for spatial autocorrelation in the analysis of species distributional data: a review. Ecography 30: 609–628.
41. Pearce J, Ferrier S (2000) Evaluating the predictive performance of habitat models developed using logistic regression. Ecological Modelling 133: 225–245.

4

Climate-Driven Effects of Fire on Winter Habitat for Caribou in the Alaskan-Yukon Arctic

David D. Gustine[1]*, Todd J. Brinkman[2], Michael A. Lindgren[2], Jennifer I. Schmidt[2], T. Scott Rupp[2], Layne G. Adams[1]

1 U. S. Geological Survey, Alaska Science Center, Anchorage, Alaska, United States of America, **2** Scenarios Network for Alaska and Arctic Planning, University of Alaska Fairbanks, Fairbanks, Alaska, United States of America

Abstract

Climatic warming has direct implications for fire-dominated disturbance patterns in northern ecosystems. A transforming wildfire regime is altering plant composition and successional patterns, thus affecting the distribution and potentially the abundance of large herbivores. Caribou (*Rangifer tarandus*) are an important subsistence resource for communities throughout the north and a species that depends on terrestrial lichen in late-successional forests and tundra systems. Projected increases in area burned and reductions in stand ages may reduce lichen availability within caribou winter ranges. Sufficient reductions in lichen abundance could alter the capacity of these areas to support caribou populations. To assess the potential role of a changing fire regime on winter habitat for caribou, we used a simulation modeling platform, two global circulation models (GCMs), and a moderate emissions scenario to project annual fire characteristics and the resulting abundance of lichen-producing vegetation types (i.e., spruce forests and tundra >60 years old) across a modeling domain that encompassed the winter ranges of the Central Arctic and Porcupine caribou herds in the Alaskan-Yukon Arctic. Fires were less numerous and smaller in tundra compared to spruce habitats throughout the 90-year projection for both GCMs. Given the more likely climate trajectory, we projected that the Porcupine caribou herd, which winters primarily in the boreal forest, could be expected to experience a greater reduction in lichen-producing winter habitats (−21%) than the Central Arctic herd that wintered primarily in the arctic tundra (−11%). Our results suggest that caribou herds wintering in boreal forest will undergo fire-driven reductions in lichen-producing habitats that will, at a minimum, alter their distribution. Range shifts of caribou resulting from fire-driven changes to winter habitat may diminish access to caribou for rural communities that reside in fire-prone areas.

Editor: Ben Bond-Lamberty, DOE Pacific Northwest National Laboratory, United States of America

Funding: This work is part of the U.S. Geological Survey's (USGS) Changing Arctic Ecosystem Initiative and is supported by funding from the Wildlife Program of the USGS Ecosystem Mission Area. The funders had no role in study design, data collection and analysis, decision to publish, or preparation of the manuscript.

Competing Interests: The authors have declared that no competing interests exist.

* Email: dgustine@usgs.gov

Introduction

Recent and projected climatic warming has direct implications for fire-dominated disturbance regimes [1], particularly at high latitudes where warming has been amplified [2]. Evidence suggests strong linkages among the increased temperatures and altered precipitation patterns associated with climate warming and increases in wildfire frequency, severity, and area burned in boreal forests of North America [3–6]. For example, half of the largest fire years in Alaska's 60-year record have occurred since 1990 and two of the three most extensive wildfire seasons have happened in the last decade [7].

A transforming wildfire regime is affecting ecosystem structure and function in the north by altering plant composition and successional patterns [8]. These shifts in vegetation alter the distribution and abundance of northern herbivores in different ways [9]. Wildfire may benefit some species, such as moose (*Alces alces*), by increasing early-successional habitats [10,11]. However, wildfire may have negative effects on species dependent on late successional habitats, such as caribou (*Rangifer tarandus*) [11–15].

Understanding impacts of a changing wildfire regime on caribou is important for social and ecological reasons. Culturally, barren-ground or migratory tundra caribou, the ecotype occurring in the North American Arctic [13,16], constitute the most important terrestrial resource for subsistence hunters throughout the region, with many indigenous groups identifying themselves as "caribou people" [17,18]. Ecologically, migratory tundra caribou are distributed across areas experiencing the most evident and unprecedented changes resulting from climate warming [19]. Of all the climate-driven factors affecting shifts in vegetation types at northern latitudes, fire has the most potential to rapidly alter the composition and distribution of plant communities that caribou use in boreal forest and tundra systems [11,12,20]. Gaining insights to how fire-related disturbance may affect caribou distribution on the landscape will aid in understanding challenges facing people who rely on caribou as a crucial subsistence resource.

Migratory tundra caribou that occur throughout the Alaskan and Canadian Arctic are highly gregarious, occur at high localized densities, aggregate on calving grounds to bear young, undergo long seasonal migrations, and have very large annual ranges. Typically, these caribou migrate to the arctic tundra for the growing season and winter in the boreal forest and the forest-tundra interface [13,21], where they subsist primarily on the terrestrial lichens (*Cladonia* spp.) [22–25] that occur in late-succession habitats [12,26]. Indeed, the availability of high energy forage lichens throughout the winter reduces the reliance of caribou on body stores [27] that are vital to survival and reproduction [28]. The loss of lichen-rich vegetative associations, such as lichen-bearing spruce (*Picea* spp.) forests and tundra communities, could alter the distribution of caribou in winter [29,30] and possibly influence population dynamics [11,31,32].

Fire exerts a profound effect on the abundance of lichen in boreal forest and tundra ecosystems [30,33]. In Alaskan forests, the past low-frequency fire regime created large patches of relatively old-aged stands. However, the frequency of large fires and the total area burned annually has been increasing [34–36]; this trend is expected to continue well into the 21st century [11]. In young, recently-burned vegetation, forage lichens are largely absent, while lichen abundance is highest in older conifer stands [12,30,37]. On the scale of centuries, large fires may enhance nutrient cycling and promote habitat diversity for caribou [30], but in the shorter term, lichen abundance is greatly reduced and remains low for decades [32]. In east-central Alaska, caribou strongly selected spruce stands that were >80 years old while stands <60 years old were rarely used and lichen biomass and recovery post-fire were strongly correlated with these patterns of use [12,38].

Although less is understood about fire history in tundra communities, the effect of fire on forage lichens is similar to that of the boreal forest. Fires in the tundra biome were rare in the past 2,000–5,000 years [6,20], but Paleo-climate records show marked increases in fire frequencies during warming periods that coincided with expansion of shrubs into tundra communities [20,39]. Changes in hydrology and temperature regimes [40] will likely increase the drying of tussock tundra throughout the growing season and increase the frequency of fires [6]. Although several factors, such as burn severity and edaphic characteristics, can influence succession following a tundra fire, burns in the tundra are typically recolonized by graminoids and shrubs [41,42]. The presence of these vascular plants, especially deciduous shrubs, may further increase fire frequency [20] and thereby preclude forage lichens from recolonizing [42,43]. Simulations in northwest Alaska suggest that further increases in temperature could double the total area burned per decade in tundra systems [11,44]. As in the boreal forest, tundra fires affect the distribution of caribou in winter for decades: caribou typically avoided burns <60 years old in northwest Alaska [14,29].

To explore the effects of the climate-fire dynamic on the availability of lichen-producing winter habitats (hereafter to referred to as winter habitat) for migratory tundra caribou through 2100, we used an established simulation modeling platform [Alaskan Frame-based Ecosystem Code (ALFRESCO)] [45], and employed two global circulation models (GCMs) that defined the range of plausible climate projections for northwestern North America along with a moderate emissions scenario, to project fire regime characteristics, the abundance of winter habitat (tundra and spruce forest stands >60 years old), and the relative flammability (defined below) of the winter ranges of the Central Arctic and Porcupine caribou herds in the Alaskan-Yukon Arctic (Fig. 1). These adjacent herds represent the two wintering behaviors observed in migratory tundra caribou populations: the Central Arctic herd primarily uses Arctic and montane tundra habitats whereas the Porcupine herd typically winters in boreal forest habitats at or south of latitudinal treeline. The first analysis of this type was conducted on the Nelchina caribou herd in Interior Alaska [15] where basic hypotheses on the response of vegetation to a changing fire regime were tested. Contrary to previous work [11], we used markedly different current and projected temperature and precipitation regimes than in western Alaska. Thus, we used different vegetative inputs and contrasted the climate-mediated changes to winter habitat of two herds in a region influenced by the continental-montane-arctic climates. Therefore, our research represents modeling advancements that provide state-of-the-science scenarios of biophysical change, expanding the relevance and application of our findings to management, research, and conservation as well as subsistence-based communities throughout northern North America.

Materials and Methods

Study area and caribou herds

The simulation domain comprised 570,112 km^2 of eastern Alaskan and western Canadian boreal and arctic ecosystems, encompassing the winter ranges of the Central Arctic and Porcupine caribou herds (named after the Porcupine River; Fig. 1). There was a large physiographic gradient (0–3,200 m) that included boreal deciduous and coniferous forests, expansive wetlands, and arctic and alpine tundra. Since the mid-1970s when regular monitoring of these herds began [46], the Central Arctic herd has increased markedly from 5,000 to 70,000, whereas the Porcupine herd increased from 100,000 to 178,000 in the late 1980s and now numbers around 169,000 [47]. Winter range extents for each herd were based on over 30 years of radiotelemetry studies. Specifically, the Central Arctic winter range was based on a combination of telemetry data and expert opinion [48], whereas the winter range of the Porcupine herd was derived via a 90% adaptive kernel algorithm applied to available telemetry data (K. Poole, Aurora Wildlife Research, unpublished analyses). The historic winter ranges of the Central Arctic (60,175 km^2) and Porcupine (176,540 km^2) caribou herds span east from the Colville River in north-central Alaska to the Northwest Territories and north from the Yukon River to the Beaufort Sea (Fig. 1). The Central Arctic herd primarily winters in tundra habitats that span the northern and southern slopes of the Brooks Range, while the Porcupine herd winters predominantly in the northern interior boreal regions of Alaska and Yukon (Fig. 1). Detailed descriptions of the ranges for each herd have been provided elsewhere [23,49,50].

Fire

We used the "raster" package [51] in the R statistical language version 2.15.2 (R Core Team, http://www.R-project.org/) to calculate historic (1950–2012) wildfire characteristics (i.e., number of fires, average fire area [km^2], area burned annually [km^2]) from the Bureau of Land Management-Alaska Fire Services and the Yukon Territorial Government's burn perimeter database records (data available at http://fire.ak.blm.gov and http://www.geomaticsyukon.ca/data/datasets) for our modeling domain. These databases contain fire perimeters dating back to 1917 and 1946, respectively.

Modeling platform

We used ALFRESCO [1,15,45] to simulate the change in winter caribou habitat (2010–2100) in response to climate projections and the resulting fire regime. ALFRESCO is a spatially-explicit, stochastic landscape succession model for sub-arctic and boreal vegetation types that operates at a 1-km resolution and an annual time-step to model the interactions between fire, a changing climate, and 4 vegetation types (white spruce, black spruce, deciduous, and tundra) in Alaska. For this

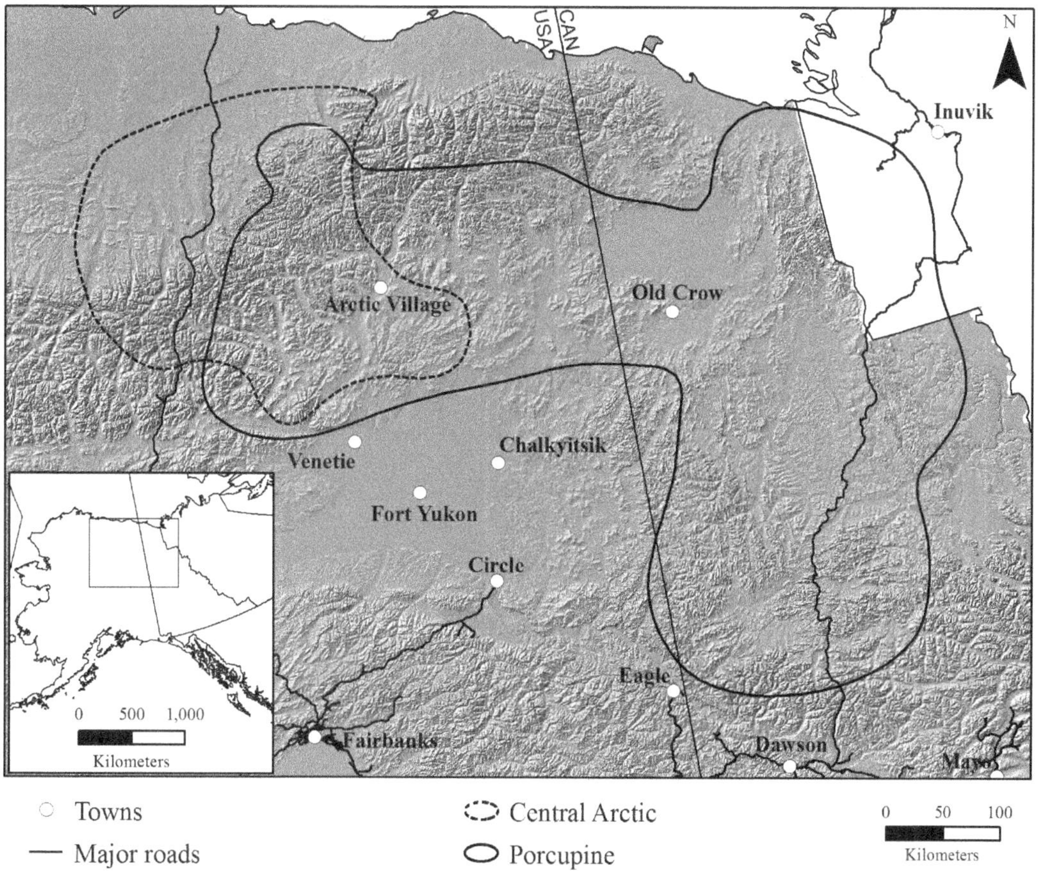

Figure 1. Simulation domain and winter ranges of the Central Arctic and Porcupine caribou herds, Alaska and Yukon.

research, the model used downscaled scenarios of temperature and precipitation from each GCM given the A1B greenhouse gas emission scenario (see below for details). Wildfires were randomly "ignited" using a cellular automata approach based on a linear regression model developed by Duffy et al. [3]. Cell ignition and fire spread (i.e., flammability) is a function of climate, vegetation state, and time since last fire. The ignition of any given cell is determined by comparing a randomly generated number against the flammability coefficient of that cell. The flammability coefficient allows for changes in flammability that occur through succession (i.e., fuel build up). Following a wildfire in ALFRESCO, general successional trajectories were as follows: burned spruce forest (white or black) transitioned into early successional deciduous forest, and burned deciduous forest and tundra self replaces. Vegetation transition times differed probabilistically between climax black and white spruce trajectories [11]. Transitional times were modeled probabilistically to represent early successional (i.e., recolonization) deciduous vegetation following wildfires in spruce and deciduous forest and to determine the amount of time, in the absence of fire, until the climax spruce stage dominates the site again. Self replacement of deciduous forest can occur when repeated burning and/or climate conditions preclude transition to climax spruce. ALFRESCO incorporates the effects of fire severity on transition times using measurements of the area of the wildfire (i.e., fire size), complex topography, and vegetation type on flat landscapes [52]. We calibrated the relationship between climate and fire by comparing model output (e.g., fire regime, stand age structure) to the corresponding historical data [15,53].

Climate models

We used two downscaled Coupled Model Intercomparison Project 3 GCMs in the ALFRESCO simulations: Canadian Center for Climate Modeling Analysis Coupled Global Climate Model 3.1 (hereafter referred to as the "warm" GCM; more info at http://www.ec.gc.ca/ccmac-cccma/default.asp?lang = En&n = 1299529F-1) and Max Planck Institute European Center-Hamburg 5 Model (hereafter referred to as the "hot" GCM; more info at http://www-pcmdi.llnl.gov/ipcc/model_documentation/ECHAM5_MPI-OM.htm). Global circulation models are mathematical representations of atmospheric and oceanic conditions and have the dual-purpose of simulating historical conditions based on known atmospheric values, such as greenhouse gas concentrations, and extrapolating those conditions to the future through the use of greenhouse gas emission scenarios. The GCMs we used represent the low and high limits of projected climate change of the five best performing GCMs for Alaska [54]. The GCMs were downscaled to 2-km resolution using the delta downscaling method by the Scenarios Network for Alaska and Arctic Planning (available at http://www.snap.uaf.edu/), and subsequently resampled to 1 km for input to the ALFRESCO model. We used the A1B emissions scenario [55], which assumed a mid-range of emissions in the future with a steady increase in carbon dioxide, however, recent climate and emission trends suggest that A1B may be conservative [56].

Vegetation

The input vegetation map is a simplified version of the North American Land Change Monitoring System land cover map [57]. This land cover map was reclassified to meet the needs of the ALFRESCO model by collapsing classes into 5 groups: rock/ice, tundra, black spruce, white spruce, and deciduous. At the time of these analyses, there were no large-scale vegetation data available in a comparable format for the Northwest Territories, so we did not include this region in our simulations. This vegetation map was used as input in the ALFRESCO "spin up" phase of approximately 90 replicates running for 1,000 years to create a simulated landscape similar to current conditions with regards to burn characteristics (e.g., stand age), and vegetation distribution and composition [11,15,53].

Analyses

We used the ALFRESCO simulations, driven by temperature and precipitation projections from each GCM, to analyze the number and size of fires and total area burned (mean ± SD), associated vegetation changes, and relative flammability of the landscape during 2010–2100 for each herds' winter range, as well as throughout the modeled spatial domain. For each GCM, ALFRESCO produced 90 simulation runs of annual series of spatial raster maps that depicted stand age and vegetation types over the 90 years of the simulation. Time (years) since fire from simulated historic (<2010) and projected (2010–2100) fire histories was used to determine age structures of vegetation communities on the landscape (i.e., pixels that burned in a previous year had an age of zero in the subsequent year). Forage lichen recovery generally occurs >60 years following fire [12,26,29], so stands of spruce or tundra in these age-classes were assumed to produce more lichen than younger stands (<60 years old), thus we classified older, lichen-producing stands as winter habitat for caribou. To find the most representative model run for each GCM, we compared the annual area burned from each run over the 90 simulated years with the median values for each year of all 90 runs and selected the run with the highest correlation coefficient (r) [53]. The annual representative runs were used to summarize fire characteristics (km^2) and the amount of winter habitat (km^2; measure of uncertainty = 5th and 95th percentiles) by decade for each GCM.

We assessed the spatial distribution of relative flammability across the landscape for each GCM. Relative flammability was defined as the likelihood of a pixel to ignite throughout spatial and temporal domain of the simulations, thus we calculated the proportion of years among all the simulations (90 simulation runs x 90 years per simulation) that each individual pixel burned. From the resulting relative flammability maps for each GCM, we determined the top 20% of the modeling domain where fires occurred most frequently throughout our simulations, then calculated the percent of each herd's winter range that occurred within this highly flammable category. To identify areas within the domain with the largest relative differences in flammability between GCMs and, thereby, a proxy for the largest uncertainty in future fire characteristics, we subtracted the continuous relative flammability maps of the warm GCM from that of the hot GCM.

Results

The annual representative runs from ALFRESCO (i.e., most correlated with median area burned of all repetitions across 90 years) that were used for subsequent analyses had correlation coefficients of 0.904 and 0.922 for the warm and hot GCMs, respectively. Throughout our simulations, fires were less numerous, smaller, and burned less area in tundra compared to spruce for both GCMs (Table 1). Fire characteristics followed general expectations for the warm and hot GCMs with the hot model projecting larger fires and more area burned. Due primarily to the increase in the size of fires, the average area of winter habitat that burned per decade was 64 and 25% higher in the hot versus warm GCM for tundra and spruce, respectively (Table 1).

The projected changes in the amount of winter habitat within each herd's range differed by vegetation type and GCM. For the Central Arctic herd, the percent of winter range that was winter habitat changed little through the nine decade simulation for the warm GCM, but decreased under the hot GCM (Fig. 2). Under the warm GCM, the extent of winter habitat was essentially the same through 2100, increasing <1% (41,009 km^2 in 2010s versus 41,300 km^2 in 2090s) with little change in composition of tundra and spruce communities (89.2 and 88.9% tundra, respectively). Conversely, under the hot GCM, winter habitat decreased by 11% throughout the same time period (41,692 to 37,092 km^2) with a shift to less old-aged spruce habitats (11.2 and 6.5% spruce in 2010s and 2090s, respectively). For the Porcupine herd, results across the 90 years for the two GCMs differed more widely, with a modest increase (+5%; 102,710 to 107, 909 km^2) in the availability of winter habitat under the warm GCM, compared to marked declines (−21%; 107,224 to 84,353 km^2) for the hot GCM (Fig. 2). Similar to the Central Arctic herd's range, the composition of winter habitat within the Porcupine herd's winter range changed little under the warm GCM (64.3% tundra in 2010s and 2090s, respectively), but for the hot GCM, the proportion comprised of spruce habitats declined markedly (36.8 to 28.6%).

In assessing the capacity of areas to burn during the 90-year simulations for both GCMs, the Porcupine herd had a larger portion of its winter range within the highest relative flammability category (Fig. 3). For the warm GCM, 0.9 and 3.5% of the Central Arctic and Porcupine herd's ranges, respectively, had the highest occurrence of fires. The amount of highly flammable area increased approximately 2–4 times for both herds under the hot GCM (Central Arctic =3.4% and Porcupine =8.9%). In examining the spatial distribution of the differences in the relative flammability between GCMs, most of the uncertainty between simulations was located on the southern slope of the Brooks Range in Alaska as well as a large portion of the simulation domain in the northern interior Yukon (Fig. 4). Conversely, two prominent regions were consistent in flammability between the GCM simulations: the north slope of the Brooks Range had low relative flammability while the Yukon Flats in Alaska, including the communities of Venetie, Chalkyitsik, and Fort Yukon, had a high relative flammability (Fig. 4).

Discussion

The projected influences of climate changes to fire regimes in northern ecosystems yielded insights into the potential availability of winter habitat for the Central Arctic and Porcupine caribou herds, with implications for other migratory tundra caribou populations throughout the 21st century. The projected availability of winter habitat and relative flammabilities depended on the GCM and coarse vegetation type within the each herd's winter range (Figs. 2 and 3). Specifically, under the hot GCM we projected declines in winter habitat for both herds with larger decreases for the Porcupine herd that wintered primarily in the boreal forest (Fig. 2). If flammability of the Arctic tundra biome does indeed increase due to vegetation changes, drying, and increase in ignition agents [6,11,58,59], losses of winter habitat for migratory tundra caribou may be amplified. Herd-specific

Table 1. Simulated average (SD) annual fire characteristics by decade within lichen-producing vegetation classes (i.e., spruce and tundra >60 y) under a moderate emissions scenario (A1B) for two global circulation models [GCM; Canadian Center for Climate Modeling Analysis Coupled Global Climate Model 3.1 (Warm GCM) and Max Planck Institute European Center-Hamburg 5 Model (Hot GCM)] within the spatial domain of the modeling effort including the winter ranges of the Central Arctic and Porcupine caribou herds in northern Alaska and Yukon.

Decade	Warm GCM						Hot GCM					
	Tundra			Spruce			Tundra			Spruce		
	Number of Fires	Fire Size (km^2)	Area Burned (km^2)	Number of Fires	Fire Size (km^2)	Area Burned (km^2)	Number of Fires	Fire Size (km^2)	Area Burned (km^2)	Number of Fires	Fire Size (km^2)	Area Burned (km^2)
2010s	8.2 (4.2)	5.8 (1.8)	50 (27.7)	21.7 (4.9)	52.6 (30.4)	1219 (745.9)	10.4 (6.1)	26.7 (55.8)	506 (1186.2)	22.8 (4.8)	68.9 (86.8)	1794 (2398.6)
2020s	5.2 (1.8)	3.9 (1.8)	21 (14.2)	20.9 (4.7)	30.2 (16.5)	614 (296.5)	9.6 (6.3)	7.0 (7.4)	104 (203.4)	22 (5.9)	56.9 (54.2)	1449 (1692.2)
2030s	8.1 (3.4)	9 (6.9)	91 (104.1)	21.8 (4.3)	60.8 (44.7)	1385 (1152.6)	9.4 (3.4)	17.8 (23.9)	205 (292.5)	25.1 (3.6)	85 (71.8)	2140 (1930.5)
2040s	7.7 (4.6)	7.2 (5.4)	64 (62.5)	24 (5.4)	55.7 (36.4)	1426 (1020.6)	9.1 (4.7)	14.3 (10.8)	151 (165.2)	24 (7.8)	64 (40.8)	1781 (1401.1)
2050s	8.9 (4)	8.5 (5.1)	86 (74.4)	26.3 (4.9)	65.4 (34)	1719 (770.7)	9.1 (5.2)	11.9 (9.4)	150 (172.1)	21.4 (6.1)	79.4 (48.9)	1853 (1434.9)
2060s	9.8 (3.2)	7 (3.1)	69 (31.7)	24.3 (5.5)	43.5 (21.5)	1070 (587.4)	9.8 (3.7)	31.8 (56.2)	431 (924.2)	21.4 (5.9)	67.1 (44.1)	1633 (1516)
2070s	8.8 (2.7)	12.1 (11.9)	119 (150.3)	24.6 (5.9)	60.5 (47.3)	1649 (1471.3)	8.3 (3.8)	13.4 (12.5)	146 (190.3)	23.2 (5)	64 (63.9)	1616 (1744.4)
2080s	9.1 (3.2)	9.8 (6.9)	86 (52.5)	23.3 (3.8)	49.2 (24.6)	1167 (603.2)	8.6 (4.6)	31.4 (52.2)	460 (1021.8)	23 (6.3)	62.8 (37.6)	1589 (1511.3)
2090s	11.1 (4.1)	18.6 (15)	230 (204.1)	23.3 (4.4)	89.8 (39.5)	2025 (724.9)	11.4 (4.3)	55.3 (60.1)	689 (750)	23.3 (5.6)	93.5 (49.8)	2240 (1388.7)

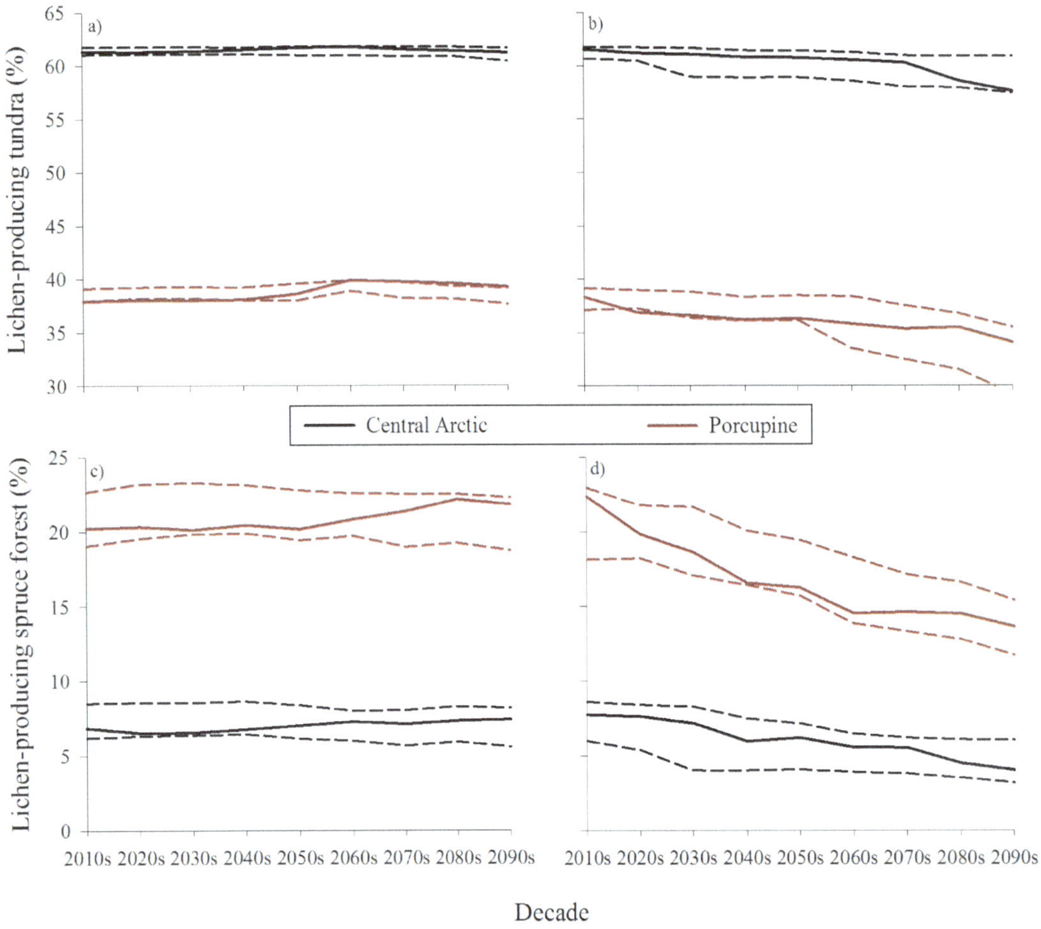

Figure 2. Simulated proportion of the winter ranges of the Central Arctic and Porcupine caribou herds in northern Alaska and Yukon that were covered in lichen-producing (>60 y) tundra (a, b) and spruce forest (c, d) under a moderate emissions scenario (A1B) for the warm [Canadian Center for Climate Modeling Analysis Coupled Global Climate Model 3.1 (a, c)] and hot [Max Planck Institute European Center-Hamburg 5 Model (b, d)] global circulation models, Alaska and Yukon. Representative runs are denoted by solid lines and the 5th and 95th percentiles are denoted by dashed lines.

simulations, however, will help to establish regionally appropriate climate-change projections for migratory tundra caribou as well as the repercussions to the communities that rely on these caribou as a subsistence resource [60].

Given current greenhouse gas emissions and climate trajectories, as well as the simplification of the successional pathways and associated flammability of tundra, our projections of winter habitat for these caribou herds were conservative. Since IPCC's fourth assessment, greenhouse gas emissions have exceeded expectations and considering recent trends, the A1B emissions scenario is likely an underestimate of what actual greenhouse gas emissions will be for the remainder of the century [19,56]. Temperature projections in particular are most sensitive to greenhouse gas emissions scenarios [19], therefore future temperature regimes in Alaska may more closely align with the hot GCM. Coupled with a moderate emissions scenario, these two GCMs identified areas of uncertainty between scenarios in flammability (Fig. 4) and established a range of possible outcomes for caribou that included increases of 0.7 and 5% (warm GCMs) and decreases of 11 and 21% (hot GCM) of winter habitat for the Central Arctic and Porcupine herds, respectively (Table 1, Fig. 2). However unlikely, if greenhouse gas emissions decrease and (or) temperature projections align with the warm GCM, than this work provides a relevant alternative regarding the availability of winter habitat for these two herds.

Accounting for spatial variation in ignition agents and the capacity of a pixel to burn and the actual successional pathways of tundra vegetation will likely increase projections of the area burned [11], and thereby magnify the projected losses of winter habitat for migratory tundra caribou. In the current version of ALFRESCO, all types of tundra habitat, such as alpine, graminoid, shrub, or wetland tundra, are treated the same with regards to flammability and successional pathways (i.e., low flammability and tundra transitions to tundra after a fire) [11,15]. However, flammability and successional pathways following a tundra fire appear to be related to initial tundra type, geography, local isotherms, fire severity, juxtaposition of seed sources, and colonization of shrubby species [20,39,61]. For example, during periods of warming, Paleo-records suggest birch-dominated shrub tundra burned as frequently as modern day boreal forest [20,39]. Thus, expansion of shrub-dominated tundra [20], as is being observed on portions of the North Slope of Alaska [62], coupled with warmer temperatures throughout the growing season may increase the flammability of vegetation in areas where fires have, until recently, been rare [6]. In an attempt to deal with this spatial variation in flammability within tundra vegetation in

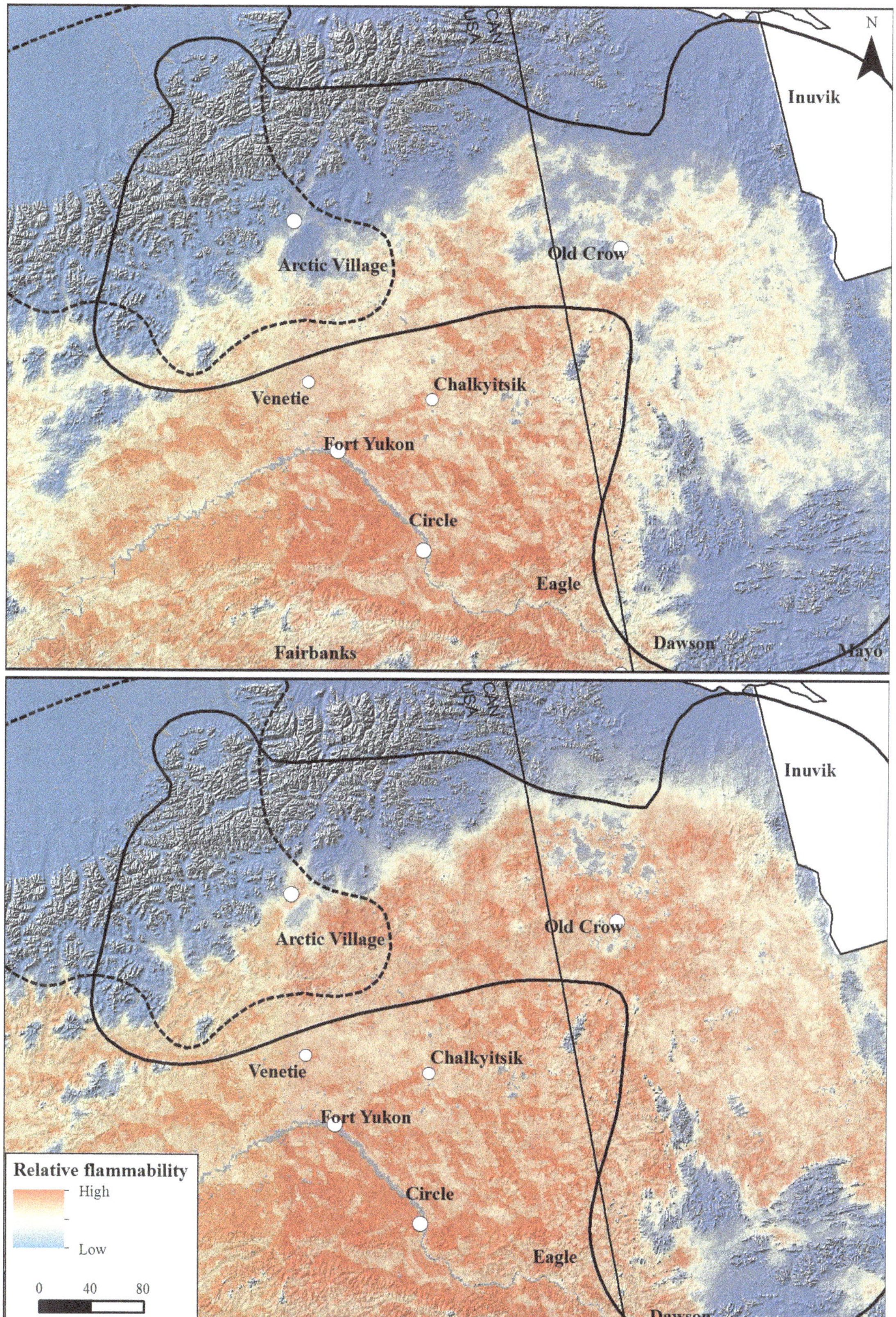

Figure 3. Relative flammability under a moderate emissions scenario (A1B) for the warm [Canadian Center for Climate Modeling Analysis Coupled Global Climate Model 3.1 (top panel)] and hot [Max Planck Institute European Center-Hamburg 5 Model (bottom panel)] global circulation models in the winter ranges of the Central Arctic and Porcupine caribou herds, Alaska and Yukon, 2010–2100.

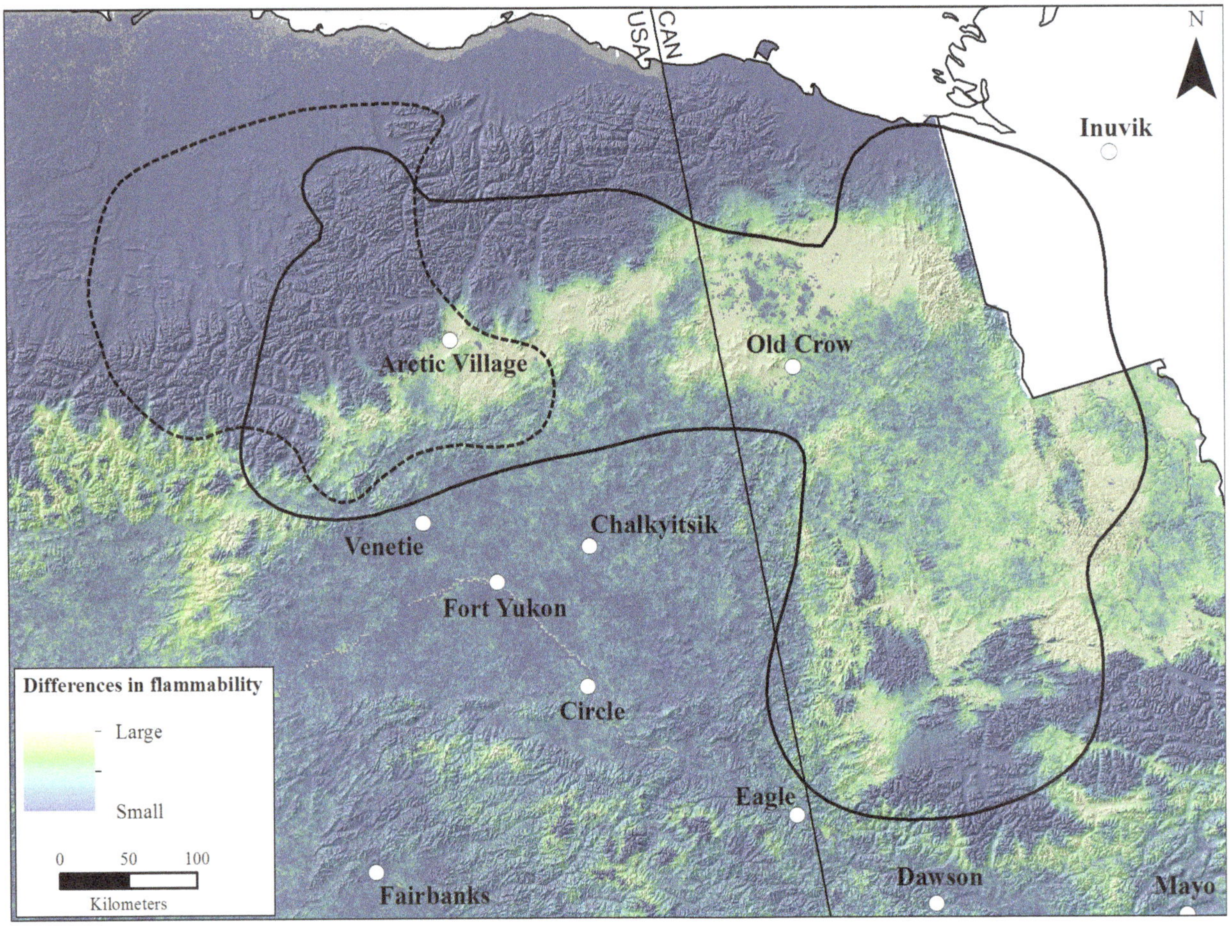

Figure 4. Differences in relative flammability of vegetation between the warm (Canadian Center for Climate Modeling Analysis Coupled Global Climate Model 3.1) and hot (Max Planck Institute European Center-Hamburg 5 Model) global circulation models under a moderate emissions scenario (A1B) in the winter ranges of the Central Arctic and Porcupine caribou herds, Alaska and Yukon, 2010–2100.

northwestern Alaska, previous work altered the flammability coefficients in ALFRESCO by physiographic region [11]. Fire return intervals in that region of Alaska (approx. 140–240 yrs) [58] are not indicative of tundra vegetation on the North Slope of Alaska where fire return intervals may exceed 1,000s years [6,20]. The fire regime in western Alaska is more similar to fire regimes of lowland tundra interspersed amongst spruce forest that is common throughout the winter range of the Porcupine caribou herd [23]. Regardless, any modification of tundra flammability would have decreased the fire return interval and increased the amount of area burned thereby amplifying the magnitude of the trends we simulated.

Despite the potential for shifting fire dynamics to influence the lichen-producing tundra, the projected availability of winter habitat for these caribou herds was driven by climate-fire dynamics in the boreal forest. Fire greatly influences ecosystem structure in the boreal forest by influencing successional trajectories [34–36]. The projected increases in temperature coupled with extended growing seasons, increased fuel buildup, and drying will reduce the mean stand age and increase the prevalence of deciduous stands [8]. Spruce forest comprised a relatively small portion of the Central Arctic herd's range (Fig. 2), and correspondingly, this herd incurred a smaller loss of winter habitat (~11%) under the hot GCM. Although the simulated effects to the Central Arctic herd appeared relatively minor under the hot GCM, if the recent heavy use of the forested southern slope of the Brooks Range in winter is indicative of a longer term trend [47], then any losses of these old-growth spruce forests may become more influential to the distribution of this herd in winter. Alternatively, old-growth spruce forest was more abundant within the winter range of the Porcupine caribou herd (Fig. 2). Comparable to simulations in western Alaska [11], we project this herd will lose 21% of winter habitat to fire by the end of the century. The majority (67%) of this loss was driven by increased flammability in spruce forests in the Yukon (Fig. 3). Thus, managers there will likely contend with relatively significant changes in the fire regime and subsequent reductions in the availability of lichen-producing vegetation. Based on the expected shifts in fire regime throughout the boreal forest [8] and the predominantly boreal-wintering behavior of migratory tundra caribou in North America, these projections for the Porcupine herd may be indicative of fire regime shifts influencing other continental caribou populations.

As noted throughout fire-caribou literature over the past 50 years, the effects of fire-driven habitat changes to caribou population dynamics is uncertain. The linkage between the availability of winter habitat, shifts in distribution, and changes in abundance of herds is difficult to establish for populations that range over large areas and demonstrate remarkable demographic plasticity in spite of austere and variable conditions. At the heart of this complexity and uncertainty are the numerous intra- and inter-annual factors that affect caribou populations and, to some degree, are influenced by climate changes. These include the influence of snow conditions on forage availability and energetic costs of movement and foraging [63], capacity to shift seasonal ranges to account for changes in range quality [14,64], vegetative shifts that may increase apparent competition [11], intra-specific competition [65], top-down effects of herbivory on vegetation communities [66,67], expansion of the growing season and increases in primary productivity that may confer nutritional benefits to reproductive females [68,69], influence of tundra fires on the quality of summer habitats [in sensu 70, 71], insect harassment in the summer [72], and anthropogenic influences [73–76]. Caribou populations can and do shift their distributions to minimize any or all of the above factors [64,77,78]. However, available wintering habitat is continuously occupied by migratory tundra caribou herds from the Bering Sea to the Hudson Bay and is limited to the north by the Arctic Ocean [13,79]. Climate-fire-winter habitat dynamics will most certainly affect the availability of winter habitats for these herds as well, but to varying degrees. Thus, the capacity for a herd to shift its range to accommodate these changes will vary with the availability of winter habitat in the adjacent herds' ranges. Despite these complexities, it is relatively clear: caribou will need to alter their distribution to avoid recently burned areas in winter due to reduced lichen presence [11,12,14,29,38,80].

Although the population-level effects may be unclear, the potential changes in caribou distribution will most certainly affect human communities that have a cultural and nutritional reliance on caribou. Unlike caribou herds, communities have limited resilience to large shifts in availability of food resources. For example, four indigenous communities occur in a region that, regardless of GCM, had a relatively high flammability (Figs. 3 and 4). One community (Old Crow, Yukon Territory) is within the traditional winter range of the Porcupine herd, while hunters from three villages (Fort Yukon, Venetie, and Chalkyitsik, Alaska) travel north each year to harvest animals from this herd. With caribou avoiding or shifting migrations away from recent burns, harvest opportunities are impacted immediately by wildfire and the effects could last for two generations of hunters [81,82]. Also, wildfire indirectly affects hunting opportunities by impeding hunter travel across the landscape [7,60,83]. Based on simulated relative flammabilities, projected increases in fire sizes under the hot GCM, and historic distribution of the Porcupine caribou herd (Figs. 3 and 4), it is unlikely that caribou will become more accessible to these aboriginal communities in winter.

Projecting the influences of climate changes to wildlife populations is a necessary but daunting task fraught with numerous ecological, climatic, and technical complexities, uncertainties, and assumptions [84–86]. As noted in other efforts using ALFRESCO, this work was not insulated from these challenges [11,15], however, certain aspects of the life-history characteristics of migratory tundra caribou and their habitats facilitated the projection climate-fire induced changes to the availability of winter habitat. These characteristics included the clear roles that vegetation, temperature, and precipitation have in structuring fire regimes, the strong influence of fire on successional pathways in northern systems, the specific stand characteristics that facilitate lichen colonization and growth in boreal and arctic systems, and the important role of lichens in the winter diets of caribou. Projected warming and greenhouse gas emission trends will indeed alter fire dynamics and shift vegetative composition and age structure to lower the availability of winter habitat to migratory tundra caribou [11,15]. Yet, linking these projected losses of winter habitat with changes in abundance of caribou, and rectifying the apparent opposing influences of other climate-driven changes to winter habitats [negative: this study, 11] and throughout the growing season [69] remain important issues in elucidating climate-induced effects to caribou populations and the communities that depend on them.

Acknowledgments

Any use of trade names is for descriptive purposes only and does not imply endorsement by the U.S. Government. We thank S. Arthur, D. Cooley, K. Poole, D. Russell, and the Porcupine Caribou Herd Technical Committee for providing the winter ranges of the Central Arctic and Porcupine caribou herds. This manuscript was improved by comments from B. Bond-Lamberty, T. Hegel, L. McNew, J. Pearce, and three anonymous reviewers.

Author Contributions

Conceived and designed the experiments: DDG TSR LGA. Analyzed the data: DDG MAL JIS. Wrote the paper: DDG TJB MAL LGA.

References

1. Rupp TS, Chen X, McGuire AD (2007) Sensitivity of simulated boreal fire dynamics to uncertainties in climate drivers. Earth Interactions 11: 1–21.
2. Field CB, Boesch DF, Chapin III FS, Gleick PH, Janetos AC, et al. (2008) National Research Council. Washington, DC: Ecological impacts of climate change. The National Academies Press. 58 p.
3. Duffy PA, Walsh JE, Graham JM, Mann DH, Rupp TS (2005) Impacts of large-scale atmospheric-ocean variability on Alaskan fire season severity. Ecological Applications 15: 1317–1330.
4. Kasischke ES, Turetsky MR (2006) Recent changes in the fire regime across the North American boreal region - spatial and temporal patterns of burning across Canada and Alaska. Geophysical Research Letters 33.
5. Flannigan M, Logan KA, Amiro BD, Skinner WR, Stocks BJ (2005) Future area burned in Canada. Climate Change 72: 1–16.
6. Hu FS, Higuera PE, Walsh JE, Chapman WL, Duffy PA, et al. (2010) Tundra burning in Alaska: linkages to climatic change and sea ice retreat. Journal of Geophysical Research: Biogeosciences 115: G04002.
7. Chapin FS, Trainor SF, Huntington O, Lovecraft AL, Zavaleta E, et al. (2008) Increasing wildfire in Alaska's boreal forest: pathways to potential solutions of a wicked problem. BioScience 58: 531–540.
8. Fauria MM, Johnson EA (2008) Climate and wildfires in the North American boreal forest. Philosophical Transactions of the Royal Society B 363: 2317–2329.
9. Klein DR, Baskin LM, Bogoslovskaya LS, Danell K, Gunn A, et al. (2005) Management and conservation of wildlife in a changing arctic environment. In: C Symon, L Arris and B Heal, editors. Arctic Climate Impact Assessment. New York, NY: Cambridge University Press. pp. 597–648.
10. Maier JAK, Hoef JMV, McGuire AD, Bowyer RT, Saperstein L, et al. (2005) Distribution and density of moose in relation to landscape characteristics: effects of scale. Canadian Journal of Forest Research 35: 2233–2243.
11. Joly K, Duffy PA, Rupp TS (2012) Simulating the effects of climate change on fire regimes in Arctic biomes: implications for caribou and moose habitat. Ecosphere 3: art36.
12. Collins WB, Dale BW, Adams LG, Mcelwain DE, Joly K (2011) Fire, grazing history, lichen abundance, and winter distribution of caribou in Alaska's taiga. Journal of Wildlife Management 75: 369–377.
13. Festa-Bianchet M, Ray JC, Boutin S, Côté SD, Gunn A (2011) Conservation of caribou (*Rangifer tarandus*) in Canada: an uncertain future. Canadian Journal of Zoology 89: 419–434.
14. Joly K, Chapin FS, Klein DR (2010) Winter habitat selection by caribou in relation to lichen abundance, wildfires, grazing, and landscape characteristics in northwest Alaska. Ecoscience 17: 321–333.
15. Rupp TS, Olson M, Adams LG, Dale BW, Joly K, et al. (2006) Simulating the influences of various fire regimes on caribou winter habitat. Ecological Applications 16: 1730–1743.
16. Hummel M, Ray JC (2008) Caribou and the north: a shared future. Toronto, Ontario: Dundurn, Press. 288 p.

17. Kofinas GP (2005) Hunters and researchers at the co-management Interface: emergent dilemmas and the problem of legitimacy. Anthropologica 47: 179–196.
18. Kofinas GP, Osherenko G, Klein D, Forbes B (2000) Research planning in the face of change: the human role in reindeer/caribou systems. Polar Research 19: 3–21.
19. IPCC (2013) Climate Change 2013: the Physical Science Basis. United Nations, International Panel on Climate Change, Working Group I. pp. 127.
20. Higuera PE, Brubaker LB, Anderson PM, Brown TA, Kennedy AT, et al. (2008) Frequent fires in ancient shrub tundra: implications of paleorecords for arctic environmental change. Plos One 3: 1–7.
21. Russell DE, Martell AM (1984) Winter range ecology of caribou (*Rangifer tarandus*). In: R Olsen, F Gedde and R Hastings, editors. Northern ecology and resource management. Edmonton, Alberta, Canada: University of Alberta Press. pp. 117–144.
22. Gustine DD, Barboza PS, Lawler JP, Adams LG, Parker KL, et al. (2012) Diversity of nitrogen isotopes and protein status in caribou: implications for monitoring northern ungulates. Journal of Mammalogy 93: 778–790.
23. Russell DE, Martell AM, Nixon WA (1993) The range ecology of the Porcupine caribou herd in Canada. Rangifer 8: 1–168.
24. Scotter GW (1967) The winter diet of barren-ground caribou in northern Canada. Canadian Field-Naturalist 81: 31–38.
25. Gustine DD, Barboza PS, Addison J, Shively R, Oliver L (2014) Isotopic nitrogen in fecal fiber as an indicator of winter diet in caribou and muskoxen. Rapid Communications in Mass Spectrometry 28: 625–634.
26. Thomas DC, Kiliaan HPL (1998) Fire-caribou relationships (IV) - Recovery of habitat after fire on winter range of Beverly herd. Technical Report Series No 312. Edmontion, AB Canada: Canadian Wildlife Service.
27. Adamczewski JZ, Hudson RJ, Gates CC (1993) Winter energy balance and activity of female caribou on Coats Island, Northwest Territories - the relative importance of foraging and body reserves. Canadian Journal of Zoology 71: 1221–1229.
28. White RG (1992) Nutrition in relation to season, lactation, and growth of north temperate deer. In: R. D Brown, editor editors. The biology of deer. New York, NY: Springer-Verlag. pp. 407–417.
29. Joly K, Bente P, Dau J (2007) Response of overwintering caribou to burned habitat in northwest Alaska. Arctic 60: 401–410.
30. Schaefer JA, Pruitt WO (1991) Fire and woodland caribou in southeastern Manitoba. Wildlife Monographs 116: 1–39.
31. Skogland T (1986) Density dependent food limitation and maximal production in wild reindeer herds. Journal of Wildlife Management 50: 314–319.
32. Klein DR (1982) Fire, lichens, and caribou. Journal of Range Management 35: 390–395.
33. Jandt R, Joly K, Meyers CR, Racine C (2008) Slow recovery of lichen on burned caribou winter range in Alaska tundra: potential influences of climate warming and other disturbance factors. Arctic Antarctic and Alpine Research 40: 89–95.
34. Stocks BJ, Mason JA, Todd JB, Bosch EM, Wotton BM, et al. (2003) Large forest fires in Canada, 1959-1997. Journal of Geophysical Research-Atmospheres 108: 1–12.
35. Barrett K, McGuire AD, Hoy EE, Kasischke ES (2011) Potential shifts in dominant forest cover in interior Alaska driven by variations in fire severity. Ecological Applications 21: 2380–2396.
36. Kasischke ES, Verbyla DL, Rupp TS, McGuire AD, Murphy KA, et al. (2010) Alaska's changing fire regime – implications for the vulnerability of its boreal forests. Canadian Journal of Forest Research 40: 1313–1324.
37. Barrier TA, Johnson CJ (2012) The influence of fire history on selection of foraging sites by barren-ground caribou. Ecoscience 19: 177–188.
38. Joly K, Dale BW, Collins WB, Adams LG (2003) Winter habitat use by female caribou in relation to wildland fires in interior Alaska. Canadian Journal of Zoology 81: 1192–1201.
39. Higuera PE, Brubaker LB, Anderson PM, Hu FS, Brown TA (2009) Vegetation mediated the impacts of postglacial climate change on fire regimes in the south-central Brooks Range, Alaska. Ecological Monographs 79: 201–219.
40. Martin PD, Jenkins JL, Adams FJ, Jorgenson MT, Matz AC, et al. (2009) Wildlife response to environmental arctic change: predicting future habitats of Arctic Alaska. Report of the Wildlife Response to Environmental Arctic Change (WildREACH). Fairbanks, AK: U. S. Fish and Wildlife Service. pp. 138.
41. Landhausser SM, Wein RW (1993) Postfire vegetation recovery and tree establishment at the Arctic treeline: climate-change-vegetation-response hypotheses. Journal of Ecology 81: 665–672.
42. Racine C, Jandt R, Meyers C, Dennis J (2004) Tundra fire and vegetation change along a hillslope on the Seward Peninsula, Alaska, U.S.A. Arctic, Antarctic, and Alpine Research 36: 1–10.
43. Racine CR (2008) Slow recovery of lichen on burned caribou winter range in Alaska tundra: potential influences of climate warming and other disturbance factors. Arctic, Antarctic, and Alpine Research 40: 89–95.
44. Rupp TS, Chapin FS, Starfield AM (2000) Response of subarctic vegetation to transient climatic change on the Seward Peninsula in north-west Alaska. Global Change Biology 6: 541–555.
45. Rupp TS, Starfield AM, Chapin FS (2000) A frame-based spatially explicit model of subarctic vegetation response to climatic change: comparison with a point model. Landscape Ecology 15: 383–400.
46. Valkenburg P (2001) Stumbling towards enlightenment: understanding caribou dynamics. Alces 37: 457–474.
47. Lenart EA (2011) Units 26B and 26C caribou. In: P Harper, editor editors. Caribou management report of survey and inventory activities 1 July 2008-30 June 2010. Juneau, Alaska: Alaska Department of Fish and Game, Juneau, AK. pp. 299–325.
48. Arthur SM, Del Vecchio PA (2009) Effects of oil field development on calf production and survival in the Central Arctic herd. Juneau, AK: Alaska Department of Fish and Game.
49. Cameron RD, Whitten KR (1979) Seasonal movements and sexual segregation of caribou determined by aerial survey. The Journal of Wildlife Management 43: 626–633.
50. Wolfe SA (2000) Habitat selection by calving caribou of the Central Arctic Herd, 1980–1995. Department of Biology and Wildlife. Fairbanks, AK: University of Alaska Fairbanks. pp. 83.
51. Hijmans RJ, van Etten J (2013) raster: Geographic data analysis and modeling. R package version 21–16.
52. Duffy PA, Epting J, Graham JM, Rupp TS, McGuire DA (2007) Analysis of Alaskan burn severity patterns using remotely sensed data. International Journal of Wildland Fire 16: 277–284.
53. Duffy PA (2006) Interactions among climate, fire, and vegetation in the Alaskan boreal forest. Fairbanks, Alaska: University of Alaska Fairbanks.
54. Walsh JE, Chapman WL, Romanovsky V, Christensen JH, Stendel M (2008) Global climate model performance over Alaska and Greenland. Journal of Climate 21: 6156–6174.
55. Nakicenovic N, Alcamo J, Davis G, de Vries B, Fenhann J, et al. (2000) Special report on emissions scenarios: a special report of working group III of the Intergovernmental Panel on Climate Change. Cambridge, U.K.: Intergovernmental Panel on Climate Change. pp. 599.
56. United Nations Environment Programme (2009) Climate change science compendium. Nairobi, Kenya.
57. U.S. Geological Survey (2005) North American Land Cover at 250-m spatial resolution.: Natural Resources Canada/Canadian Center for Remote Sensing (NRCan/CCRS), U.S. Geological Survey (USGS); Insituto Nacional de Estadística y Geografía (INEGI), Comisión Nacional para el Conocimiento y Uso de la Biodiversidad (CONABIO) and Comisión Nacional Forestal (CONAFOR).
58. Higuera PE, Chipman ML, Barnes JL, Urban MA, Hu FS (2011) Variability of tundra fire regimes in Arctic Alaska: millennial-scale patterns and ecological implications. Ecological Applications 21: 3211–3226.
59. Kochtubajda B, Flannigan MD, Gyakum JR, Stewart RE, Logan KA, et al. (2006) Lightning and fires in the Northwest Territories and responses to future climate change. Arctic 59: 211–221.
60. Nelson JL, Zavaleta E, Chapin III FS (2008) Boreal fire effects on subsistence resources: landscape diversity as a critical component of rural livelihoods in Alaska and adjacent Canada. Ecosystems 11: 156–171.
61. Barrett CM, Kelly R, Higuera PE, Hu FS (2013) Climatic and land cover influences on the spatiotemporal dynamics of Holocene boreal fire regimes. Ecology 94: 389–402.
62. Tape K, Sturm M, Racine C (2006) The evidence for shrub expansion in Northern Alaska and the Pan-Arctic. Global Change Biology 12: 686–702.
63. Tyler NC, Forchhammer MC, Oritsland NA (2008) Nonlinear effects of climate and density in the dynamics of a fluctuating population of reindeer. Ecology 89: 1675–1686.
64. Skoog R (1968) Ecology of the caribou (*Rangifer tarandus granti*) in Alaska. Unpublished PhD thesis, University of California, Berkeley.
65. Skogland T (1985) The effects of density-dependent resource limitations on the demography of wild reindeer. Journal of Animal Ecology 54: 359–374.
66. Zamin TJ, Grogan P (2012) Birch shrub growth in the low Arctic: the relative importance of experimental warming, enhanced nutrient availability, snow depth and caribou exclusion. Environmental Research Letters 7.
67. Cahoon SMP, Sullivan PF, Post E, Welker JM (2012) Large herbivores limit CO_2 uptake and suppress carbon cycle responses to warming in West Greenland. Global Change Biology 18: 469–479.
68. Tews J, Ferguson MAD, Fahrig L (2007) Potential net effects of climate change on High Arctic Peary caribou: lessons from a spatially explicit simulation model. Ecological Modelling 207: 85–98.
69. Tveraa T, Stien A, Bårdsen B-J, Fauchald P (2013) Population densities, vegetation green-up, and plant productivity: impacts on reproductive success and juvenile body mass in reindeer. Plos One 8: e56450.
70. Bret-Harte MS, Mack MC, Shaver GR, Huebner DC, Johnston M, et al. (2013) The response of Arctic vegetation and soils following an unusually severe tundra fire. Philosophical Transactions of the Royal Society B: Biological Sciences 368.
71. Barrett K, Rocha AV, van de Weg MJ, Shaver G (2012) Vegetation shifts observed in arctic tundra 17 years after fire. Remote Sensing Letters 3: 729–736.
72. Witter LA, Johnson CJ, Croft B, Gunn A, Poirier LM (2012) Gauging climate change effects at local scales: weather-based indices to monitor insect harassment in caribou. Ecological Applications 22: 1838–1851.
73. Johnson CJ, Boyce MS, Case RL, Cluff HD, Gau RJ, et al. (2005) Cumulative effects of human developments on arctic wildlife. Wildlife Monographs 160: 1–36.
74. Cameron RD, Smith WT, White RG, Griffith B (2005) Central Arctic caribou and petroleum development: distributional, nutritional, and reproductive implications. Arctic 58: 1–9.

75. Northrup JM, Wittemyer G (2013) Characterising the impacts of emerging energy development on wildlife, with an eye towards mitigation. Ecology Letters 16: 112–125.
76. McDonald RI, Fargione J, Kiesecker J, Miller WM, Powell J (2009) Energy sprawl or energy efficiency: climate policy impacts on natural habitat for the United States of America. Plos One 4: e6802.
77. Hinkes MT, Collins GH, Van Daele LJ, Kovach SD, Aderman AR, et al. (2005) Influence of population growth on caribou herd identity, calving ground fidelity, and behavior. Journal of Wildlife Management 69: 1147–1162.
78. Nagy JA, Johnson DL, Larter NC, Campbell MW, Derocher AE, et al. (2011) Subpopulation structure of caribou (*Rangifer tarandus*) in arctic and subarctic Canada. Ecological Applications 21: 2334–2348.
79. Joly K, Klein DR, Verbyla DL, Rupp TS, Chapin FS (2011) Linkages between large-scale climate patterns and the dynamics of Arctic caribou populations. Ecography 34: 345–352.
80. Thomas DC, Kiliaan HPL, Trottier TWP (1998) Fire-caribou relationships (III) - Movement patterns of the Beverly herd in relation to burns and snow. Technical Report Series No 311. Edmontion, AB Canada: Canadian Wildlife Service.
81. Kofinas GP, Chapin III FS, Burnsilver S, Schmidt JI, Fresco NL, et al. (2010) Resilience of Athabascan subsistence systems to interior Alaska's changing climate. Canadian Journal of Forestry Research 40: 1347–1359.
82. Brinkman TJ, Kofinas G, Hansen WD, Chapin III FS, Rupp TS (2013) A new framework to manage hunting: why we should shift focus from abundance to availability. The Wildlife Professional 7: 38–43.
83. McNeeley SM, Shulski MD (2011) Anatomy of a closing window: vulnerability to changing seasonality in Interior Alaska. Global Environmental Change 21: 464–473.
84. Walther G-R, Post E, Convey P, Menzel A, Parmesan C, et al. (2002) Ecological responses to recent climate change. Nature 416: 389–395.
85. Lawler JJ, Shafer SL, White D, Kareiva P, Maurer EP, et al. (2009) Projected climate-induced faunal change in the Western Hemisphere. Ecology 90: 588–597.
86. Travis JMJ, Delgado M, Bocedi G, Baguette M, Bartoń K, et al. (2013) Dispersal and species' responses to climate change. Oikos: 1532–1540.

5

Accounting for Biomass Carbon Stock Change Due to Wildfire in Temperate Forest Landscapes in Australia

Heather Keith[1]*, David B. Lindenmayer[1], Brendan G. Mackey[2], David Blair[1], Lauren Carter[1], Lachlan McBurney[1], Sachiko Okada[1], Tomoko Konishi-Nagano[3]

1 The Fenner School of Environment and Society, Australian National University, Building 48, Canberra, ACT, Australia, **2** Griffith Climate Change Response Program, Griffith University, Queensland, Australia, **3** Fujitsu Laboratories Ltd., Kawasaki, Japan

Abstract

Carbon stock change due to forest management and disturbance must be accounted for in UNFCCC national inventory reports and for signatories to the Kyoto Protocol. Impacts of disturbance on greenhouse gas (GHG) inventories are important for many countries with large forest estates prone to wildfires. Our objective was to measure changes in carbon stocks due to short-term combustion and to simulate longer-term carbon stock dynamics resulting from redistribution among biomass components following wildfire. We studied the impacts of a wildfire in 2009 that burnt temperate forest of tall, wet eucalypts in south-eastern Australia. Biomass combusted ranged from 40 to 58 tC ha^{-1}, which represented 6–7% and 9–14% in low- and high-severity fire, respectively, of the pre-fire total biomass carbon stock. Pre-fire total stock ranged from 400 to 1040 tC ha^{-1} depending on forest age and disturbance history. An estimated 3.9 TgC was emitted from the 2009 fire within the forest region, representing 8.5% of total biomass carbon stock across the landscape. Carbon losses from combustion were large over hours to days during the wildfire, but from an ecosystem dynamics perspective, the proportion of total carbon stock combusted was relatively small. Furthermore, more than half the stock losses from combustion were derived from biomass components with short lifetimes. Most biomass remained on-site, although redistributed from living to dead components. Decomposition of these components and new regeneration constituted the greatest changes in carbon stocks over ensuing decades. A critical issue for carbon accounting policy arises because the timeframes of ecological processes of carbon stock change are longer than the periods for reporting GHG inventories for national emissions reductions targets. Carbon accounts should be comprehensive of all stock changes, but reporting against targets should be based on human-induced changes in carbon stocks to incentivise mitigation activities.

Editor: Benjamin Poulter, Montana State University, United States of America

Funding: The funder provided support in the form of salaries for the authors [HK, SO], but did not have any additional role in the study design, data collection and analysis, decision to publish, or preparation of the manuscript. The specific roles of these authors are articulated in the 'author contributions' section. The author affiliated with the funding organization contributed to the conceptual framework of the study.

Competing Interests: The authors declare that one of the authors [TKN] is affiliated with one of the funding organizations of this research study, Fujitsu Laboratories Ltd. This affiliation does not alter the authors' adherence to PLOS ONE policies on sharing data and materials.

* Email: Heather.Keith@anu.edu.au

Introduction

Stabilising the carbon stock in the atmosphere to prevent dangerous anthropogenic interference with the climate system is the objective of the United Nations Framework Convention on Climate Change (UNFCCC) [1]. Forest ecosystems naturally exchange carbon dioxide (CO_2) with the atmosphere, but losses of carbon stocks from forests due to land management activities is a significant source of greenhouse gas (GHG) emissions [2]. Losses of carbon as emissions of CO_2 from forest ecosystems to the atmosphere occur due to biological decomposition processes, natural disturbance events like fire, as well as human activities. Removals of CO_2 from the atmosphere to forest ecosystem carbon stocks occur by the uptake of carbon in growing vegetation.

Under the UNFCCC and Kyoto Protocol, each country constructs GHG inventories and reports on their net annual emissions, that is, flows of CO_2 and other GHGs, due to specific activities and sectors including Land Use, Land Use Change and Forestry. Flow-based inventories in the land sector, however, obscure important differences between ecosystems types and between the impacts of human activities and natural disturbances. A more comprehensive, stock-based approach to accounting can address these problems and provide information that complements flow-based accounts (Ajani et al. 2013). The flow-based accounts are currently used for national reporting against emission reduction targets over short time periods, such as annual or 5 to 8 year commitment periods, as determined by political negotiations. However, this short time period exposes nations to high variability in emissions due to natural disturbance events, especially wildfires, which are outside human control, yet create large GHG fluctuations. The resulting variability in national net emissions due to natural disturbance can be much greater than the changes due to land use impacts, thus confounding attempts to provide incentives for reducing emissions in the land sector through improved forest management.

Losses of carbon due to wildfire are a significant component of global annual GHG balances and so it is important that they are included in national inventories. Seasonal losses of carbon from wildfires are detected as high concentrations of trace gases in the troposphere [3,4]. Estimates of gross losses of carbon from forest fires globally range from 2.0–2.5 Pg C yr^{-1} [5,6] to 3.8–4.3 Pg C yr^{-1} [7] as annual averages. As a comparison of the magnitude of this flux, the global net losses are 1.1 PgC yr^{-1} from land-use change and 7.9 PgC yr^{-1} from fossil fuel emissions [8]. For nations with large areas of forest that are subject to periodic wildfires, such as Australia, Canada, USA, Russia and South Africa [9,10,11], the treatment of wildfires in carbon accounting for forest management is an important consideration. The rules and methodologies are complex, often controversial, and raise many issues in international accounting systems.

Carbon stock changes due to wildfires occur at different temporal scales that should be included in comprehensive stock-based accounts. Large losses occur over periods of hours to days or weeks due to combustion and are highly variable each year due to occurrence of fire. Uptake through forest regrowth and losses through decomposition of dead biomass occur over decades to centuries. These changes in carbon stocks are balanced in comprehensive accounts, and in the accumulated concentration of CO_2 in the atmosphere, when assessed over a sufficiently long time period, assuming the fire regime is stationary. The following information is required to enable inclusion of wildfire disturbance in comprehensive stock-based national carbon accounts: carbon stock loss due to combustion, redistribution of carbon stocks between living and dead biomass components; subsequent rates of decomposition; and carbon uptake by regenerating vegetation. These estimates of carbon stock change need to be calculated on a landscape-wide basis and integrated over appropriate time periods [12] with data from a range of forest types. However, methodologies for measuring carbon stock change due to wildfire are problematic for a number of reasons, including: (i) wildfires typically extend over large areas, within which fire severity varies spatially; (ii) initial carbon stocks vary across the landscape in response to differences in forest type, age, disturbance history and environmental factors; (iii) biomass components are combusted with differing efficiencies; (iv) fire occurrence is stochastic [13,14]; and (v) the dynamics of carbon stocks occur over long time periods in response to wildfire.

Of these methodological issues, the parameter with the greatest uncertainty is the amount of carbon combusted, calculated as either an amount or a proportion of the fuel load or total stock, and then upscaled from individual components to sites and landscapes [5,15]. Combustion efficiency varies due to characteristics of individual fuel materials, such as logs of varying size, wood density, state of decay and moisture content [16], and characteristics of the fire, such as intensity and residence time, which determine the oxidation conditions of flaming or smouldering combustion [7,17,18]. At the landscape scale, the proportion combusted depends on the areas burnt within the fire boundary, including small scale heterogeneity in fire occurrence and severity. The high degree of uncertainty created by the limited existing data is demonstrated when aggregated spatially in regional and global inventories. For example, estimates of carbon stock losses from wildfires in Australia derived from global analyses differ by up to six-fold [6,19,20,21,22].

The objectives of this study were to investigate the magnitude and timeframe of carbon stock changes due to wildfire and to evaluate the results in terms of implications for national carbon accounts.

(i) We assessed the magnitude of carbon stock losses due to combustion in a wildfire in terms of the amount and proportion of the total forest ecosystem carbon stock. We then considered the significance of these carbon losses compared with potential impacts from human activities.

(ii) We considered the impact of fire on the ecosystem carbon stock in terms of the timeframe of stock changes. The mean residence times of carbon stocks change in response to redistribution of carbon among biomass components post-fire and their subsequent relative rates of losses and gains. We investigated the dynamics of carbon stock change within the current disturbance regime and not predicted future regimes or climate.

(iii) We evaluated our results about the magnitude and timeframe of carbon stock changes in terms of their significance for the rules and methodologies governing national carbon accounting for forest management. Now is an important time to contribute quantitative information to the development of accounting guidelines for post-2020 agreements which include forest management and disturbance events under the Durban Platform for Enhanced Action [23].

To address these issues about accounting for carbon stock changes due to wildfire, we examined changes in carbon stocks in a temperate eucalypt forest in south-east Australia that was subjected to a wildfire in 2009. We selected a temperate forest because they contain high carbon stocks and contribute 34% to the global forest carbon sink; a proportion that has been increasing over the last two decades [24]. Changes in carbon stocks due to combustion during the fire were assessed empirically and we also investigated the long-term dynamics of biomass carbon pools post-fire using a simulation model, both at the site level and up-scaled across the landscape.

Methods

Description of study region

Access to field sites was granted by Parks Victoria and the Department of Environment and Primary Industries Victoria. Our study region (2,326 km^2) was the tall, wet sclerophyll montane ash forests (*Eucalyptus regnans*, *E. delegatensis* and *E. nitens*) in the Central Highlands of Victoria (Figure 1). These temperate forests are among the most carbon-dense in the world and protection of these carbon stocks offers the potential for mitigation [25]. These forests are subject to a disturbance regime of infrequent, high intensity wildfires [26,27].

Wildfire disturbance regime

The known historical fire frequency has been one in the 1700s, three in the 1800s, nine in the 1900s, and three in the 2000s [27,28], and their spatial extent was mapped across the study region (Figure 1). Occurrence of these fires does not necessarily mean that each conflagration occurred at the same locations or at the same intensity. The mean interval for fires that kill ash trees at a site has been estimated to be 75 to 150 years [29,30]. However, an estimate that approximately half the trees survive within the boundary of a wildfire [30] means that only every second fire in any one location is of sufficient intensity to kill ash trees. Hence, the mean interval of all fires is estimated to be 37 to 75 years. A maximum fire interval of 350–400 years was estimated based on maximum age of the montane eucalypt species [31].

Unlike most eucalypt species, *E. regnans* and *E. delegatensis* are obligate seeders and usually killed if their canopies are predominantly scorched by fire [32,33,34]. However, wildfires are highly

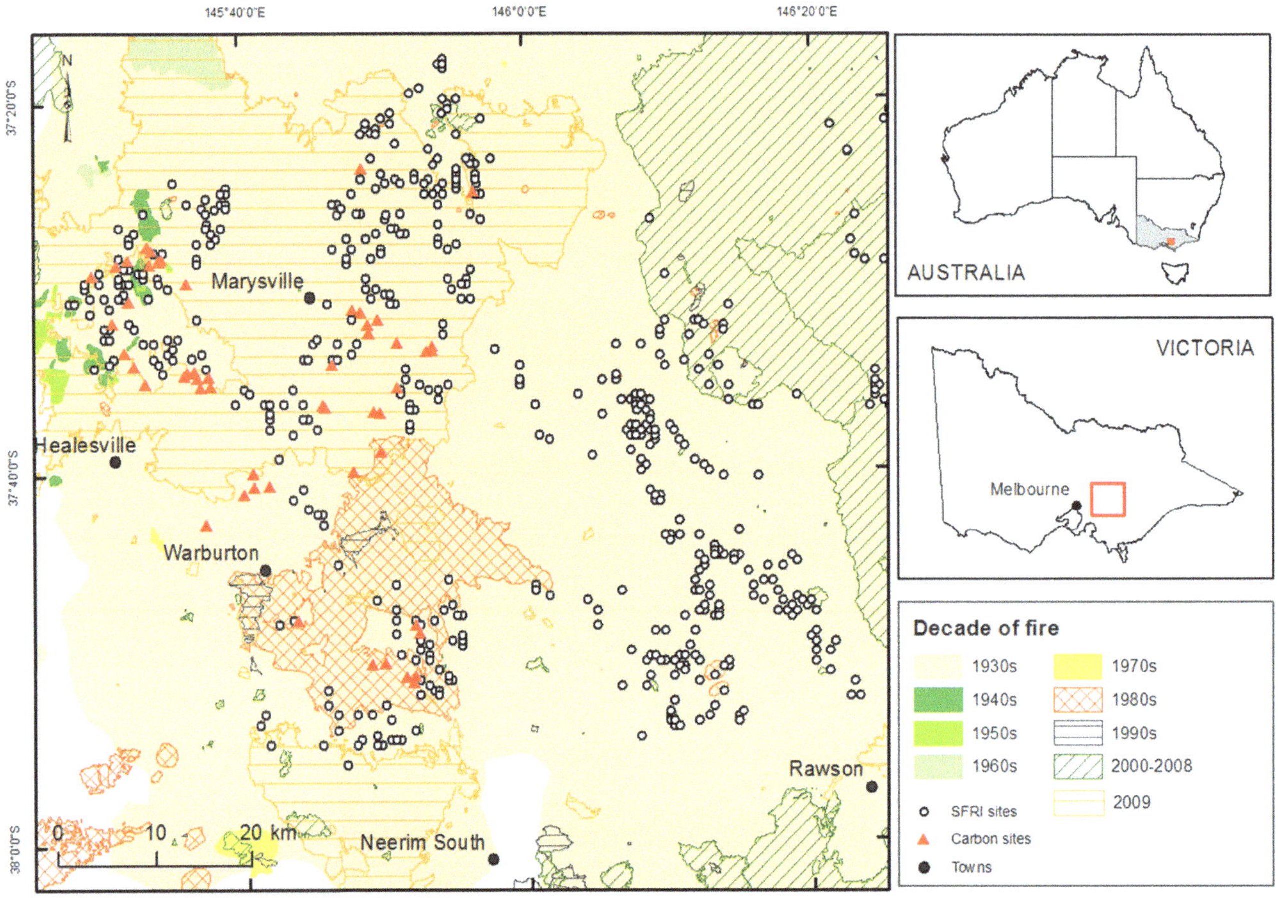

Figure 1. Location of the study region in the Central Highlands of Victoria, Australia. Location of field sites is marked, including long-term ecological monitoring sites (n = 54) where carbon stocks in biomass components were measured, and inventory sites (n = 876) where carbon stocks were estimated and used for upscaling. The spatial extent of wildfires shown represents the outer boundary of the burnt area.

variable in extent and severity, which is reflected in current stand age structures of a mosaic of even-aged and multi-aged forests [26,35,36]. Even-aged stands result from widespread mortality and regeneration following high-severity wildfire. Multi-aged stands result from partial tree death and subsequent regeneration alongside surviving trees [29,37,38]. Hence, the variability of historical fires is important for determining current stand structures. For example, our entire study area is within the boundary of the 1939 fire. However, evidence that individual trees and forest stands survived is seen by the current age structure of patches of old-growth forest where many living trees have fire scars from the 1939 fire event [39].

A wildfire started on 7th February 2009 that burnt over 450,000 ha in Victoria under extreme hot, dry weather conditions (46.4 °C in the nearby city of Melbourne), with winds up to 100 km hr^{-1} [40,41]. Within our study region of montane ash forest, 485 km^2 burnt at low-severity and 283 km^2 burnt at high-severity (Figure 2). Fire severity was mapped by the Victorian Department of Environment and Primary Industries using air photo interpretation and categorised by degree of crown scorch and subsequent tree survival or mortality.

Field sites

We selected 54 sites within the montane ash forest region of the Central Highlands of Victoria (Figure 1) from an existing network of 170 long-term ecological monitoring sites [27] to provide sites with a range of conditions of forest age and severity of the 2009 wildfire. The site categories of forest age were 1983 and 1939 regrowth after wildfire followed by salvage logging or clearcut logging, and old growth. Salvage logging was common practice after wildfire to remove the timber resource [42,43,44,45]. Selection of these regrowth sites provided reasonably even-aged stands of similar structure, although some residual older trees occur [46]. Old growth forest consisted of dominant trees greater than 120 years old but mostly 250 years old that had regenerated following a fire in approximately 1750 [39]. The site categories of severity of the 2009 wildfire were unburnt, low-severity and high-severity burns. Our distinction between low- and high-severity fire was based on site observations of full scorch of the canopy and mortality of *E. regnans and E. delegatensis* trees. The three forest age and three fire severity categories produced a matrix of nine categories, each with six replicate sites.

While the 54 sites covered a range of forest stand structures and disturbance histories, the assumption that the sites had comparable pre-fire biomass carbon stocks, disturbance history, and local biophysical conditions did not necessarily hold. Their carbon stocks likely varied because of the natural heterogeneity of the landscape and the fact that fires are less likely to burn in wet, highly productive mature and old growth forest areas [47,48,49]. Furthermore, the monitoring protocols used prior to the 2009 fire at the sites were aimed at biodiversity assessments, and did not include all the measurements required to quantify ecosystem

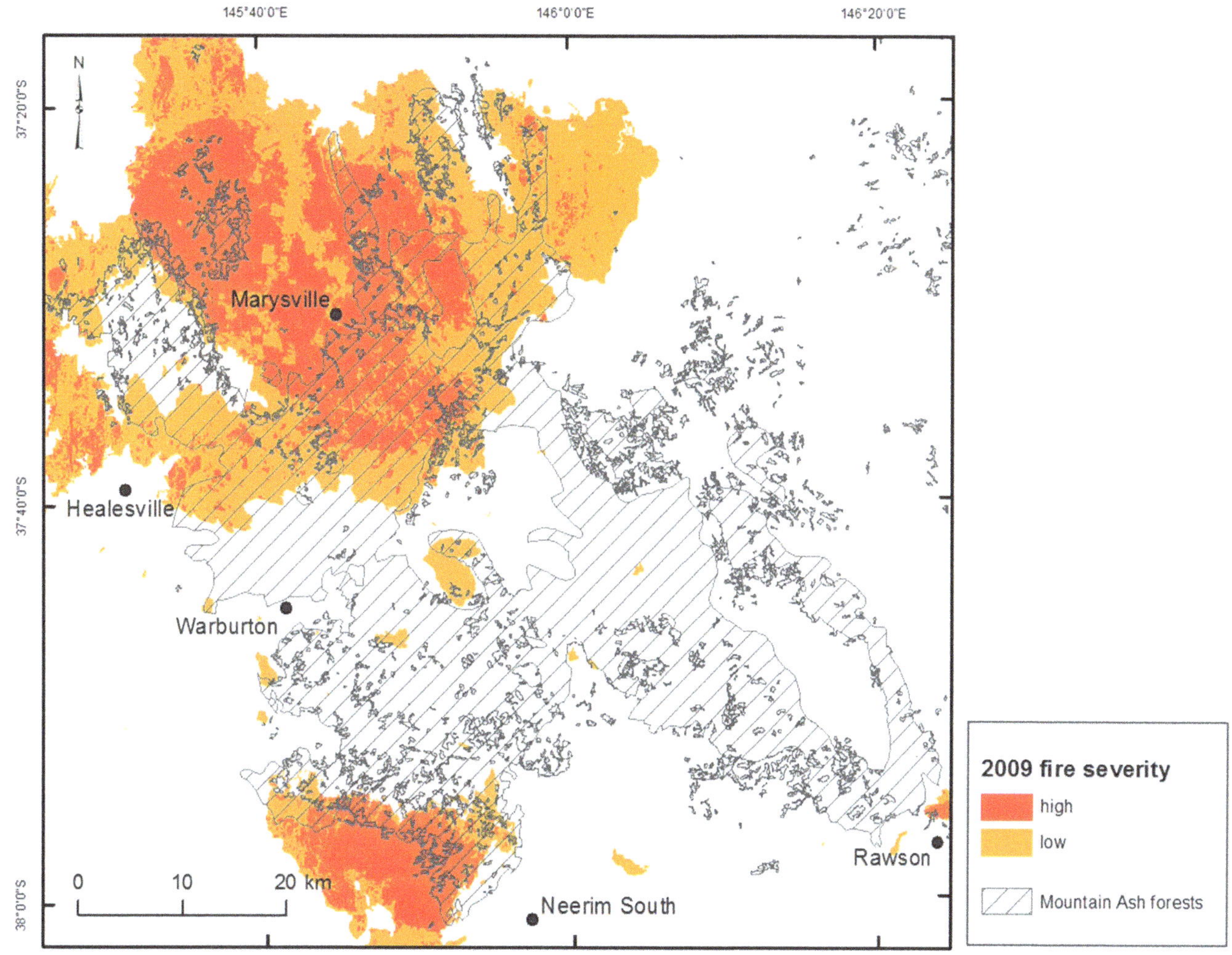

Figure 2. Spatial extent of the 2009 wildfire within the study region. The fire was categorised as low-severity where most trees survived, and high-severity where most trees were killed.

carbon stocks. Therefore, we undertook additional field survey to ensure there was a consistent set of measurements across sites.

Measurement of biomass carbon stocks and stock loss due to fire

We measured carbon stocks in all aboveground biomass components in the forest (living vegetation, dead standing trees, coarse woody debris (CWD or dead and downed woody debris), and litter) at the selected 54 sites in 2010. We defined biomass as all intact organic components of the ecosystem, both living and dead, above- and below-ground, but excluding soil organic matter. Biomass components were classified for ease of measurement and to track movement of carbon between ecosystem components. Classification was based on the status of living or dead, size and vertical location. We used different sampling strategies for each biomass component within 1 ha sites to maximise accuracy and efficiency of measurements of different sizes, densities and distributions of components (Table 1 and Figure 3).

All woody stems, both living and dead, greater than 2 m height and less than 100 cm diameter (diameter at breast height, DBH) were assessed in height and diameter categories for each species within three 10 m×10 m plots (0.03 ha). Trees greater than 100 cm DBH were measured in two perpendicular and intersecting 100 m transects by 30 m width (0.51 ha). In large trees that form buttresses with fluted stems [50], there is a cross-sectional area deficit that was accounted for by converting to a 'functional' diameter [51]. The allometric equation for *E. regnans* derived by Sillett et al. [51] was used to calculate stem and branch volume, and multiplied by wood density and carbon concentration to derive aboveground biomass carbon content. Average stem wood density was 0.520 g cm^{-3} [52,53,54,55] and branch wood density was 0.677 g cm^{-3} [51]. A carbon concentration of 0.5 gC g^{-1} was used for all biomass components [56,57]. Internal decay or hollows in stems were accounted for in the calculation of biomass derived from stem volume [58,59]. Equations to predict occurrence and volume of decay related to tree size were derived from a subset of inventory data (734 trees) [60] where dimensions of defect in sawlogs had been measured. Allometric equations for other vegetation components included *Acacia* spp as mid-storey trees [61], a general rainforest equation for other mid- and understorey species [62], and treeferns were calculated as a cylindrical volume and measured wood density. Root mass in *E. regnans* forests has not been measured, hence an average root: shoot ratio for eucalypt forests of 0.25 [63] was used to convert aboveground biomass to total tree biomass for living and dead trees.

Table 1. Methods for measuring carbon stocks in biomass components within 1 ha sites post-fire at burnt and unburnt sites.

Biomass component	Sampling strategy	Measurement
Trees (living and dead) eucalypt and rainforest species		
<100 cm diameter	Three 10 m×10 m plots (0.03 ha)	Estimated height and diameter in size categories
>100 cm diameter	Two perpendicular intersecting 100 m×30 m transects (0.51 ha)	Measured DBH and height.
Mid- and Under-storey (living and dead) non-eucalypt species		
<20 m height	Three 10 m×10 m plots (0.03 ha)	Estimated height and diameter in size categories
Coarse woody debris		
<60 cm diameter	Line intersect method along 6×10 m transects	Log diameter, decay class, hollows, charcoal and bulk density [64,65]
>60 cm diameter	2×100 m transects	
Litter layer		
<2.5 cm diameter	15 points along transects per site. 30 quadrats randomly located at sites to cover range in litter depths	Measured litter depth.
		Measured litter dry mass.

Coarse woody debris consisted of all woody material ≥25 mm diameter on the ground, and was measured using the line intersect method [64,65]. Logs less than 60 cm diameter were measured along 6×10 m transects within the site and logs greater than 60 cm diameter were measured along 2×100 m perpendicular transects. Each piece of CWD was assessed for hollows and degree of decay in three categories. Bulk density was measured for each category of decay to determine biomass. The litter layer consisted of organic material less than 25 mm diameter including leaves, twigs, insect detritus, animal scats, and comminuted material that was recognisable as organic material. Litter depth was measured at 15 points per site on a compressed litter pack to standardise the quantity of loose material. Depth was converted to biomass using a relationship derived from 30 quadrats sampled for dry weight of litter. Our results for carbon stocks refer to the total biomass of living and dead, above- and below-ground components.

We estimated biomass carbon stock loss by combustion in the 2009 wildfire from the following lines of evidence: (i) field observations and photographs taken immediately after the fire (within 2 months) [66], (ii) measurements of some components of the forest that were measured pre-fire compared with post-fire, and (iii) measurements of biomass components post-fire taken at burnt and unburnt sites. Biomass components combusted in the fire were measured specifically; including, hollow trees, decorticating and rough bark, canopy, shrub biomass, CWD and litter (Table 2 and Figure 4).

Hollow trees, both living and dead, are highly susceptible to combustion because fire is funnelled through the pipe [67]. Data for individual tree heights and diameters from 2005 and 2011 [67] were compared to assess the change pre- and post-2009 fire. The difference in carbon stock between years represented a combination of natural mortality, decay and collapse of trees plus combustion at the burnt sites. Carbon stock loss was calculated as a proportion of the initial stock and the difference between burnt and unburnt used to distinguish the effect of natural decay and combustion.

Biomass of decorticating or ribbon bark hanging from the upper stem and branches was related to tree DBH by a non-destructive sampling method using estimated standard units [68]. All decorticating bark was combusted in the fire. Biomass of rough or fibrous bark on the lower stem was estimated from measurements of bark thickness and bulk density related to tree DBH. Combustion of rough bark was determined from the difference between burnt and unburnt sites. Both decorticating and rough bark mass were calculated for all the inventory trees at the 54 sites.

Canopy leaf biomass was estimated to be approximately 7 tC ha^{-1}, derived from annual litterfall of 4.7 tC ha^{-1} yr^{-1} in ash forests [69,70] and longevity in the canopy of about 18 months [50,71]. Twig (diameter <4 mm) biomass is a similar amount to the leaves [72]. The proportion of the canopy combusted was estimated from photographs after the fire of the leaves remaining in the canopy and the scorched leaves that had fallen to the ground.

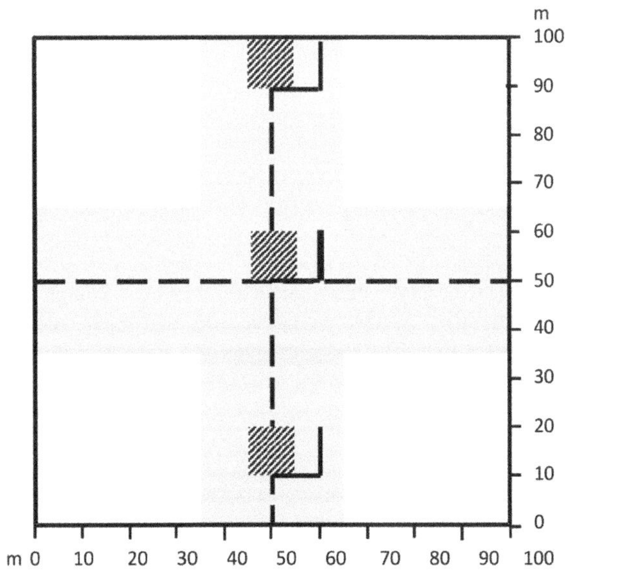

Figure 3. Design of 1 ha experimental sites. Sites of 100 m×100 m included central and perpendicular transects (dashed lines), 10 m×10 m plots (hatched), 10 m transects (black lines), and area sampled for large trees 2×30 m×100 m = 0.51 ha) (grey shaded area). Different biomass components were sampled in different parts of the site.

Table 2. Methods for measuring changes in carbon stocks of biomass components after wildfire, within 1 ha sites.

Biomass component	Sampling time	Sampling strategy	Measurement	Proportion of biomass combusted	
				Low-severity fire	High-severity fire
Hollow trees	Measured in 2005 and 2011	All trees in 1 ha	Measured DBH and height.	0.12 to 0.29	0.16 to 0.56
Decorticating bark	Post-fire, burnt and unburnt sites	Representative size range of trees.	Measured relationship between tree DBH and bark mass.	1.0	1.0
Rough bark	Post-fire, burnt and unburnt sites	Representative size range of trees.	Measured relationship between tree DBH and bark thickness plus bulk density.	0.3 (for trees >200 cm)	0.6 (for trees >200 cm)
Canopy leaves	Post-fire, burnt and unburnt sites	Observations and photos	Estimated biomass from litterfall and leaf longevity	0.5	1.0 (mostly)
Canopy twigs	Post-fire, burnt and unburnt sites	Observations and photos	Relationship between leaf and twig mass.	0.5	1.0 (mostly)
Shrubs	Post-fire, burnt sites	Representative range of shrub sizes and species.	Minimum size of tips remaining on stems after fire. Calculated biomass in each stem diameter class in plots.	0.25 to 0.5 (from branch diameter <5 cm)	0.3 to 0.75 (from branch diameter <5 cm)
Coarse woody debris	Post-fire burnt sites	Line intersect transects	Proportions of logs charred. Lines of charcoal remaining post-fire.	Highly variable dependent on log size	Highly variable dependent on log size
Litter	Photos soon after the fire	Average from all low- and high-severity burnt sites.	Proportion of litter layer combusted.	1.0	1.0

Combustion of shrub biomass in woody stems, branches and twigs was assessed using measurements of: (i) minimum sizes of tips remaining on stems after the fire, (ii) the proportion of biomass lost from a shrub when different tip sizes were removed, and (iii) the amount of biomass in each stem diameter size class of shrubs from the inventory data. A range of species and sizes of understorey shrubs was sampled.

Combustion of CWD was determined on each piece along the measurement transects by the proportion of the piece charred, or by the number of lines of charcoal remaining on the ground. The initial carbon stock of the logs that were completely converted to charcoal was assumed to be represented by the mean size of logs at the site. CWD was too variable between burnt and unburnt sites within stand age categories to allow a comparison to be indicative of the amount combusted as the mean of a category. The proportion of the litter layer combusted was estimated from photographs of quadrats taken soon after the fire in low- and high-severity burnt areas and compared with litter biomass in unburnt sites.

Spatial estimation of carbon stocks

Spatial up-scaling of the site data to the landscape level followed the statistical modelling approach of Keith et al. [73]. Calibration data for the model used the pre-2009 carbon stocks from the 54 sites described above, as well as an additional 876 sites with inventory data [60]. The statistical model regressed site carbon stocks against the following explanatory variables at a 250 m grid resolution: climate variables of precipitation, water availability index, temperature (minimum, mean, maximum) and radiation; topographic variables of topographic position, elevation, aspect and slope; substrate variables of soil parent material, lithology and soil organic carbon content; gross primary productivity (GPP, derived from remote sensing); forest type; forest management area; and disturbance history. Methods used to derive spatial estimates of the environmental and GPP variables are documented in Mackey et al. and Berry et al. [26,74,75]. Disturbance history included wildfires and logging under different silvicultural systems, both of which have complex effects on forest age structure. The statistical regression model was used to generate spatial estimates of carbon stocks across the study region.

Impacts of the wildfire on ecosystem carbon stocks were quantified by determining first, the loss in carbon stock of each biomass component combusted, and second, the redistribution of carbon among biomass components within the ecosystem. Spatial estimation of the carbon stock post-2009 fire was based on measured reductions in the stock according to the fire severity category in the burnt areas and forest age.

Modelling change in carbon stocks

Changes in carbon stocks were simulated for each forest age category after high-severity fire at the average historical return time of 112 years. Initial conditions for the model used site data of the current carbon stock in each age category and the proportion of the stock combusted. After the fire, carbon stocks were redistributed among living and dead components according to the proportions measured in each age category. For the second fire cycle, the initial condition of total biomass at the time of the fire was the stock simulated for year 112, and the proportions of the stock combusted and redistributed were taken as the average of the site data from the 1939 regrowth and old growth forests. Changes in carbon stocks over time were simulated using functions to describe the processes of regeneration, mortality, collapse of dead trees, and decomposition of CWD derived from existing data and observations for montane ash forest (Table 3).

The rate of regeneration was based on carbon stocks at sites with different times since stand-replacing disturbance, derived from the current study and inventory data [60]. Mortality due to self-thinning of the regeneration was estimated from changes in stem density with age, based on data from the literature [76,77]. Mortality of living trees that survived the fire was based on measured rates of mortality over 28 years [67] and the average longevity of *E. regnans* of at least 250 years [78], but up to 500 years [79]. The rate of collapse of dead standing trees was described by a logistic function, with a slow initial rate that

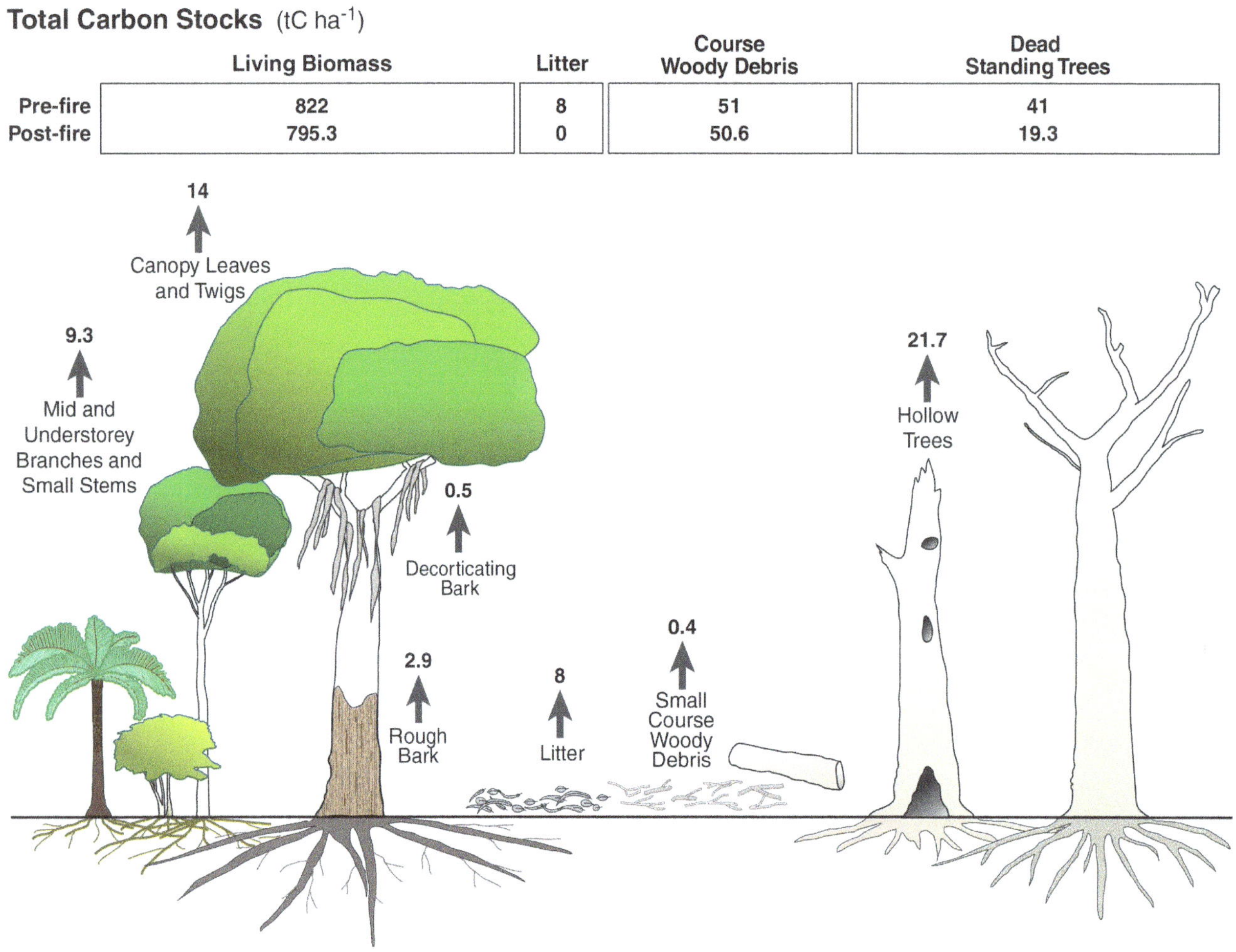

Figure 4. Schematic of carbon stocks in the forest ecosystem and stock changes resulting from fire.

increased to a maximum and then a slow final rate when few trees remain. This pattern was based on observations and evidence that dead trees remain standing for approximately 10 to 75 years [66,80]. The biomass in dead trees was then transferred to the CWD pool. The rate of decomposition of CWD and dead root biomass used in the model was 0.07 yr^{-1}, which had been derived from empirical data for *E. regnans* [55]. This rate was intermediate among published rate constants [80,81,82], and produced amounts of CWD that were consistent with observed amounts in montane ash forest.

Results

Current carbon stock

Biomass density varied substantially among sites within the montane ash forest due to the disturbance history and subsequent age distribution of the forest. Biomass density (mean ±standard error, n = 6 sites) in unburnt stands was 405±33 tC ha^{-1} in 1983 regrowth, 603±74 tC ha^{-1} in 1939 regrowth, and 1039±44 tC ha^{-1} in old growth sites. The structure of these forest stands is illustrated in terms of the distribution of their carbon stock by tree sizes (Figure 5). These tree sizes demonstrate that the regrowth stands were not necessarily even-aged and included some larger residual trees which contained high, but variable, levels of biomass.

CWD biomass was highly variable among sites in the age/fire categories, ranging from 15 to 186 tC ha^{-1}, with a mean of 67 tC ha^{-1}. There was no significant difference in CWD biomass between forest age categories, reflecting the varied disturbance history of these forests where CWD remained after logging and wildfire, and its biomass was not related to the age of the stands. Litter biomass in the unburnt sites ranged from 6 tC ha^{-1} in the 1983 regrowth to 9 tC ha^{-1} in mature and old growth stands. This litter biomass corresponded to average site depth of litter when compressed of 2 cm and 3 cm, respectively, with the majority of measured points less than 5 cm and only a few points with 5–10 cm depth.

Biomass components combusted

In areas burnt by low-severity fire, most of the over-storey trees survived as well as variable proportions of mid- and under-storey vegetation with green, scorched or combusted leaves. In areas burnt by high-severity fire, most of the leaves, twigs and decorticating bark in all canopy strata were combusted, although scorched leaves remained in patches. Some montane ash forests included a few *E. nitens* trees that regenerated by epicormic growth. Woody stems of most shrubs and mid-storey trees remained, but some were entirely combusted. The average branch size combusted from shrubs was 4 mm (maximum 19 mm) in low-severity fire, and 8 mm (maximum 42 mm) in high-severity fire.

Table 3. Input data and equations to predict biomass carbon stock (B in tC ha^{-1}) as functions of time (t) since disturbance in the model of change in carbon stocks after a high-severity wildfire in old growth montane ash forest.

Process	Function	Data	Source
Regeneration (= gain in carbon)	B_{regen} (t) = 1200×(1−exp (−0.0045×t))$^{0.7}$	Inventory data from 99 sites	46,60
	$Log_{10}(SD_t)$ = 1.28+3.16×0.913^t +1.9×0.99^t Log_{10} (SD_t) = 11.61−1.624 log_{10} (t) (average from the two functions used)	Double exponential function fitted to chronosequence site data for *E.regnans*. Function derived from site data	76,77
Mortality of regeneration (= input to dead trees)	$B_{dead}(t)$ = [(SD(t −1) − SD(t))/SD(t)]×$B_{regen}(t)$	Empirically derived from stem density changes	Current
Branch fall (= input to CWD)	$B_{branch}(t)$ = B_{regen} (t)×0.005	Rate constant derived empirically to produce CWD biomass in the range observed (3–255 tC ha^{-1}).	96
Mortality of surviving trees (= input to dead trees)	$B_{dead}(t)$ = 135 exp (−0.015×t)	Rate of mortality estimated from site data.	67
Collapse of dead trees (= input to CWD)	$B_{CWD\_in}(t)$ = 687/(1+ exp(0.1×t −5)	Logistic function derived to fit observations that dead trees remain standing for 10 to 75 years.	66
Decomposition of CWD (= loss of carbon)	$B_{CWD\_loss}(t)$ = 124 exp (−0.07×t)	Rate constant. Modelled range in CWD consistent with site data (mean of 51 tC ha^{-1} and max. of 255 tC ha^{-1})	55,96 Current

The carbon stock losses from each biomass component and the proportion of the total biomass combusted differed with age class of the forest (Table 4). The distribution of readily combusted components within the ecosystem influenced the proportion of the stock emitted during the fire. Much of the observed change due to fire was combustion of shrubs in the understorey. However, this represented a small proportion of the total biomass in the ecosystem. Combustion of 7.8–16.2 tC ha^{-1} in the small branches on shrubs of all stem diameter categories in the understorey (Table 4) represented 1.6–4% of the total living biomass (above- and below-ground) in low-severity fire, and 1.9–6.7% in high-severity fire (range from young regrowth to old growth stands).

Some CWD was combusted during the fire, but both the absolute amount and proportion of initial biomass were highly variable spatially. Higher amounts of CWD were combusted in the 1983 and 1939 regrowth stands (range 1.37–2.41 tC ha^{-1}) than in the old growth stands (0.16–0.43 tC ha^{-1}) (Table 4). Large amounts of CWD occur in these regrowth stands due to self-thinning and residual material from previous logging and fire events. The high rates of combustion in regrowth stands occur

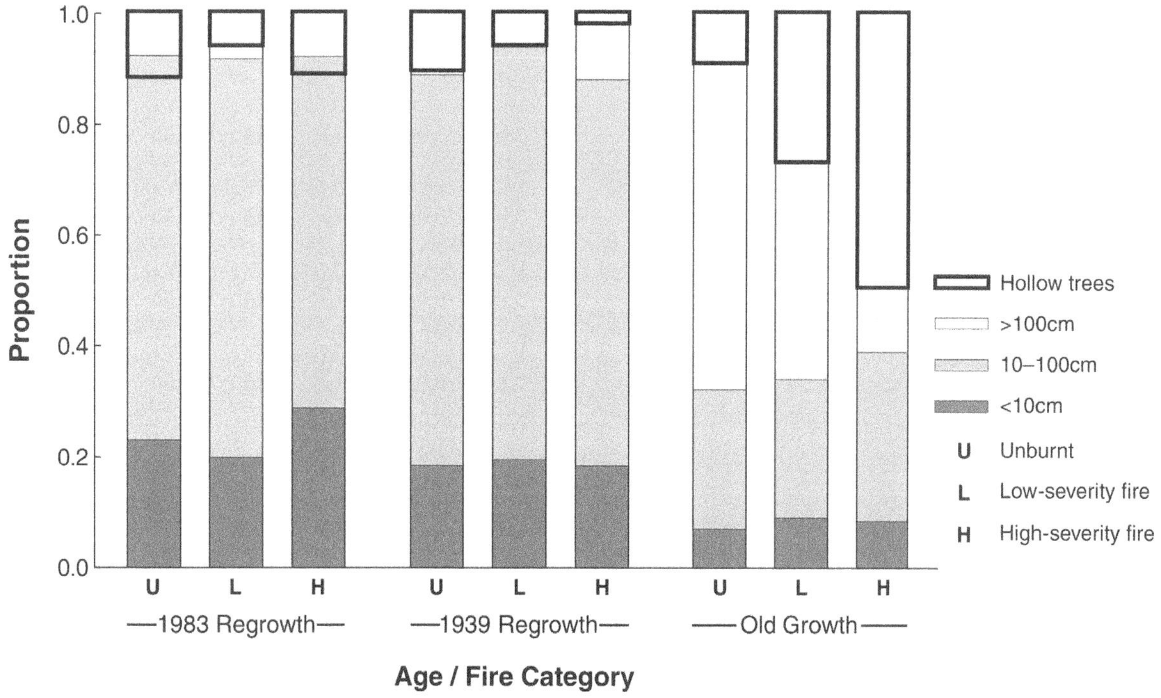

Figure 5. Distribution of carbon stocks in trees according to their size. Carbon stock is shown as the proportion in stem size classes of DHB and trees with hollows, averaged for the sites in each age/fire category (n = 6 sites per category).

Table 4. Summary of biomass components combusted by low- and high-severity fires.

	Mass of carbon[1] combusted (tC ha^{-1})					
	1983 regrowth		**1939 regrowth**		**Old growth**	
Low-severity fire	**mean**	**SE**	**mean**	**SE**	**mean**	**SE**
Litter	6	1.0	9	1.0	9	1.0
Canopy leaves	3.5		3.5		3.5	
Small stems	1.5	0.6	0.6	0.4	0.2	0.2
Small branches	11.9	1.2	11.8	2.7	7.6	3.4
Decorticating bark	0.3	0.02	0.3	0.06	0.5	0.07
Rough bark	0.6	0.2	0.6	0.1	1.3	0.3
CWD	1.4	0.6	1.8	0.7	0.2	0.2
Dead standing trees	3.8	0.9	9	1.7	17.6	5.2
Total biomass[2] combusted[2]	29	4.5	36.6	6.7	39.9	10.4
Total carbon stock[3]	433	35	475	91	564	88
Proportion combusted[4]	0.063		0.072		0.066	
High-severity fire						
Litter	6	1.0	9	1.0	9	1.0
Canopy leaves	7		7		7	
Canopy twigs	7		7		7	
Small stems	2.2	0.5	0.8	0.3	0.3	0.2
Small branches	14	3.0	15	4.0	9	2.0
Decorticating bark	0.3	0.05	0.4	0.1	0.5	0.1
Rough bark	0.8	0.1	0.7	0.2	2.9	0.4
CWD	2.4	1.7	1.7	0.5	0.4	0.3
Dead standing trees	7	1.9	14.8	1.2	21.7	6.2
Total biomass[2] combusted[2]	46.7	8.3	56.4	7.9	57.8	10.2
Total carbon stock[3]	28	46	459	102	589	63
Proportion combusted[4]	0.140		0.109		0.089	

[1]Mass of carbon (mean ±standard error, n = 6 sites).
[2]Total biomass combusted is the sum of all components combusted.
[3]Total carbon stock (above- and below-ground) pre-fire is current carbon stock before the fire.
[4]Proportion combusted is the total biomass combusted divided by the pre-fire carbon stock.

because large proportions of the CWD are small logs and drier logs (decay class 1), compared with the old growth stands. Hollow or damaged trees, either living or dead, were a major source of carbon stock loss (3.8–21.7 tC ha^{-1}) (Table 4). Where these trees had hollows in the centre of the stem, flames funnel up like a chimney and ignite exposed dry wood. In regrowth stands, hollow trees comprised 2–12% of the total tree biomass, and contained a carbon stock ranging from 7–50 tC ha^{-1}. In old-growth stands, trees with hollows generally consisted of the older cohort of trees but also included damaged younger trees. They represented a highly variable proportion of the total biomass, from 9–49%, and contained a carbon stock of 45–241 tC ha^{-1}.

The average carbon stock loss from biomass components was 40 tC ha^{-1} and 58 tC ha^{-1} in low- and high-severity fires, respectively, and this represented an estimated maximum amount (Table 4). As a proportion of the total biomass carbon stock (above- and below-ground), this loss represented 6–7% in low-severity fires and 9–14% in high-severity fires.

Distribution of carbon among components due to fire

The largest impact of fire on the ecosystem carbon stock was a shift in distribution among biomass components. The proportion of total biomass that was in living vegetation in unburnt sites was 72%, 71% and 89%, respectively in the 1983, 1939 regrowth and old growth sites. The shift from living to dead biomass was 3%, 6% and 19% due to low-severity fire, and 85%, 97% and 84% due to high-severity fire, respectively in the age categories (Figure 6). CWD increased by only a small proportion in the burnt sites; some logs were combusted and a similar amount was transferred into the CWD component.

Carbon stock dynamics

Combustion of biomass during the fire resulted in an initial loss of some carbon, but most of the biomass remained at the site. Our simulation model predicted that carbon stocks would increase for several decades after the fire due to the combination of regeneration of living biomass and persistence of dead biomass components (Figure 7). Mortality of trees during the fire produced large amounts of dead biomass, but the trees only collapsed slowly and then contributed to the CWD pool, from where carbon stock loss by decomposition occurred over many decades. The lowest carbon stock in an old growth forest occurred nearly a century after a single fire when most of the dead biomass produced by the fire had decomposed. The maximum fluctuation in carbon stock

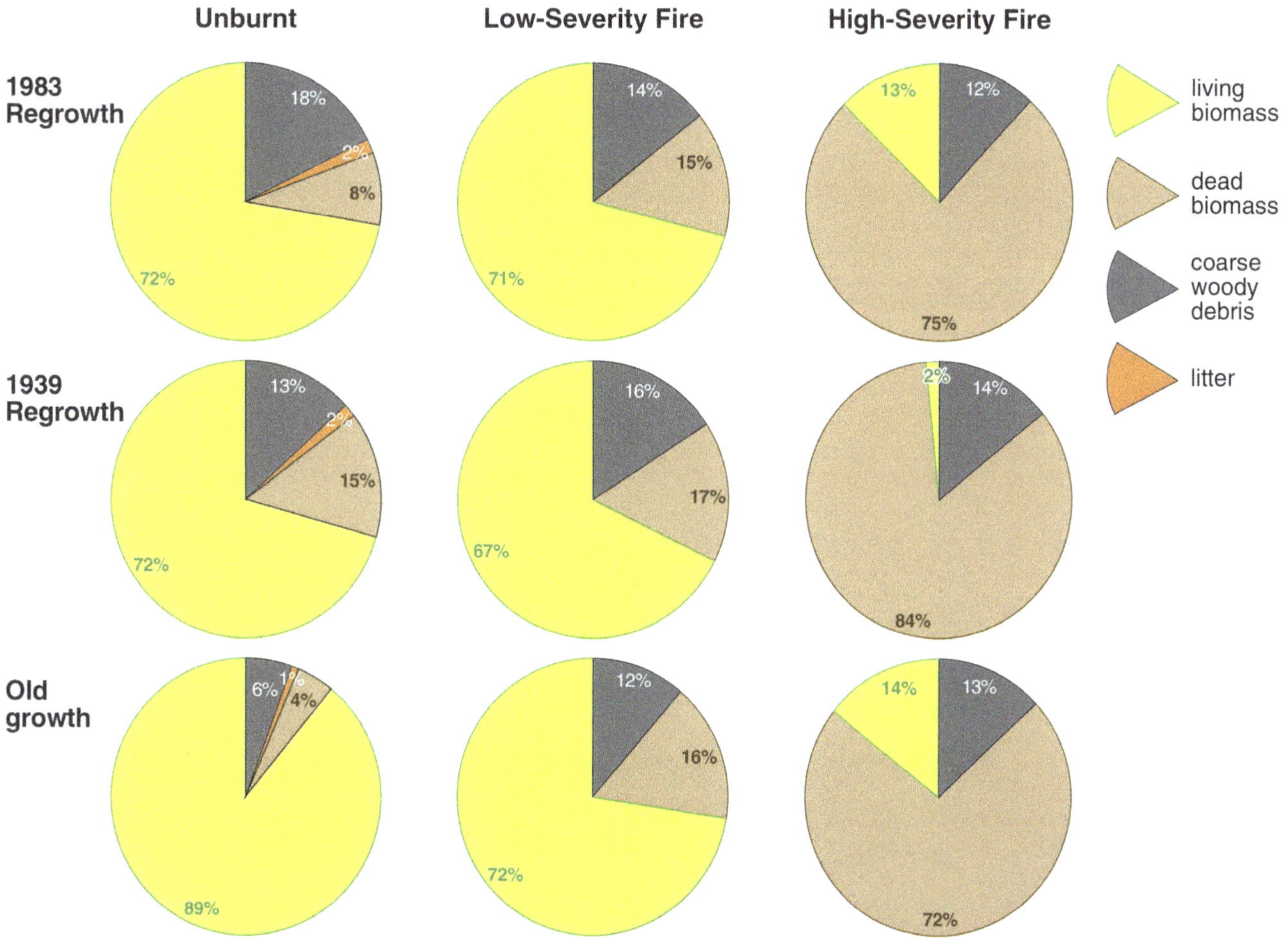

Figure 6. Distribution of carbon stocks among components in relation to stand age and fire severity. Distribution is shown as the proportion of carbon in each biomass component, including living vegetation, dead standing trees, litter and coarse woody debris.

over time as a result of fire was estimated to be 20% from the scenario of a single high-severity fire in old growth forest where there was no trend in total biomass due to changes in forest age (Figure 7a).

The carbon stock dynamics in the forests of initially different ages reflect the differences in the total biomass stock, its distribution among living and dead trees and CWD, the proportion combusted, and the proportions remaining in each of the components after the fire. The old growth forest had the largest increase in total biomass with the combined regeneration plus remaining dead trees, but subsequently the largest decline in total biomass as the dead trees collapsed and decomposed (Figure 7b). In the regrowth forests from 1939 and 1983, there was less dead biomass after the fire and the regeneration resulted in increasing total biomass carbon stock over most of the century following the fire (Figure 7c and d).

Under a regime of repeated fires on a 112-year return time, trends occurred in the total biomass carbon stock because forest age changed from that of the initial forest. The old growth forest decreased in carbon stock because of the more frequent disturbance, whereas the 1983 and 1939 regrowth forests increased in carbon stock because they remained undisturbed for a longer period than at their initial age. These scenarios of carbon stock dynamics represent an average forest stand of a given initial age subject to the average wildfire regime. A range of forest age classes occur across the landscape, including forests much older than 112 years. Forests that remain undisturbed for several hundred years have the capacity to continue accumulating carbon stocks.

Spatial estimation of carbon stocks post-2009 wildfire

The carbon stock loss post-2009 fire was estimated spatially from the distribution of biomass carbon density in the unburnt montane ash forest and the proportion of biomass combusted in low- and high-severity fire for each forest age category (Figure 8). The total carbon stock in the montane ash forest region (2,326 km^2) was 112.8 TgC pre-2009 and an estimated 3.9 TgC was emitted from biomass combustion during the 2009 wildfire (burnt area of 768 km^2 or 33% of the area of montane ash forest). The carbon emitted represented 8.5% of the total biomass carbon stock in the area of montane ash forest that was burnt.

Discussion

Although the total amount of carbon stock loss across the landscape burnt during the wildfire was large as an annual flow in the national inventory, we found that the proportion of total ecosystem carbon stock combusted was relatively small, both at site and landscape scales. The major impact of wildfire was the redistribution of carbon among biomass components, from living to dead biomass and CWD, and their longer-term rates of change due

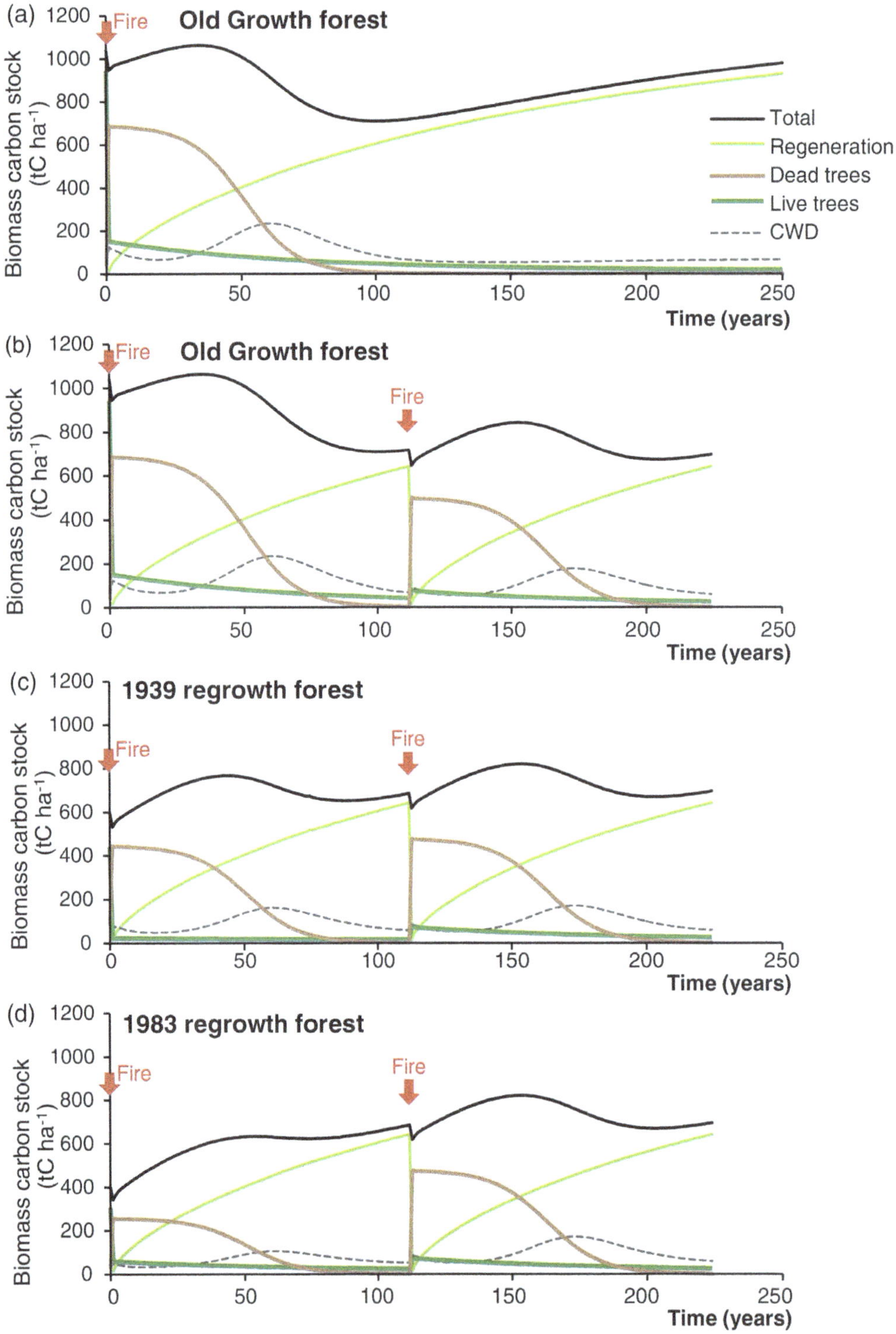

Figure 7. Simulated carbon stock change in biomass components of forests of different ages after high-severity wildfire. Initial fire occurred in year 1 and the change over time in each biomass component, plus the total biomass, is simulated; (a) old growth forest initially that is burnt in year 1 and recovery is simulated over 250 years assuming no other disturbance; (b) old growth forest initially that is burnt in years 1 and 113, according to the average fire frequency, and recovery simulated over two fire cycles; (c) 1939 regrowth forest initially that is simulated over two fire cycles; (d) 1983 regrowth forest initially that is simulated over two fire cycles.

to decomposition and regeneration. Assessing the carbon stock loss by combustion in relation to the total stock, rather than fuel load, meant that changes in stock could be tracked over time and compared between different human activities and disturbance types.

Amount and proportion of forest ecosystem carbon stock combusted

We found that the proportion of total ecosystem biomass carbon stock combusted during a wildfire in the montane ash forest was

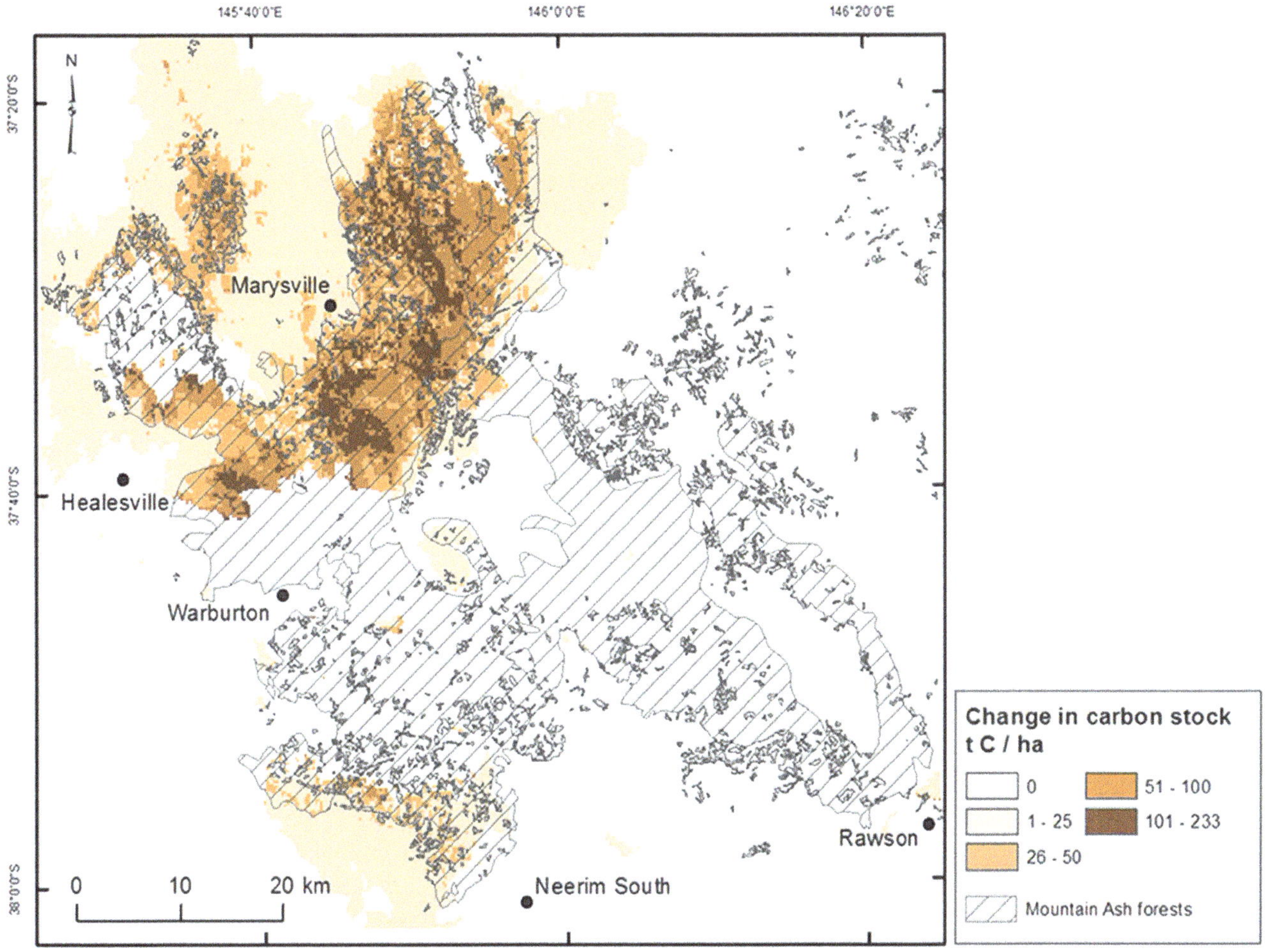

Figure 8. Spatially modelled change in carbon density (tC ha^{-1}) across the montane ash forest region. Change in carbon stock was calculated as the difference between pre- and post-2009 wildfire, including total biomass of living and dead trees, above- and below-ground, shrubs, litter and coarse woody debris.

6–14% at the site scale depending on fire severity and forest age (27 years to old growth), with an average of 8.5% across the landscape that was burnt. This is a relatively small proportion compared with impacts of human disturbance events, such as logging that removes biomass off-site [46,83,84]. Our results are similar to studies in other forest ecosystems (boreal, temperate mixed conifer, savannah woodlands, and other temperate wet sclerophyll) that estimated the relationship between carbon stock loss with total stock, but are lower than estimates of proportion of biomass combusted related to fine and coarse fuel loads (data and references in Table S1).

The estimated loss of only a small proportion of ecosystem carbon stock combusted in a fire concords with simulation results of net losses of carbon by 4 to 8% over 180 years under a fire regime scenario of increased frequency and intensity in south-east Australia [85]. Under this scenario, the loss was due to increased mortality leading to a younger age distribution of trees. Similarly, a sensitivity analysis of the carbon balance model for Canadian forests showed that a three-fold increase in area burned decreased the total forest carbon stock by 6% over 100 years [86]. These predicted losses of carbon stocks are relatively small over a century or two compared with the impact of many human activities.

Differences in the proportion of fine fuel combusted in different ecosystems are insignificant in terms of the total ecosystem carbon stock, representing 0.5 to 1% or 0.5 to 3 tC ha^{-1}. The proportion of coarse fuel combusted varies greatly depending on the size of material, which is influenced not only by forest age but also by disturbance history. Although proportions of biomass combusted were lower in montane ash forest than in other forest types, the amounts combusted in these fuel size classes were greater (Table S1). Amounts of carbon emitted during combustion in different ecosystems depend on vegetation type, past disturbances resulting in dead biomass, and fire weather conditions [87].

We conclude from our methodology that estimating carbon stock loss as a proportion of the total ecosystem carbon stock (based on tracking each biomass component) is a useful metric to compare with other impacts on ecosystem carbon stocks and other sectors of the national GHG inventory. This metric is thus more appropriate for national accounts than the current default reporting of emissions as the product of a combustion factor and fuel load in the IPCC guidelines and Australian national inventory report [6,15,82,88,89]. In our method, all biomass components are assessed without making assumptions about which components constitute fuel.

Our values of carbon stock loss in a wildfire may be small overestimates because some of the biomass combusted is not volatilised and therefore not emitted to the atmosphere. Under conditions of reduced oxygen, partial combustion forms char that remains on the ground, or particulate carbon that is transported in smoke but settles to the ground elsewhere. The proportion of char produced in wildfires has been estimated to be 1–10% of the biomass combusted [90]. Both these forms of carbon are highly resistant to decay and form a long-term stable component of soil carbon stocks [90].

Changes in forest biomass carbon dynamics

The major impact of wildfire in the montane ash forest was the redistribution of carbon among biomass components, from living to dead, and subsequent long-term stock losses from decomposition and gains by regeneration. The relative rates of these processes and subsequent mean residence times of the biomass components determine the long-term dynamics of carbon stock in the ecosystem. These long-term dynamics resulted in changes of up to 20% about the mean stock, whereas direct losses from combustion represented 6–14% of the total biomass stock.

Mean residence time of the biomass components combusted is critical for determining the impact of wildfire on longer-term carbon dynamics. More than half the stock losses from combustion were derived from biomass components with short lifetimes; in the canopy, bark, litter layer and shrubs. These components are described as fine fuel, and this amount remains reasonably constant after about 10 years post-fire in a range of forest types [91]. For example, fine litter in montane ash forest has a mean residence time of 1–4 years [50,70], and 85–95% of the litter mass has accumulated within a decade after fire [91]. Recovery of the carbon stock within a decade in the montane ash forest is within the range of up to 10–20 years for recovery of rates of carbon uptake estimated from eddy covariance measurements of net ecosystem carbon exchange at chronosequence sites across a range of forest types in North America [92].

The components combusted with longer mean residence times, such as CWD and large hollow trees, have variable stocks and proportions combusted depending on forest type, conditions for decomposition, logging and fire history. This coarse material contributed 18 to 45% of the carbon combusted in the montane ash forest, with an increasing proportion in older stands. The residence time of this coarse material is in the order of decades to centuries [67,93].

Rates of carbon accumulation in regenerating stands are difficult to estimate due to limited site data of known forest age, particularly in older aged stands. We compared carbon uptake in regenerated living biomass at a given age estimated from our site data, for example, 3.8 tC ha^{-1} yr^{-1} at 80 years, with the net ecosystem exchange (NEE) estimated from eddy covariance measurements. An average NEE over 10 years of measurements was 5.8 tC ha^{-1} yr^{-1} and the maximum was 9.3 tC ha^{-1} yr^{-1} in an 80 year old stand of *Eucalyptus delegatensis* or Alpine Ash [94]. This average NEE is comparable to the 6 tC ha^{-1} yr^{-1} in a temperate forest in Oregon, which is the highest value reported for forests of a similar age in North America [92]. Carbon accumulated in living biomass does not include uptake of carbon that is then transferred to dead biomass pools and soil organic matter.

Our simulated carbon stocks for each biomass component were within the range observed in the field. Actual magnitudes of stocks at different ages were difficult to verify because of the confounding disturbances of wildfire and salvage logging that have occurred in these forests. Among our sites, there were no younger stands that had been burnt but not logged and such stands would be rare in the landscape. Our simulated turnover times appeared to be slower than in some other ecosystems, such as boreal forests [95]. However, observations of large logs remaining after 60–90 years in the montane ash forest [50,96] suggested that even slower rates may be characteristic of large biomass components. Uncertainty about the magnitude of the changes and their timing in the long-term dynamics of carbon occurred mainly because of limited information about rates of tree fall and decomposition.

Inclusion of wildfire in national carbon accounts

Changes in carbon stocks due to natural disturbances, such as wildfire, must be included in national carbon accounts. The current UNFCCC carbon accounting guidelines for forest management (Article 3 paragraph 4 clause 33) address carbon stock change due to natural disturbances under the '*force majeure*' provision (Decision 2/CMP.7, CoP 17 Durban Climate Conference) [97]. Natural disturbances, such as wildfire, outbreaks of pests and diseases, extreme weather events and geological disturbances, are considered as extraordinary events that are outside human control. An average net stock change due to natural disturbances is calculated as a background level within the forest management reference level. These accounts for forest management and natural disturbances must be included in the 2015 inventory reports from signatory nations to the Kyoto Protocol (38 of the 43 Annex 1 nations), which are assessed for compliance against emissions reduction targets [98].

The current negotiations under the Durban Platform for Enhanced Action [23] represent a critical time to provide quantitative data on carbon stock changes due to disturbance and to interpret the implications of the inventory for maximising outcomes of mitigation activities. Nations can develop and choose methodologies, within the UNFCCC guidelines, that are appropriate for their accounts [99]. The default method is calculated as an average of net stock change from natural disturbances from 1990 to 2009, plus a margin of twice the standard deviation. Annual stock changes that are above this background level are excluded from the accounts. When this exclusion occurs, the area of land affected is removed from the inventory so that subsequent uptake by the regenerating vegetation is not counted, until pre-disturbance carbon stocks are restored [99]. The UNFCCC guidelines provide some flexibility, but the methodology should adhere to the objectives of the *force majeure* provision of minimising annual variability in the reported net stock changes due to natural disturbance events. The rationale is that this variability should not negate reductions in emissions due to human mitigation activities over the timeframe of the reporting [97].

Our results about the magnitude and timeframes of carbon stock changes due to wildfires suggest that carbon accounts need to be comprehensive and include all stocks and flows. Such accounts enable differentiation of stock changes due to natural disturbances and human activities. There are four reasons why constructing carbon accounts which differentiate carbon stock changes are beneficial.

First, our results show that in terms of national annual net change in carbon stocks, the magnitude of carbon stock loss due to combustion in a wildfire is sufficient to mask reduced emissions due to mitigation activities. Our estimated carbon stock loss was 3.9 TgC from a single (albeit large) fire in 2009 within the montane ash forest. The annual average for all fires in Australia is estimated to be 8.7 TgC yr^{-1} [100]. These single and average wildfire emissions represent about 27% and 59%, respectively, of Australia's annual emissions reduction target of 14.7 TgC yr^{-1} [100]. Similar magnitudes of emissions from individual wildfires have been quantified elsewhere, for example 2.9 TgC from the 2003 Californian fires [101]. Hence, differentiation of carbon stock changes due to natural disturbances and human activities is important when national GHG inventory reports are used for assessing progress in meeting emission reduction targets over short time periods.

Second, we demonstrated that the changes in carbon stocks in response to a disturbance event occur over the following decades. Hence, derivation of a background level of net carbon stocks over a two decade period (1990–2009) is not adequate to account for the variations in carbon dynamics within a disturbed area. However, the carbon stock at the landscape scale represents the sum of forest stands that have been burnt at different times and intensities, thus producing a mosaic of stand age classes, amounts of debris, and stages of recovery from disturbance. Under a stable disturbance regime, the carbon stock at the landscape scale remains stable when accounted for over sufficiently long time periods. Hence, carbon stock losses from combustion of biomass components do not increase the accumulated atmospheric CO_2 concentration under a stable fire regime. Carbon stock dynamics may alter at the landscape level if fire regimes change due to climate change, or due to the effects of growth from a forest age class distribution affected by historical activities [9,85,86, 102,103,104].

Third, the increases and decreases in carbon stock post-fire resulting from differential rates of regeneration, mortality and decomposition mean that excluding burnt areas from the national inventory until carbon stocks are restored to pre-fire levels, is not an adequate method for excluding the effect of natural disturbances on net changes in carbon stocks. In fact, the increase in carbon stocks for some decades post-fire would produce artificial credits, but later decreases due to decomposition would produce artificial debits in the national inventory, which are not related to human activities. Additionally, the high levels of uncertainty in these long-term dynamics could produce errors in the national inventory.

Fourth, an averaging period of net stock changes from a disturbance regime over two decades, or within the time period of adequate records, does not encompass the temporal scale of natural variability in wildfire regimes in many temperate forest ecosystems [105], such as montane ash forest with a return interval of many decades [30,31]. The frequency and intensity of wildfire is influenced by interactions of large-scale features of atmospheric circulation, such as the El Niña-Southern Oscillation, Indian Ocean Dipole, Southern Annular Mode and the Sub-Tropical Ridge, which drive multi-decadal climate variability [13,106,107]. If a background level is to be used in an accounting system, then the preferred method would be to re-calculate the actual incidence of disturbance events at the end of the reporting period [108,109,110].

Methodological and data uncertainties

Our results suggest that data from ecological monitoring sites can be used *a posteriori* to estimate carbon stock losses from wildfire, when supplemented with additional field data and ancillary data sources. Our approach used the basic framework of estimating carbon stock changes due to wildfire from the area burned, fuel load and combustion efficiency [87,90] However, our sampling also incorporated spatial variability in fire severity and initial carbon stocks at the landscape level. This provided a more representative range in carbon stock losses at the regional scale. Covering this range is important even though a balanced experimental design is difficult to obtain in studies of stochastic disturbance events where field sites may not be comparable because of uncertainties in disturbance histories and stand ages. Previous studies in Australia have used either experimental fires imposed on a planned design of sites and pre-fire measurements, where the fires were either lower severity than wildfire or regeneration burns of logging slash [89,111,112,113], or post-fire comparisons of burnt and unburnt sites albeit with limited replicate sites [41,114].

Improved assessment of fire severity as a continuous variable at the landscape scale is important to allow more accurate estimates of carbon stock changes spatially [115]. New techniques in remote sensing are providing quantitative information about fire radiative power that can be related to factors that control fire severity [116,117,118]. However, adequate field calibration of the remotely sensed metric currently limits their application [116].

The results of up-scaling the site data to estimate carbon stocks and stock losses across our study region revealed that the effect of disturbance history on tree age structure and CWD was the major source of uncertainty in spatial estimates of biomass carbon stocks. Other studies have drawn similar conclusions that the determination of forest stand age is the main factor contributing to uncertainty in assessment of regional carbon stocks [5].

Our results suggest that current national accounting that uses default values (Tier 2 methodology) for fuel combustion efficiency and fuel load can be improved by using site data from different ecosystems. Data from existing state forest inventories need to be augmented by comprehensive sampling of forest age structure, including old growth forests, and all components of the carbon stock, including dead wood. The methods used here are generic and would constitute Tier 3 methodology that is appropriate for national and project-based accounting. The sampling methods were non-destructive and feasible to apply to a large number of sites.

Conclusions

Carbon stocks in forest ecosystems change in response to wildfire in the short-term by combustion and in the long-term by redistribution among biomass components and their subsequent differential rates of decomposition and regeneration. It is important that carbon accounts include these processes because all stock changes, activities and land areas contribute to quantifying the global carbon cycle. However, the timeframes of the ecological processes that determine carbon stock changes are longer than the time periods for reporting of national inventories for compliance with emissions reductions targets. This problem of incongruous timeframes means that reporting against targets should be based on net changes in carbon stocks from human activities to incentivise mitigation activities. If the carbon accounts differentiate stock changes between the sources from natural and human-induced disturbances, then the net stock change from human activities can be extracted from the accounts for use in reporting. This distinction is currently made between the comprehensive UNFCCC accounts and the reports against targets for the Kyoto Protocol, except that the guidelines for the Protocol include net stock changes from natural disturbances that can be discounted above an averaged background level. However, our results indicate that the spatial and temporal variability in carbon dynamics is too high and the uncertainty of estimates too great to enable a realistic averaged background level that would prevent distortions of the inventory by stock changes that were outside human control.

Acknowledgments

We wish to thank Clive Hilliker for expert assistance with the figures and Steve Meacher for assistance with field measurements of shrub biomass. Helpful discussions about fire ecology with Malcolm Gill, Geoff Cary and Michael Doherty are gratefully acknowledged. The inventory plot data and spatial data for the Central Highlands were obtained from the Victorian Department of Sustainability and Environment under data access licence agreement DALA_2009_082. We thank anonymous reviewers for comments that have improved the manuscript.

Author Contributions

Conceived and designed the experiments: HK DL BM. Performed the experiments: HK DB LM SO. Analyzed the data: HK LC. Contributed to the writing of the manuscript: HK DL BM. Developed the conceptual framework and global context: HK DL BM TKN.

References

1. United Nations Framework Convention on Climate Change 1992. Available: http://unfccc.int/resource/docs/convkp/conveng.pdf. Accessed 2013 Oct 9.
2. Cubasch U, Wuebbles D, Chen D, Facchini MC, Frame D, et al. (2013) Introduction. In: Stocker TF, Qin D, Plattner G-K, Tignor M, Allen SK, et al., editors. Climate Change 2013: The Physical Science Basis. Contribution of Working Group I to the Fifth Assessment Report of the Intergovernmental Panel on Climate Change. Cambridge University Press, Cambridge, United Kingdom and New York, NY, USA. pp. 119–158.
3. Watson CE, Fishman J, Reichle HG (1990) The significance of biomass burning as a source of carbon monoxide and ozone in the southern hemisphere tropics: a satellite analysis. J Geophys Res 95: 16444–16450.
4. Cahoon DR, Stocks BJ, Levine JS, Cofer WR, O'Neill KP (1992) Seasonal distribution of African savanna fires. Nature 359: 812–815.
5. Andreae MO, Merlet P (2001) Emission of trace gases and aerosols from biomass burning. Global Biogeochem Cycles 15(4): 955–966.
6. van der Werf GR, Randerson JT, Giglio L, Collatz GJ, Kasibhatla PS, et al. (2006) Interannual variability in global biomass burning emissions from 1997 to 2004. Atmosph Chem Phys 6: 3423–3441.
7. Kasischke ES, Bruhwiler LP (2003) Emissions of carbon dioxide, carbon monoxide, and methane from boreal forest fires in 1998. J Geophys Res 108: D1,8146.
8. Le Quéré C, Raupach MR, Canadell JG, Marland G, Bopp L, et al. (2009) Trends in the sources and sinks of carbon dioxide. Nature Geosci doi: 10.1038/ngeo689.
9. Stocks BJ, Mason JA, Todd JB, Bosch EM, Wotton BM, et al. (2003) Large forest fires in Canada, 1959–1997. J Geophys Res 108:(D1), 8149.
10. Kurz WA, Stinson G, Rampley GJ, Dymond CC, Neilson ET (2008) Risk of natural disturbances makes future contribution of Canada's forests to the global carbon cycle highly uncertain. Proc Natl Acad Sci USA 105: 15551–1555.
11. Gill AM, Stephens SL, Cary GJ (2013) The worldwide "wildfire" problem. Ecol Appl 23(2): 438–454.
12. Hurst DF, Griffith DWT, Cook GD (1994) Trace gas emissions from biomass burning in tropical Australian savannas. J Geophys Res 99(D8): 16,441–6,456.
13. van der Werf GR, Randerson JT, Collatz GJ (2004) Continental-scale partitioning of fire emissions during the 1997 to 2001 El Niño/La Niña period. Science 303: 73–76.
14. Gill AM and Allan G (2008) Large fires, fire effects and the fire-regime concept. International J Wildland Fire 17: 688–695.
15. de Groot WJ, Pritchard JM, Lynham TJ (2009) Forest floor fuel consumption and carbon emissions in Canadian boreal forest fires. Can J For Res 39: 367–382.
16. Hollis JJ, Matthews S, Ottmar RD, Prichard SJ, Slijepcevic A, et al. (2010) Testing woody fuel consumption models for application in Australian southern eucalypt forest fires. For Ecol Manag 260: 948–964.
17. Scholes RJ, Kendall J, Justice CO (1996a) The quantity of biomass burned in southern Africa. J Geophys Res 101, D19: 23667–23676.
18. Scholes RJ, Ward DE, Justice CO (1996b) Emissions of trace gases and aerosol particles due to vegetation burning in southern hemisphere Africa. J Geophys Res 101: D19: 23677–23682.
19. Horowitz LW, Walters S, Mauzerall DL, Emmons LK, Rasch PJ, et al. (2003) A global simulation of tropospheric ozone and related tracers: Description and evaluation of MOZART, version 2. J Geophys Res 108: D24,4784.
20. Shirai T, Blake DR, Meinardi S, Rowland FS, Russell-Smith J, et al. (2003) Emission estimates of selected volatile organic compounds from tropical savannah burning in northern Australia. J Geophys Res 108: D3, 8406.
21. Hoelzemann J, Schulz MG, Brasseur GP, Granier C (2004) Global wildland fire emission model (GWEM): Evaluating the use of global area burnt satellite data. J Geophys Res 109: D14S04.
22. Ito A, Penner JE (2004) Global estimates of biomass burning emissions based on satellite imagery for the year 2000. J Geophys Res Atm 109: D14S05.
23. UNFCCC (2012) Ad Hoc Working Group on the Durban Platform for Enhanced Action. Available: http://unfccc.int/bodies/body/6645.php Accessed 8 August 2013.
24. Pan Y, Birdsey RA, Fang J, Houghton R, Kauppi PE, et al. (2011) A large and persistent carbon sink in the world's forests, 1990–2007. Science 333: 988–993.
25. Keith H, Mackey BG, Lindenmayer DB (2009) Re-evaluation of forest biomass carbon stocks and lessons from the world's most carbon dense forests. Proc Natl Acad Sci USA 106(28): 11635–11640.
26. Mackey BG, Lindenmayer DB, Gill AM, McCarthy AM, Lindesay JA (2002) Wildfire, Fire and Future Climate: A forest ecosystem analysis. CSIRO Publ., Melbourne.
27. Lindenmayer DB (2009) Forest Pattern and Ecological Processes: A Synthesis of 25 Years of Research. CSIRO Publishing, Collingwood, Australia.
28. Department of Sustainability and Environment (1996) Study of old growth forest in Victoria's Central Highlands. Available: http://www.dse.vic.gov.au/forests/publications/research-reports/study-of-old-growth-forests. Accessed 2013 Aug 8.
29. McCarthy MA, Lindenmayer DB (1998) Multi-aged mountain ash forest, wildlife conservation and timber harvesting. For Ecol Manag 104: 43–56.
30. McCarthy MA, Gill AM, Lindenmayer DB (1999) Fire regimes in mountain ash forest: evidence from forest age structure, extinction models and wildlife habitat. For Ecol Manag 124: 193–203.
31. McCarthy MA, Gill AM, Bradstock RA (2001) Theoretical fire interval distributions. Int J Wildland Fire 10: 73–77.
32. Ashton DH (1975a) The root and shoot development of *Eucalyptus regnans* F. Muell. Aust J Bot 23: 867–887.
33. Flinn D, Squire R, Waring K (2007) Victoria. In: Raison RJ, Squire RO, editors. Forest Management in Australia: Implications for Carbon Budgets. National Carbon Accounting System Technical Report No. 32, Australian Greenhouse Office, Canberra. Ch 3, 103–146.
34. Tng DYP, Williamson GJ, Jordan GJ, Bowman DMJS (2012) Giant eucalypts – globally unique fire – adapted rainforest trees? New Phytol doi: 10.1111/j.1469-8137.2012.04359.x
35. Simkin R, Baker PJ (2008) Disturbance history and stand dynamics in tall open forest and riparian rainforest in the Central Highlands of Victoria. Austral Ecol 33: 747–760.
36. Turner PAM, Balmer J, Kirkpatrick JP (2009) Stand-replacing wildfires? The incidence of multi-cohort and single-cohort *Eucalyptus regnans* and *E. obliqua* forests in southern Tasmania. For Ecol Manag 258: 366–373.
37. Banks JCG (1993) Tree ring analysis of two Mountain Ash trees from the Watts and O'Shannassy catchments, Central Highlands, Victoria. Report to the Central Highlands Old Growth Forest Project, Department of Conservation and Natural Resources, Melbourne.
38. Vivian LM, Cary GJ, Bradstock RA, Gill AM (2008) Influence of fire severity on the regeneration, recruitment and distribution of eucalypts in the Cotter River Catchment ACT. Austral Ecol 33: 55–67.
39. Lindenmayer DB, Cunningham RB, Donnelly CF, Franklin JF (2000) Structural features of old growth Australian montane ash forests. For Ecol Manag 134: 189–204.
40. Victorian Bushfires Royal Commission 2009, Australia. Available: http://www.royalcommission.vic.gov.au/Documents/HR-Chapters/. Accessed 2013 Aug 8.
41. Cruz MG, Sullivan AL, Gould JS, Sims NC, Bannister AJ, et al. (2012) Anatomy of a catastrophic wildfire: The Black Saturday Kilmore East fire in Victoria, Australia. For Ecol Manag 284: 269–285.
42. Noble WS (1977) Ordeal by Fire: The Week a State Burned Up. Hawthorn Press, Melbourne, Australia.
43. Ough K (2001) Regeneration of wet forest flora a decade after clear-felling or wildfire – is there a difference? Aust J Bot 49: 645–664.
44. Lindenmayer DB, Ough K (2006) Salvage harvesting in the montane ash forests of the Central Highlands and its potential impacts on biodiversity. Conserv Biol 20: 1005–1015.
45. Lindenmayer DB, Burton PJ, Franklin JF (2008) Salvage Logging and its Ecological Consequences. CSIRO Publishing, Collingwood, Australia.
46. Keith H, Lindenmayer DB, Mackey B, Blair D, Carter L, et al. (2014) Managing temperate forests for carbon storage: impacts of logging versus forest protection on carbon stocks. Ecosphere 5(6): 75 doi 10.1890
47. Jackson WD (1968) Fire, air, water and earth – an elemental ecology of Tasmania. Proc Ecol Soc Aust 3: 9–16.
48. Lindenmayer DB, Hobbs RJ, Likens GE, Krebs CJ, Banks SC (2011) Newly discovered landscape traps produce regime shifts in wet forests. Proc Natl Acad Sci USA doi 10.1073
49. Taylor C, McCarthy MA, Lindenmayer DB (2014) Non-linear effects of stand age on fire severity. Conserv Letters doi: 10.1111/conl.12122
50. Ashton DH (1975b) Studies in litter in *Eucalyptus regnans* forests. Aust J Bot 23: 413–433.

51. Sillett SC, van Pelt R, Koch GW, Ambrose AR, Carroll AL, et al. (2010) Increasing wood production through old age in tall trees. For Ecol Manag 259: 976–994.
52. Illic J, Boland D, McDonald M, Downes G, Blakemore P (2000) Woody density phase 1– state of knowledge. National Carbon Accounting System Technical Report No. 18, Australian Greenhouse Office, Canberra.
53. Bootle KR (2005) Wood in Australia: types, properties and uses. McGraw-Hill, Sydney.
54. Chafe SC (1985) The distribution and interrelationship of collapse, volumetric shrinkage, moisture content and density in trees of *Eucalyptus regnans* F. Muell. Wood Sci Tech 19: 329–345.
55. Mackensen J, Bauhus J (1999) The decay of coarse woody debris. National Carbon Accounting System, Technical Report No. 6, Australian Greenhouse Office, Canberra.
56. Gifford RM (2000) Carbon contents of above-ground tissues of forest and woodland trees. National Carbon Accounting System Technical Report No. 22, Australian Greenhouse Office, Canberra.
57. Keith H, Leuning R, Jacobsen KL, Cleugh HA, van Gorsel E, et al. (2009a) Multiple measurements constrain estimates of net carbon exchange by a *Eucalyptus* forest. Agric For Meteorol 149: 535–558.
58. Mackowski CW (1987) Wildlife hollows and timber management in Blackbutt forest. Masters' Thesis, University of New England, Armidale, NSW.
59. Gibbons P, Lindenmayer DB (2002) Tree hollows and wildlife conservation in Australia. CSIRO Publications, Collingwood.
60. Department of Sustainability and Environment (2007) Victoria's Statewide Forest Resource Inventory – Central, Central Gippsland and Dandenong Forest Management Areas, Natural Resources Report Series 07–1, Victorian Government.
61. Feller MC (1980) Biomass and nutrient distribution in two eucalypt forest ecosystems. Aust J Ecol 5: 309–333.
62. Keith H, Barrett D, Keenan R (2000) Review of allometric relationships for estimating woody biomass for New South Wales, the Australian Capital Territory, Victoria, Tasmania and South Australia. National Carbon Accounting System Technical Report No. 5b, Australian Greenhouse Office, Canberra.
63. Snowdon P, Eamus D, Gibbons P, Khanna P, Keith H, et al. (2000) Synthesis of allometrics, review of root biomass and design of future woody biomass sampling strategies. National Carbon Accounting System Technical report No. 17, Australian Greenhouse Office, Canberra.
64. Van Wagner CE (1968) The line intersect method in forest fuel sampling. For Sci 14: 20–26.
65. McKenzie N, Ryan P, Fogarty P, Wood J (2001) Sampling, measurement and analytical protocols for carbon estimation in soil, litter ad coarse woody debris. National Carbon Accounting System Technical Report No. 14, Australian Greenhouse Office, Canberra.
66. Lindenmayer DB, Blair D, McBurney L, Banks S (2010) Forest Phoenix: How a Great Forest Recovers after Wildfire. CSIRO Publishing, Melbourne. 128 pp.
67. Lindenmayer DB, Blanchard W, McBurney L, Blair D, Banks S, et al. (2012) Interacting factors driving a major loss of large trees with cavities in a forest ecosystem. PLoS 1: 7(10), e41864.
68. Andrew MH, Noble IR, Lange RT (1979) A non-destructive method for estimating the weight of forage on shrubs. Aust Range J 1(3): 225–231.
69. Ashton D H (1975a) Studies of litter in *Eucalyptus regnans* forests. Aust J Bot 23: 413–433.
70. Polglase PJ, Attiwill PM (1992) Nitrogen and phosphorus cycling in relation to stand age of *Eucalyptus regnans* F. Muell. I. Return from plant to soil in litterfall. Pl Soil 142: 157–166.
71. Jacobs M (1955) Growth Habits of the Eucalypts. Forestry and Timber Bureau, Canberra.
72. Keith H (1991) Effects of fire and fertilization on nitrogen cycling and tree growth in a subalpine eucalypt forest. PhD Thesis, Australian National University, Canberra. 369 pp.
73. Keith H, Mackey BG, Berry S, Lindenmayer D, Gibbons P (2010) Estimating carbon carrying capacity in natural forest ecosystems across heterogeneous landscapes: addressing sources of error. Glob Change Biol 16: 2971–2989.
74. Mackey B, Berry S, Brown T (2008) Reconciling approaches to biogeographic regionalization: a systematic and generic framework examined with a case study of the Australian continent. J Biogeog 35: 213–229.
75. Berry S, Mackey B, Brown T (2007) Potential applications of remotely sensed vegetation greenness to habitat analysis and the conservation of dispersive fauna. Pac Conserv Biol 13(2): 120–127.
76. Ashton DH (1976) The development of even-aged stand of *Eucalyptus regnans* F. Muell. in Central Victoria. Aust J Bot 24: 397–414.
77. Watson FGR, Vertessy RA (1996) Estimating leaf area index from stem diameter measurements in Mountain Ash forest. Cooperative Research Centre for Catchment Hydrology, Report 96/7.
78. Attiwill PM (1994) Ecological disturbance and the conservative management of eucalypt forests in Australia. For Ecol Manag 63: 301–346.
79. Wood SW, Hua Q, Allen KJ, Bowman DMJS (2010) Age and growth of a fire prone Tasmanian temperate old-growth forest stand dominated by *Eucalyptus regnans*, the world's tallest angiosperm. For Ecol Manag 260, 438–447.
80. Grove SJ, Stamm L, Barry C (2009) Log decomposition rates in Tasmanian *Eucalyptus obliqua* determined using an indirect chronosequence approach. For Ecol Manag 258: 389–397.
81. IPCC (2006) Guidelines for National Greenhouse Gas Inventories. Eggleston S, Buendia L, Miwa K, Ngara T, Tanabe K, editors. Institute for Global Environmental Strategies, Hayama, Japan. Available: http://www.ipcc-nggip.iges.or.jp/public/2006gl/index.html. Accessed 2013 Oct 30.
82. DCCEE (2012) Department of Climate Change and Energy Efficiency, Australian National Greenhouse Gas Accounts, National Inventory Report 2010, Australian Government Submission to the United National Framework Convention on Climate Change, April 2012.
83. Harmon ME, Ferrell WK, Franklin JF (1990) Effects on carbon storage of conversion of old growth forests to young forests. Science 247, 699–702.
84. Gough CM, Vogel CS, Harrold KH, George K, Curtis PS (2007) The legacy of harvest and fire on ecosystem carbon storage in a north temperate forest. Glob Change Biol 13, 1935–1949.
85. King KJ, deLigt RM, Cary GJ (2011) Fire and carbon dynamics under climate change in south-eastern Australia: insights from FullCAM and FIRESCAPE modelling. Int J Wildland Fire 20: 563–577.
86. Kurz WA, Apps MJ (1999) A 70-year retrospective analysis of carbon fluxes in the Canadian forest sector. Ecol Appl 9(2): 526–547.
87. French NHF, de Groot WJ, Jenkins LK, Rogers BM, Alvarado E, et al. (2011) Model comparisons for estimating carbon emissions from North American wildland fire. J Geophys Res 116: G00K05.
88. IPCC (2003) Good practice guidance for land use, land use change and forestry. Penman J et al., editors. National Greenhouse Gas Inventories Programme, Institute for Global Environmental Strategies, Kanagawa, Japan. Available: http://www.ipcc-nggip.iges.or.jp/public/gpglulucf/gpglulucf-files. Accessed 2013 Oct 30.
89. Gould JS, Cheney NP, Hutchings PT, Cheney S (1996) Final Report on Prediction of Bushfire Spread for Australian Co-ordination Committee International Decade of Natural disaster Reduction. Project 4/95. CSIRO Forestry and Forest Products.
90. Seiler W, Crutzen PJ (1980) Estimates of gross and net fluxes of carbon between the biosphere and the atmosphere from biomass burning. Clim Change 2: 207–247.
91. Gould JS, Cheney NP (2007) Fire management in Australian forests. In: Forest management in Australia: implications for carbon budgets. Raison RJ and Squire RO, editors. 341–371. National Carbon Accounting System Technical Report No.32.
92. Amiro BD, Barr AG, Barr JG, Black TA, Bracho R, et al. (2010) Ecosystem carbon dioxide fluxes after disturbance in forests of North America. J Geophys Res 115: G00K02.
93. Lindenmayer DB, Wood JT (2010) Long-term patterns in the decay, collapse and abundance of trees with hollows in the mountain ash (*Eucalyptus regnans*) forests of Victoria, southeastern Australia. Can J For Res 40: 48–54.
94. van Gorsel E, Berni JAJ, Briggs P, Cabello-Leblic A, Chasmer L, et al. (2013) Primary and secondary effects of climate variability on net ecosystem carbon exchange in an evergreen *Eucalyptus* forest. Agric For Meteorol 182–183: 248–256.
95. Manies KL, Harden JW, Bond-Lamberty BP, O'Neill KP, (2005) Woody debris along an upland chronosequence in boreal Manitoba and its impact on long-term carbon storage. Can J For Res 35: 472–482.
96. Lindenmayer DB, Incoll RD, Cunningham RB, Donnelly CF (1999) Attributes of logs on the floor of Australian Mountain Ash (*Eucalyptus regnans*) forests of different ages. For Ecol Manag 123: 195–203.
97. UNFCCC (2011) 17th Conference of the Parties to the UNFCCC in Durban, CMP.7 Land Use, Land Use Change and Forestry. Available: http://unfccc.int/files/meetings/durban_nov_2011/decisions/application/pdf/awgkp_lulucf.pdf. Accessed 2013 Aug 8.
98. UNFCCC (2013b) National Inventory Submissions. Available: http://unfccc.int/national_reports/annex_i_ghg_inventories/national_inventories_submissions/. Accessed 2013 Aug 8.
99. UNFCCC (2013a) Revised Supplementary Methods and Good Practice guidance arising from the Kyoto Protocol. Good Practice Guidance. Available: http://www.ipcc-nggip.iges.or.jp/home/docs/kpsg/KP_Supplement_precopyedit.pdf. Accessed 2013 Aug 8.
100. Australian Treasury (2011) Strong growth, low pollution: modelling a carbon price. Australian Government, Canberra.
101. Mühle J, Lueker TJ, Su Y, Miller BR, Prather KA, et al. (2007) Trace gas and particulate emissions from the 2003 southern Californian wildfires. J Geophys Res 112,D03307.
102. Williams A, Karoly D, Tapper N (2001) The sensitivity of Australian fire danger to climate change. Clim Change 49: 171–191.
103. Pitman A, Narisma G, McAneney J (2007) The impact of climate change on the risk of forest and grassland fires in Australia. Clim Change 84: 383–401.
104. Clarke H, Smith P, Pitman A (2011) Regional signatures of future fire weather over eastern Australia from global climate models. Int J Wildland Fire 20: 550–562.
105. Macintosh A (2012a) Are forest management reference levels incompatible with robust climate outcomes? A case study on Australia. Carbon Manag 2(6): 691–707.
106. Esplin B, Gill AM, Enright N (2003) Report of the Inquiry into the 2002–2003 Victorian Bushfires. Office of the Emergency Services Commissioner, State

Government of Victoria. Available: http://www.oesc.vic.gov.au/home/reviews+and+inquiries/report+of+the+inquiry+into+the+2002-2003+Victorian+bushfires. Accessed 2013 Oct 15.

107. Murphy B, Timbal B (2008) A review of recent climate variability and climate change in southeastern Australia. Int J Clim 28: 859–879.

108. Canadell J, Kirschbaum M, Kurz W, Sanz M, Schlamadinger B, et al. (2007) Factoring out natural and indirect human effects on terrestrial carbon sources and sinks. Environ Sci Policy 10, 370–384.

109. Schlamadinger B, Johns T, Ciccarese L, Braun M, Sato A, et al. (2007) Options for including land use in a climate agreement post-2012: improving the Kyoto Protocol approach. Environ Sci Policy 10, 295–305.

110. Macintosh A (2011) Potential carbon credits from reducing native forest harvesting in Australia. ANU Centre for Climate Law and Policy, Working Paper Series 2011/1.

111. Marsden-Smedley JB, Slijepcevic A (2001) Fuel characteristics and low intensity burning in *Eucalyptus obliqua* wet forest at the Warra LTER site. Tasforests 13(2): 261–277.

112. Slijepcevic A (2001) Loss of carbon during controlled regeneration burns in *Eucalyptus obliqua* forest. Tasforests 13(2): 281–290.

113. Tolhurst KG, Anderson WR, Gould JS (2006) Woody fuel consumption experiments in an undisturbed forest. In: Conference Proceedings of the V International Conference on Forest Fire Research. Viegas DX, editor.

114. Hollis JJ, Anderson WR, McCaw WL, Cruz MG, Burrows ND, et al. (2011) The effect of fireline intensity on woody fuel consumption in southern Australian eucalypt forest fires. Aust For 74: 81–96.

115. Turetsky MR, Kane ES, Harden JW, Ottmar RD, Manies KL, et al. (2011) Recent acceleration of biomass burning and carbon losses in Alaskan forests and peatlands. Nature Geosciences 4: 27–31.

116. Campbell J, Donato D, Azuma D, Law B (2007) Pyrogenis carbon emissions from a large wildfire in Oregon, United States. J Geophys Res 122: G04014.

117. Barrett K, Kasischke ES, McGuire AD, Turetsky MR, Kane ES (2010) Modelling fire severity in black spruce stands in the Alaskan boreal forest using spectral and non-spectral geospatial data. Remote Sens Environ 114: 1494–1503.

118. Barrett K, Kasischke ES (2013) Controls on variation in MODIS fire radiative power in Alaskan boreal forests: Implications for fire severity conditions. Remote Sens Environ 130: 171–181.

6

Size Dependency of Post-Disturbance Recovery of Multi-Stemmed Resprouting Trees

Jennifer L. Schafer*¤, Michael G. Just

Department of Plant and Microbial Biology, North Carolina State University, Raleigh, North Carolina, United States of America

Abstract

In frequently burned ecosystems, many plants persist by repeated resprouting from basal or belowground buds. This strategy requires that plants reach a balance between biomass loss and recovery, which depends on the shape of the relationship between pre- and post-fire size. Previous analyses of this relationship, however, have focused on the size of the largest stem, which ignores the importance of the multi-stem growth habit that is common in pyrogenic ecosystems. We hypothesized that the presence of multiple stems causes a substantial shift in the relationship between pre- and post-fire size and in the relationship between pre-fire size and size recovery. We measured the height and basal diameter, then calculated volume and biomass, of all stems of six tree species before and nine months after complete removal of aboveground biomass via coppicing. The number of resprouts was correlated with the original number of stems for four species. For all species, the relationship between pre-coppicing and resprout size fit a positive curvilinear function, and the shape of this curve did not differ for maximum and total stem size. Smaller individuals recovered a larger proportion of their pre-coppicing size than larger individuals, but the shape of the size recovery curves were the same regardless of whether the analysis was performed with all stems or only the largest stem. Our results indicate that measuring only the largest stem of multi-stemmed individuals is sufficient to assess the ability of individuals to recover after complete loss of aboveground biomass and persist under frequent burning.

Editor: Christopher Carcaillet, Ecole Pratique des Hautes Etudes, France

Funding: This research was supported by a cooperative agreement between the US Army Engineer Research and Development Center and North Carolina State University (W9132T-11-2-0007 to W. Hoffmann). The funders had no role in study design, data collection and analysis, decision to publish, or preparation of the manuscript.

Competing Interests: The authors have declared that no competing interests exist.

* Email: schaferj@william.jewell.edu

¤ Current address: Department of Biology, William Jewell College, Liberty, Missouri, United States of America

Introduction

Resprouting provides resilience to fire and allows plants to persist in pyrogenic ecosystems. When aboveground stems are killed by fire (i.e., topkilled), species that are able to resprout generate new biomass from plant parts that survive fire [1,2] such as basal buds, lignotubers, rhizomes, or the root collar [3,4]. Resprouting ability and resprout biomass [5–7] are influenced by the size of the belowground bud bank [8], the pool of belowground resources (e.g., carbohydrates and nutrients [9–14]), and pre-fire plant size [15,16].

In frequently burned ecosystems, resprouting species are subjected to repeated cycles of topkill and resprouting [17], so persistence depends on the ability of plants to recover their pre-fire size to maintain a balance between biomass loss and recovery [13,18]. Resprout height and diameter are positively correlated with pre-fire stem height and diameter [16,19,20], with the relationship between pre- and post-fire size fitting a curvilinear scaling function [18]. This "resprout curve" illustrates the balance between biomass loss and recovery and determines the equilibrium size (i.e., where pre-fire and post-fire size are equal) upon which plants will converge over multiple fire cycles ([18]; Figure 1A). Although resprout size is correlated with pre-fire size [16,19,20], large plants often recover their pre-fire size more slowly than small plants [18,21]. This "recovery curve" is a negative curvilinear relationship between pre-fire size and the ratio of post- to pre-fire size (Figure 1B).

Studies on the relationship between pre- and post-fire size and the size dependency of post-fire recovery, however, often focus only on the largest pre-fire stem and largest resprout [16,18,19] even though many resprouting species are multi-stemmed before and/or after fire (e.g., [22–24]). In fact, the number of resprouts is correlated with the number of stems pre-fire [20,25,26]. Allocation of biomass to multiple stems, rather than one stem, may be beneficial due to limitations on maximum stem height and growth rates [27–29] and the improvement in competitive success conferred by a large crown volume [30]. If the curvilinear nature of resprout and recovery curves is a consequence of limitations on maximum stem growth rates [28,31], then this limitation could be overcome by producing multiple stems.

Accounting for all stems of multi-stemmed resprouting species, therefore, may cause an upward shift in resprout (Figure 1A) and recovery (Figure 1B) curves. An upward shift in the resprout curve would indicate that individual plants are able to maintain a greater biomass (i.e., a greater equilibrium plant size) with frequent burning. Consequently, larger individuals would be able to recover

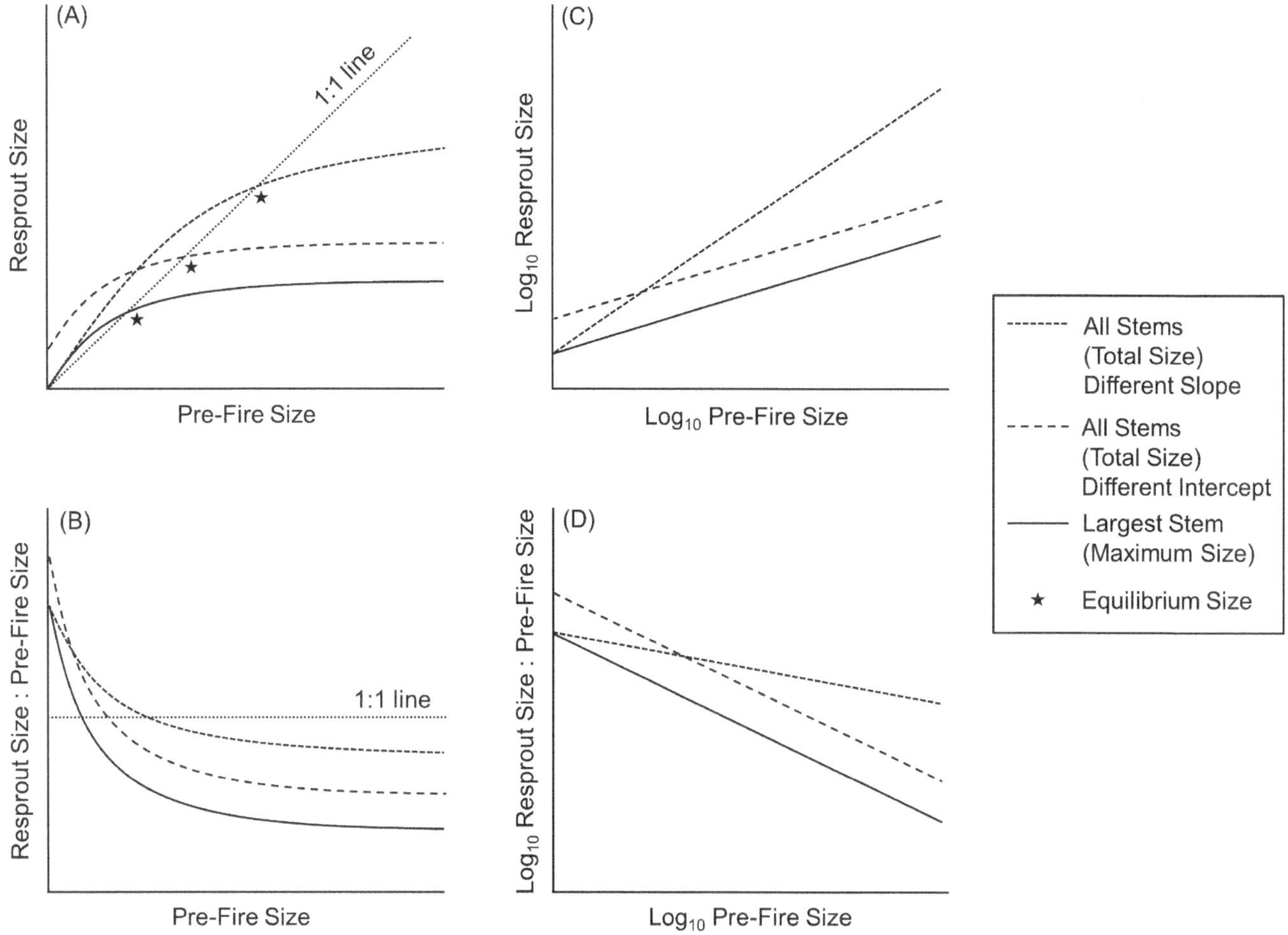

Figure 1. Hypothesized shifts in resprout and recovery curves resulting from inclusion of all stems. (A) Differences in resprout curves that could arise from inclusion of all stems of multi-stemmed trees. Stars indicate the equilibrium size that develops over multiple fire cycles that corresponds to the point at which biomass loss is equal to biomass recovery (i.e., intersects with the 1:1 line; following [18]). (B) Recovery curves that correspond with resprout curves. (C) Illustration of the transformation of resprout curves to a logarithmic scale. (D) Illustration of the transformation of recovery curves to a logarithmic scale. We assessed shifts in resprout and recovery curves by testing for differences in the slopes and y-intercepts of the log_{10}-transformed relationships between maximum and total size and size recovery.

their pre-fire size. In this case, production of multiple stems could increase the ability of plants to escape a suppressed state of repeated topkill and resprouting [17,32,33] during a longer fire free interval. Alternatively, accounting for all stems could lead to a change from a curvilinear to linear relationship between pre- and post-fire size and size recovery, indicating that curvilinearity is not a fundamental property of resprouting. Regardless, understanding the impact of multiple stems on resprout and recovery curves is important because the ability of individual plants to recover biomass lost during fire allows for persistence with repeated burning [13].

We assessed resprouting success and the size dependency of volume and biomass recovery after complete loss of aboveground biomass. Specifically, we coppiced aboveground stems – as has been done in other studies to simulate fire-induced topkill [6,14,34,35] – of six tree species that occur in the pyrogenic longleaf pine savannas and adjacent stream-head pocosins of the southeastern United States [36,37]. To test the hypothesis that accounting for all stems of multi-stemmed resprouting species causes a shift in resprout and recovery curves, we measured all stems pre-coppicing and all resprouts. We assessed possible shifts in resprout and recovery curves by testing for differences in the slopes and y-intercepts of the log-transformed relationships between pre-coppicing and resprout size (i.e., volume and biomass) of the largest stem (maximum size) and all stems (total size; Figures 1C and 1D).

Materials and Methods

Ethics Statement

We obtained approval for data collection from the Endangered Species Branch at Fort Bragg Military Installation. Data was collected on publicly owned land. No protected species were used in this study.

Study site and species

We conducted our study at Fort Bragg, which is located in the Sandhills region of North Carolina (35° 07′N, 79° 10′W). Longleaf pine (*Pinus palustris* Mill.) savanna (e.g., upland pine/scrub oak sandhill *sensu* [36]) is the most widespread vegetation type on the installation. Pine needles and wiregrass (*Aristida stricta* Michx.) accumulate quickly and facilitate frequent fire; the average historical fire return interval is approximately 2 years [38]. For the lowland stream-head pocosins (i.e., wetlands)

Table 1. General characteristics of individuals included in the study.

Species[a]	# Individuals Coppiced[b]	# Individuals Resprouted	# of Stems Pre-Coppicing	# of Resprouts	Maximum Stem Height (cm)[c]	Maximum Basal Diameter (mm)[c]
Quercus laevis	29	26	1-6	0-8	60-330	6.99-61.18
Diospyros virginiana	26	26	1-2	1-5	58-285	7.74-40.01
Liquidambar styraciflua	32	31	1-5	0-17	12-399	2.38-41.20
Liriodendron tulipifera	29	27	1-8	0-37	37-610	5.12-56.92
Persea palustris	31	31	1-4	1-7	53-289	5.46-34.33
Acer rubrum	30	28	1-6	0-9	38-382	4.27-30.13

[a]Species are listed in order of their position along the savanna-to-pocosin gradient.
[b]Does not include individuals burned in the June 2013 wildfire.
[c]For stems pre-coppicing.

embedded within the savanna matrix [37], higher moisture content and differences in species composition contribute to a longer fire return interval of 7–50 years [39]. Soil moisture increases along the gradient from upland savanna to lowland pocosin, and soils are classified as entisols, inceptosols, or ultisols [40]. Mean annual precipitation is 1275 mm, and summer is the wettest season [37]. Fort Bragg is divided into discrete landscape units locally referred to as burn blocks, within which prescribed fire is applied approximately every 3 years [41]. Prescribed fires at Fort Bragg are conducted during the dormant season (December–March) and the growing season (April–July) [42]. Lightning

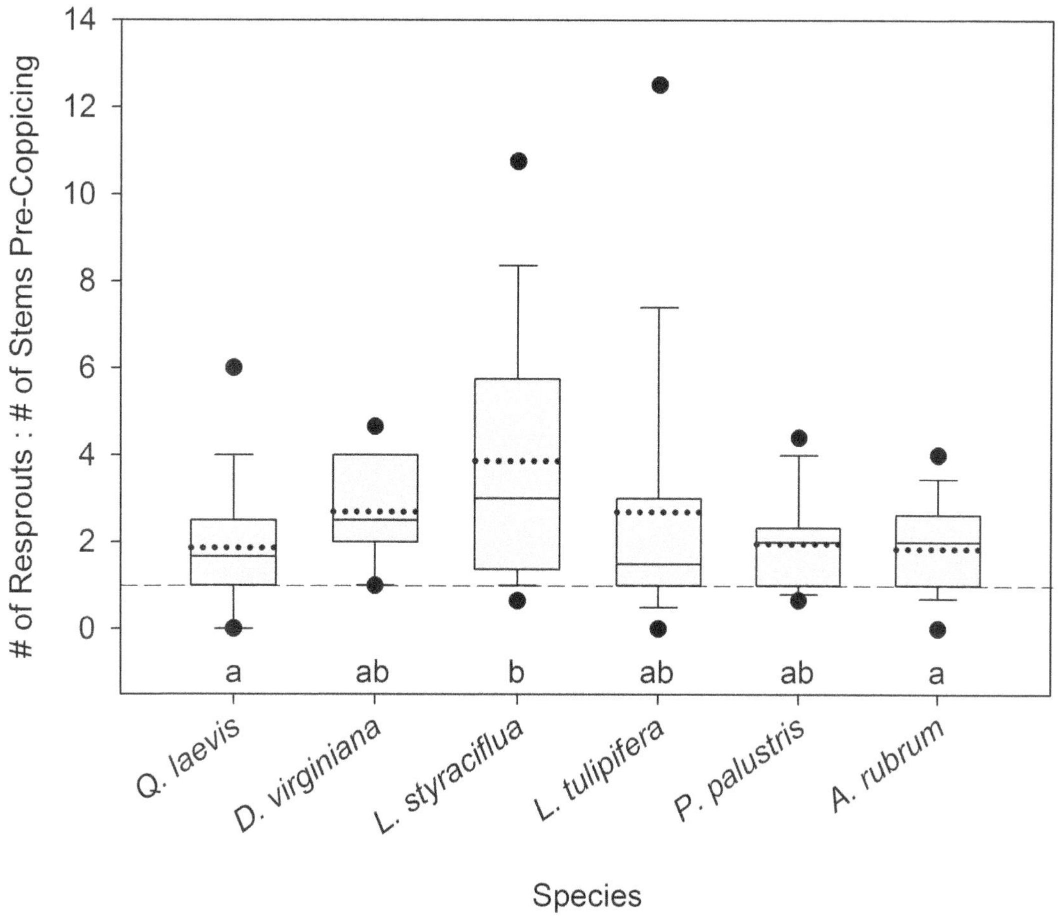

Figure 2. Ratio of resprout stem number to pre-coppicing stem number. In the boxplots, the solid and dotted bars represent the median and mean, respectively; the lower and upper bars represent the 25th and 75th percentiles, respectively. The lower and upper "whiskers" show the largest and smallest values that are not outliers and the lower and upper dots show the 5th and 95th percentiles. Kruskal-Wallis $X^2 = 20.03$, $P = 0.001$; different letters indicate significant differences among species. The dashed line indicates where the number of resprouts equals the number of stems pre-coppicing.

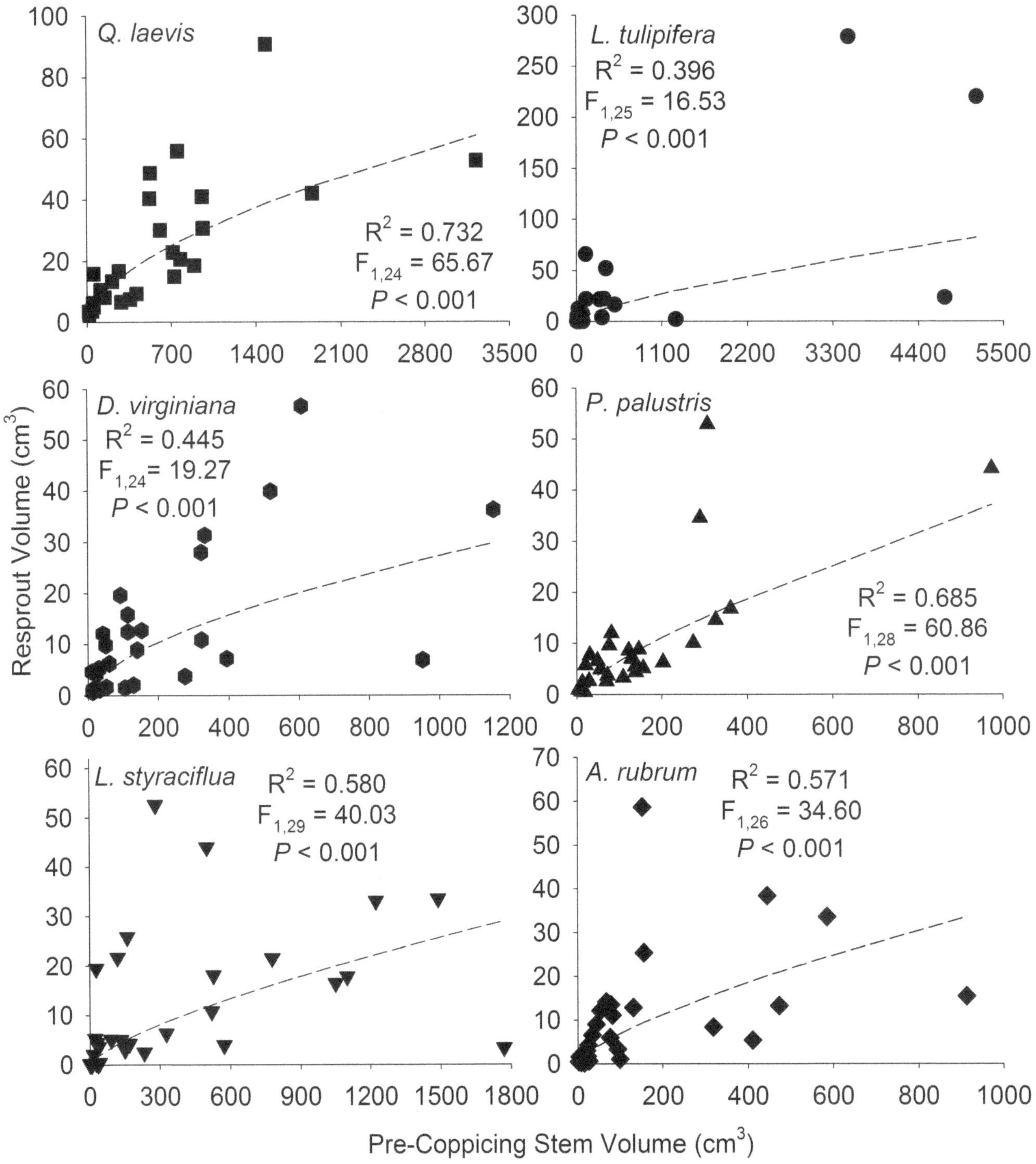

Figure 3. Curvilinear relationships between total pre-coppicing stem volume and total resprout stem volume.

ignited fires in the southeastern US typically occur during the spring and summer (April–September) [43,44].

We selected six focal species that differ in their distribution along the savanna-to-pocosin gradient: *Quercus laevis* Walter (turkey oak), *Diospyros virginiana* L. (persimmon), *Liquidambar styraciflua* L. (sweetgum), *Liriodendron tulipifera* L. (tulip poplar), *Persea palustris* (Raf.) Sarg. (swamp bay), and *Acer rubrum* L. (red maple; Table 1; nomenclature follows The PLANTS Database (US Department of Agriculture, Natural Resources Conservation Service; http://plants.usda.gov/java/). *Quercus laevis* is a savanna species, and *D. virginiana* occurs in the savanna and the ecotone between savanna and pocosin. *Liquidambar styraciflua* is most common in the ecotone, and *L. tulipifera*, *P. palustris*, and *A. rubrum* are restricted to the pocosin and ecotone. All study species typically resprout from basal or belowground buds after topkill via fire or other damage.

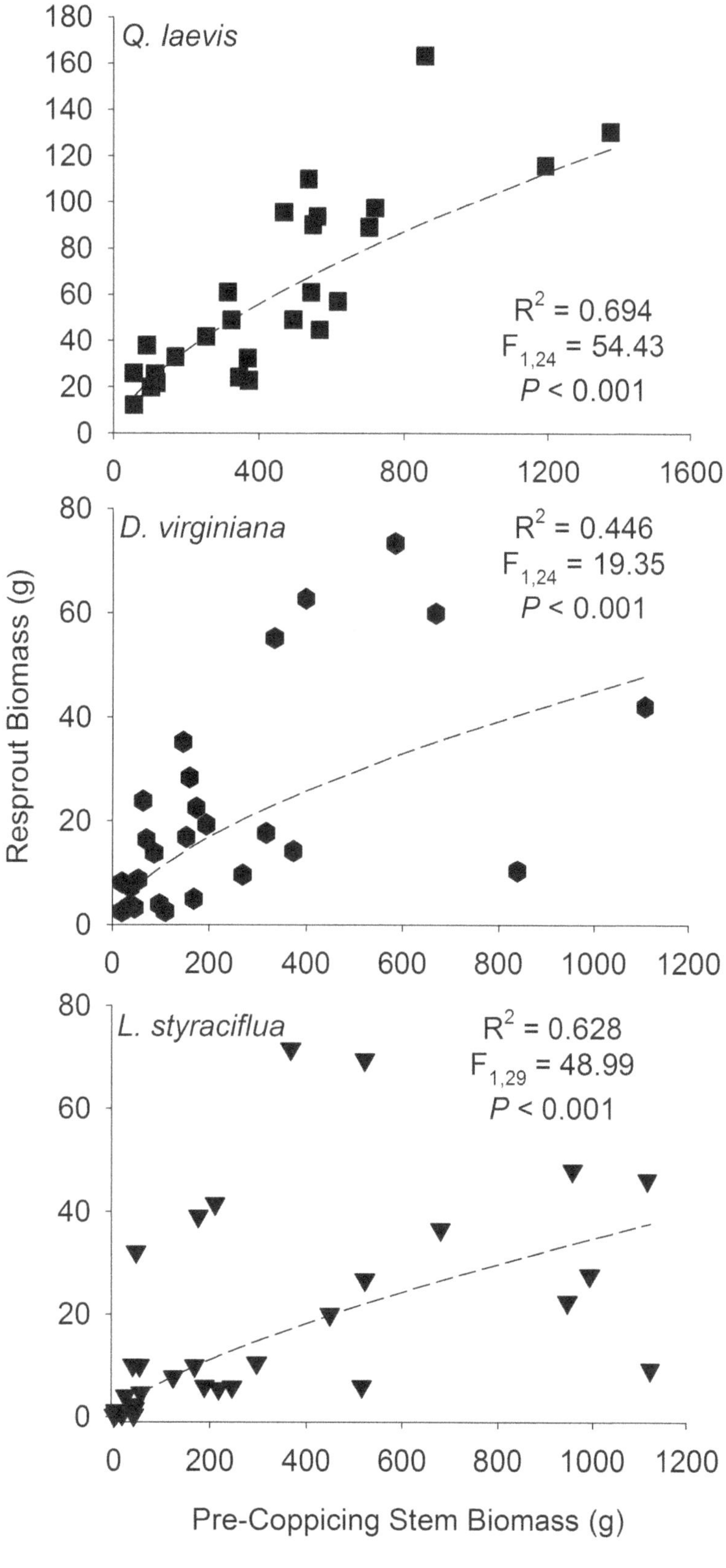

Figure 4. Curvilinear relationships between total pre-coppicing stem biomass and total resprout biomass.

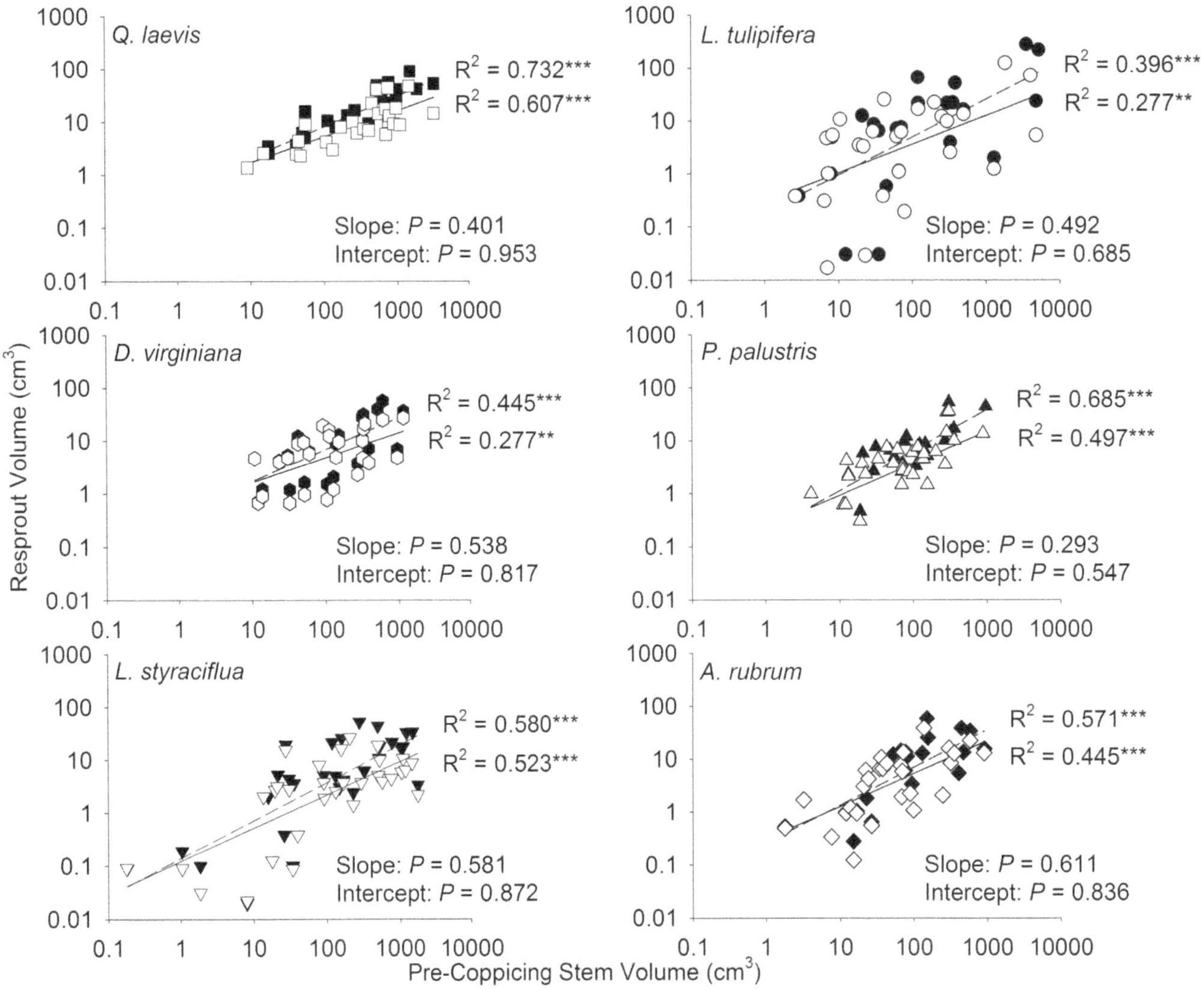

Figure 5. Relationships between pre-coppicing stem volume and resprout stem volume on a logarithmic scale. Total stem volume is denoted with filled symbols and dashed lines. Maximum stem volume is denoted with open symbols and solid lines. ** $P<0.01$, *** $P<0.001$ for regressions. *P* values for differences in slopes and intercepts between maximum and total volume are given.

Field measurements and calculations

In October 2012, we selected 30–32 individuals of each species that spanned a range of maximum stem heights and diameters (Table 1); even the largest individuals were considered saplings. Individuals were found throughout their distribution along the savanna-to-pocosin gradient in multiple burn blocks ($n = 6$) that were burned 3–4 years previously. We measured the height and basal diameter (within 2 cm of ground level) of all stems of each individual and then, to simulate topkill, coppiced all stems at ~2 cm above ground level. Stems were coppiced in October, at the beginning of the dormant season and after the typical wildfire season [43,44]. Coppicing is commonly used as a surrogate for disturbance-induced topkill (e.g., [34,35]), and resprouting success is similar between burned and coppiced individuals [6,15,45]. A wildfire in June 2013 burned a small section of one of our study sites, which reduced our sample size of all species except *L. styraciflua* (Table 1). In July 2013, 9 months after coppicing and near the end of the growing season, we measured the height and basal diameter of all resprouts of each individual.

For individuals that resprouted, we calculated the conical volume of each stem from measurements of stem height and diameter. We determined the maximum stem volume (i.e., volume of the largest stem) and total stem volume (i.e., the sum of volumes of all stems) of each individual pre-coppicing and after resprouting. For three species – *Q. laevis*, *D. virginiana*, and *L. styraciflua* – we used allometric equations from Robertson and Ostertag [46] to calculate maximum and total stem biomass of each individual pre-coppicing and after resprouting. Except for large pre-coppiced *Q. laevis* and small *L. styraciflua* resprouts, the majority of our stems were within the range of diameters used to develop the allometric equations [46].

Statistical analyses

For each species, we analyzed the relationship between pre-coppicing stem number and the number of resprouts using Kendall's tau. To analyze differences among species in resprouting success (i.e., production of resprouts), we calculated the ratio of resprouts to pre-coppicing stems for all individuals and used a Kruskal-Wallis test, with post-hoc pair-wise significance tests

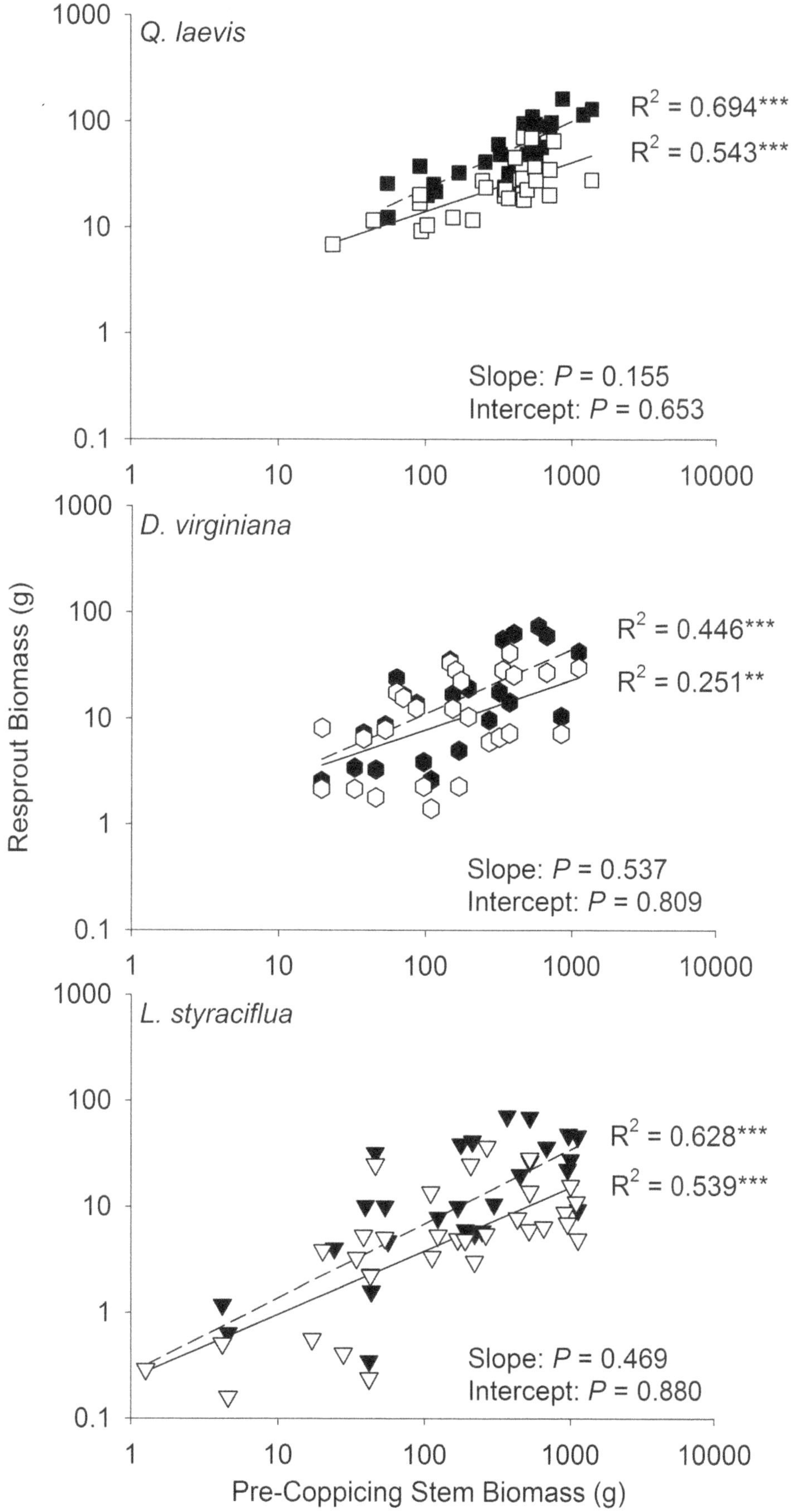
Q. laevis
R2 = 0.694***
R2 = 0.543***
Slope: P = 0.155
Intercept: P = 0.653
D. virginiana
R2 = 0.446***
R2 = 0.251**
Slope: P = 0.537
Intercept: P = 0.809
L. styraciflua
R2 = 0.628***
R2 = 0.539***
Slope: P = 0.469
Intercept: P = 0.880
Resprout Biomass (g)
Pre-Coppicing Stem Biomass (g)
1000
100
10
1
0.1
1
10
100
1000
10000

Figure 6. Relationships between pre-coppicing stem biomass and resprout stem biomass on a logarithmic scale. Total stem biomass is denoted with filled symbols and dashed lines. Maximum stem biomass is denoted with open symbols and solid lines. ** *P*<0.01, *** *P*<0.001 for regressions. *P* values for differences in slopes and intercepts between maximum and total biomass are given.

adjusted for multiple comparisons. To determine if there was a significant curvilinear relationship between total resprout size and total pre-coppicing size, we used the curve estimation function in SPSS version 19.0 (IBM Corporation, Armonk, NY, USA). Specifically, we fit power functions to the relationships between total pre-coppicing stem volume and total resprout stem volume (for all species) and total pre-coppicing stem biomass and total resprout biomass (for *Q. laevis*, *D. virginiana*, and *L. styraciflua*). We used a regression model with $\log_{10}$ resprout size (i.e., volume or biomass) as the dependent variable and $\log_{10}$ pre-coppicing size (i.e., volume or biomass), data type (i.e., maximum or total size, coded as 1 and 0, respectively), and an interaction term ($\log_{10}$ pre-coppicing size * data type) as independent variables entered into the model to determine if there was a significant difference between the slopes and y-intercepts of the relationships between: (1) pre-coppicing and resprout maximum and total stem size (Figure 1C) and (2) pre-coppicing size and recovery of maximum and total stem size (Figure 1D) for each species.

Results

For all species, at least 90% of individuals resprouted after coppicing (Table 1). The number of resprouts was positively correlated with the number of coppiced stems for *Q. laevis* ($\tau = 0.339$, $P = 0.025$), *L. styraciflua* ($\tau = 0.296$, $P = 0.042$), *P. palustris* ($\tau = 0.296$, $P = 0.058$), and *A. rubrum* ($\tau = 0.491$, $P = 0.001$). The number of resprouts of *D. virginiana* ($\tau = 0.251$, $P = 0.239$) and *L. tulipifera* ($\tau = 0.224$, $P = 0.139$) was not correlated with the number of coppiced stems. The number of

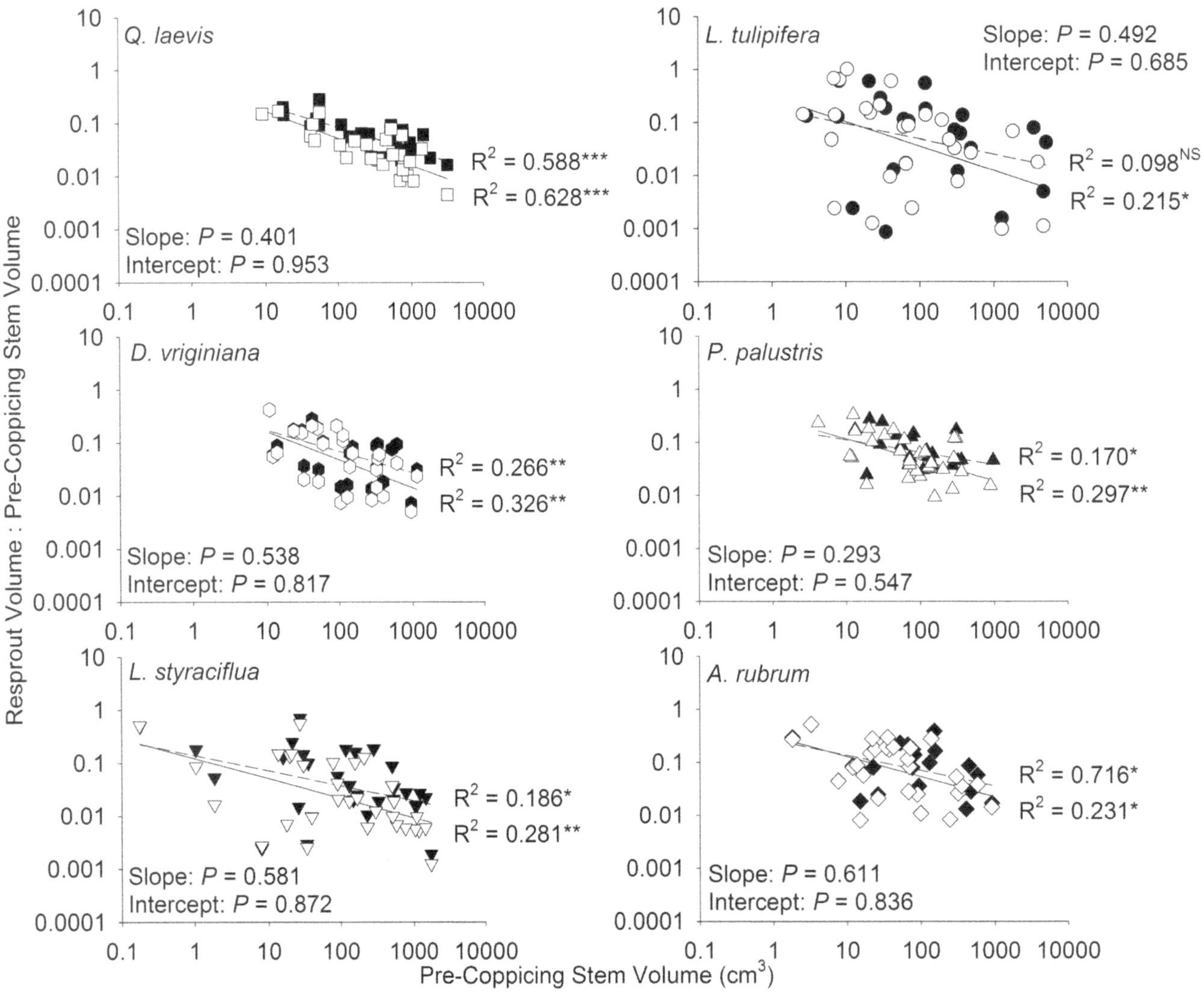

Figure 7. Relationships between pre-coppicing stem volume and the ratio of resprout volume to pre-coppicing volume. Data is shown on a logarithmic scale. Total stem volume is denoted with filled symbols and dashed lines. Maximum stem volume is denoted with open symbols and solid lines. * *P*<0.05, ** *P*<0.01, *** *P*<0.001 for regressions; NS indicates not significant. *P* values for differences in slopes and intercepts between maximum and total volume are given.

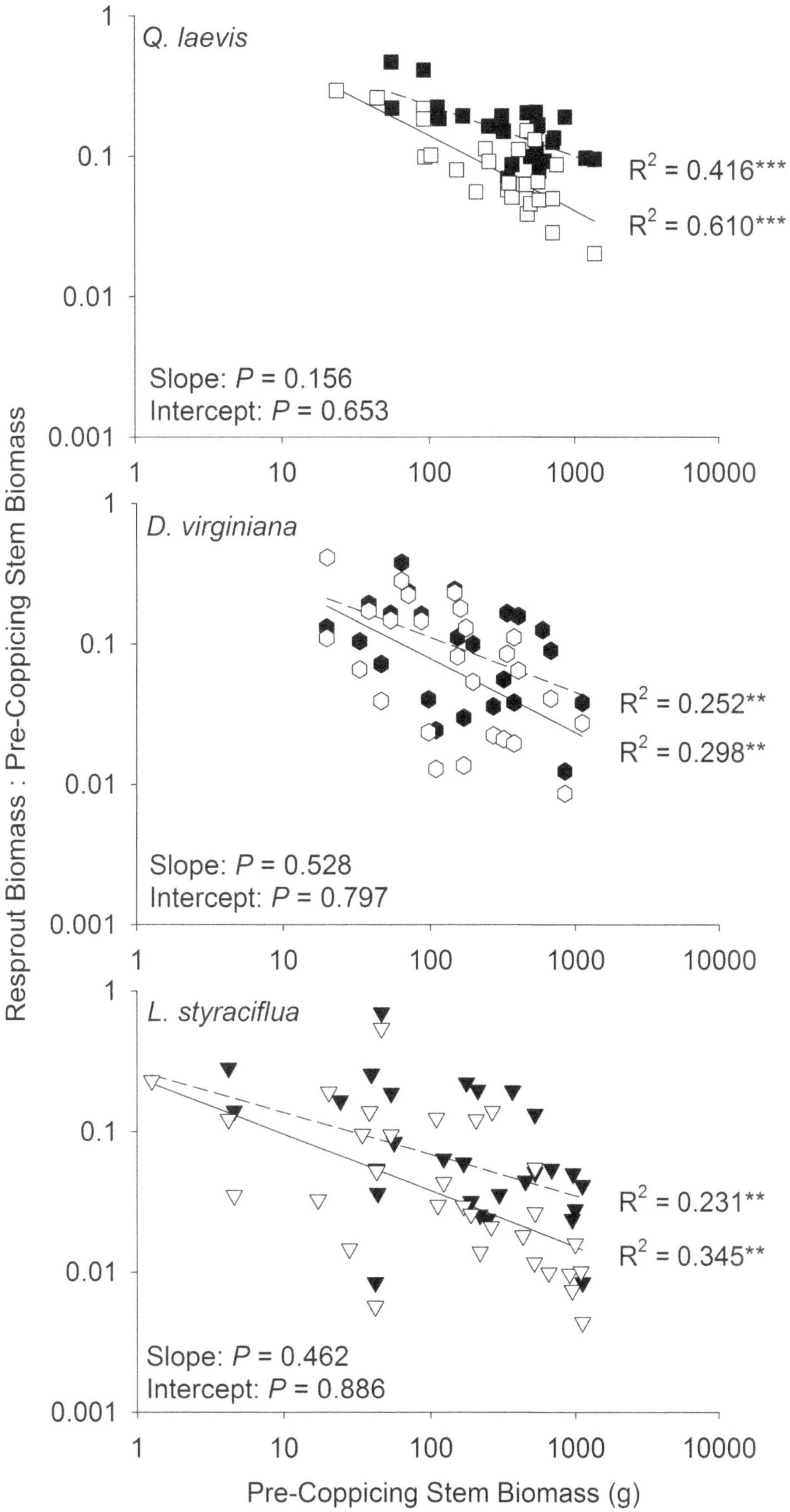
Q. laevis
R^2 = 0.416***
R^2 = 0.610***
Slope: P = 0.156
Intercept: P = 0.653
D. virginiana
R^2 = 0.252**
R^2 = 0.298**
Slope: P = 0.528
Intercept: P = 0.797
L. styraciflua
R^2 = 0.231**
R^2 = 0.345**
Slope: P = 0.462
Intercept: P = 0.886
1
0.1
0.01
0.001
1
10
100
1000
10000
Resprout Biomass : Pre-Coppicing Stem Biomass
Pre-Coppicing Stem Biomass (g)

Figure 8. Relationships between pre-coppicing stem biomass and the ratio of resprout biomass to pre-coppicing biomass. Data is shown on a logarithmic scale. Total stem biomass is denoted with filled symbols and dashed lines. Maximum stem biomass is denoted with open symbols and solid lines. ** $P<0.01$, *** $P<0.001$ for regressions. P values for differences in slopes and intercepts between maximum and total biomass are given.

resprouts of all species tended to be equal to or greater than the number of coppiced stems (Figure 2); 73% of individuals that resprouted had more resprouts than coppiced stems.

There was a positive curvilinear relationship between total resprout and pre-coppicing stem volume (Figure 3) and biomass (Figure 4) for all species studied. There was no difference between the slopes or intercepts of the relationships between pre-coppicing and resprout maximum and total volume (Figure 5) and pre-coppicing and resprout maximum and total biomass (Figure 6). In other words, for all species, resprout curves were similar for maximum and total size.

Recovery of volume and biomass was negatively correlated with pre-coppicing volume and biomass, regardless of whether all stems were included in the analysis. For all species, there was no difference between the slopes or intercepts of the relationships between recovery of maximum and total volume (Figure 7) and recovery of maximum and total biomass (Figure 8). Most individuals recovered less than 55% of their maximum stem volume.

Discussion

The ability of plants to resprout after topkill contributes to their persistence in pyrogenic ecosystems. Across species in our study, 95% of individuals resprouted after complete removal of aboveground biomass (Table 1). Resprout number has been found to be positively correlated with the number of stems present before fire [20]; this was the case for four of our six study species. Individuals with more stems pre-fire may have larger storage organs, which translate to larger pools of carbohydrates and buds to support resprouting [7,47,48]. For the two species in our study for which resprout number was not positively correlated with pre-coppicing stem number, *D. virginiana* and *L. tulipifera*, individuals with only one stem pre-coppicing produced 1 to 5 and 1 to 14 resprouts, respectively. Factors such as bud activation and proximity of buds to the soil surface [49], as well as pre-fire stem size [15,16], may have affected resprout number.

We found that total resprout size was positively correlated with pre-coppicing total size and fit a curvilinear function (Figures 3 and 4). Although maximum resprout size may be limited by growth rates [28,31], the ability to increase mechanical strength to support height growth [50], or physiological changes that alter allocation of photosynthates [51], we found no difference in the resprout curves for maximum and total size (Figures 5 and 6; determined by analyzing $\log_{10}$-transformed data). Accounting for all stems did not cause a significant shift in the resprout curves, and thus, production of multiple stems does not change the size at which individuals persist in frequently burned ecosystems. One explanation for the lack of upward shift in resprout curves could be related to the concomitant increase in resprout and pre-coppicing volume and/or biomass – individuals in our study had up to eight stems pre-coppicing – such that inclusion of all stems of multi-stemmed individuals affected the location of an individual on the curve rather than the shape of the curve. In addition, intraspecific variation in pre- and post-fire sizes could be related to resource availability because plants in high-resource environments are larger after fire than plants of the same initial size in low-resource environments [18].

Similar to Grady and Hoffmann [18], when we accounted for only the largest stem, larger individuals recovered a smaller fraction of their pre-coppicing size than smaller individuals. Contrary to our hypothesis, there was no difference in the shape of the recovery curves of maximum and total size (Figures 7 and 8; determined by analyzing $\log_{10}$-transformed data). Accounting for all stems, rather than only the largest stem, does not affect which individuals, in terms of pre-coppicing size, are able to recover all biomass lost during fire. Although 73% of the resprouting individuals in our study experienced an increase in stem number after complete removal of aboveground biomass (Figure 2), our results suggest that the potential benefits of producing multiple stems do not extend to post-disturbance biomass recovery.

The shape of resprout and recovery curves may be influenced by resprout age. Historically, the fire return interval in longleaf pine savannas ranged from 0.5 to 12 years [38], and savannas at our study site are currently burned every 3 years, on average, with stream-head pocosins generally burning less frequently [39]. The number of resprouts per clump is higher in more recently burned sites than longer unburned sites [52,53], suggesting that self-thinning of resprouts can occur over time such that total size may converge on maximum size as the number of stems decreases. Furthermore, large individuals recovered a lower proportion of their size regardless of whether resprouts were 9 months (Figures 7 and 8) or 3 years old [18].

The shape of resprout and recovery curves may also be influenced by fire season. In the southeastern US, plants burned during the dormant season have greater post-fire stem densities [4,54] and aboveground biomass [55] than plants burned during the growing season. Differences in resprout number should have little or no effect on the persistence equilibrium since there is no difference between resprout and recovery curves of maximum and total stem size (Figures 5-8). Greater biomass [55] and a larger increase in growth rates [56] after dormant season fires suggests that resprout curves of plants burned during the dormant season could be shifted upward from plants burned during the growing season. Nonetheless, any effects of fire season on resprout and recovery curves should be consistent for maximum and total stem size.

Our study assessed the importance of considering all stems, not just the largest stems, when assessing size recovery after complete removal of aboveground biomass. The ability of plants to produce multiple resprouts did not allow individuals to reach a larger equilibrium size or recover a greater proportion of their pre-coppicing size. Production of multiple stems, therefore, does not appear to affect persistence in frequently burned ecosystems in regard to the balance between biomass loss and recovery. Our study species are all trees that resprouted from the root crown; it is not clear how differences in allometric constraints on resprout allocation (e.g., shrubs vs. trees; [57]) or the belowground structure from which resprouts are produced (e.g., lignotubers or rhizomes; [3]) influence relationships between maximum and total stem size. Nevertheless, accounting for only the largest pre-fire stem and largest resprout appears to be an adequate predictor of species' equilibrium size and their ability to recovery their pre-fire size and should not lead to misinterpretation of persistence ability over multiple fire cycles.

Acknowledgments

We thank Alicia Ballard, Spencer Bell, Bradley Breslow, and Ashley McGuigan for help in the field. Alice Broadhead, Renée Marchin, Matthew Hohmann, and William Hoffmann provided helpful comments on the manuscript.

Author Contributions

Conceived and designed the experiments: JLS MGJ. Performed the experiments: JLS MGJ. Analyzed the data: JLS MGJ. Contributed to the writing of the manuscript: JLS MGJ.

References

1. Bellingham PJ, Sparrow AD (2000) Resprouting as a life history strategy in woody plant communities. Oikos 89: 409–416.
2. Hoffman WA, Moreira AG (2002) The role of fire in the population dynamics of woody plants. In: Oliveira PS, Marquis RJ, editors. The cerrados of Brazil: ecology and natural history of a Neotropical savanna. New York: Columbia University Press. pp. 159–177.
3. Clarke PJ, Lawes MJ, Midgley JJ, Lamont BB, Ojeda F, et al. (2013) Resprouting as a key functional trait: how buds, protection and resources drive persistence after fire. New Phytol 197: 19–35.
4. Drewa PB, Platt WJ, Moser EB (2002) Fire effects on resprouting shrubs in headwaters of southeastern longleaf pine savannas. Ecology 83: 755–767.
5. Moreno JM, Oechel WC (1991) Fire intensity and herbivory effects on postfire resprouting of *Adenostoma fasciculatum* in southern California chaparral. Oecologia 85: 429–433.
6. Lloret F, López-Soria L (1993) Resprouting of *Erica multiflora* after experimental fire treatments. J Veg Sci 4: 367–374.
7. Cruz A, Pérez B, Quintana J, Moreno J (2002) Resprouting in the Mediterranean-type shrub *Erica australis* affected by soil resource availability. J Veg Sci 13: 641–650.
8. Klimešová J, Klimeš L (2007) Bud banks and their role in vegetative regeneration – A literature review and proposal for simple classification and assessment. Perspect Plant Ecol 8: 115–129.
9. Canadell J, López-Soria L (1998) Lignotuber reserves support regrowth following clipping of two Mediterranean shrubs. Funct Ecol 12: 31–38.
10. El Omari B, Aranda X, Verdaguer D, Pascual G, Fleck I (2003) Resource remobilization in *Quercus ilex* L. resprouts. Plant Soil 252: 349–357.
11. Kabeya D, Sakai S (2005) The relative importance of carbohydrate and nitrogen for the resprouting ability of *Quercus crispula* seedlings. Ann Bot 96: 479–488.
12. Paula S, Ojeda F (2009) Belowground starch consumption after recurrent severe disturbance in three resprouter species of the genus *Erica*. Botany 87: 253–259.
13. Schutz AEN, Bond WJ, Cramer MD (2009) Juggling carbon: allocation patterns of a dominant tree in a fire-prone savanna. Oecologia 160: 235–246.
14. Moreira B, Tormo J, Pausas JG (2012) To resprout or not to resprout: factors driving intraspecific variability in resprouting. Oikos 121: 1577–1584.
15. Bonfil C, Cortés P, Espelta JP, Retana J (2004) The role of disturbance in the co-existence of the evergreen *Quercus ilex* and the deciduous *Quercus cerrioides*. J Veg Sci 15: 423–430.
16. Konstantinidis P, Tsiourlis G, Xofis P (2006) Effect of fire season, aspect and pre-fire plant size on the growth of *Arbutus unedo* L. (strawberry tree) resprouts. Forest Ecol Manag 225: 359–367.
17. Bond WJ, Midgley JJ (2001) Ecology of sprouting in woody plants: the persistence niche. Trends Ecol Evol 16: 45–51.
18. Grady JM, Hoffmann WA (2012) Caught in a fire trap: Recurring fire creates stable size equilibria in woody resprouters. Ecology 93: 2052–2060.
19. Kennedy LM, Horn SP (2008) Postfire vegetation recovery in highland pine forests of the Dominican Republic. Biotropica 40: 412–421.
20. Dacy EC, Fulbright TE (2009) Survival of sprouting shrubs following summer fire: effects of morphological and spatial characteristics. Rangeland Ecol Manag 62: 179–185.
21. Hoffmann WA, Solbrig OT (2003) The role of topkill in the differential response of savanna woody species to fire. Forest Ecol Manag 180: 273–286.
22. Masaka K, Ohno Y, Yamada K (2000) Fire tolerance and the fire-related sprouting characteristics of two cool-temperate broad-leaved tree species. Ann Bot 85: 137–142.
23. Schafer JL (2010) Effects of fire on nutrient availability and limitation in Florida scrub ecosystems. Ph.D. Dissertation, Department of Biology, University of Florida, Gainesville, FL. pp. 257.
24. Maguire AJ, Menges ES (2011) Post-fire growth strategies of resprouting Florida scrub vegetation. Fire Ecol 7: 12–25.
25. Malanson GP, Trabaud L (1988) Vigour of post-fire resprouting by *Quercus coccifera* L. J Ecol 76: 351–365.
26. Olson MS, Platt WJ (1995) Effects of habitat and growing season fires on resprouting of shrubs in longleaf pine savannas. Vegetatio 119: 101–118.
27. Falster DS, Westoby M (2005) Tradeoffs between height growth rate, stem persistence and maximum height among plant species in a post-fire succession. Oikos 111: 57–66.
28. Niklas KJ, Cobb ED (2005) N, P, and C stoichiometry of *Eranthis hyemalis* (Ranunculaceae) and the allometry of plant growth. Am J Bot 92: 1256–1263.
29. Niklas KJ (2007) Maximum plant height and the biophysical factors that limit it. Tree Physiol 27: 433–440.
30. Kozovits AR, Matyssek R, Winkler JB, Göttlein A, Blaschke H, et al. (2005) Above-ground space sequestration determines competitive success in juvenile beech and spruce trees. New Phytol 167: 181–196.
31. Niklas KJ (2006) Plant allometry, leaf nitrogen and phosphorus stoichiometry, and interspecific trends in annual growth rates. Ann Bot 97: 155–163.
32. Waldrop TA, White DL, Jones SM (1992) Fire regimes for pine-grassland communities in the southeastern United States. Forest Ecol Manag 47: 195–210.
33. Higgins SI, Bond WJ, February EC, Bronn A, Euston-Brown DIW, et al. (2007) Effects of four decades of fire manipulation on woody vegetation structure in savanna. Ecology 88: 1119–1125.
34. Cruz A, Pérez B, Moreno JM (2003) Plant stored reserves do not drive resprouting of the lignotuberous shrub *Erica australis*. New Phytol 157: 251–261.
35. Goorman R, Bartual A, Paula S, Ojeda F (2011) Enhancement of photosynthesis in post-disturbance resprouts of two co-occuring Mediterranean *Erica* species. Plant Ecol 212: 2023–2033.
36. Schafale MP, Weakley AS (1990) Classification of the Natural Communities of North Carolina: Third Approximation. North Carolina Natural Heritage Program, Division of Parks and Recreation, Department of Environment, Health, and Natural Resources, Raleigh, North Carolina. pp. 321.
37. Sorrie BA, Bracey Gray J, Crutchfield PJ (2006) The vascular flora of the longleaf pine ecosystem of Fort Bragg and Weymouth Woods, North Carolina. Castanea 71: 129–161.
38. Stambaugh MC, Guyette RP, Marschall JM (2011) Longleaf pine (*Pinus palustris* Mill.) fire scars reveal new details of a frequent fire regime. J Veg Sci 22: 1094–1104.
39. Frost CC (1995) Presettlement fire regimes in southeastern marshes, peatlands, and swamps. In: Cerulean SI, Engstrom RT, editors. Proceeding of the Tall Timbers Fire Ecology Conference No 19. Tallahassee: Tall Timbers Research Station. pp 39–60.
40. U.S. Department of Agriculture, National Resources Conservation Service (2007) Soil Survey Geographic (SSURGO) database for Fort Bragg, North Carolina, Cumberland and Hoke Counties.
41. Lashley MA, Chitwood MC, Prince A, Elfelt MB, Kilburg EL, et al. (2014) Subtle effects of a managed fire regime: A case study in the longleaf pine ecosystem. Ecol Indic 38: 212–217.
42. Krall JS, Hohmann MG, Fraterrigo JM (2014) Contingent fire effects on granivore removal of exotic woody plant seeds in longleaf pine savannas. Biol Invasions 16: 1055–1068.
43. Fowler C, Konopik E (2007) The history of fire in the Southern United States. Human Ecol Rev 14: 165–176.
44. Fill JM, Welch SM, Waldron JL, Mousseau TA (2012) The reproductive response of an endemic bunchgrass indicates historical timing of a keystone process. Ecosphere 3: art61.
45. Vesk PA, Westoby M (2004) Sprouting ability across diverse disturbances and vegetation types worldwide. J Ecol 92: 310-320.
46. Robertson KM, Ostertag TE (2009) Biomass equations for hardwood resprouts in fire-maintained pinelands in the southeastern United States. South J App For 33: 121–128.
47. Bowen BJ, Pate JS (1993) The significance of root starch in post-fire shoot recovery of the resprouter *Stirlingia latifolia* R. Br. (Proteaceae). Ann Bot 72: 7–16.
48. Olano JM, Menges ES, Martínez E (2006) Carbohydrate storage in five resprouting Florida scrub plants across a fire chronosequence. New Phytol 170: 99–106.
49. Noble JC (2001) Lignotubers and meristem dependence in mallee (*Eucalyptus* spp.) coppicing after fire. Aust J Bot 49: 31–41.
50. King DA (1986) Tree form, height growth, and susceptibility to wind damage in *Acer saccharum*. Ecology 67: 980–990.
51. Fang X-W, Turner NC, Xu D-H, Jin Y, He J, et al. (2013) Limits to the height growth of *Caragana korshinskii* resprouts. Tree Physiol 33: 275–284.
52. Davies SJ, Myerscough PJ (1991) Post-fire demography of the Wet-mallee *Eucalyptus leuhmanniana* F.Muell. (Myrtaceae). Aust J Bot 39: 459–66.

53. Schafer JL, Mack MC (2014) Growth, biomass, and allometry of resprouting shrubs after fire in scrubby flatwoods. Am Midl Nat In press.
54. Drewa PB, Thaxton JM, Platt WJ (2006) Responses of root-crown bearing shrubs to differences in fire regimes in Pinus palustris (longleaf pine) savannas: Exploring old-growth questions in second-growth systems. Appl Veg Sci 9: 27–36.
55. Hmielowski TL, Robertson KM, Platt WJ (2014) Influence of season and method of topkill on resprouting characteristics and biomass of Quercus nigra saplings from a southeastern U.S. pine-grassland ecosystem. Plant Ecol DOI:10.1007/s11258-014-0380-5.
56. Robertson KM, Hmielowski TL (2014) Effects of fire frequency and season on resprouting of woody plants in southeastern US pine-grassland communities. Oecologia 174: 765–776.
57. Hermann J-M, Huag S, Pillar VD, Pfadenhauer J (2012) Shrubs versus 'gullivers': woody species coping with disturbance in grasslands. Plant Ecol 213: 1757–1768.

7

Dynamic Analysis and Pattern Visualization of Forest Fires

António M. Lopes[1]*, J. A. Tenreiro Machado[2]

1 Institute of Mechanical Engineering, Faculty of Engineering, University of Porto, Porto, Portugal, **2** Institute of Engineering, Polytechnic of Porto, Porto, Portugal

Abstract

This paper analyses forest fires in the perspective of dynamical systems. Forest fires exhibit complex correlations in size, space and time, revealing features often present in complex systems, such as the absence of a characteristic length-scale, or the emergence of long range correlations and persistent memory. This study addresses a public domain forest fires catalogue, containing information of events for Portugal, during the period from 1980 up to 2012. The data is analysed in an annual basis, modelling the occurrences as sequences of Dirac impulses with amplitude proportional to the burnt area. First, we consider mutual information to correlate annual patterns. We use visualization trees, generated by hierarchical clustering algorithms, in order to compare and to extract relationships among the data. Second, we adopt the Multidimensional Scaling (MDS) visualization tool. MDS generates maps where each object corresponds to a point. Objects that are perceived to be similar to each other are placed on the map forming clusters. The results are analysed in order to extract relationships among the data and to identify forest fire patterns.

Editor: Jesus Gomez-Gardenes, Universidad de Zarazoga, Spain

Funding: The authors have no support or funding to report.

Competing Interests: The authors have declared that no competing interests exist.

* Email: aml@fe.up.pt

Introduction

Forest fires are a major concern in many countries, like United States, Australia, Russia, Brazil, China and Mediterranean Basin European regions [1–3]. Every year forest fires consume vast areas of vegetation, compromising ecosystems and contributing to the carbon dioxide emissions that are changing Earth's climate [4–5]. Besides the long-term economic implications associated to the climate change, forest fires have direct impact upon economy due to the destruction of public and private property and infrastructures [6]. Fires are mainly caused by natural factors, human negligence, or even human intent. Fire propagation and burnt area depend on many natural factors and conditions, not only on the terrain orography and the type of vegetation, but also on the efficacy of detection and suppression strategies. Moreover, fires caused by incendiaries contribute to increase the complexity of the phenomena. Understanding the underlying patterns of forest fires in terms of their size and spatiotemporal distributions may help the decision makers to take preventive measures beforehand, identifying possible hazards and deciding strategies for fire prevention, detection and suppression [7–8].

Forest fires have been studied using classical statistical tools. However, those methods reveal limitations, both in capturing all characteristics underneath forest fires dynamics, and the evolution along years [9]. Forest fires dynamics exhibits correlations in size, space and time. Size-frequency distributions unveil long range memory, which is typical in complex systems. Correlation between data is characterized by self-similarity and absence of characteristic length-scale, meaning that forest fires exhibit power-law (PL) behaviour [10–13].

Several studies have been published during the last years about this topic [14–17]. In references [18–19] it is shown that forest fires exhibit PL frequency-size relationship over many orders of magnitude and that such behaviour seems consistent with the self-organized criticality arising in complex systems. The most important practical implication of such results is that the frequency-size distribution of small and medium fires can be used to quantify the risk of large fires [19]. Nevertheless, some authors [15] suggest that a simple PL distribution of sizes may be too simple to describe the distributions of forest fires over their full range.

In reference [20] the time dynamics of forest fires is investigated and it is shown that forest fires exhibit time-clustering phenomena. More recently, the fractality of the forest fires was addressed in [21] using spatial and temporal fractal tools. The authors prove that these phenomena exhibit space–time clustering behaviour.

In this paper we look at forest fires from the perspective of dynamical systems. A public domain forest fires catalogue containing data of events occurred in Portugal, in the period 1980 up to 2012, is addressed. The data is analysed in an annual basis, modelling the occurrences as a sequence of Dirac impulses. Therefore, instead of modelling individual forest fires, we are describing the global dynamics along several decades. In this perspective, mutual information and visualization trees, generated by hierarchical clustering algorithms, are used. The Multidimensional Scaling (MDS) tool is adopted in order to compare and to

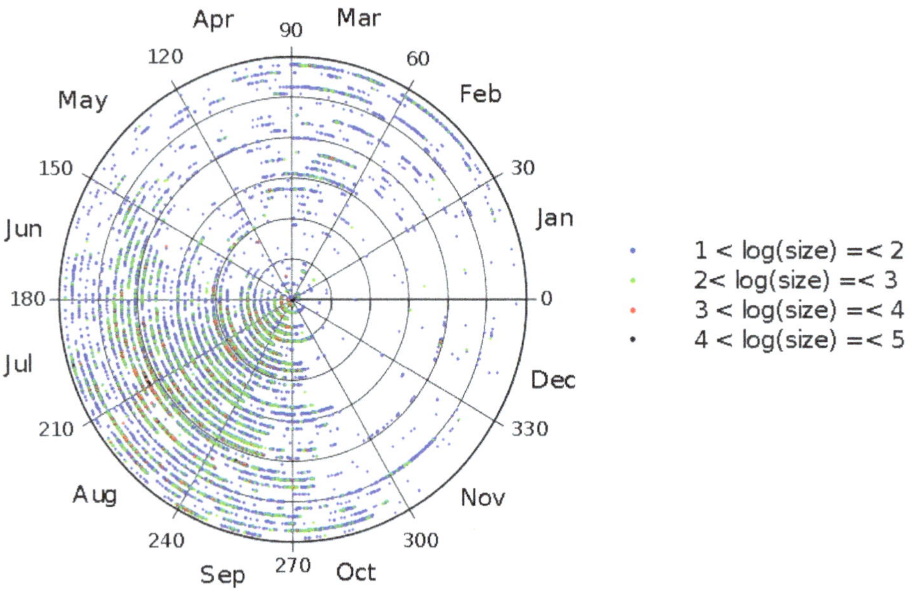

Figure 1. Temporal evolution and size (in log units) of forest fires registered in Portugal in the time period 1980–2012, with burnt area larger than A_{min} = 10 ha. Each (r, θ) point represents the time of the event and the color represents the size.

extract relationships among the data. Finally, we propose an amplitude-space embedding technique that produces a clear fire pattern classification.

Characterization of the Dataset

Data from forest fires is available online at the Portuguese Institute of Nature and Forest Conservation (INCF), http://www.icnf.pt/portal/florestas/dfci/inc/estatisticas, and the catalogue contains events since 1980 up to 2012. Ignitions might have different sources, as natural causes, human negligence or human intentionality, among others. The data analysed in this paper was retrieved in December, 2013. Each record contains information about the events date, time (with one minute resolution), geographic location and size (in terms of burnt area). We decided to discard small size events, as those are prone to measurement errors, by adopting a cutoff threshold value of A_{min} = 10 hectares.

Fig. 1 illustrates the temporal evolution and size of the events occurred in Portugal, during 1980–2012 and meeting the cutoff threshold criterion. We tackle the concept of 'circular time' (since

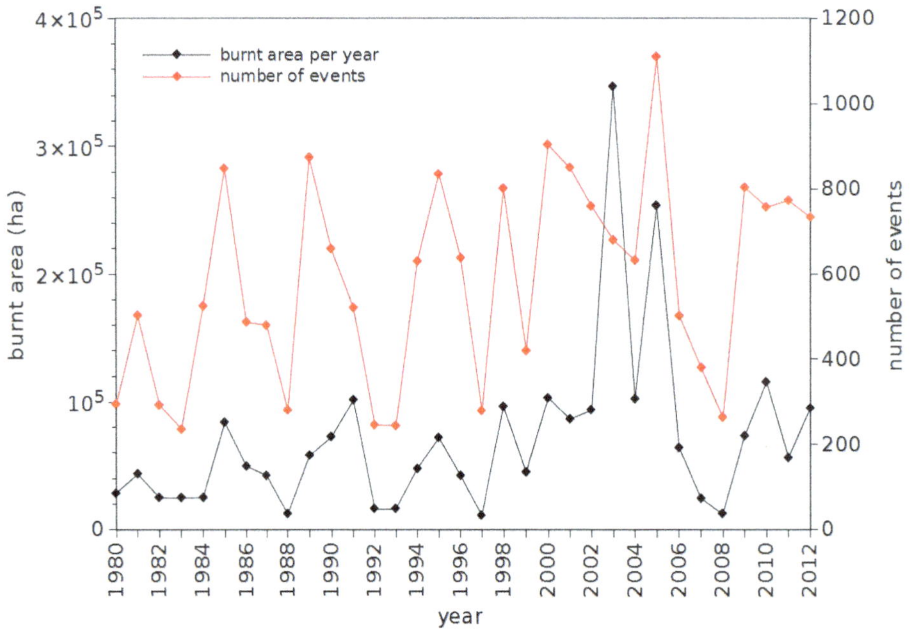

Figure 2. Evolution of burnt area per year and number of occurrences registered in Portugal in the time period 1980–2012 (are considered events with burnt area larger than A_{min} = 10 ha).

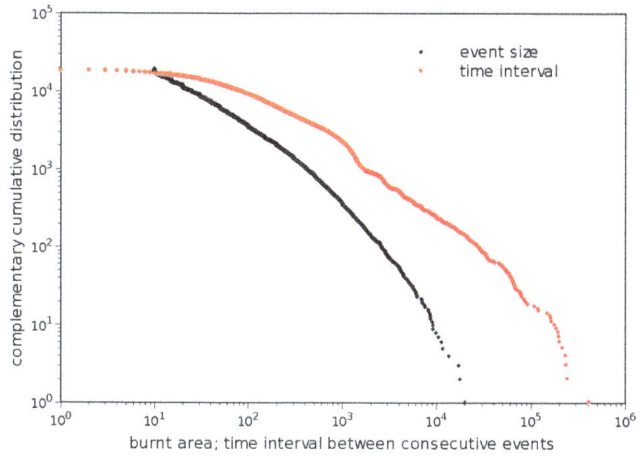

Figure 3. Complimentary cumulative distribution of events size and time interval (in minutes) between consecutive events, corresponding to occurrences registered in Portugal in the time period 1980–2012, with burnt area larger than A_{min} = 10 ha.

there is a kind of one-year periodicity, with December close to January and not the opposite, as a Cartesian scale implicitly assumes). The (circular) time scale evolves along an Archimedean spiral, with origin at the center of the circumferences, given by:

$$\theta = 2\pi \cdot \left(\frac{t}{T} + i\right) \qquad (1)$$

$$r = p + q \cdot \theta \qquad (2)$$

where (r, θ) denotes the radius and angle coordinates, respectively, $i = 0, \ldots, 32$, represents the year and $p = q = 1$. The burnt area is expressed in logarithmic units and is related to the color of the marks. We can note two annual cycles: the first is weaker and includes the months of February and March; the second is stronger and is due to the major incidence of fires during summer [22].

In Fig. 2 we depict the evolution of the burnt area per year and number of occurrences versus year. It is visible the increasing number of events as well as the strong activity verified around the middle of the decade 2000–2009. Nevertheless, the charts reveal a large volatility and pose difficulties to capture some trend. We observe minimal values for years 1983, 1988, …, 2008, and maximum values for 2003 and 2005, but no straightforward method to correlate data points. Fig. 3 represents the complementary cumulative distributions of the events size and the time interval between consecutive events.

The results shown above illustrate through simple statistics the increasing importance of understanding the behavior of forest fires and characterizing the spatiotemporal distributions unveiled by such a complex phenomenon. For that purpose, in the next sections we adopt several complementary mathematical tools.

Mutual Information Analysis

In this section we adopt the mutual information to correlate forest fires annual patterns. First we compute the mutual information, based on events size (i.e., burnt area), for each pair of years in the time period 1980–2012. Second, we use a hierarchical clustering algorithm to find relationships among the

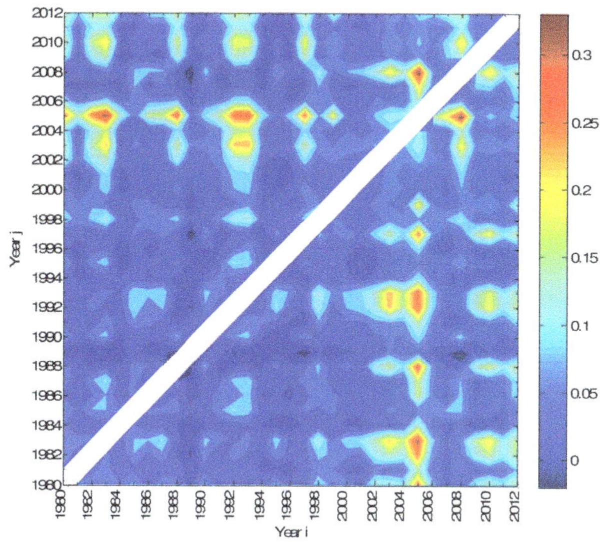

Figure 4. Contour map of the mutual information, $I_M(X_i, X_j)$, between occurrences registered in Portugal during 1980–2012. The cutoff threshold value A_{min} = 10 ha is adopted.

data. Visualization trees are used to highlight the interpretation of the results.

Mutual Information

The mutual information is a measure of the statistical dependence between two random variables, giving the amount of information that one variable "contains" about the other. If X_i and X_j are two discrete random variables, then the mutual information, $I(X_i, X_j)$, is given by:

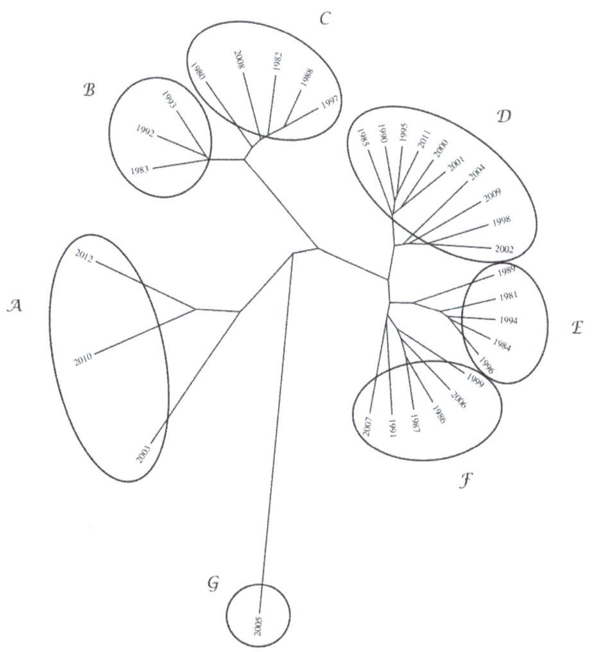

Figure 5. Tree representing mutual information, $I_M(X_i, X_j)$, between occurrences registered in Portugal during 1980–2012. The cutoff threshold value A_{min} = 10 ha is adopted.

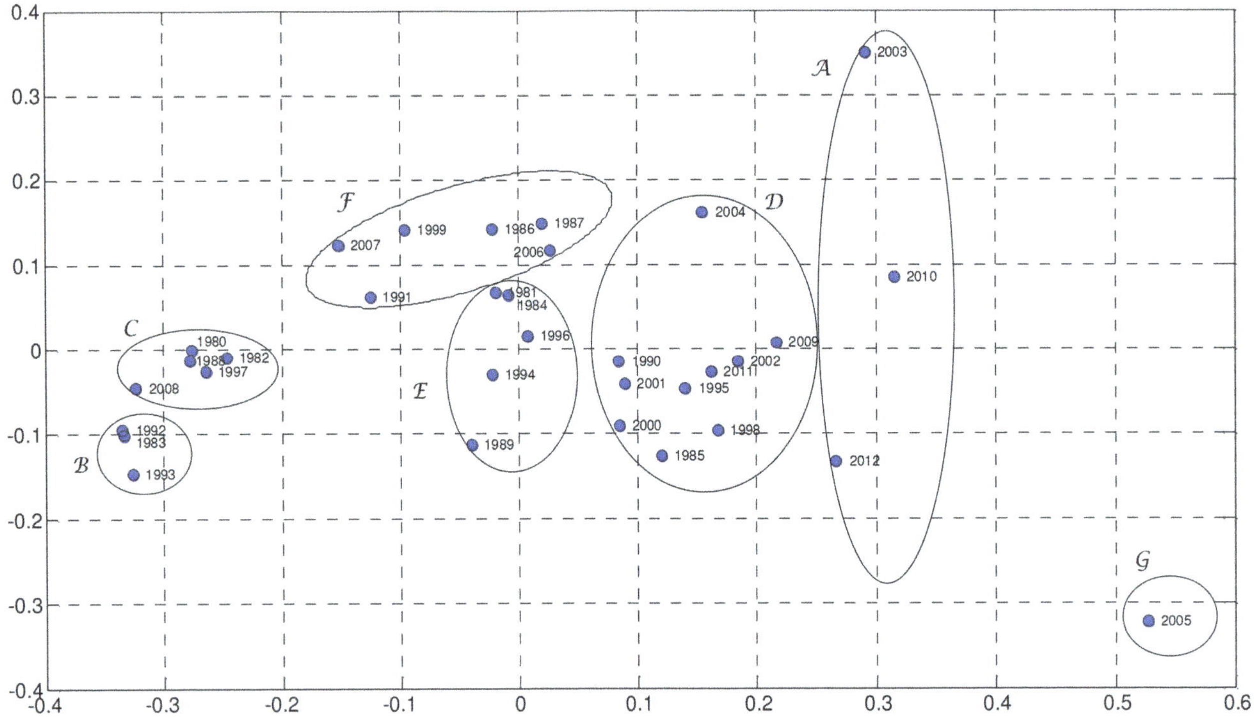

Figure 6. MDS map based on matrix D, for visualization space with dimension $m = 2$.

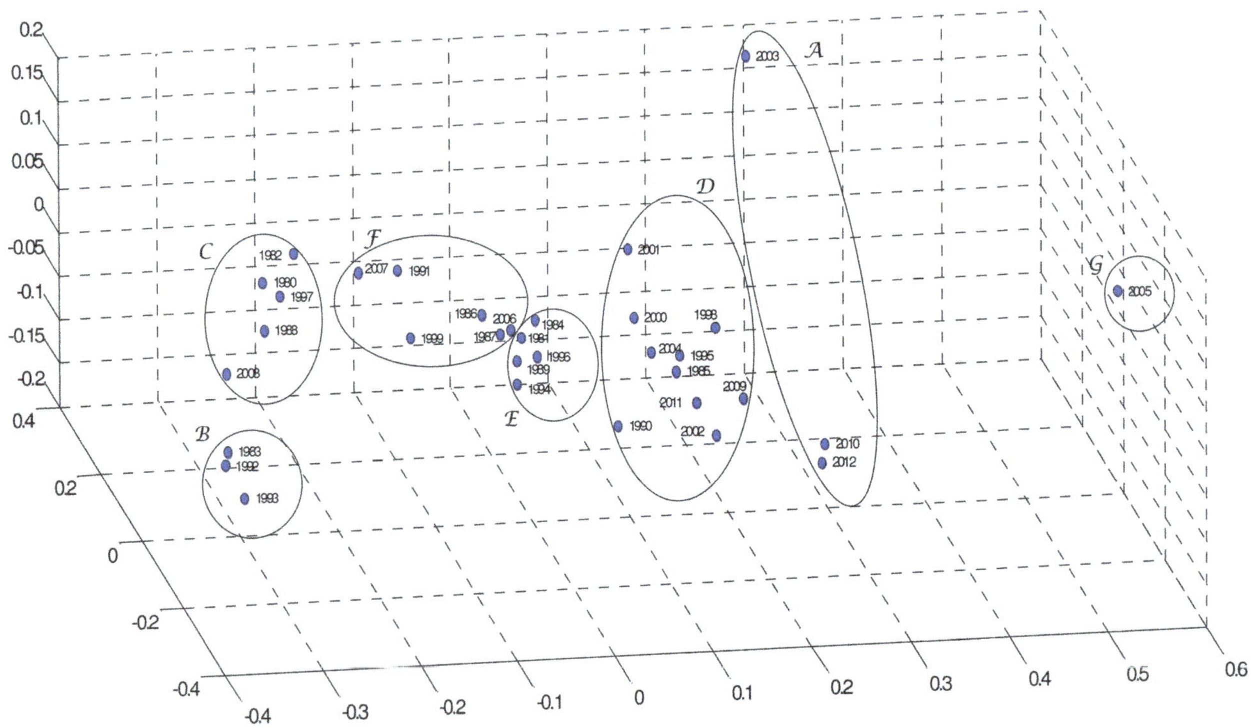

Figure 7. MDS map based on matrix D, for visualization space with dimension $m = 3$.

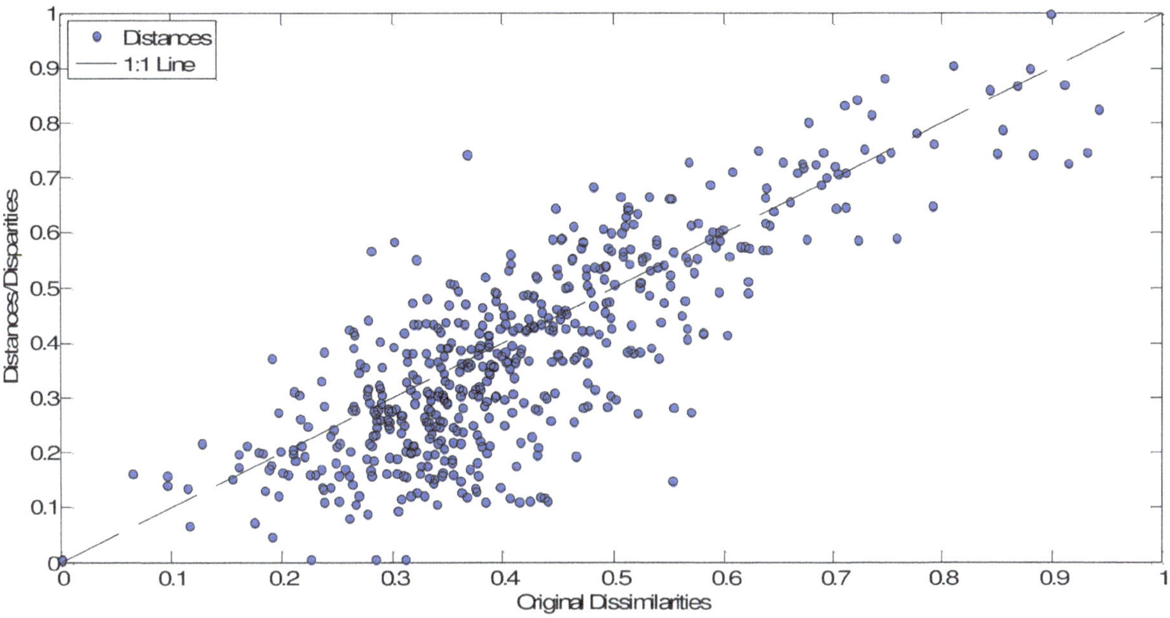

Figure 8. Shepard plot of the MDS map based on matrix D, for visualization space with dimension $m = 2$.

$$I(X_i, X_j) = \sum_{x_j \in X_j} \sum_{x_i \in X_i} p(x_i, x_j) \cdot \log\left(\frac{p(x_i, x_j)}{p(x_i) \cdot p(x_j)}\right) \quad (3)$$

where $p(x_i, x_j)$ is the joint probability distribution function of (X_i, X_j), and $p(x_i)$ and $p(x_j)$ are the marginal probability distribution functions of X_i and X_j, respectively.

The concept of mutual information comes from the information theory [23] and has been adopted in the study of complex systems from diverse fields, namely in experimental time series analysis, in DNA and symbol sequencing and in providing a theoretical basis for the notion of complexity [24–30].

In this section, instead of expression (3), we use the normalized mutual information, $I_N(X_i, X_j)$, given by [31]:

$$I_N(X_i, X_j) = \frac{I(X_i, X_j)}{H(X_i, X_j)} \quad (4)$$

where $H(X_i, X_j)$ represents the joint entropy between X_i and X_j:

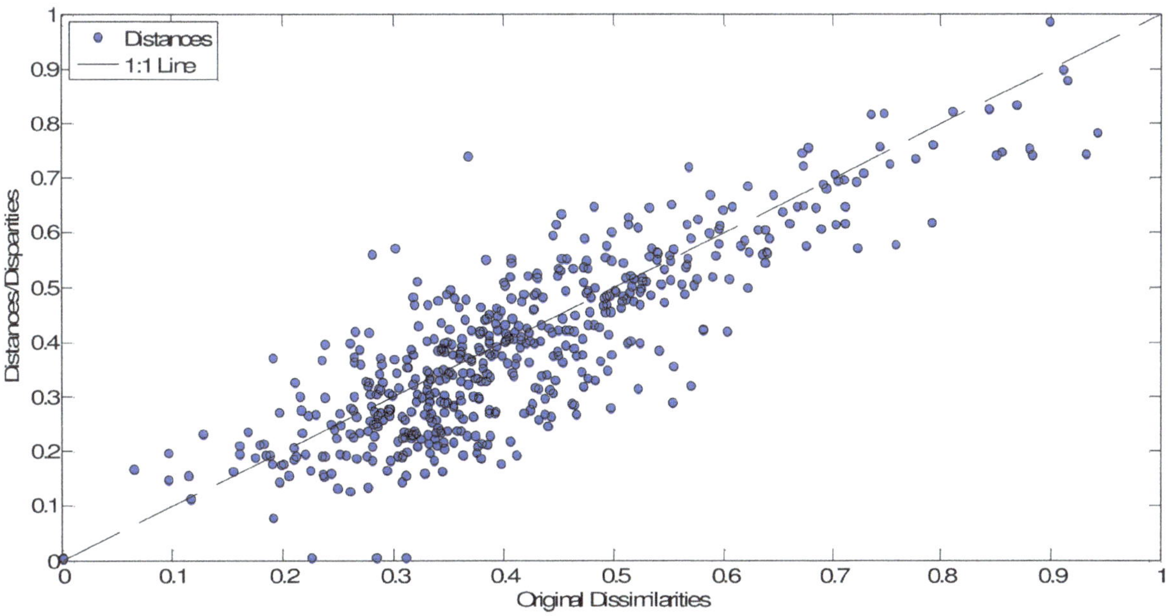

Figure 9. Shepard plot of the MDS map based on matrix D, for visualization space with dimension $m = 3$.

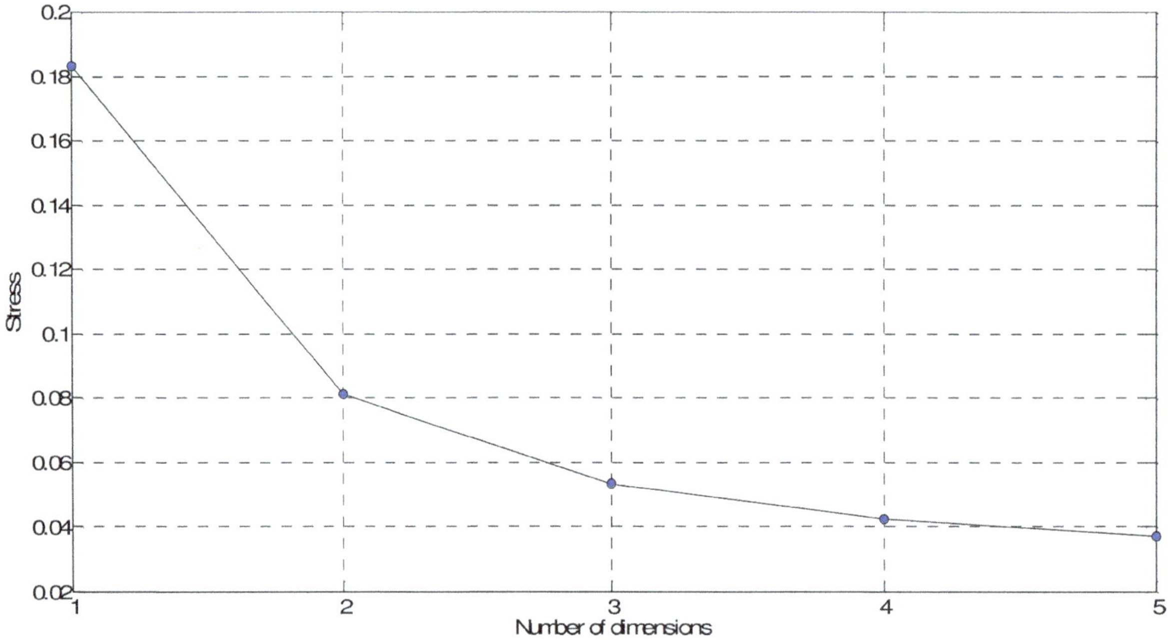

Figure 10. Stress plot for the MDS based on matrix D.

$$H(X_i,X_j) = -\sum_{x_j \in X_j}\sum_{x_i \in X_i} p(x_i,x_j)\cdot\log\left[p(x_i,x_j)\right] \quad (5)$$

The normalized mutual information $I_N(X_i, X_j) \in [0, 1]$ simplifies comparison across different conditions and improves sensitivity.

Forest fires are analysed in an annual basis. For each year, $i = 0, \ldots, 32$, in the period 1980–2012 the events are represented by:

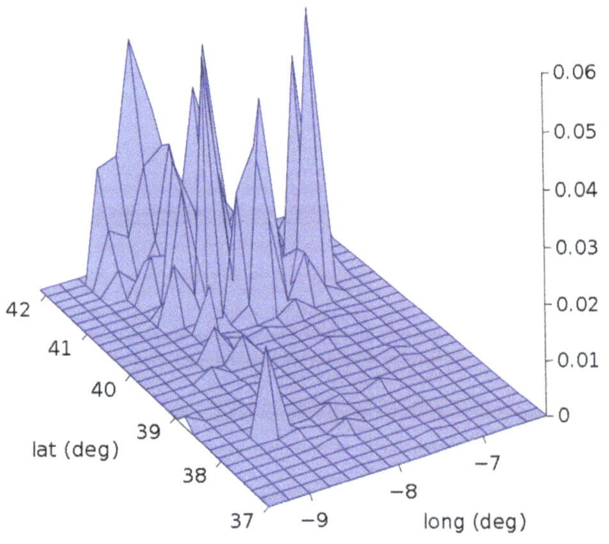

Figure 11. Bidimensional histogram of forest fires versus latitude and longitude, in Portugal, for year 2010.

$$x_i(t) = \sum_{k=1}^{T} A_k \delta(t - t_k), \quad i = 0, \cdots, 32 \quad (6)$$

leading to 33 one-year length time series. This means that the events are modelled as Dirac impulses, where A_k represents fire size (i.e., burnt area), t_k is the instant of occurrence (with one minute resolution), t represents time and T denotes the time period of one year.

The signals $x_i(t)$ are then normalized according to (7):

$$\tilde{x}_i(t) = \frac{x_i(t) - \mu}{\sigma} \quad (7)$$

where μ and σ represent the mean and standard deviation values of all events listed during 1980–2012, with magnitude larger than $A_{min} = 10$ ha. The mutual information is calculated to correlate events occurred in different years of the analysed time period.

Fig. 4 depicts in a contour map the mutual information, $I_N(X_i, X_j)$, between every pair of years $i, j = 0, \ldots, 32$. The probabilities for calculating the mutual information are estimated from the histograms of amplitudes A_k, constructed considering 476 bins, each one having width equal to 0.1 ha. To facilitate the comparison the cases $i = j$ (i.e., those with maximum value of mutual information) are removed from the graph.

The map reveals strong correlations between certain years, corresponding to higher values of mutual information. This is well noted for the years a = {2003, 1983}, b = {2003, 1993}, c = {2005, 1980}, d = {2005, 1983}, e = {2005, 1988}, f = {2005, 1993}, g = {2008, 2005} and h = {2010, 1983}. Nevertheless, the analysis is not totally assertive and requires multiple comparisons.

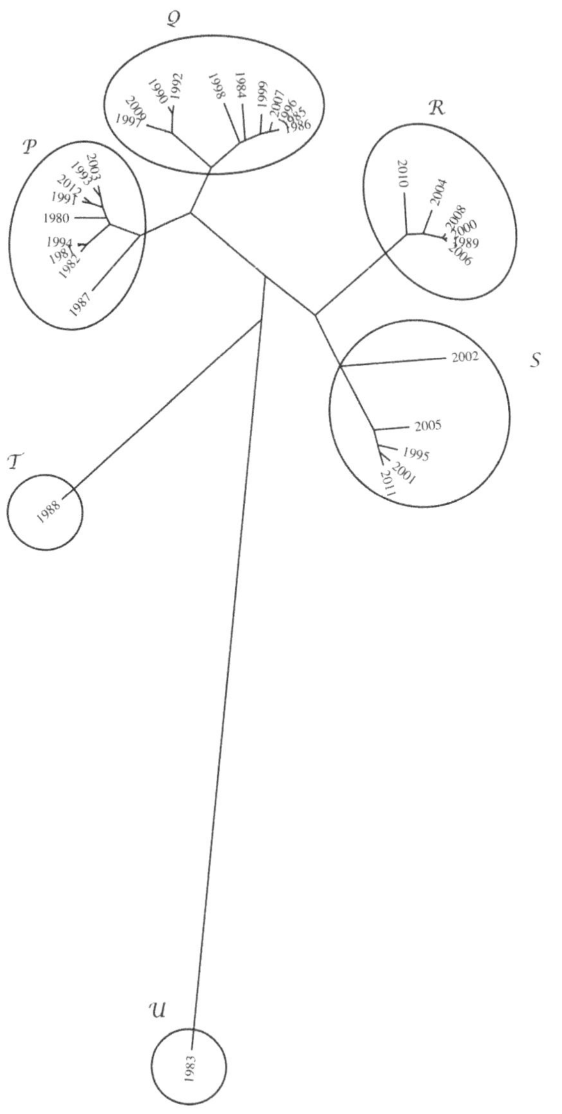

Figure 12. Tree comparing the 33 bidimensional histograms based on index d_{ij}.

Hierarchical clustering

Having in mind an efficient method to visualize and to compare results, a hierarchical clustering algorithm is adopted, based on the mutual information, $I_N(X_i, X_j)$, between pairs of objects.

The goal of hierarchical clustering is to build a hierarchy of clusters, in such a way that objects in the same cluster are, in some sense, similar to each other [30,32–33]. Based on a measure of dissimilarity between clusters, those are combined (or, alternatively, split) for agglomerative (or, alternatively, divisive) clustering. This is achieved by using an appropriate metric, quantifying the distance between pairs of objects, and a linkage criterion, defining the dissimilarity between clusters as a function of the pairwise distances between objects. The results of hierarchical clustering are presented in a phylogenetic tree adopting the successive (agglomerative) clustering and average-linkage method (Fig. 5). The software PHYLIP was used for generating both graphs (http://evolution.genetics.washington.edu/phylip.html).

Fig. 5 unveils groups of objects (years) in such a way that objects in the same group (cluster) are more similar to each other than to

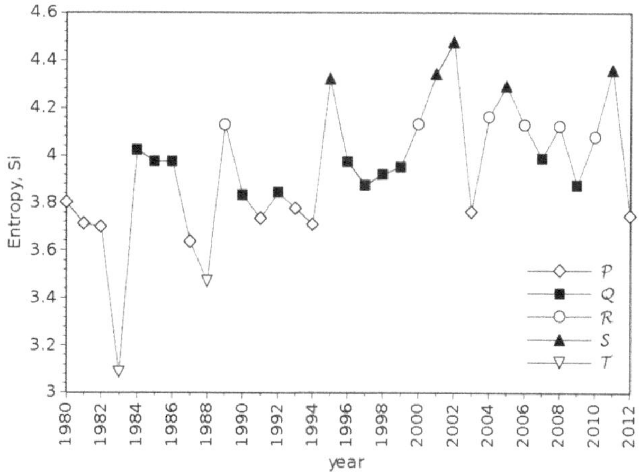

Figure 13. Entropy, S_i, versus year, during the time period 1980–2012.

those in other groups. For example, we can easily identify clusters composed by years A = {2003, 2010, 2012}, B = {1983, 1992, 1993} and C = {1980, 1982, 1988, 1997, 2008,}. On the contrary, year G = {2005} is far aside, meaning that it is different from all the rest. Both representations of Fig. 4 and Fig. 5 can be used to visualise and to compare the events, in an annual basis. Fig. 5 leads to a result easier to interpret than Fig. 4, as it identifies groups of objects that are similar.

MDS Analysis and Visualization

In this section we adopt the MDS tools to handle information and the relationships embedded into the data.

MDS is a statistical technique for visualizing data that can reveal similarities between objects. The algorithm requires the definition of a similarity measure (or, inversely, of a distance) and the construction of a $s \times s$ symmetric matrix **D** of similarities (or distances) between each pair of s objects. MDS assigns a point to each object in a m-dimensional space and arranges the set in order to reproduce the observed similarities. A shorter (larger) distance between two points means that the corresponding objects are more similar (distinct). For $m = 2$ or $m = 3$ dimensions the resulting locations may be displayed in a "map" that can be visualized [34–39].

In our case, we obtain **D** (33×33 dimensional) by means of the mutual information (4). Fig. 6 and Fig. 7 show the MDS maps for $m = 2$ and $m = 3$, respectively. The Shepard and the stress plots assess the quality of the MDS maps. The Shepard diagrams (Fig. 8 and Fig. 9) show an acceptable distribution of points around the 45 degree line, which means a good fit of the distances to the dissimilarities. On the other hand, the stress plot reveals that a three dimensional space describes adequately the data (Fig. 10). This can be concluded by observing the stress line, which diminishes strongly until the dimensionality is two, moderately towards dimensionality three and weakly from then on. Often, the maximum curvature point of the stress line is adopted as the criterion for deciding the dimensionality of the MDS map.

The MDS maps of Fig. 6 and Fig. 7 confirm the groups previously identified by the hierarchical clustering and, consequently, the relationships between the corresponding yearly patterns. Comparing Fig. 5 with Fig. 6 and Fig. 7, we conclude that all allow an easy interpretation of the results. The MDS maps, in particular the 3D plot, are more intuitive than the phylogenetic

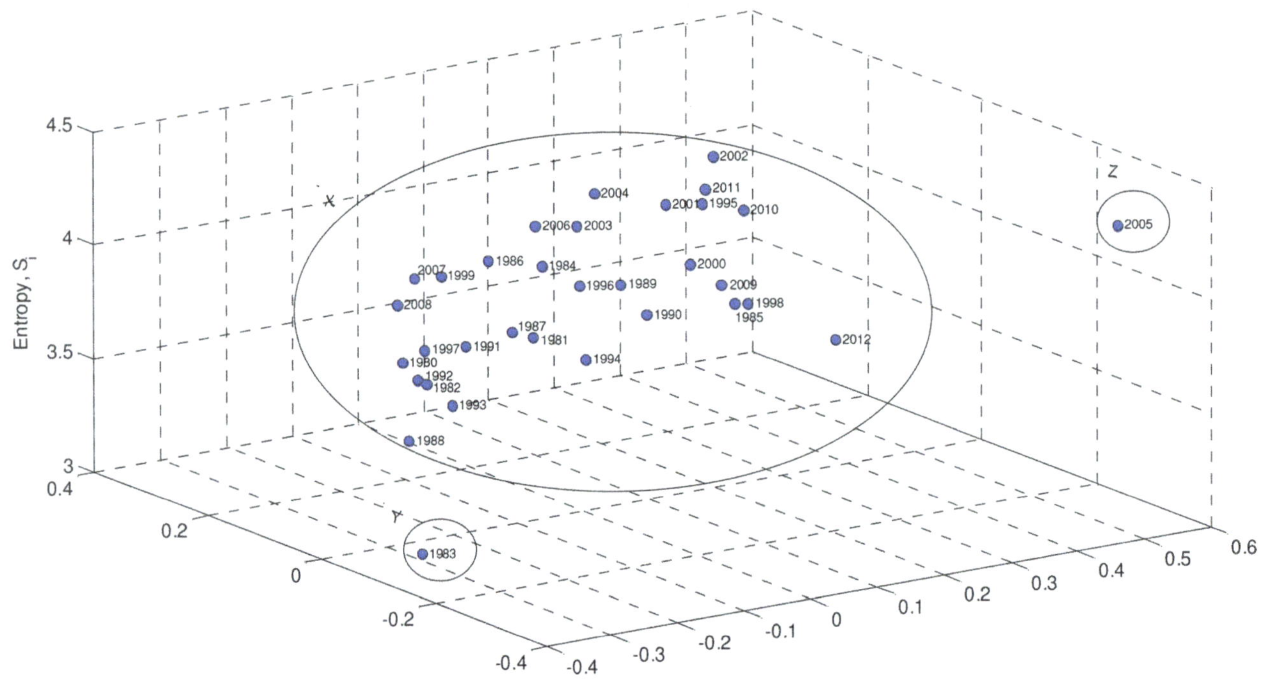

Figure 14. MDS 2D plot with the vertical axis representing Shannon entropy.

tree. Moreover, most software for MDS analysis allows the user to rotate and visualise the maps from different perspectives, easing the identification of clusters. This is useful especially when dealing with large amounts of data.

Forest Fires Spatial Patterns

In this section we study forest fires in a complementary line of thought, namely by considering spatial information. First, we divide the geographic territory under study (i.e., 36.95°≤ lat ≤ 42.15°; −9.50°≤ lon ≤−6.19°), using a $M \times N$ ($M = 30$, $N = 15$) rectangular grid, and we determine the 33 bidimensional histograms of relative frequencies for all years in the period 1980–2012. Second, for characterizing the histograms, we calculate the Shannon entropy, S_i, given by:

$$S_i = -\sum_{m=1}^{M}\sum_{n=1}^{N} p_i(m,n)\cdot\log[p_i(m,n)], \quad i = 0, \cdots, 32 \qquad (8)$$

where the probabilities $p_i(m, n)$ are approximated by the relative frequencies.

In Fig. 11, for example, we depict the bidimensional histogram for year 2010. The corresponding entropy is $S_i = 4.08$ ($i = 30$).

For histogram comparison we calculate a $M \times N$ symmetric matrix $\mathbf{D} = [d_{ij}]$, where

$$d_{ij} = |S_i - S_j|, \quad i,j = 0, \cdots, 32 \qquad (9)$$

The results are visualized in the phylogenetic tree of Fig. 12. We can observe six clusters: P= {1980, 1981, 1982, 1987, 1991, 1993, 1994, 2003, 2012}; Q= {1984, 1985, 1986, 1990, 1992, 1996, 1997, 1998, 1999, 2007, 2009}; R= {1989, 2000, 2004, 2006, 2008, 2010}; S= {1995, 2001, 2002, 2005, 2011}; T= {1983} and U= {1988}.

The evolution of S_i versus year is represented in Fig. 13, where the clusters shown in Fig. 12 are identified. In this chart is clear a large volatility and apparently some increase of entropy along time.

In a more global perspective, we verify that amplitude and space data lead to distinct observations. The conclusions are 'decoupled' and reveal that both directions must be explored, with more data, in order to include all information in a global tool of analysis.

In this line of though, we embed amplitude and space data into a single graph by adding to the bidimensional MDS plot of Fig. 6 a vertical axis representing the Shannon entropy (Fig. 14).

We note that only two years, Y= {1983} and Z= {2005}, have now a clearly distinct separation from the main cluster, X. In Fig. 13 we observed them to be located at near extreme values, but, as mentioned, it is difficult to get idea due to large volatility. The embedding of amplitude-space techniques produced a clear classification pattern.

Conclusions

We analysed forest fires from the perspective of dynamical systems. Data from a public domain forest fires catalogue, containing information of events for Portugal, during the period 1980–2012, was studied in an annual basis. Mutual information to correlate annual patterns was considered. Phylogenetic trees generated by hierarchical clustering algorithms and MDS visualization tools were used to compare to extract relationships among the data and to identify forest fire patterns. Those tools allow different perspectives over forest fires that may be used to better understand the dynamics emerging in the plethora of phenomena that occur in forest fires.

Author Contributions

Conceived and designed the experiments: AML JTM. Performed the experiments: AML. Analyzed the data: AML JTM. Contributed reagents/

materials/analysis tools: AML JTM. Contributed to the writing of the manuscript: AML JTM.

References

1. Di Bella CM, Jobbágy EG, Paruelo JM, Pinnock S (2006) Continental fire density patterns in South America. Global Ecology and Biogeography 15, 192–199.
2. Bradstock RA (2008) Effects of large fires on biodiversity in south-eastern Australia: disaster or template for diversity?. International Journal of Wildland Fire 17, 809–822.
3. Hanson CT, Odion DC (2013) Is fire severity increasing in the Sierra Nevada, California, USA?. International Journal of Wildland Fire 23, 1–8.
4. Flannigan MD, Krawchuk M, de Groot W, Wotton B, Gowman L (2009) Implications of changing climate for global wildland fire. International Journal of Wildland Fire 18, 483–507.
5. Zumbrunnen T, Pezzatti GB, Menéndez P, Bugmann H, Bürgi M, et al. (2011) Weather and human impacts on forest fires: 100 years of fire history in two climatic regions of Switzerland. Forest Ecology and Management, 261, 2188–2199.
6. Silva FR, González-Cabán A (2010) 'SINAMI': a tool for the economic evaluation of forest fire management programs in Mediterranean ecosystems. International journal of wildland fire, 19 927–936.
7. Zamora R, Molina-Martínez JR, Herrera MA, Rodríguez y Silva F (2010) A model for wildfire prevention planning in game resources. Ecological Modelling 221, 19–26.
8. Preisler HK, Westerling AL, Gebert KM, Munoz-Arriola F, Holmes TP (2011) Spatially explicit forecasts of large wildland fire probability and suppression costs for California. International Journal of Wildland Fire 20, 508–517.
9. Alvarado E, Sandberg DV, Pickford SG (1998) Modeling Large Forest Fires as Extreme Events. Northwest Science 72, 66–75.
10. Bak P, Chen K, Tang C (1990) A forest-fire model and some thoughts on turbulence. Physics Letters A 147, 297–300.
11. Barford P, Bestavros A, Bradley A, Crovella M (1999) Changes in Web client access patterns: characteristics and caching implications. World Wide Web 2, 15–28.
12. Barabási AL (2005) The origin of bursts and heavy tails in human dynamics. Nature 435, 207–211.
13. Pinto CMA, Lopes AM, Tenreiro Machado JA (2012) A review of power laws in real life phenomena. Communications in Nonlinear Science and Numerical Simulations 17, 3558–3578.
14. Ricotta C, Avena G, Marchetti M (1999) The flaming sandpile: self-organized criticality and wildfires. Ecological Modelling 119, 73–77.
15. Reed W, McKelvey K (2002) Power-law behaviour and parametric models for the size-distribution of forest fires. Ecological Modelling 150, 239–254.
16. Prentice IC, Kelley DI, Foster PN, Friedlingstein P, Harrison SP, et al. (2011) Modeling fire and the terrestrial carbon balance. Global Biogeochemical Cycles. 25, GB3005
17. Fletcher IN, Aragão LE, Lima A, Shimabukuro Y, Friedlingstein P (2013) Fractal properties of forest fires in Amazonia as a basis for modelling pan-tropical burned area. Biogeosciences Discussions 10, 14141–14167.
18. Drossel B, Schwabl F (1992) Self-organized critical forest-fire model. Physical Review Letters 69, 1629–1632.
19. Malamud B, Morein G, Turcotte D (1998) Forest Fires: An Example of Self-Organized Critical Behavior. Science 281, 1840–1842.
20. Telesca L, Amatulli G, Lasaponara R, Lovallo M, Santulli A (2005) Time-scaling properties in forest-fire sequences observed in Gargano area (southern Italy). Ecological Modelling 185, 531–544.
21. Telesca L, Amatucci G, Lasaponara R, Lovallo M, Rodrigues MJ (2007) Space–time fractal properties of the forest-fire series in central Italy. Communications in Nonlinear Science and Numerical Simulation 12, 1326–1333.
22. Marques S, Borges JG, Garcia-Gonzalo J, Moreira F, Carreiras JMB, et al. (2011) Characterization of wildfires in Portugal. European Journal of Forest Research 130, 775–784.
23. Shannon CE (1948) A Mathematical Theory of Communication. Bell System Technical Journal 27, 379–423, 623–656.
24. Herzel H, Schmitt AO, Ebeling W (1994) Finite Sample effects in Sequence Analysis. Chaos, Solitons & Fractals 4, 97–113.
25. Mori T, Kudo K, Tamagawa Y, Nakamura R, Yamakawa O, et al. (1998) Edge of chaos in rule-changing cellular automata. Physica D 116, 275–282.
26. Matsuda H (2000) Physical nature of higher-order mutual information: Intrinsic correlations and frustration. Physical Review E 62, 3096–3102.
27. Posadas A, Hirata T, Vidal F, Correig A (2000) Spatio-temporal seismicity patterns using mutual information application to southern Iberian peninsula (Spain) earthquakes. Physics of the Earth and Planetary Interiors 122, 269–276.
28. Telesca L (2011) Tsallis-Based Nonextensive Analysis of the Southern California Seismicity. Entropy 13, 1267–1280.
29. Mohajeri N, Gudmundsson A (2012) Entropies and Scaling Exponents of Street and Fracture Networks. Entropy 14, 800–833.
30. Tenreiro Machado JA, Lopes AM (2013) Analysis and Visualization of Seismic Data using Mutual Information. Entropy 15, 3892–3909.
31. Kvalseth TO (1987) Entropy and correlation: Some comments. IEEE Transactions on Systems, Man and Cybernetics 17, 517–519.
32. Jain A, Dubes R (1988) 'Algorithms for Clustering Data.' (Prentice-Hall: Englewood Cliffs, NJ).
33. Johnson SC (1996) Hierarchical clustering schemes. Psychometrika 32, 241–54.
34. Shepard R (1962) The analysis of proximities: multidimensional scaling with an unknown distance function. Psychometrika 27, 219–246.
35. Kruskal J, Wish M (1978) 'Multidimensional Scaling.' (Sage Publications: Newbury Park, CA).
36. Cox T, Cox M (2001) 'Multidimensional scaling.' (Chapman & Hall/CRC).
37. Martinez W, Martinez A (2005) 'Exploratory Data Analysis with MATLAB.' (Chapman & Hall/CRC Press: UK).
38. Tzagarakis C, Jerde TA, Lewis SM, Ugurbil K, Georgopoulos AP (2009) Cerebral cortical mechanisms of copying geometrical shapes: a multidimensional scaling analysis of FMRI patterns of activation, Experimental Brain Research 194, 369–380.
39. Costa AM, Tenreiro Machado JA, Quelhas MD (2011) Histogram-based DNA analysis for the visualization of chromosome, genome and species information. Bioinformatics 27, 1207–1214.

8

Altered Community Flammability in Florida's Apalachicola Ravines and Implications for the Persistence of the Endangered Conifer *Torreya taxifolia*

John M. Mola[1,4]*, J. Morgan Varner[2¤a], Erik S. Jules[1], Tova Spector[3¤b]

1 Department of Biological Sciences, Humboldt State University, Arcata, California, United States of America, **2** Department of Forestry, Forest & Wildlife Research Center, Mississippi State University, Mississippi State, Mississippi, United States of America, **3** Florida Park Service, Department of Environmental Protection, Panama City, Florida, United States of America, **4** Graduate Group in Ecology, University of California Davis, Davis, California, United States of America

Abstract

Plant species and communities often reflect historic fire regimes via ecological and evolutionary responses to recurrent fires. Plant communities of the southeastern USA experience a wide array of fire regimes, perhaps nowhere more marked than the juxtaposition of fire-prone uplands and adjacent mesic ravines along Florida's Apalachicola River. The ravines contain many endemic and disjunct species, most notably the endangered endemic conifer *Torreya taxifolia*. A rapid decline in *T. taxifolia* over the past 60 years has been associated with widespread replacement by other tree species. To understand the changes accompanying the shift in ravine composition, we compared leaf litter flammability of nine historic and contemporary species. We measured maximum flame height, flame duration, smoldering duration, mass loss, absorptive capacity, and drying rate. Ordination and perMANOVA suggest the nine species segregated into three distinct groups: the fire-impeding *T. taxifolia* and *Taxus floridana*; an intermediate group of three deciduous angiosperms; and a mixed cluster of four flammable species. Results suggest *T. taxifolia* and *T. floridana* were fire-impeding species in these communities, while contemporary dominants burn similarly to the upslope pyric species. The increasing presence of fire-facilitating species may portend a shifting fire regime that further imperils *T. taxifolia* and other rare species in the formerly fire-safe ravines.

Editor: Ben Bond-Lamberty, DOE Pacific Northwest National Laboratory, United States of America

Funding: Funding for this research was provided by the National Science Foundation (NSF) Research Experience for Undergraduates Program (REU 0755466 to S.F. Craig and M.D. Johnson), the NSF Graduate Research Fellowship Program (Grant 1049702 to J.M.M.) and the Mississippi Forest & Wildlife Research Center (http://fwrc.msstate.edu/). The funders had no role in study design, data collection and analysis, decision to publish, or preparation of the manuscript.

Competing Interests: The authors have declared that no competing interests exist.

* Email: jmmola@ucdavis.edu

¤a Current address: Department of Forest Resources & Environmental Conservation, Virginia Polytechnic Institute & State University, Blacksburg, Virginia, United States of America
¤b Current address: US Fish & Wildlife Service, Utah Ecological Services, West Valley City, Utah, United States of America

Introduction

Fire is a dominant driver of ecological processes in fire-prone environments worldwide [1]. Flammability of senesced leaves, cones, or sloughed bark [2–4] can facilitate the spread and intensity of fires in some plant communities and impede or dampen fire in others [5–7]. Flammability has been quantified as a combination of ignitibility of fuel (energy required for combustion), combustibility (intensity, flame height, energy output), sustainability (duration of flaming and smoldering), and consumability [mass loss; 8]. Flammable plants promote conditions favoring fire, resulting in the persistence of plant communities with high fire tolerance [2], while plant communities composed of species intolerant of fire tend to burn with lower flammability and dampen the intensity or spread of fire [4,5]. Understanding the flammability of individual species can provide insight into community fire regimes [9].

Changes in plant community composition that alter fire regimes can initiate feedbacks that govern ecological processes [1,10,11]. For example, widespread shifts from pyrophytic to mesophytic tree species in woodlands and forests of eastern North America (termed "mesophication"; 10) have resulted in a positive feedback cycle where the increases in fire-intolerant, less flammable species reduce the intensity and spread of fires in formerly fire-prone environments. Negative feedbacks can also occur where species with low flammability are replaced by more flammable species, as in transitional Amazonian tropical rainforests [e.g., 12,13]. These shifts in fuels may have extreme consequences for fire regimes, affecting ecosystem processes broadly and species of conservation concern more specifically.

In this paper, we explore the possibility that the distribution and recovery of a rare tree, *Torreya taxifolia*, is further imperiled due to shifts in the woody plant community and their litter fuels. *Torreya taxifolia* is a federally endangered conifer that has undergone a precipitous decline in the numbers and health of remnant populations [14,15]. A 35 km corridor along Florida's Apalachicola River [14] harbors ravines that shelter mesophytic forests and a rich diversity of endemic and disjunct plant species as many taxa found refuge in the cool, moist ravines during the warming interglacial period [Fig. 1; 16]. These ravines hold the only wild populations of the rare conifers *T. taxifolia* and *Taxus floridana* [17]. As recently as the 1950s, *T. taxifolia* comprised ~15% of the dominant trees within the ravines totaling ca. 300,000 to 650,000 individuals [18,19]. The current population includes fewer than 2,000 small trees, with few reproductive individuals observed since 1962, representing <1% of the trees in the ravines [14,20]. Mechanisms proposed to explain the decline of *T. taxifolia* include fungal pathogens, water stress, regional warming, hydrologic changes, and altered fire regimes [21,22]. While the roles of pathogens, water stress, and warming climate have been addressed in previous work [21,22], the role of fire in the decline and potential recovery of *T. taxifolia* has not been studied.

It remains unknown how fire affected the decline, or may affect the future recovery, of *T. taxifolia*. It has been proposed that fire frequency of the steephead ravines has increased, thereby making it difficult for the thin-barked, fire-susceptible *T. taxifolia* to recruit new individuals [15,21]. It has also been proposed that smoke from historic fires in the adjacent uplands helped to suppress fungal loads that currently plague remnant trees in the ravines [21]. The Apalachicola ravines have been described as some of the most "fire-proof" areas of Florida [18,23], while others contend that fire-caused canopy gaps may have helped sustain *T. taxifolia* prior to fire exclusion [15,21]. In spite of this disagreement, no studies have been conducted on the flammability of *T. taxifolia* or its associated species. Anecdotal evidence suggests fire spreads into the ravines during upslope prescribed burns and has been observed to creep through the ravines for several days (J. McKenzie, personal communication; D. Prentiss, personal communication). These fires may result in the eventual resprouting or mortality of *T. taxifolia* [24; E. Johnson, personal communication]. It is unclear, however, if these ravine fires are consistent with historical patterns, or if they are novel due to the changing species and resulting fuel composition of the ravines. The lack of studies on fire in the ravines is especially noteworthy given the extensive attention given the role of fire and flammability in the pine-oak ecosystems immediately upslope [e.g., 5,6,25,26].

If fire played a dominant role in the ravines historically, a necessary condition would have been the presence of flammable fuels. Following the logic of Mutch [2], we proposed that if the flammability of historic ravine species is high, then fire may have played a persistent role in ravine plant dynamics. If, however, contemporary fuels are more flammable than historic fuels in *T. taxifolia* -dominated communities, then this shift would suggest negative fire feedbacks, as described for wet Neotropical forests [12,13,27]. To better understand the role of fire in ravine communities past and present, we evaluated the flammability of ravine species, comparing historic and contemporary tree community composition. Results will help clarify the role and potential effects of fire in the ravine ecosystems and, more broadly, help shed light on the effects of shifting species composition on fire regimes.

Figure 1. The mesic ravines along the Apalachicola River in northern Florida harbor a variety of rare plant and animal species. 20th century declines in the formerly dominant endemic conifer Torreya taxifolia have resulted in shifting tree community composition.

70
60
50
40
30

Figure 2. A laboratory flammability experiment with *Magnolia grandiflora*, a native tree species in Florida's Apalachicola ravines. *M. grandiflora* is one of several tree species that have increased following the demise of the endemic conifer *Torreya taxifolia*. The resulting shift in species composition has cascading effects on community flammability and consequences for the recovery of *Torreya taxifolia* and perhaps a wider suite of fire-sensitive ravine taxa.

Methods

We collected surface litter from nine currently or formerly dominant tree species in the Apalachicola ravines. We collected eight species from Torreya State Park, Florida: *T. taxifolia*, *T. floridana*, *Liquidambar styraciflua*, *Liriodendron tulipifera*, *Ilex opaca*, *Magnolia grandiflora*, *Fagus grandifolia*, and *Ostrya virginiana*. Due to its rarity, *T. taxifolia* litter was also collected at Maclay Gardens State Park and Three Rivers State Park, Florida. We also used previously collected *Pinus glabra* litter collected from the Tuskegee National Forest, Alabama, though the species is found in the Apalachicola ravines. Litter from all species was collected in May and June, during the typical fire season in the region [28]. For each species, we collected ~200 g of surface litter (dry weight; ~20 g from ten separate trees of each of the nine species). Access and permits for collecting litter were provided by the Florida Division of Recreation and Parks. Samples were transported to the laboratory with less than seven days between field collection and lab preparation.

In the lab, litter was sorted to remove other species and any woody debris. Five to seven replicates of each of the nine species were oven-dried at 70°C for 48 hours. After drying, 15±0.1 g (±0.5 g for the large-leaved *M. grandiflora*) of litter were burned using standard methods [3,5]. We distributed 15 g of litter across the inner 20×20 cm area of a 35×35 cm grid of eight xylene-soaked cotton strings on a stainless steel platform beneath a 2.75×2.75 m ventilation hood with constant draw. Strings were ignited and a timer started once the litter ignited. Flame height was visually estimated using a 125 cm ruler (with 1 cm graduations) behind the fuel (Fig. 2). Timers captured the point when flames extinguished (flaming time, sec). The duration of smoldering (sec) was measured until glowing stopped. The remaining ash and unburned litter were collected from the platform and weighed to calculate mass loss (%).

We coupled our burning data with data from litter drying experiments on the same nine species from our sites [7; and our own unpublished data; See Appendix S1 for full data]. In humid, punctuated wet-dry climates like those in northern Florida [23], "dry flammability" measures such as ours can be misleading, since oven-dry conditions are rare on landscapes that receive abundant or frequent precipitation and have otherwise high relative humidity. To account for these factors, we used two additional metrics of litter flammability: drying time [hours, as measured by time lag; 7] and their initial moisture content (%) following prolonged saturation [or absorptive capacity; 7]. Detailed methods are presented elsewhere [7] but entailed saturating four 15 g litter samples for 24 hours, measuring their water absorption (following agitation to remove excess water), and tracking their moisture loss over seven days in a constant drying environment (air temperature 23° to 24°C, 45 to 50% relative humidity). We used data for five species from Kreye et al. [7] and unpublished data for *T. taxifolia*, *T. floridana*, *O. virginiana*, and *P. glabra* measured in the same experiments.

Because our burning and drying response variables were highly correlated, we used principal components analysis (PCA) to assess differences in multiple flammability measures among our nine taxa. We included the four burning metrics (flame height, flame duration, smoldering time, and mass loss) and the two drying metrics (time lag and initial moisture content) in our PCA, and we relativized each variable based on the number of standard deviations from the mean. To determine if species could be segregated into discrete groups based on the burning and drying characteristics, we used *k*-means cluster analysis. Next, we tested for differences in burning and drying characteristics among the groups identified in the cluster analysis using perMANOVA. We used the non-parametric perMANOVA because we could not meet the assumptions of normality of MANOVA. The PCA and cluster analysis were performed with NCSS [29] and the perMANOVA was done using the adonis function in the vegan package for R version 3.01 [30].

Results

Litter from the nine ravine tree species differed in several aspects of flammability (Table 1). Mean maximum flame height differed among the nine species burned. *Liriodendron tulipifera* and *F. grandifolia* burned with the tallest flames with means of 84.6 cm and 81 cm, respectively. *Torreya taxifolia* had the lowest mean flame height at only 26.2 cm. Fuel consumption for the nine species segregated into two distinct groups with most species having means ranging from 73 to 89% consumption and *T. taxifolia* and *T. floridana* being different from all others with only 44% and 41% mass loss, respectively. *Torreya taxifolia* and *T. floridana* had the longest duration flaming at 155 and 145 sec, respectively. *Pinus glabra* had the longest smoldering duration at 278 sec and *T. taxifolia* had the shortest duration at 128 sec.

Combining the four burning and two drying flammability metrics in the PCA explained 54%, 18%, and 16% of the variance for the first, second, and third axes, respectively (88% cumulative). Four flammability metrics were strongly correlated with the first axis (Table 2): those taxa with low axis 1 scores burned with tall flames ($r = -0.832$), brief flaming duration ($r = 0.912$), substantial mass loss ($r = -0.957$), and dried rapidly ($r = 0.864$; Fig. 3). There were strong positive correlations between axis 2 and initial litter moisture content ($r = 0.899$) and modest positive correlations with the duration of smoldering ($r = 0.404$). The third axis was strongly and negatively correlated with smoldering duration ($r = -0.896$); *T. taxifolia* and *T. floridana* differed from other taxa with particularly brief smoldering.

The *k*-means cluster analysis segregated the species into three distinct groups explaining 38.95% of the variation in the burning and drying characteristics (Fig. 3), and the perMANOVA demonstrated that the groups differed significantly in these characteristics ($F = 4.312$, df = 46, $P = 0.021$; Fig. 3). The rare conifers *T. taxifolia* and *T. floridana* group with short flame heights, brief flaming and smoldering durations, minor fuel consumption, and high moisture retention. In contrast, three deciduous hardwood trees *O. virginiana*, *F. grandifolia*, and *L. tulipifera* burned with tall flames, flamed and smoldered for long durations, and had heavy fuel consumption while losing their high initial moisture rapidly. The remaining species, evergreens *P. glabra*, *I. opaca*, *M. virginiana*, and deciduous *L. styraciflua* grouped with tall flames, long durations of flaming and smoldering, with heavy fuel consumption while taking on little initial moisture and losing it rapidly.

Table 1. Litter flammability of ravine and upland tree species, sorted by maximum flame height.

Species	Community	Flame height (cm)	Flame duration (sec)	Mass loss (%)	Smoldering time (s)	Drying time (hr)	Initial moisture content (g)
Liriodendron tulipifera	Ravine	84.6	48.6	0.89	173.4	4.81	386.83
Quercus laevis[a]	Upland	81.4	50.4	0.90	350.7	-	-
Fagus grandifolia	Ravine	81.0	50.8	0.86	177.6	6.86	414.67
Pinus glabra	Ravine	78.3	53.1	0.88	278.7	4.95	127.33
Magnolia grandiflora	Ravine	76.4	55.2	0.87	123.2	6.42	226.50
Quercus falcata[a]	Upland	75.0	52.0	0.87	350.7	-	-
Ilex opaca	Ravine	68.6	74.4	0.89	144.8	7.30	188.50
Quercus stellata[a]	Upland	68.4	52.1	0.85	216.3	-	-
Quercus margaretta[a]	Upland	68.1	66.1	0.79	286.5	-	-
Quercus nigra[a]	Ravine	57.3	77.1	0.78	268.8	-	-
Liquidambar styraciflua	Ravine	54.4	48.2	0.87	288.6	2.67	217.17
Quercus incana[a]	Upland	52.1	77.3	0.86	341.8	-	-
Taxus floridana	Ravine	46.0	149.4	0.42	267.4	12.33	309.33
Ostrya virginiana	Ravine	43.2	69.8	0.74	278.4	8.08	466.33
Quercus hemisphaerica[a]	Ravine	40.6	91.4	0.61	348.0	-	-
Quercus virginiana[a]	Upland	33.6	84.5	0.70	230.7	-	-
Torreya taxifolia	Ravine	26.2	155.4	0.44	128.2	9.59	248.00

[a]Moisture metrics for the ravine species were collected in part from Kreye et al. (2013), moisture metrics for upland species were not collected as part of the Kane et al. (2008) study and so are unavailable.
[b]Ravine species' flammability data were collected as part of this study, upland species flammability data are from Kane et al. (2008).

Table 2. Factor loadings describing the contribution to three principal components (PC) axes of burning characteristics and drying characteristics for litter of nine woody plant taxa collected in the Apalachicola Ravine, Florida.

Factor	PC 1	PC 2	PC 3
Flame Height	−0.834	0.072	0.107
Flame Duration	0.912	−0.157	−0.149
Smolder Duration	−0.181	0.404	−0.896
Mass Loss	−0.957	0.053	0.014
Drying Time	0.864	0.173	0.004
Initial Moisture Content	0.199	0.899	0.367

Discussion

The observed magnitude of interspecific differences in flammability suggests that major changes in community flammability have transpired in Florida's Apalachicola ravines. With the decline of *T. taxifolia*, a species with very low flammability (Fig. 3, and the relative increases in relatively flammable species, a dramatic shift in community flammability has occurred. Prior to the decline of *T. taxifolia*, it is likely that *T. taxifolia* litter dampened the intensity and extent of surface fires that spread from the flammable pine-oak savannas upslope [31]. Extant *T. taxifolia* contributes little to the litter fuels because it has declined in both abundance and stature, with the average tree height near only 1 m [14, T.S. unpublished manuscript]. As species such as *I. opaca, F. grandifolia,* and *M. grandiflora* have replaced forest area formerly occupied by *T. taxifolia* (Table 3) [19], the flammability and therefore susceptibility of the ravines to fire has increased. The higher flammability of litter of these successor species may promote the spread and increased intensity of fires within the ravines, further suppressing the recovery of *T. taxifolia* in a negative feedback. Similar fire feedbacks have been found in transitional forests of the tropics [12,13] and savannas in the temperate zone [11] where these feedbacks can act as stabilizing (negative feedback) or destabilizing (positive feedback) forces. In the ravines, the shift from non-flammable to flammable species has apparently resulted in a negative fire feedback that may further destabilize these formerly fire-safe sites.

Not only have more flammable species replaced *T. taxifolia* within the ravines, but also the flammability of replacement species is somewhat comparable to pyric species in adjacent *Pinus palustris* [3] and *Quercus* spp. [5] uplands. These replacement species burned with flame heights comparable to the upland *Quercus* (*Q. laevis*, *Q. falcata*, *Q. margaretta*, and *Q. stellata*) and *Pinus palustris* [3,5; Table 1]. Among increasing species, *L. styraciflua*, *I. opaca*, and *M. grandiflora* also dry rapidly [7], enabling them to burn under the mesic conditions that characterize the Apalachicola ravines. The flammability of these species shows that, given favorable fire weather, the spread of fires in contemporary communities are more probable than with the low-flammability litter that formerly prevailed in the ravines. Inferences from our results may be limited since the relationship examined here is on the flammability of litter and not necessarily related to species abundance *per se*, but such data cannot be obtained for pre-decline ravine forests. Furthermore, previous research demonstrates that often the most flammable species determines the overall flammability of a given fuel mixture [32].

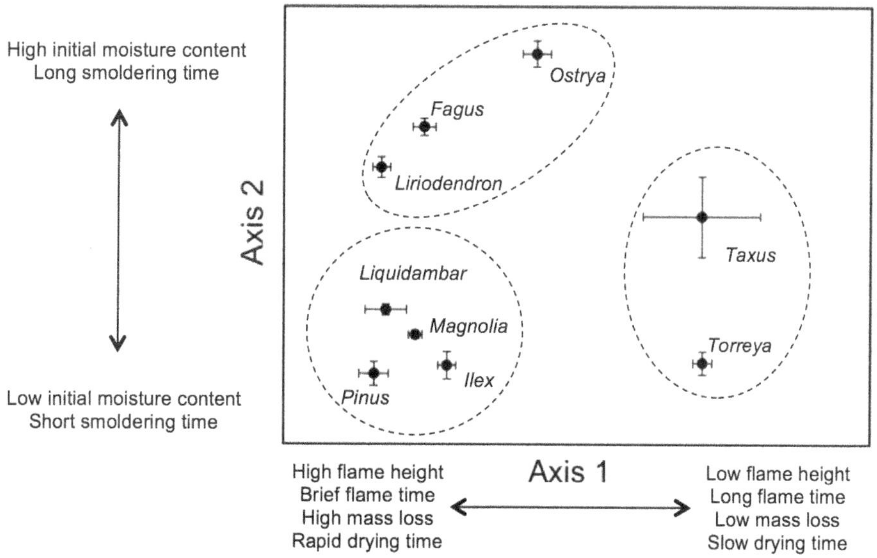

Figure 3. Results of principal components analysis using burning characteristics and drying characteristics for litter of nine woody plant taxa found in the Apalachicola ravines, Florida. The average PCA score (±1.0 S.E.) is shown for each taxa. Circles around taxa are groupings determined using *k*-means clustering analysis.

Table 3. Historic and post-decline dominant tree densities, sorted by percent change in density, and their maximum flame height as determined from our study.

Species	Historic density[a]	Post-decline density[b]	% Density change[c]	Max flame height (cm)
Fagus grandiflora	15.3%	25.9%	+69.2%	81.0
Ilex opaca	13.1%	18.5%	+41.2%	68.6
Magnolia grandiflora	35.5%	43.5%	+22.5%	76.4
Pinus glabra	20.9%	12.0%	−42.6%	78.3
Torreya taxifolia	14.9%	<0.001%	∼−100%	26.2

[a]From Harper (1914).
[b]From Schwartz (2000), only trees with a diameter >15 cm are included.
[c]Caution should be used in interpreting these results due to the different sampling methods.

As a result, the replacement of *T. taxifolia* with more flammable species could have a disproportionate role in increasing community flammability. As such, our results highlight the importance of understanding how declines in dominant species can shift community flammability.

The mechanisms for the shifts in community flammability are unclear. Variation among species, even those phylogenetically similar, has been found elsewhere [3,33,34]. In several species, burning and drying rates were linked to leaf traits (long leaves that curled) that generated porous, well-ventilated fuels [7,33,34]. In other genera, chemical differences (terpene content and identity) drove interspecific differences in burning [35]. The species that have replaced *T. taxifolia* and the similarly flammable *T. floridana* have different leaf morphologies and presumably divergent chemical compositions. Future research should evaluate traits responsible for interspecific differences in flammability in these species and in other ecosystems where species and resulting fire regimes may be changing.

While it is evident that upland fires creeping into the ravines is not the original cause of *T. taxifolia* decline, our results demonstrate that accounting for the increased fire susceptibility of the ravines will be necessary in parallel with reestablishment efforts. If *T. taxifolia* populations can be reestablished in the wild, understanding the flammability of the ravine community will be crucial in order to sustain those populations until the fuels of this ecosystem recover their former fire-suppressing characteristics. The increased prevalence of fire in the ravines may further imperil *T. taxifolia*, *T. floridana* and other rare ravine taxa (Appendix S2). Some evidence shows that fire spread was impeded in ∼25% of *T. taxifolia*'s range over the past few decades (M. Schwartz, personal communication). In these cases, fire was apparently impeded due to previous upslope management, which included the creation of windrows and the establishment of fire-intolerant species upslope (T. S., personal observations). Upslope and ecotonal areas that were not so heavily altered by such ground disturbance and recent upslope restoration efforts may provide the bridge for recent fires to spread into the ravines (J. McKenzie, personal communication, D. Prentiss, personal communication) and supports our concern that future fires may complicate *T. taxifolia* population recovery.

The Apalachicola ravines are a biodiversity hotspot especially for disjunct and rare flora [36]; ∼48 rare plant species occur in the ravines, of which 41 are considered endangered and 6 threatened in Florida (Appendix S2). Many of the rare taxa found in the ravines depend on the cool, stable temperatures, moist soils, and humid microclimates [16,31]. In addition to the general fire sensitivity of these taxa, fire regime changes in the ravines could alter the microclimate by opening the canopy, increasing winds and decreasing humidity potentially having cascading effects on the rare ravine taxa.

As individual species decline due to habitat loss, the introduction of invasive species, and the spread of emerging pathogens, examples of the negative feedback phenomenon we describe may become more common. For example, the functional extinction of the flammable tanoak (*Notholithocarpus densiflorus*) in coastal California and Oregon forests [37] or the loss of the non-flammable, foundation tree eastern hemlock (*Tsuga canadensis*) in eastern US forests [38] may portend shifts in flammability and future fire regimes for remnant communities. Understanding the nuances of individual species interactions with ecosystem processes will be important; replacing species may result in negative, positive, or neutral feedbacks that have the potential to have cascading effects on surviving individuals or on other remnant species [37]. As ecological changes transpire in other ecosystems [39], our task will be to determine where potential fire regime shifts may occur and how, or if, we can intervene.

Supporting Information

Appendix S1 Data used for analysis.

Appendix S2 Rare plant taxa found in the steephead ravines of the Apalachicola River watershed. Conservation status is shown for each taxa (E = Endangered, T = Threatened). FNAI rank refers to the ranking system used by Florida Natural Areas Inventory (FNAI 2010).

Acknowledgments

Florida Division of Recreation and Parks provided access and permits for collecting litter. Laboratory assistance was provided by K. Williams and J. Garcia and help with analyses was provided by M. DeSiervo and J. Kreye. Discussions and comments from E. Engber, M. Schwartz, and S.F. Craig improved the manuscript.

Author Contributions

Conceived and designed the experiments: JM JMV TS. Performed the experiments: JM. Analyzed the data: ESJ JMV JM. Contributed reagents/materials/analysis tools: JMV. Contributed to the writing of the manuscript: JM JMV ESJ TS. Directed the collection of leaf litter: TS.

References

1. Bond WJ, van Wilgen BW. (1996) Fire and Plants. Springer.
2. Mutch RW. (1970) Wildland fires and ecosystems–a hypothesis. Ecology 51: 1046–1051.
3. Fonda RW. (2001) Burning characteristics of needles from eight pine species. Forest Science 47: 390–396.
4. Gagnon PR, Passmore HA, Platt WJ, Myers JA, Paine CET, et al. (2010) Does pyrogenicity protect burning plants? Ecology 91: 3481–3486.
5. Kane JM, Varner JM, Hiers JK. (2008) The burning characteristics of southeastern oaks: discriminating fire facilitators from fire impeders. Forest Ecology and Management 256: 2039–2045.
6. Ellair DP, Platt WJ. (2013) Fuel composition influences fire characteristics and understorey hardwoods in pine savanna. Journal of Ecology 101: 192–201.
7. Kreye JK, Varner JM, Hiers JK, Mola JM. (2013) Toward a mechanism for eastern North American forest mesophication: the role of litter drying. Ecological Applications 23: 1976–1986.
8. Anderson HE. (1970) Forest fuel ignitibility. Fire Technology 6: 312–319.
9. Schwilk DW, Caprio AC. (2011) Scaling from leaf traits to fire behaviour: community composition predicts fire severity in a temperate forest. Journal of Ecology 99: 970–980.
10. Nowacki GJ, Abrams MD. (2008) The demise of fire and "mesophication" of forests in the eastern United States. BioScience 58: 123–138.
11. Beckage B, Platt WJ, Gross LJ. (2009) Vegetation, fire, and feedbacks: A disturbance-mediated model of savannas. The American Naturalist 174: 805–818.
12. Cochrane MA, Alencar A, Schulze MD, Souza CM, Nepstad DC, et al. (1999) Positive feedbacks in the fire dynamic of closed canopy tropical forests. Science 284: 1832–1835.
13. Balch JK, Nepstad DC, Brando PM, Curran LM, Portela O, et al. (2008) Negative fire feedback in a transitional forest of southeastern Amazonia. Global Change Biology 14: 2276–2287.
14. Schwartz MW, Hermann SM. (1993) The continuing population decline of *Torreya taxifolia* Arn. Bulletin of the Torrey Botanical Club 120: 275–286.
15. Schwartz MW, Hermann SM. (1999) Is slow growth of the endangered *Torreya taxifolia* (Arn.) normal? Journal of the Torrey Botanical Society 126: 307–312.
16. Delcourt HR, Delcourt PA. (1977) Presettlement magnolia-beech climax of the Gulf Coastal Plain: quantitative evidence from the Apalachicola River Bluffs, North-Central Florida. Ecology 58: 1085–1093.
17. Godfrey RK, Kurz H. (1962) The Florida torreya destined for extinction. Science 136: 900–902.
18. Harper RM. (1914) Geography and vegetation of northern Florida. Pages 163–451. Florida Geological Survey, Sixth Annual Report, Tallahassee, FL.
19. Schwartz MW, Hermann SM, van Mantgem PJ. (2000) Population persistence in Florida torreya: comparing modeled projections of a declining coniferous tree. Conservation Biology 14: 1023–1033.
20. US Fish and Wildlife Service (USFWS) (1986) Florida torreya (*Torreya taxifolia*) recovery plan. US Fish and Wildlife Service, Atlanta, GA.
21. Schwartz MW, Hermann SM, Vogel CS. (1995) The catastrophic loss of *Torreya taxifolia*: Assessing environmental induction of disease hypotheses. Ecological Applications 5: 501–516.
22. Smith JA, O'Donnell K, Mount LL, Shin K, Peacock K, et al. (2011) A novel *Fusarium* species causes a canker disease of the critically endangered conifer, *Torreya taxifolia*. Plant Disease 95: 633–639.
23. Chen E, Gerber JF. (1990) Climate. In: Myers, R.L and J.J Ewel, eds. Ecosystems of Florida. University of Central Florida Press, Orlando, FL. 11–34.
24. Spector T. (2009) Florida Park Service management plan for *Torreya taxifolia* and *Taxus floridana*. Florida Park Service, Tallahassee, FL.
25. Glitzenstein JS, Platt WJ, Streng DR. (1995) Effects of fire regime and habitat on tree dynamics in north Florida longleaf pine savannas. Ecological Monographs 65: 441–476.
26. Hiers JK, Walters JW, Mitchell RJ, Varner JM, Conner LM, et al. (2014) Ecological value of retaining pyrophytic oaks in longleaf pine ecosystems. Journal of Wildlife Management 78: 383–393.
27. Hoffmann WA, Geiger EL, Gotsch SG, Rossatto DR, Silva LC, et al. (2012) Ecological thresholds at the savanna-forest boundary: how plant traits, resources and fire govern the distribution of tropical biomes. Ecology Letters 15: 759–768.
28. Robbins LE, Myers RL. (1992) Seasonal effects of prescribed burning in Florida: A review. Tall Timbers Research Station miscellaneous publication No. 8. Tall Timbers Research Station, Tallahassee, FL.
29. Hintze JL, (2006) Number Crunching Statistical Systems Version 2006, Kaysville, UT.
30. R Core Team (2013). R: A language and environment for statistical computing. R Foundation for Statistical Computing, Vienna, Austria.
31. Kwit C, Schwartz MW, Platt WJ, Geaghan JP. (1998) The distribution of tree species in steepheads of the Apalachicola River Bluffs, Florida. Journal of the Torrey Botanical Society 125: 309–318.
32. de Magalhães, RMQ, Schwilk DW. (2012) Leaf traits and litter flammability: Evidence for non-additive mixture effects in a temperate forest. Journal of Ecology 100: 1153–1163.
33. Scarff FR, Westoby M. (2006) Leaf litter flammability in some semi-arid Australian woodlands. Functional Ecology 20: 745–752.
34. Engber EA, Varner JM. (2012) Patterns of flammability of the California oaks: the role of leaf traits. Canadian Journal of Forest Research 42: 1965–1975.
35. Ormeño E, Céspedes B, Sánchez IA, Velasco-García A, Moreno JM, et al. (2009) The relationship between terpenes and flammability of leaf litter. Forest Ecology and Management 257: 471–482.
36. Stein BA, Kutner LS, Adams JS, editors. (2000) Precious Heritage: The Status of Biodiversity in the United States. Oxford University Press, USA, New York.
37. Metz MR, Varner JM, Frangioso KM, Meentemeyer RK, Rizzo DM. (2013) Unexpected redwood mortality from synergies between wildfire and an emerging infectious disease. Ecology 94: 2152–2159.
38. Ellison AM, Bank MS, Clinton BD, Colburn EA, Elliott K, et al. (2005) Loss of foundation species: consequences for the structure and dynamics of forested ecosystems. Frontiers in Ecology and the Environment 3: 479–486.
39. Fisher MC, Henk DA, Briggs CJ, Brownstein JS, Madoff LC, et al. (2012) Emerging fungal threats to animal, plant and ecosystem health. Nature 484: 186–194.

9

Trace Metal Inventories and Lead Isotopic Composition Chronicle a Forest Fire's Remobilization of Industrial Contaminants Deposited in the Angeles National Forest

Kingsley O. Odigie*, A. Russell Flegal

WIGS Laboratory, Environmental Toxicology, University of California Santa Cruz, Santa Cruz, California, United States of America

Abstract

The amounts of labile trace metals: [Co] (3 to 11 µg g^{-1}), [Cu] (15 to 69 µg g^{-1}), [Ni] (6 to 15 µg g^{-1}), [Pb] (7 to 42 µg g^{-1}), and [Zn] (65 to 500 µg g^{-1}) in ash collected from the 2012 Williams Fire in Los Angeles, California attest to the role of fires in remobilizing industrial metals deposited in forests. These remobilized trace metals may be dispersed by winds, increasing human exposures, and they may be deposited in water bodies, increasing exposures in aquatic ecosystems. Correlations between the concentrations of these trace metals, normalized to Fe, in ash from the fire suggest that Co, Cu, and Ni in most of those samples were predominantly from natural sources, whereas Pb and Zn were enriched in some ash samples. The predominantly anthropogenic source of excess Pb in the ash was further demonstrated by its isotopic ratios ($^{208}Pb/^{207}Pb$: $^{206}Pb/^{207}Pb$) that fell between those of natural Pb and leaded gasoline sold in California during the previous century. These analyses substantiate current human and environmental health concerns with the pyrogenic remobilization of toxic metals, which are compounded by projections of increases in the intensity and frequency of wildfires associated with climate change.

Editor: Ben Bond-Lamberty, DOE Pacific Northwest National Laboratory, United States of America

Funding: This work was supported by Flegal Research Fund and the Department of Energy (DOE) Office of Science Graduate Fellowship Program. All opinions expressed in this paper are the authors' and do not necessarily reflect the policies and views of DOE Office of Science. The funders had no role in study design, data collection and analysis, decision to publish, or preparation of the manuscript.

Competing Interests: The authors have declared that no competing interests exist.

* Email: kodigie@ucsc.edu

Introduction

Historic Emissions of Trace Metals

Extensive anthropogenic emissions of environmentally persistent contaminants have substantially altered the natural biogeochemical cycles of some trace metals in the biosphere over the past century [1–3]. For example, ~90% and 75% of atmospheric lead (Pb) in California and the world, respectively, during the second half of the previous century were attributed to anthropogenic sources, predominantly the combustion of leaded gasoline [1].

These industrial emissions extensively contaminated aerosols and sediments in Southern California, where enormous amounts of leaded gasoline were combusted during the previous century. For example, ~8.9% (412,300 metric tons) of the estimated 4,639,000 metric tons of Pb additives used in gasoline in the US between 1950 and 1982 was in California [4]. Subsequently, Pb concentrations in surface soil in Los Angeles Metropolitan area increased from 16±0.5 µg/g (mean ± SD) between 1919 and 1933 to 79±23 µg/g between 1967 and 1970 [5]; and the flux of Pb to sediments in the coastal San Pedro Basin, which abuts the Los Angeles Metropolitan area, increased by ~7-fold, compared to the natural rate, following the introduction of alkyl-lead gasoline additives in Southern California [6]. Similar temporal increases in fluxes of Pb in age-dated sediments that were attributed to the use of leaded gasoline have been documented in adjacent areas both north and south of Southern California, albeit at lower levels [7,8]. Substantial temporal increases in the fluxes of other trace metals, e.g., zinc (Zn), associated with anthropogenic inputs have also been measured in Southern California [9].

However, efforts to reduce industrial metal contamination, especially Pb contamination, over the past half century have been quite successful in much of the world, notably the US and specifically Los Angeles. For example, the burden of emissions from the past combustion of leaded gasoline on humans is shown by the relatively high average blood Pb levels (PbB) of children (ages 1–5 years) in the US, which declined from 15 µg/dL in 1976–1980 to 3.6 µg/dL in 1988–1991, during the phase-out and subsequent elimination of leaded gasoline in the US, and then to the current level of 1.3 µg/dL [10,11]. Despite these marked improvements, the current average PbB level in the US is still about 80-fold greater than the estimated natural PbB (0.016 µg/dL) level in humans [12] and health impairments are now being reported at Pb levels that were previously considered harmless [11,13]. Similarly, sublethal toxicities of other trace metals (e.g., Cu and Zn) continue to be documented at increasingly lower levels [14].

Moreover, the legacy of historic anthropogenic Pb emissions is evident in terrestrial and aquatic sediments in California and elsewhere [5,7]. Ten years after Pb was phased out from gasoline in California, concentrations of Pb (77±19 µg/g) in soil in Pasadena, California were still 6-fold greater than the natural (baseline) Pb level (~12.5 µg/g) of crustal rock [5]. This persistence of industrial Pb deposits was summarized by Steding

Figure 1. Photograph of the 2012 Williams Fire in Angeles National Forest, California showing the relatively remote and mountainous terrain where the fire occurred. Photo by Freddie W. Duncan, US Forest Service.

et al. [15] and Dunlap et al. [1], who predicted that it could take decades to centuries to flush leaded gasoline depositions from surface sediments in California's Central Valley to San Francisco Bay. Similarly, Semlali et al. [16], based on a study in France, estimated that it would take at least 700 years to reduce the amount of anthropogenic Pb accumulated in soil by 50%.

Remobilization of Trace Metals by Forest Fires

Wildfires remobilize legacy trace metal contaminants deposited in grasslands and forests. Metals are sequestered in sediment, soil organic matter, and vegetation in forests, where they are relatively immobile [17–19]. However, the combustion of vegetation and soil organic matter releases these trace metals in more labile forms [17,20]. In addition, forest fires increase erosion rates by as much as 100-fold and facilitate the rapid transport of remobilized trace metals to draining water bodies, raising contamination levels in aquatic ecosystems [21,22].

While there have been several investigations of the remobilization of mercury by forest fires [23,24], there have been few similar studies on the remobilization of Pb and other trace metals (e.g., Co and Zn). The latter include a recent study in the relatively unpopulated and relatively pristine Santa Barbara area north of Los Angeles [25] and a similar study in Australia [26]. In contrast to those studies, this study was conducted in the Angeles National Forest, which is adjacent to the Los Angeles Metropolitan area, which has a population of ~17 million [27], has many large sources of industrial contaminants [28–30], and ranks as one of the nation's most contaminated metropolitan areas [31]. Moreover, "car is king" in Los Angeles, where leaded gasoline emissions were the dominant source of atmospheric Pb during the previous century and resuspension of that legacy pollution continues to be the principal component of atmospheric Pb today, as documented in mass balance study of atmospheric Pb contamination in the Los Angeles Air Basin [5]. Therefore, we hypothesized that forest fires within that basin would remobilize relatively large amounts of industrial Pb and other industrial metals, and that remobilization would add to the already large amounts of legacy gasoline Pb deposits that now constitute the major source of atmospheric Pb in Los Angeles.

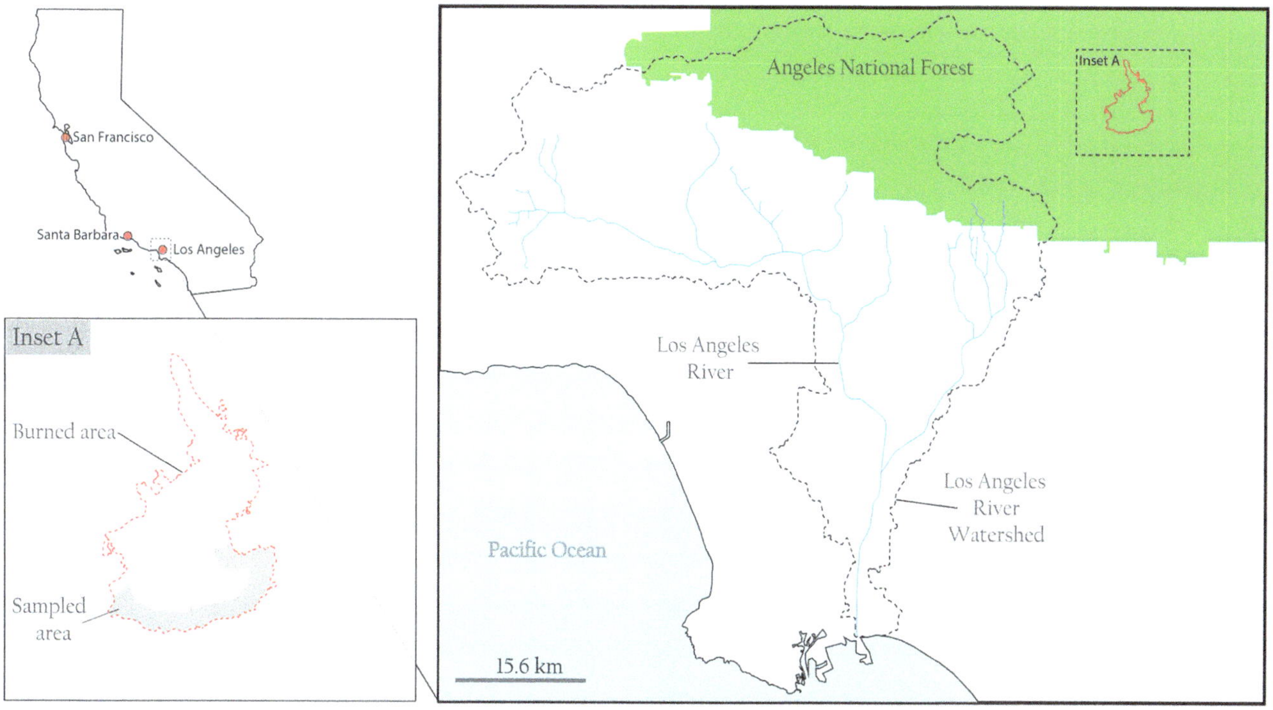

Figure 2. Map showing approximate samples collection sites within the boundary of the 2012 Williams Fire in Angeles National Forest, California.

Table 1. Detection limits of analytes.

	Detection Limit (3 x SD, Standard Deviation of 3 blanks)						
	(ng/g)					(µg/g)	
Element	[Co]	[Cu]	[Ni]	[Pb]	[Zn]	[Al]	[Fe]
Detection Limit	0.002	0.010	0.002	0.008	0.148	0.015	0.035

Table 2. Digestion recoveries (concentrations) for elements in standard reference materials (SRMs).

	% Recovery (Mean ± SD, n = 3)						
SRM	**Co**	**Cu**	**Ni**	**Pb**	**Zn**	**Al**	**Fe**
NIST SRM 1547 (Peach Leaves)	82.6±3.0	89.3±0.7	74.5±2.8	104±0.8	98.7±1.2	65.7±5.2	86.7±4.7
NIST SRM 2709 (San Joaquin Soil)	52.0±0.4	49.1±0.6	49.1±0.4	77.5±0.7	53.2±0.7	55.6±1.1	93.7±0.5
NIST SRM 1633b (Coal Fly Ash)	43.7±0.7	38.6±0.5	42.6±0.6	62.5±0.3	41.2±0.0	25.6±0.8	76.7±1.9

Table 3. Analytical (instrumental) precision.

SRM	% RSD (n = 5)					% RSD (n = 4)	
	[Co]	[Cu]	[Ni]	[Pb]	[Zn]	[Al]	[Fe]
SRM 1640a	0.9	1.1	0.9	0.5	1.6	-	-
Consistency Standard A	2.3	1.6	2.3	0.2	0.9	0.3	0.5
Consistency Standard B	-	-	-	-	-	0.3	0.6

Materials and Methods

Samples Collection Sites

The Angeles National Forest, which covers ~ 283,280 hectares (700,000 acres), abuts the Los Angeles Metropolitan Area [32]. The September 2012 Williams Fire scorched about 1,696 hectares (4,192 acres) of land in Angeles National Forest from approximately September 2nd to 13th (Fig. 1). The fire occurred in a very steep (~ 30–80% slope) area of the forest that is covered with medium to heavy brush. The fuels, mostly chaparral, brush, and mixed conifers, were approximately 15 to 20 years old [33]. Fourteen ash and 6 soil (from unburned patches) samples were collected from different accessible locations within the burned perimeter of the fire site (Figs. 1 and 2) using established trace metal clean techniques [25]. The samples were then transported in plastic containers to the University of California, Santa Cruz for processing. The collection location (GPS coordinates) for each sample is presented in Table S1 in File S1. No specific permissions were required to access and collect samples from the fire site, and the sampling did not involve interactions with endangered or protected species.

Analytical Methods

Trace metal or ultra-pure (2x sub-boiling quartz distilled) grade reagents and high purity (18.2 MΩ cm) water (Milli-Q) were used for cleaning labware and processing samples. Fourteen ash and 6 soil samples, 3 procedural blanks, and triplicates of 3 National Institute of Standards and Technology (NIST) Standard Reference Materials (SRMs), SRM 1547 (peach leaves), SRM 2709 (San Joaquin soil), and SRM 1633b (coal fly ash), were processed concurrently in a Class-100 HEPA-filtered trace metal clean laboratory. Three ash samples (CWA-1, CWA-8, and CWA-14) and 2 soil samples (CWS-1 and CWS-6) were processed in triplicate.

A concentrated aqua regia (HNO_3: HCl, 1:3 v/v) digestion was used to extract the acid-leachable (labile) trace (Co, Cu, Ni, Pb, and Zn) and major (Al and Fe) metals in the samples and reference materials [34,35] using established methods [8]. These metals were selected to represent a suite of inorganic contaminants emitted from anthropogenic sources at different – from relatively low (e.g., Ni) to high (e.g., Pb) – levels in Los Angeles [29,36,37]. Briefly, approximately 0.3 to 0.5 g of ash, soil or reference material was transferred to a clean Teflon digestion vial, dried overnight at 65°C, refluxed in hot (~130°C) aqua regia for about 6 hours, dried, weighed and redissolved in 10 mL of 1M ultrapure HNO_3.

The digests were analyzed for trace metal (Co, Cu, Ni, Pb, and Zn) concentrations with a Finnigan ELEMENT 2 high resolution inductively coupled plasma – mass spectrometer (HR ICP-MS), and for Al and Fe concentrations with a PerkinElmer Optima 4300 DV inductively coupled plasma – optical emission spectrometer (ICP-OES). The concentrations of Fe were used to compute the enrichment factors of trace metals in the samples. The samples were then analyzed for stable lead isotopic composition after optimizing their concentrations with a Finnigan ELEMENT 2 HR ICP-MS. Internal standards were used to correct for instrumental drifts during analysis, and NIST SRM 981 (common lead) was used as standard to correct for instrumental fractionation of lead isotopic composition. The detection limit (3 x standard deviation of procedural blanks), reproducibility, and recoveries of each analyte are presented in Tables 1–4. The recoveries of metals in the NIST SRM 1547 (peach leaves) are assumed to be most representative of the metals

Table 4. Analytical (instrumental) precision for lead isotopic compositions.

	% RSD (n = 5)		
Sample	$^{208}Pb/^{207}Pb$	$^{206}Pb/^{207}Pb$	$^{207}Pb/^{204}Pb$
CWA-2	0.07	0.04	0.09

in the ash samples since they were derived predominantly from vegetation and soil organic matter.

Results and Discussion

Metal Concentrations

The "acid-leachable" concentrations of Al, Co, Cu, Fe, Ni, Pb, and Zn, and lead isotopic ratios ($^{208}Pb/^{207}Pb$, $^{206}Pb/^{207}Pb$, and $^{207}Pb/^{204}Pb$) of the ash and soil samples are listed in Table 5. The enrichment factors of the trace metals (Co, Cu, Ni, Pb, and Zn) normalized to Fe are listed in Table S2 in File S1. These concentrations and enrichment factors are considered conservative because they may not include metals volatilized during the fire and those contained in refractory aluminosilicate lattices, which require dissolution with concentrated HF [38,39].

Concentrations of Co (3.1 to 11 $\mu g\ g^{-1}$), Cu (15 to 69 $\mu g\ g^{-1}$), Ni (5.9 to 15 $\mu g\ g^{-1}$), Pb (7.4 to 42 $\mu g\ g^{-1}$), and Zn (65 to 500 $\mu g\ g^{-1}$) in the ash varied substantially. The ranges of these concentrations are consistent with those measured in ash collected from the 2009 Jesusita Forest Fire site in Santa Barbara, California (Table 5). The concentrations are also comparable to concentrations of these metals in ash sourced primarily from the experimental combustion of ponderosa pines in Montana [40]. The relatively high Zn levels potentially originated from anthropogenic sources, which have typically included fossil fuel combustion exhaust, galvanized parts and railings, break lining wear, and tire wear [41,42].

Variations in the amounts of these remobilized metals presumably reflect variations in baseline concentrations of the metals, anthropogenic input, and levels of volatilization of organic matter during the fire [43,44]. The concentrations of Co (6.3 to 13 $\mu g\ g^{-1}$), Cu (17 to 32 $\mu g\ g^{-1}$), Ni (6.0 to 21 $\mu g\ g^{-1}$), Pb (8.6 to 140 $\mu g\ g^{-1}$), and Zn (32 to 61 $\mu g\ g^{-1}$) in the soil samples are also comparable to previously reported values of soils in Southern California [45–47]. The concentrations of some metals (e.g., Co and Ni) in the soils approximate their baseline concentrations in Southern California, but the >2-fold increase in Pb concentrations (e.g., 140 $\mu g\ g^{-1}$) indicated that the Angeles National Forest site had received substantial Pb inputs from anthropogenic sources.

Enrichment Factors

The amounts of anthropogenic trace metals in the ash are indicated by the enrichment factors of the metals normalized to Fe: Co (0.9 to 2.5), Cu (0.9 to 6.2), Ni (0.3 to 1.0), Pb (1.9 to 29), and Zn (0.9 to 19) (Table S2 in File S1).

$$EF = \frac{\frac{[X]Sample}{[Fe]Sample}}{\frac{[X]Baseline}{[Fe]Baseline}}$$

We calculated the enrichment factor of each metal by normalizing the concentrations of the metal ($[X]_{Sample}$) and iron ($[Fe]_{Sample}$) in the sample to the baseline concentrations of each metal ($[X]_{Baseline}$) and iron ($[Fe]_{Baseline}$) of ^{210}Pb-dated preindustrial sediment deposited in San Pedro Basin, which drains parts of the Los Angeles Metropolitan Area. Normalization of trace metals to Fe is considered more appropriate than normalization to Al for samples that were not treated with HF because of the recalcitrance of aluminosilicates as indicated by the difference in the recoveries of these metals in the NIST SRMs (Table 2). Using the enrichment classification proposed by Sutherland [48], the ratios indicate that: Co was not measurably enriched (EF <2) in 10 samples and moderately enriched (EF = 2–5) in 4 samples; Cu was not measurably enriched in 5 samples, moderately enriched in 7 samples, and significantly enriched (EF = 5–20) in 2 samples; Ni was not measurably enriched in all 14 ash samples; Pb was not measurably enriched in 1 sample, moderately enriched in 7 samples, significantly enriched in 5 samples, and very highly enriched (EF = 20–40) in 1 sample; and Zn was not measurably enriched in 4 samples, moderately enriched in 5 samples, and significantly enriched in 5 samples. These industrial enrichments were further indicated by the simple linear correlations between the concentrations of the metals and the concentrations of Fe: [Co] ($r = 0.576$, $p = 0.031$, $n = 14$), [Cu] ($r = 0.058$, $p = 0.844$, $n = 14$), [Ni] ($r = 0.612$, $p = 0.020$, $n = 14$), [Pb] ($r = -0.158$, $p = 0.590$, $n = 14$), and [Zn] ($r = -0.221$, $p = 0.448$, $n = 14$) in the ash. Metals that are primarily from natural origin, e.g., [Al] and [Fe] ($r = 0.909$, $p = 0.000007$, $n = 14$), which are relatively abundant and not easy to contaminate, correlate better than those enriched by anthropogenic inputs [49]. The addition of anthropogenic metal, e.g., Pb, to a system often changes its relative abundance to normalizing major elements as indicated by the correlation of [Pb] and [Fe] above. These varying enrichments are also generally consistent with those of these metals in ash collected from the 2009 Jesusita Fire in Santa Barbara, California [e.g., 25].

Lead Isotopic Compositions

The apparent anthropogenic enrichment of Pb in some of the samples is further demonstrated in Figure 3, which is a plot of the lead isotopic ratios ($^{208}Pb/^{207}Pb$: $^{206}Pb/^{207}Pb$) of the ash and soil samples. The plot also includes the lead isotopic compositions of pre-industrial and contaminated sediment deposited in the San Pedro Basin, which drains parts of the Los Angeles Metropolitan area as previously mentioned, and those of leaded gasoline sold in California during the second half of the last century.

The plot indicates that the Pb in the ash is mainly from anthropogenic sources, predominantly the past combustion of leaded gasoline. The plot also indicates that the portion of anthropogenic Pb from the leaded gasoline used between 1964 to 1979 was substantially larger than that from the 1980s leaded gasoline in the ash, which contrasts the results from the Santa Barbara study that indicate that the predominant source of anthropogenic Pb was the combustion of the 1980s leaded gasoline [25]. The predominantly alkyl-lead in the samples is further corroborated by the isotopic compositions of Pb in most (4) of the soil samples (Fig. 3). Two soil samples (CWS-4 and CWS-5) have distinct Pb isotopic compositions compared to the other samples and are assumed to be representative of mixtures between baseline and anthropogenic Pb or variants of baseline Pb isotopic compositions.

Table 5. Concentrations (μg g^{-1}) of trace metals: Co, Cu, Ni, Pb, and Zn; concentrations (mg g^{-1}) of Al and Fe; and lead isotopic composition of ash (CWA) and soil (CWS) samples collected from the 2012 Williams Fire in Angeles National Forest in Southern California.

Sample ID	Concentrations [mean ± *(standard deviation, SD)*]							Lead Isotopic Ratios [mean ± *(SD)*]		
	[Co] (μg g^{-1})	[Cu] (μg g^{-1})	[Ni] (μg g^{-1})	[Pb] (μg g^{-1})	[Zn] (μg g^{-1})	[Al] (mg g^{-1})	[Fe] (mg g^{-1})	$^{208}Pb/^{207}Pb$	$^{206}Pb/^{207}Pb$	$^{207}Pb/^{204}Pb$
CWA-1*	3.90	15.4	6.77	16.0	216	13.8	10.9	2.454	1.189	15.673
	(0.05)	*(0.24)*	*(0.25)*	*(0.04)*	*(2.92)*	*(0.55)*	*(0.29)*	*(0.001)*	*(0.001)*	*(0.004)*
CWA-2	7.54	68.5	10.6	24.0	128	18.4	25.0	2.461	1.185	15.635
CWA-3	3.59	41.3	9.21	42.2	83.0	5.79	6.94	2.438	1.178	15.656
CWA-4	6.76	30.8	10.6	14.8	85.0	18.7	24.4	2.457	1.190	15.633
CWA-5	3.81	38.6	9.72	20.8	65.3	7.86	10.8	2.454	1.192	15.659
CWA-6	8.75	56.0	12.5	27.2	414	13.6	16.2	2.463	1.213	15.684
CWA-7	5.17	65.5	14.3	38.5	203	14.3	13.8	2.451	1.195	15.711
CWA-8*	10.9	58.2	13.3	8.95	374	13.0	17.0	2.457	1.193	15.660
	(0.74)	*(3.24)*	*(0.52)*	*(1.02)*	*(15.7)*	*(0.36)*	*(0.55)*	*(0.001)*	*(0.002)*	*(0.011)*
CWA-9	4.32	17.3	10.5	9.68	96.5	14.3	16.2	2.471	1.186	15.662
CWA-10	7.47	26.3	14.6	20.0	73.6	24.4	31.5	2.458	1.187	15.672
CWA-11	3.13	17.0	5.86	8.23	74.2	7.42	8.71	2.450	1.189	15.733
CWA-12	6.05	20.5	11.2	17.3	65.5	18.6	18.1	2.462	1.183	15.669
CWA-13	6.30	56.8	9.17	11.5	503	7.85	10.7	2.472	1.188	15.664
CWA-14*	6.66	43.3	12.9	7.38	66.6	20.7	19.2	2.450	1.189	15.686
	(0.27)	*(0.23)*	*(0.15)*	*(0.25)*	*(0.68)*	*(0.53)*	*(0.39)*	*(0.001)*	*(0.001)*	*(0.017)*
CWS-1*	7.31	20.0	6.04	20.7	44.5	29.2	52.4	2.459	1.181	15.635
	(0.10)	*(3.89)*	*(0.14)*	*(0.76)*	*(0.46)*	*(2.11)*	*(1.84)*	*(0.002)*	*(0.000)*	*(0.011)*
CWS-2	8.65	19.6	21.1	143	61.2	35.1	46.8	2.461	1.204	15.647
CWS-3	12.5	32.1	16.8	20.3	41.5	32.1	43.5	2.455	1.178	15.614
CWS-4	6.26	25.9	11.8	8.62	32.0	20.7	28.1	2.506	1.196	15.684
CWS-5	8.82	17.0	15.0	10.9	37.3	34.9	40.1	2.491	1.185	15.643
CWS-6*	9.98	20.0	13.1	16.4	44.2	33.5	46.4	2.443	1.160	15.581
	(0.67)	*(0.89)*	*(1.21)*	*(1.81)*	*(0.65)*	*(5.00)*	*(1.37)*	*(0.001)*	*(0.006)*	*(0.024)*
Jesusita Fire ash, Santa Barbara, California (Data Summary, n = 30)										
Mean, [x̄]	4.46	–	22.7	18.8‡	151	16.7‡	13.9‡	2.466‡	1.205‡	15.635‡
Range (Min	1.95	–	8.88	4.29‡	41.6	5.70‡	5.00‡	2.436‡	1.161‡	15.563‡
to	–	–	–	–	–	–	–	–	–	–
Max)	7.23	–	49.1	51.0‡	433	36.0‡	31.1‡	2.481‡	1.218‡	15.716‡

Samples marked with (*) were processed in triplicates and are reported as mean ± (standard deviation, SD). Also listed in the table are the ranges and mean concentrations of metals and lead isotopic compositions of ash collected from the 2009 Jesusita Fire in Santa Barbara, California. Data marked with (‡) were summarized from Odigie and Flegal [25].

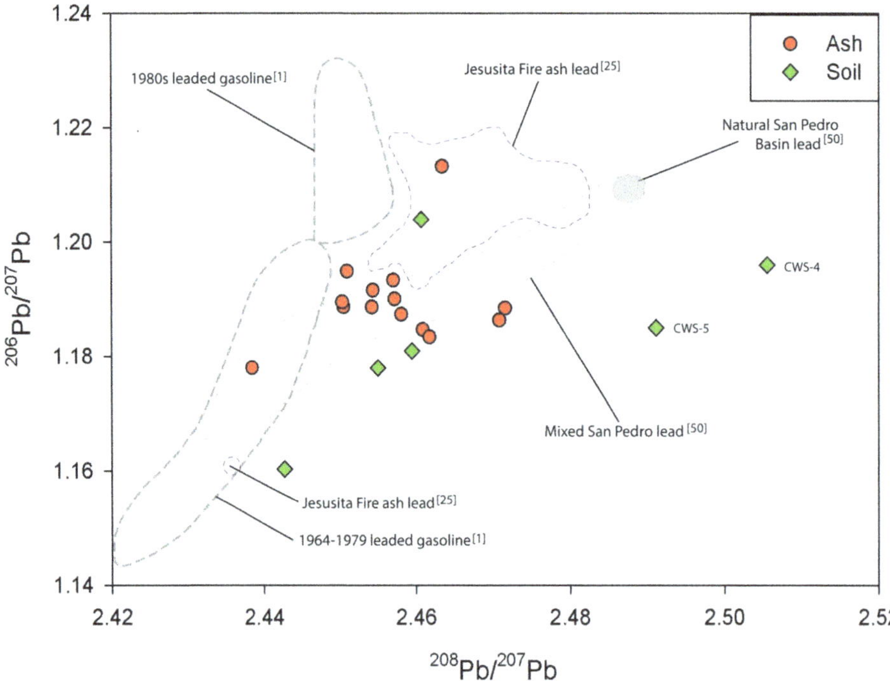

Figure 3. Lead isotopic composition ($^{208}Pb/^{207}Pb$: $^{206}Pb/^{207}Pb$) of ash and soil samples collected from the 2012 Williams Fire site in Southern California. The plot also includes the ranges of isotopic composition of alkyl-lead used in California and those of ^{210}Pb dated sediment cores in the San Pedro Basin. Also included in the plot are the ranges of Pb isotopic composition of ash collected from the 2009 Jesusita Fire in Santa Barbara, California.

The apparent leaded gasoline signature of Pb in the samples attests to the persistence of industrial Pb deposits in the Los Angeles Basin that is consistent with the results of the mass balance study by Harris and Davidson [5]. It is also consistent with previous reports on the persistence of industrial Pb deposits in California's Central Valley, which is north of Los Angeles, based on both mass balance and lead isotopic composition analyses [1,15]. The latter has been principally traced to atmospheric depositions from the combustion of leaded gasoline in California [1].

This attribution is based on the temporal changes in the isotopic compositions of leaded gasoline in California in the previous century. Before 1960, alkyl-lead in the US was primarily manufactured with ores from Australia, Canada, Mexico, and Peru, which had $^{206}Pb/^{207}Pb$ ratios of 1.037, 1.064, 1.192, and 1.200, respectively. Then, the proportion of Pb sourced from the Missouri deposits ($^{206}Pb/^{207}Pb = 1.28$ to 1.33) increased from 9% in 1962 to 82% in 1976 [50]. This change in the sources of Pb in alkyl-lead was subsequently reflected in the Pb isotopic compositions of aerosols in Southern California. There, the $^{206}Pb/^{207}Pb$ ratio of aerosols, which was ~1.15 before 1967, increased to 1.20 by 1974, and then to 1.23 by 1977 [51].

Accelerated Transport of Contaminants to Water Bodies and Atmosphere

In addition to atmospheric contamination, wildfires increase the load of contaminants in aquatic ecosystems. This was recently documented in a study [22] that examined the impact of wildfires on contaminant loadings in Southern California – the same area where the Williams Fire occurred. The results from that study show that the mean fluxes of Cu, Pb, Ni, and Zn from burned sites were 110-fold, 740-fold, 82-fold, and 110-fold greater compared to their respective fluxes at similar unburned (control) sites. That study also found that the mean post-fire concentrations of Cu, Pb, Ni, and Zn in the first stormwater collected from an area that was ~20 km from the closest fire site were each 3 times their pre-fire levels, which they attributed to ash fallout. Similarly, Sabin et al. [52] reported 4-fold, 6-fold, 8-fold, and 13-fold post-fire increases in fluxes of Cu, Zn, Pb, and Ni, respectively, over their mean fluxes in an unburned site in Southern California, which were likewise attributed to forest fires in nearby mountains. Those attributions are consistent with results from a study by Young and Jan [53], who showed that trace metals (Cu, Ni, Pb, and Zn) fallout increased by up to 3-fold in a 10,000 km^2 unburned area that was impacted by smoke plumes and ash from wildfires in the Angeles National Forest in Southern California.

Potential Impact of Climate Change

The mobilization of contaminants by wildfires is a concern, especially in the western US where climate change is expected to dramatically expand fire regimes [54]. Recent studies have linked increases in fire intensity, frequency, and burned area to climate change [55–57]. For example, wildfire frequency has been positively linked ($r = 0.76$, $p < 0.001$, $n = 34$) to regional spring and summer temperatures in western US [54]. The frequency of wildfires and total area burned in western US in the mid-1980s were ~4 times and ~7 times, respectively, higher than their mean values from 1970 to 1986, and this interval was marked by a rise in temperature of <0.9°C [54]. Furthermore, the Intergovernmental Panel on Climate Change has projected that global mean surface temperature could increase by up to 4.8°C by the end of this century (2081–2100) relative to 1986–2005, and researchers have predicted a corresponding increase in the frequency and intensity of wildfires [58–61]. Consequently, it is hypothesized that the pyrogenic remobilization and volatilization of contaminants from forests and grasslands will increase in response to projected climate change. That expected increase is a concern, especially in forests contaminated with radioactive pollutants and at sites used for nuclear waste storage, as recently highlighted for Chernobyl, Ukraine and Los Alamos National Laboratory, US, respectively [62–64]. The results from this and complementary studies provide empirical evidence that wildfires are already remobilizing and volatilizing other contaminants – albeit more pedestrian ones – that constitute a greater global health problem (e.g., Pb).

Supporting Information

File S1 Tables S1–S2. Table S1: Collection locations (coordinates) for ash (CWA) and soil (CWS) samples collected from the 2012 Williams Fire site in the Angeles National Forest in California. Table S2: Enrichment factors (*f*-Fe, normalized to Fe) of trace metals in ash (CWA) samples collected from the 2012 Williams Fire site in the Angeles National Forest in California.

Acknowledgments

Special thanks to Aradhna Tripati and John Mering of UCLA, Martha Thomas, and Priya Ganguli for help with sample collection; Rob Franks of UCSC for analytical support; and all members of the WIGS laboratory, especially Sharon Hibdon, for their support with this project.

Author Contributions

Conceived and designed the experiments: KO ARF. Performed the experiments: KO. Analyzed the data: KO. Contributed reagents/materials/analysis tools: KO ARF. Wrote the paper: KO ARF.

References

1. Dunlap CE, Alpers CN, Bouse R, Taylor HE, Unruh DM, et al. (2008) The persistence of lead from past gasoline emissions and mining drainage in a large riparian system: Evidence from lead isotopes in the Sacramento River, California. Geochim Cosmochim Acta 72: 5935–5948.
2. Nriagu JO (1996) A history of global metal pollution. Science 272: 223–224.
3. Flegal AR, Gallon C, Ganguli PM, Conaway CH (2013) All the Lead in China. Crit Rev Environ Sci Technol 43: 1869–1944.
4. Mielke HW, Laidlaw MAS, Gonzales C (2010) Lead (Pb) legacy from vehicle traffic in eight California urbanized areas: Continuing influence of lead dust on children's health. Sci Total Environ 408: 3965–3975.
5. Harris AR, Davidson CI (2005) The role of resuspended soil in lead flows in the California South Coast Air Basin. Environ Sci Technol 39: 7410–7415.
6. Chow TJ, Bruland KW, Bertine K, Soutar A, Koide M, et al. (1973) Lead pollution: Records in Southern California coastal sediments. Science 181: 551–552.
7. Ritson PI, Bouse RM, Flegal AR, Luoma SN (1999) Stable lead isotopic analyses of historic and contemporary lead contamination of San Francisco Bay estuary. Mar Chem 64: 71–83.
8. Soto-Jimenez MF, Hibdon SA, Rankin CW, Aggarawl J, Ruiz-Fernandez AC, et al. (2006) Chronicling a century of lead pollution in Mexico: Stable lead isotopic composition analyses of dated sediment cores. Environ Sci Technol 40: 764–770.
9. Finney B, Huh C-A (1989) High resolution sedimentary records of heavy metals from the Santa Monica and San Pedro Basins, California. Mar Pollut Bull 20: 181–187.
10. Pirkle JL, Brody DJ, Gunter EW, Kramer RA, Paschal DC, et al. (1994) The decline in blood lead levels in the United States-The National Health and Nutrition Examination Surveys (NHANES). Jama-J Am Med Assoc 272: 284–291.
11. Wheeler W, Brown MJ (2013) Blood lead levels in children aged 1–5 years – United States, 1999–2010. CDC MMWR 62: 245–248.
12. Flegal AR, Smith DR (1992) Lead levels in preindustrial humans. N Engl J Med 326: 1293–1294.
13. WHO (2010) Childhood lead poisoning. Geneva, Switzerland, 74: WHO Document Production Services. 74 p.
14. McGeer JC, Szebedinszky C, McDonald DG, Wood CM (2000) Effects of chronic sublethal exposure to waterborne Cu, Cd or Zn in rainbow trout. 1: Iono-regulatory disturbance and metabolic costs. Aquat Toxicol 50: 231–243.
15. Steding DJ, Dunlap CE, Flegal AR (2000) New isotopic evidence for chronic lead contamination in the San Francisco Bay estuary system: Implications for the persistence of past industrial lead emissions in the biosphere. Proc Natl Acad Sci USA 97: 11181–11186.
16. Semlali RM, Dessogne JB, Monna F, Bolte J, Azimi S, et al. (2004) Modeling lead input and output in soils using lead isotopic geochemistry. Environ Sci Technol 38: 1513–1521.
17. Biswas A, Blum JD, Klaue B, Keeler GJ (2007) Release of mercury from Rocky Mountain forest fires. Global Biogeochem Cycles 21: DOI: 10.1029/2006GB002696.
18. Obrist D, Moosmuller H, Schurmann R, Chen LWA, Kreidenweis SM (2008) Particulate-phase and gaseous elemental mercury emissions during biomass combustion: Controlling factors and correlation with particulate matter emissions. Environ Sci Technol 42: 721–727.
19. Friedli HR, Arellano AF, Cinnirella S, Pirrone N (2009) Initial estimates of mercury emissions to the atmosphere from global biomass burning. Environ Sci Technol 43: 3507–3513.
20. Wiedinmyer C, Friedli H (2007) Mercury emission estimates from fires: An initial inventory for the United States. Environ Sci Technol 41: 8092–8098.
21. Cohen AS, Palacios-Fest MR, McGill J, Swarzenski PW, Verschuren D, et al. (2005) Paleolimnological investigations of anthropogenic environmental change in Lake Tanganyika: I. An introduction to the project. J Paleolimnol 34: 1–18.
22. Stein ED, Brown JS, Hogue TS, Burke MP, Kinoshita A (2012) Stormwater contaminant loading following southern California wildfires. Environ Toxicol Chem 31: 2625–2638.
23. Witt EL, Kolka RK, Nater EA, Wickman TR (2009) Forest fire effects on mercury deposition in the boreal forest. Environ Sci Technol 43: 1776–1782.
24. Biswas A, Blum JD, Keeler GJ (2008) Mercury storage in surface soils in a central Washington forest and estimated release during the 2001 Rex Creek Fire. Sci Tot Environ 404: 129–138.
25. Odigie KO, Flegal AR (2011) Pyrogenic remobilization of historic industrial lead depositions. Environ Sci Technol 45: 6290–6295.
26. Kristensen LJ, Taylor MP, Odigie KO, Hibdon SA, Flegal AR (2014) Lead isotopic compositions of ash sourced from Australian bushfires. Environ Pollut 190: 159–165.
27. US Census (2013) US Census Bureau 2012 population estimate. Available: http://quickfacts.census.gov/qfd/states/06/06037.html. Accessed 2013 May 22.
28. Smail EA, Webb EA, Franks RP, Bruland KW, Sanudo-Wilhelmy SA (2012) Status of metal contamination in surface waters of the coastal ocean off Los Angeles, California since the implementation of the Clean Water Act. Environ Sci Technol 46: 4304–4311.
29. Singh M, Jaques PA, Sioutas C (2002) Size distribution and diurnal characteristics of particle-bound metals in source and receptor sites of the Los Angeles Basin. Atmos Environ 36: 1675–1689.
30. Sardar SB, Fine PM, Sioutas C (2005) Seasonal and spatial variability of the size-resolved chemical composition of particulate matter (PM10) in the Los Angeles Basin. J Geophys Res-Atmos 110: D07S08.
31. Su JG, Jerrett M, Beckerman B, Wilhelm M, Ghosh JK, et al. (2009) Predicting traffic-related air pollution in Los Angeles using a distance decay regression selection strategy. Environ Res 109: 657–670.
32. USDA (2013) Angeles National Forest. Available: http://www.fs.usda.gov/main/angeles/about-forest. Accessed 2013 Feb 19.
33. InciWeb (2012) Camp Williams Fire. Available: http://www.inciweb.org/incident/3230/. Accessed 2013 Feb 19.
34. Kober B, Wessels M, Bollhofer A, Mangini A (1999) Pb isotopes in sediments of Lake Constance, Central Europe constrain the heavy metal pathways and the pollution history of the catchment, the lake and the regional atmosphere. Geochim Cosmochim Acta 63: 1293–1303.
35. Emmanuel S, Erel Y (2002) Implications from concentrations and isotopic data for Pb partitioning processes in soils. Geochim Cosmochim Acta 66: 2517–2527.
36. Bruland KW, Bertine K, Koide M, Goldberg ED (1974) History of metal pollution in Southern California Coastal Zone. Environ Sci Technol 8: 425–432.
37. Hu S, Herner JD, Shafer M, Robertson W, Schauer JJ, et al. (2009) Metals emitted from heavy-duty diesel vehicles equipped with advanced PM and NO_X emission controls. Atmos Environ 43: 2950–2959.
38. Hornberger MI, Luoma SN, van Geen A, Fuller C, Anima R (1999) Historical trends of metals in the sediments of San Francisco Bay, California. Mar Chem 64: 39–55.
39. Schiff KC, Weisberg SB (1999) Iron as a reference element for determining trace metal enrichment in Southern California coastal shelf sediments. Mar Environ Res 48: 161–176.
40. Gabet EJ, Bookter A (2011) Physical, chemical and hydrological properties of Ponderosa pine ash. Int J Wildland Fire 20: 443–452.
41. Councell TB, Duckenfield KU, Landa ER, Callender E (2004) Tire-wear particles as a source of zinc to the environment. Environ Sci Technol 38: 4206–4214.
42. Pakbin P, Ning Z, Shafer MM, Schauer JJ, Sioutas C (2011) Seasonal and Spatial Coarse Particle Elemental Concentrations in the Los Angeles Area. Aerosol Sci Technol 45: 949–U156.

43. Ranali AJ (2004) A summary of the scientific literature on the effects of fire on the concentration of nutrients in surface waters. USGS Open-File Report 2004–1296.
44. Simpson IJ, Akagi SK, Barletta B, Blake NJ, Choi Y, et al. (2011) Boreal forest fire emissions in fresh Canadian smoke plumes: C-1-C-10 volatile organic compounds (VOCs), CO_2, CO, NO_2, NO, HCN and CH_3CN. Atmos Chem Phys 11: 6445–6463.
45. Wu J, Edwards R, He XQ, Liu Z, Kleinman M (2010) Spatial analysis of bioavailable soil lead concentrations in Los Angeles, California. Environ Res 110: 309–317.
46. Holmgren GGS, Meyer MW, Chaney RL, Daniels RB (1993) Cadmium, lead, zinc, copper, and nickel in agricultural soils of the United States of America. J Environ Qual 22: 335–348.
47. Shacklette HT, Boerngen JG (1984) Element concentrations in soils and other surficial materials of the conterminous United States. U.S. Geological Survey Professional Paper 1270.
48. Sutherland RA (2000) Bed sediment-associated trace metals in an urban stream, Oahu, Hawaii. Environ Geol 39: 611–627.
49. Daskalakis KD, O'Connor TP (1995) Normalization and elemental sediment contamination in the Coastal United States. Environ Sci Technol 29: 470–477.
50. Ng A, Patterson CC (1982) Changes of lead and barium with time in California off-shore basin sediments. Geochim Cosmochim Acta 46: 2307–2321.
51. Shirahata H, Elias RW, Patterson CC, Koide M (1980) Chronological variations in concentrations and isotopic compositions of anthropogenic atmospheric lead in sediments of a remote subalpine pond. Geochim Cosmochim Acta 44: 149–162.
52. Sabin LD, Lim JH, Stolzenbach KD, Schiff KC (2005) Contribution of trace metals from atmospheric deposition to stormwater runoff in a small impervious urban catchment. Water Res 39: 3929–3937.
53. Young DR, Jan TK (1977) Fire fallout of metals off California. Mar Pollut Bull 8: 109–112.
54. Westerling AL, Hidalgo HG, Cayan DR, Swetnam TW (2006) Warming and earlier spring increase western US forest wildfire activity. Science 313: 940–943.
55. Gillett NP, Weaver AJ, Zwiers FW, Flannigan MD (2004) Detecting the effect of climate change on Canadian forest fires. Geophys Res Lett 31: DOI: 10.1029/2004GL020876.
56. Westerling AL, Turner MG, Smithwick EAH, Romme WH, Ryan MG (2011) Continued warming could transform Greater Yellowstone fire regimes by mid-21st century. Proc Natl Acad Sci USA 108: 13165–13170.
57. Wang YH, Flannigan M, Anderson K (2010) Correlations between forest fires in British Columbia, Canada, and sea surface temperature of the Pacific Ocean. Ecol Model 221: 122–129.
58. IPCC (2013) Climate Change 2013: The Physical Science Basis: Cambridge Univ. Press, New York.
59. McKenzie D, Gedalof Z, Peterson DL, Mote P (2004) Climatic change, wildfire, and conservation. Conserv Biol 18: 890–902.
60. Schuur EAG, Bockheim J, Canadell JG, Euskirchen E, Field CB, et al. (2008) Vulnerability of permafrost carbon to climate change: Implications for the global carbon cycle. Bioscience 58: 701–714.
61. Flannigan MD, Logan KA, Amiro BD, Skinner WR, Stocks BJ (2005) Future area burned in Canada. Clim Change 72: 1–16.
62. Charles D (2010) Russia's forest fires ignite concerns about Chornobyl's legacy. Science 329: 738–739.
63. Gilbert N (2010) Russia counts environmental cost of wildfires. Nature DOI: 10.1038/news.2010.404.
64. Weinhold B (2011) Fields and forests in flames: Vegetation, smoke & human health. Environ Health Perspect 119: A386–A393.

10

Historical, Observed, and Modeled Wildfire Severity in Montane Forests of the Colorado Front Range

Rosemary L. Sherriff[1]*, Rutherford V. Platt[2], Thomas T. Veblen[3], Tania L. Schoennagel[3], Meredith H. Gartner[3]

1 Department of Geography, Humboldt State University, Arcata, California, United States of America, **2** Department of Environmental Studies, Gettysburg College, Gettysburg, Pennsylvania, United States of America, **3** Department of Geography, University of Colorado, Boulder, Colorado, United States of America

Abstract

Large recent fires in the western U.S. have contributed to a perception that fire exclusion has caused an unprecedented occurrence of uncharacteristically severe fires, particularly in lower elevation dry pine forests. In the absence of long-term fire severity records, it is unknown how short-term trends compare to fire severity prior to 20th century fire exclusion. This study compares historical (i.e. pre-1920) fire severity with observed modern fire severity and modeled potential fire behavior across 564,413 ha of montane forests of the Colorado Front Range. We used forest structure and tree-ring fire history to characterize fire severity at 232 sites and then modeled historical fire-severity across the entire study area using biophysical variables. Eighteen (7.8%) sites were characterized by low-severity fires and 214 (92.2%) by mixed-severity fires (i.e. including moderate- or high-severity fires). Difference in area of historical versus observed low-severity fire within nine recent (post-1999) large fire perimeters was greatest in lower montane forests. Only 16% of the study area recorded a shift from historical low severity to a higher potential for crown fire today. An historical fire regime of more frequent and low-severity fires at low elevations (<2260 m) supports a convergence of management goals of ecological restoration and fire hazard mitigation in those habitats. In contrast, at higher elevations mixed-severity fires were predominant historically and continue to be so today. Thinning treatments at higher elevations of the montane zone will not return the fire regime to an historic low-severity regime, and are of questionable effectiveness in preventing severe wildfires. Based on present-day fuels, predicted fire behavior under extreme fire weather continues to indicate a mixed-severity fire regime throughout most of the montane forest zone. Recent large wildfires in the Front Range are not fundamentally different from similar events that occurred historically under extreme weather conditions.

Editor: Ben Bond-Lamberty, DOE Pacific Northwest National Laboratory, United States of America

Funding: Data collection and analysis were funded by the National Science Foundation (awards BCS-0540928, 0541480, 0802667, 0541594 to TTV, RLS and RVP) and Humboldt State University Sponsored Programs Foundation and College of Arts, Humanities, and Social Sciences (awards to RLS). The funders had no role in the study design, decision to publish, or preparation of the manuscript.

Competing Interests: The authors have declared that no competing interests exist.

* Email: sherriff@humboldt.edu

Introduction

The social, environmental and fiscal costs of wildfire have escalated dramatically over the last few decades [1–2]. The costs associated with recent wildfires are particularly high in the arid mountain West, where residential structures abut or intermingle with wildland vegetation (Wildland-Urban Interface - WUI) and the exurban population has grown rapidly in recent decades [3]. Large fire events in the 1990s and early 2000s in the western U.S., particularly in lower elevation, relatively dry-pine forests, have contributed to widespread concern that fire exclusion has caused an unprecedented threat of uncharacteristically severe fires in these ecosystems [4–5]. Broad-scale monitoring of fire severity from satellite imagery since ca. 1984 shows a significant trend towards increased severity only in parts of the Southwest [6], yet findings are varied in other studies depending on the spatial scale, selected data types, and the location of study (e.g. Pacific West) [7–12]. In the absence of longer broad-scale records of fire severity, it is unknown how such short-term trends compare to fire severity prior to fire exclusion. In the context of debate about the potential effects of fire exclusion on modern fire regimes and their departures from historical fire regimes (i.e. prior to fire exclusion) [4,13–15], there is a critical need for research on whether the severity and other characteristics of modern fires depart from historical fire regimes for specific ecosystem types at broad landscape scales. The current study compares reconstructed historical fire severity, observed fire severity in recent fires and modeled fire severity from current fuel structures, and discusses the consequences of fire regime changes for management options under expected future climate in the Colorado Front Range.

Development of management goals and adaptation options in fire-prone ecosystems interfacing with the WUI is inextricably related to quantifying the range of variation of a set of ecological patterns and processes exhibited naturally or under human influences during a specified historical period [16–17]. This widely used approach, called the historical range of variability (HRV) framework uses historical ecological data to test hypotheses about the drivers and mechanisms of contemporary and future ecological change [18]. Management using the HRV framework assumes that ecosystem resilience (the ability to recover quickly) is reflected in observed ranges of past vegetation and fire dynamics. Retrospective ecological studies are considered essential for understanding likely consequences of climate change on future fire and landscape dynamics [18–19]. While there has been a shift away from using reference conditions as default for fire and forest management goals in relation to future climate change [19–20], HRV still provides the most viable framework for understanding the sensitivity of ecosystem resilience and ecological integrity to changes in fire regimes [21]. Ecological integrity has been defined as a measure of the composition, structure, and function of an ecosystem in relation to the system's natural or historical range of variation, as well as changes caused by humans [22–23]. A key research objective in the context of fire-prone landscapes in the U.S. West is the assessment of the historical role of fires of varying severities on the resilience and integrity of current ecosystems and their potential consequences for ecosystem services valued by humans [21].

Fuel reduction is currently the dominant management tool for reducing the likelihood of high-severity fire in the contexts of ecological restoration and mitigation of climate change impacts [24–26]. This dual approach (forest restoration and fire mitigation) assumes that the probability of severe fire occurrence has increased to uncharacteristic levels during decades of fire suppression in western forests [4,27–34]. Guided by this assumption, over 190 million acres of public land have been identified as "unnaturally dense" with an increased likelihood of catastrophic wildfires [24], [26]. In response to such concerns, both forest managers and communities have begun to develop strategies to alleviate the potential impacts of wildfire [35], and millions of hectares of forest lands have been treated in recent decades [36]. The goals of fuels reduction to decrease the likelihood of severe wildfires and restore historical forest structure and species composition are complementary in ecosystems where fuels and fire severity have increased, yet are incompatible elsewhere and threaten ecosystem integrity and ecosystem services [37–43].

In the Colorado Front Range, tree-ring evidence, historical landscape photographs, and General Land Office surveys demonstrate that the historical (i.e. pre-1920) fire regime of ponderosa pine and mixed-conifer forests included low-severity fires (i.e. nonlethal to large fire-resistant trees) as well as high-severity fires (i.e. killing >70% of canopy trees) [13,44–54]. There is a broad consensus that most of the montane zone of Colorado was characterized by fire regimes of mixed severity, including some component of high-severity fires [55]. However, a better understanding of which habitats across this heterogeneous landscape were explicitly affected predominantly by low-severity or higher-severity fires, or a combination of both, is necessary for determining where woody fuels have become uncharacteristically abundant as a consequence of fire exclusion [37,56].

Thus, in this research we examine changes in fire regimes across 564,413 ha of the montane forest zone of the Colorado Front Range (Figure 1). We compile and analyze new and existing datasets to refine the spatial resolution and expand the geographic scope of retrodicted (reconstructed) historical fire regimes in comparison with present wildfire potential. We address two main questions about the montane forests of the Colorado Front Range: 1) What areas and landscape characteristics were characterized by an historical fire regime of predominantly low-severity or mixed-severity (which includes moderate- or high-severity fires)? 2) How does historical fire severity compare with observed severity of large wildfires (since 2000) and modeled potential wildfire behavior across the landscape? Finally, we discuss how these comparisons can inform fire mitigation and ecological restoration under expected climate change.

Methods

Study area

The study area is located in the north-central Front Range of Colorado, bounded by six counties and the extent of the montane zone between 1800–3000 m to the east and west (Figure 1). Within the lower to upper montane zones, the mean annual precipitation ranges are ~35.6 to 51 cm and the mean annual temperature ranges are 11°C to 2.4°C, respectively [57]. The lower montane zone (~1800 to 2200 m) comprises primarily pure ponderosa pine (*Pinus ponderosa*) on south-facing slopes and a mixture of ponderosa pine and Douglas-fir (*Pseudotsuga menziesii*) on north-facing slopes. The upper montane zone (~2200 to 3000 m) is composed of ponderosa pine stands on south-facing slopes and more dense stands of ponderosa pine and Douglas-fir on north-facing slopes along with lodgepole pine (*Pinus contorta*), aspen (*Populus tremuloides*) and dispersed limber pine (*Pinus flexilis*) trees at higher elevations. The LANDFIRE existing vegetation type (EVT) layer was used to delineate the montane study area (564,413 ha) within 1800–3000 m because it represents the most up-to-date and detailed cover type classification available across land ownerships in the region, which is proportionally represented by the following EVT types: ponderosa pine (24.2%); lodgepole pine (23.1%); mixed montane conifer forest and woodland types (49.2%); and intermixed with pixels classified as lower montane-foothill grassland (<0.1%) or shrubland (3.5%) (Figure 1; EVT types included in the study are listed in Table S1). However, the accuracy of the LANDFIRE EVT classification at field sites was relatively low (see RESULTS).

During the period from 1980 to 2011, 48 fires burned over 100,000 ha along the Front Range causing severe damage to property and infrastructure in the WUI [2,58–59]. These large fires can occur any time of year and typically burn when wind speeds are high and weather conditions are dry, which is common along the Front Range. During these types of conditions, fire suppression is typically ineffective and fires often escape initial suppression efforts [2].

Field sampling

We used a combination of existing datasets (141 sites) [52–53,60–62] and newly sampled datasets (91 sites) for a total of 232 sites with information on forest structure (tree age structure) and fire history (tree-ring fire-scar records) in montane forest types.

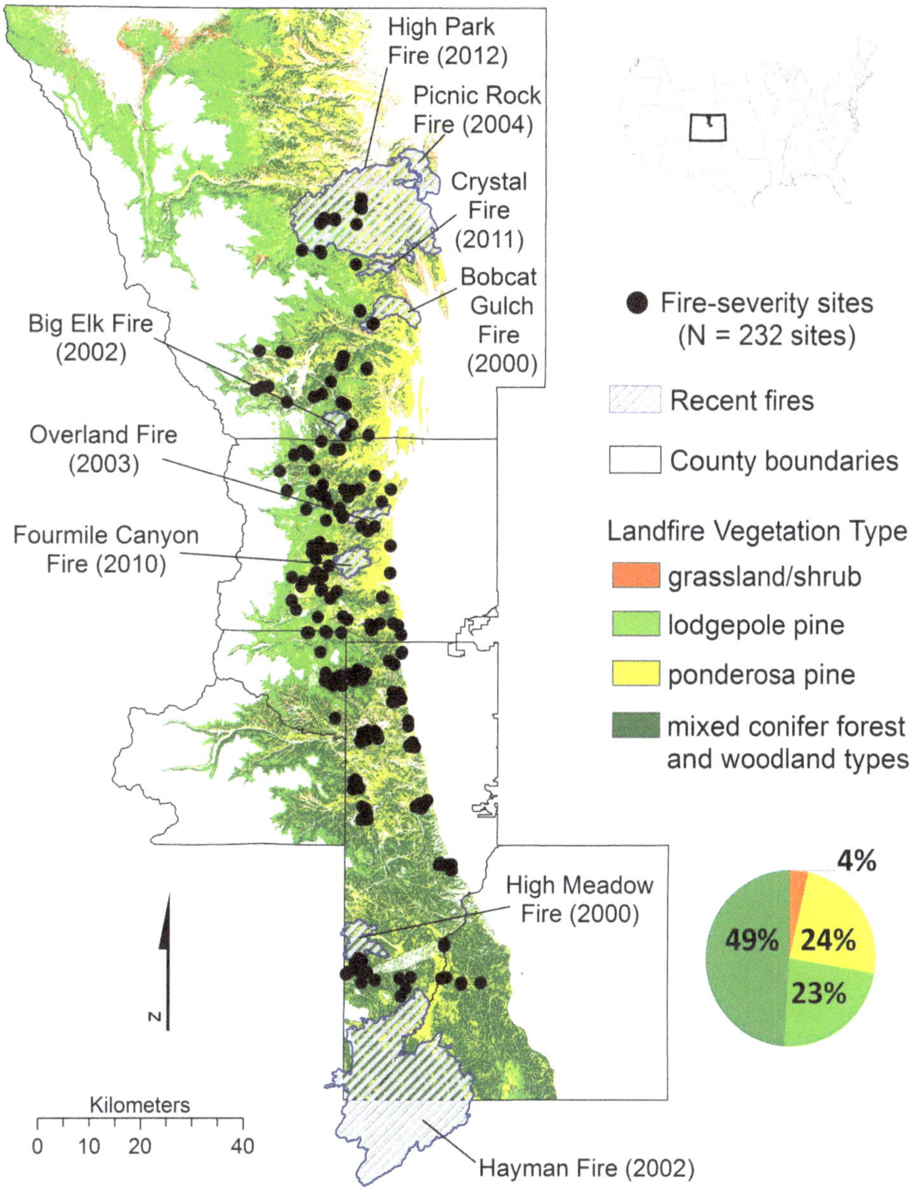

Figure 1. Study area in north-central Colorado, USA. The map includes the montane zone by cover type (LANDFIRE Existing Vegetation Type), 232 sites sampled for historical fire severity, and recent wildfires used for comparison and verification of fire behavior modeling. County boundaries (solid black lines) from north to south are of Gilpin and Clear Creek counties and the mountainous western regions of Larimer, Boulder, and Douglas counties, and Jefferson County.

The majority of sites were sampled in a stratified-random design and all sites were selected in relative proportion to their cover type within the study area, allowing inferences from our site-level datasets to reflect trends across the 564,413 ha montane study area. Because sites were dispersed over the broad regional area of montane forests we do not extrapolate the spatial spread of fire severity between sites, or determine if fires in the same year at multiple sites were from a single ignition or multiple ignition sources. We rejected sites with evidence of past logging or signs of other major anthropogenic disturbance (i.e. mining). Our sampling goal, similar to previous site-level studies of historic fire regimes in the region [51,53], was to identify the predominant structural influence of historical fires at each site (fire-severity regime as explained below). Understanding the historical fire regime and the effects on forest structure without major Euro-American influences provides the natural range of conditions. The interactive effects of logging and/or other anthropogenic disturbances are important, but beyond the scope of our study. Laboratory (tree-ring) analysis of new sites followed the same procedures used in prior studies for analyzing fire-scar samples, stand age and size structure [51–53,60–62]. The criteria used to characterize the historical fire severity at each site were derived from the results of these prior studies and are described below (*Site-level classification of historical fire severity*).

Stand-level sites. At 120 of the 232 sites [51–53,60–62] (and unpublished new data), we sampled in areas of relatively consistent (unvarying) forest structure and physical environment with the extent of sampling area varying in size from 10 to 232 ha according to tree density and the extent of the homogeneous stand structure. At 44 of the stand-level sites, fire-scarred trees were systematically sampled [60,61], fire-scarred trees and fire scars on each fire-scarred tree were tallied, and forest structure was evaluated based on the same protocol used by Sherriff and Veblen [52] (see Tables 1–2). At the remaining 76 stand-level sites, the sampling goal was more intensive. In addition to systematically sampling fire-scarred trees throughout the site, at least 50 of the closest trees (≥4 cm dbh) at a constant distance along randomly located transects were cored, and at least five of the largest and oldest characteristic live or dead trees were selectively cored to ensure large (and the oldest characteristic) trees were represented in our sample. In addition, forest structure (size, decay class, standing/down, distance from transect point), seedling (<30 cm height) and sapling (>30 cm height and <4 cm diameter) data were tallied by species along belt transects. To estimate the ages of trees too small to core (<4 cm in diameter) and to estimate the number of rings missed due to coring height (approximately 20 cm above the root-shoot boundary), 341 juvenile trees, reflecting the range of tree species within the sampling sites (in order of abundance - ponderosa pine, Douglas-fir, lodgepole pine, aspen and limber pine), were cut at sites of different elevation and aspect. Relatively open sites were selected to mimic post-fire growth conditions. Permits for sampling were issued by Rocky Mountain National Park, USDA Forest Service, and county and city open space land management agencies. Sampling did not involve protected or endangered species. For sites where seedlings were not collected, the median age-to-coring height was used from a site of similar elevation and aspect (across all sampled sites: median of 4 years – ponderosa pine; 9 years – Douglas-fir; 8 years – lodgepole pine; 1 year – aspen; 9 years – limber pine). To correct for years missed due to missing the pith, we followed Duncan's [63] procedure. Cores with more than 20 years estimated to pith were excluded.

Plot-level sites. The remaining 112 sites with information on forest structure and fire were smaller in size (3-ha plots) and randomly located from a 1-km grid throughout the CFR study area [52] (and new unpublished data). Sites with abundant logging or other human disturbances were rejected and an adjacent location was sampled. At each random plot-level site the number of fire-scarred trees and fire scars on each fire-scarred tree were tallied, fire-scarred trees were cored to estimate fire dates [64], forest structure was documented using an existing protocol [52] (see Tables 1–2), and approximately 10 cores were taken for reconstructing tree ages.

Analytical procedures

Site-level classification of historical fire severity. For this analysis, we use common and broad definitions of fire-severity regimes for montane forests of the Colorado Front Range that are relevant to forest structure and management (Table 1). Low-severity fire regimes are dominated by frequent (Mean Fire Interval – MFI <30 years), non-stand replacing fires within a stand (~100 ha) that leave multiple fire scars on individual trees throughout the stand and kill young seedlings and subcanopy trees while maintaining open, low-density stands of fire-resistant canopy trees. Mixed-severity regimes have varied fire effects that include low-severity, non-stand replacing fire to high-severity, stand- (or canopy) replacing fire both within stands and across landscapes, often in relation to topography [40,47–48,51–53,65,66]. High-severity fires occur less frequently (MFI >35–100+ years) than non-stand replacing fires, leave only small or no patches of pre-fire remnant trees, result in few to no fire scars on individual trees throughout the stand, and often initiate recruitment of a new cohort of canopy trees [51–53,55].

First, we classified the severity of fires at each (stand- and plot-level) site based on their influence on forest structure. We focused on fire dates recorded by fire scars on at least two trees per site (spreading fires) in the same year prior to 1915 (ca. effective fire exclusion). Next we used two primary metrics, along with the dates and frequency of spreading fires, for characterizing the severity of fires at each site – the percentage of remnant trees (proportion of the trees older than a spreading fire date) and the percentage of tree establishment (proportion of the trees that established within 40 years after a spreading fire). These criteria follow the approach described in prior studies in the region [51,53], correspond to classification of fire severity in other studies [65–66], and are summarized in Table 1. Specifically, individual fires were classified as low severity when ≥80% of the trees that were sampled had survived the fire (based on remnant tree ages) and ≤20% of the trees established following the fire. Fires were classified as moderate severity when 21–79% of the trees survived/established following the fire. Fires were classified as high severity when < 20% of the trees that were sampled had survived the fire and > 80% of the trees established following the fire. We excluded fires where no tree establishment followed within 40 years, assuming there was no structural impact or burning within the site [53]. Individual fires that occurred prior to subsequent mixed-severity fire, and were not associated with a pulse of establishment (>20% trees/site), were not included in the analysis of fire severity. We recognize that the estimate of fire effects for fires that occurred prior to a mixed-severity fire is less reliable because of subsequent

Table 1. Definitions of historical fire severity terms.

Term	Definition of fire effects
Historical high-severity fire	A fire that had high mortality of live, standing vegetation (<20% of the sampled trees survived the fire) and high tree establishment (>80% of the sampled trees) following the fire.
Historical low-severity fire	A fire that had low to no mortality of live, standing vegetation (>80% of the sampled trees survived) and low to no establishment (<20% of the sampled trees established following the fire).
Historical low-severity fire regime	Dominated by frequent (Mean Fire Interval <30 years), non-stand replacing fires within a stand (~100 ha) that leave multiple fire scars on individual trees throughout the stand and kill young seedlings and subcanopy trees while maintaining open, low-density stands of fire-resistant canopy trees.
Historical moderate-severity fire	A fire that had effects that were intermediate between low and high severity.
Historical mixed-severity fire regime	Varied fire effects that included low-severity, non-stand replacing fire to high-severity, stand- (or canopy) replacing fire both within stands and across landscapes, often in relation to topography.

These terms may have different meanings in the literature depending on the context in which they are used.

Table 2. Large fires in the montane study area from 2000–2012.

Fire	Year	Size (ha)*	Fire Severity (%)* Unburned/Low	Low	Moderate	High
Bobcat Fire	2000	3688	30.5	16.8	22.6	30.1
High Meadows Fire	2000	3884	21.7	37.8	32.0	8.5
Big Elk Fire	2002	1741	42.1	20.0	17.1	20.8
Hayman Fire	2002	52,016	21.1	12.7	22.5	43.7
Overland Fire	2003	1244	27.9	12.9	22.6	36.6
Picnic Rock Fire	2004	3626	27.6	37.7	22.6	12.1
Four Mile Fire	2010	2285	11.0	25.6	34.3	29.1
Crystal Fire	2011	914	22.9	38.5	26.9	11.7
High Park Fire	2012	34,905	20.8	29.7	26.9	22.5

* Total area burned and fire severity percentages within each fire perimeter (MTBS values 1–4; 1 = unburned/low, 2 = low, 3 = moderate, and 4 = high). See Figure 1 for large fire perimeters used to compare historical and observed fire severities within the study area.

mortality of older trees with time and/or subsequent disturbance. Post-fire tree establishment is not a direct measure of fire severity, and can vary significantly among species and in relation to seed availability and climate conditions, but it provides corroborative evidence of fire severity in combination with remnant trees, which provide a more direct measure of fire severity. Both metrics, pre-fire remnant trees and post-fire tree establishment, have uncertainties for evaluating fire severity and are not a perfect complement, but used in conjunction with fire dates and integrated into measureable metrics represent a robust way to reconstruct historical fire severity in montane forests (see [53] for more on methodological limitations and recommendations).

Based on the fire severity classifications, we assigned a predominant fire regime at each (stand- and plot-level) site based on the following criteria of cumulative fire effects over time. Sites characterized by frequent, non-stand replacing fires (MFI <30 years) that left multiple fire scars on individual trees throughout the site and without evidence of higher-severity fire were classified as having only low-severity fire regimes. Low-severity sites identified without historical evidence of moderate- or high-severity fire are assumed to have had only low-intensity (surface) fire and low to no canopy mortality. Sites characterized by varied fire effects that included canopy-replacing fire within sites were classified as having mixed-severity regimes (see [51–53] for within site sampling protocols). Mixed-severity fire regimes include low-, moderate- and high-severity fires where sites may illustrate different severities over time or space depending on weather, climate and stand conditions. Individual sites were differentiated based on evidence of low- and moderate-severity fire effects (no evidence of high-severity fire with <20% the stand age preceding a single fire) or low- to high-severity fire effects. Sites identified with either (low- and moderate-severity fire effects or low- to high-severity fire effects) were classified as a mixed-severity fire regime. Mixed-severity sites identified with moderate- or high-severity fire are assumed to have had components of intense fire behavior and canopy mortality, but varied in the extent of canopy mortality.

Landscape-level modeling of historical fire severity. Based on the fire severity classifications at the 232 field sites, we modeled historical fire-severity regime across the study area using potential predictor variables derived from two sources: 1) terrain variables from a 30-m digital elevation model (DEM); and 2) LANDFIRE biophysical variables. Environmental conditions at each of the 232 sites were classified in terms of the median elevation, slope steepness, aspect, cover type, distance to grassland, distance to ravine and the associated fire regime type. Elevation, slope steepness, aspect (arcsine and sine transformations) and slope curvature were all derived from the 30-m DEM. Slope curvature (concavity to convexity) and hillslope position are related to soil depth, texture and potential soil moisture [67]. Ravine drainages were delineated from the 30-m DEM using Arc GRID (ESRI 2002) hydrologic terrain modeling [68] and distance to ravine drainage was calculated. Distance to grassland was calculated as the distance to the edge of grassland areas of at least 0.1 ha.

Classification and regression trees (CART; $n = 5000$ trees) were used to identify explanatory variables with the most power to predict fire severity type (RandomForests Software, Salford Systems). CART has been increasingly used in ecological studies that require nonparametric techniques to explore complex, hierarchical interactions among variables [69–71]. We evaluated predictor variables for three response variables (historical low-, moderate- and high-severity fire) and a binary response variable (historical low- or mixed-severity fire) using all 232 sites to identify complex and non-linear interactions between variables that could be mapped across the study area (e.g. [71]). A 10-fold cross-validation process was used to avoid over fitting. The regression tree size with the smallest error was recorded for each run. The average tree size was used to prune the original tree (from the complete dataset), which tends to minimize complexity and results in a more generalizable tree [72]. There was strong overlap in the range of environmental conditions at sites with evidence of historical moderate and high fire severities. For this reason, we used the results evaluating the binary response variable (historical low- or mixed-severity fire) across all 232 sites for this study to produce a predictive map of the historical fire severity across the CFR study area in a GIS (ESRI ArcGIS 10.1). We identified the optimal model as the one with the fewest variables that best predicted the occurrence of sites with historical low and mixed severities. Model performance was evaluated using the overall percentage correctly classified (PCC) and area under the receiver operating characteristic curve (AUC) statistic. An AUC value of 0.5 indicates the prediction accuracy of sites picked at random, whereas 1.0 indicates perfect classification accuracy. AUC values of 0.7, 0.8, and 0.9 indicate fair, good, or excellent accuracy, respectively.

Comparison of historical fire severity to observed fire severity. Observed fire severities from nine large fires that occurred since 2000 (Table 2; Figure 1), including a range of fire severity, were compared with the spatially-explicit map of historical fire severity (Table 3). The 'observed' severity data came from the Monitoring Trends in Burn Severity (MTBS) program, which provides maps of the perimeters and severity of all fires larger than 1000 ha in the western United States from 1984-present [73]. MTBS thematic burn severity classification maps for each fire were used to evaluate the observed fire severity (unburned/low, low, moderate, and severe) and compared with the spatially-explicit map of historical fire severity within the study area (ESRI ArcGIS 10.1). We recognize these two datasets (observed fire severity and historical fire severity) are derived from different sources of information, each with their own limitations. Thus, our analysis focused on a summary comparison of the proportion of area burned by observed fire severity type compared to historical fire severity of the same area as a broad index of departure from historical fire regime reflecting landscape-level changes rather than a fine-scale (pixel-by-pixel) comparison. For each of the nine fires, we tested for differences (in the number of pixels) between the observed amount of low-severity fire (MTBS unburned/low and low severity classes) compared to the expected (historical low-severity) using χ^2 tests (IBM SPSS 20).

Comparison of historical fire severity to modeled fire severity. The spatially-explicit model of historical fire severity was overlaid with a model of fire potential under extreme (99th percentile) weather conditions to identify possible landscape-scale changes of historical low- and mixed-fire severity in the entire montane study area (564,413 ha; Table 3). We modeled potential fire behavior using FlamMap 3.0, a fire modeling system that predicts instantaneous fire behavior under fixed weather and fuel conditions (i.e. it is a static rather than dynamic model). FlamMap requires the following inputs: elevation, slope, aspect (derived from a 30 m USGS DEM); and percent canopy cover, stand height, canopy base height, canopy bulk density, and Scott and Burgan [74] fuel models (all from the LANDFIRE project [75]). Canopy cover and canopy base height values were adjusted downward on the recommendation of the LANDFIRE project [75] and a comparison of field-derived fuel layers in the study area [76]. The modeling procedure takes place in two steps (described in detail in [77–78] and summarized here). First, fuels are conditioned based on specified weather conditions. We used wind/weather conditions during the primary fire season (June–September) derived from four Remote Automated Weather Stations (RAWS) that span the study area and have the longest record: Redfeather (station 050505; 1964–2007), Estes Park (station 050507; 1964–2007), Corral Creek (station 051804; 1968–2007) and Bailey (station 052001; 1970–2007). In the second step, FlamMap calculates fireline intensity and crown fire activity using a suite of surface and crown fire models [79–82]. Fireline intensity is a measure of energy released per unit length along the flaming front of a fire (kW/m). Crown activity (a binary variable representing surface fire or active/passive crown fire) was calculated using the Scott and Reinhardt [83] method. To address how historical fire severity compares with modeled potential wildfire behavior, we overlaid the areas modeled as historical low-severity with potential wildfire behavior (fireline intensity and crown fire) assuming that areas of higher fireline intensity (e.g.>33,000 kW/m) and crown fire today indicate a shift from the historical fire regime to one of higher potential for crown fire and canopy mortality (>33,000 kW/m and >40,000 kW/m are the approximate median fireline intensity values for moderate and high severity fire, respectively, for all verification fires with variable severity described in the next section; Table S1). A crown fire may or may not be lethal to all dominant vegetation, and a crown fire may be continuous or may occur in patches within a mixed-severity burn. Mixed-severity fires, by definition, have components of intense fire behavior and canopy mortality, and high fireline intensity and active crown fire would be within the HRV. Our focus is on the areas of greatest change that may have shifted from low severity to potential high fire intensity or crown fire today. These fire behavior outputs were chosen because they are commonly used, interpretable, and can be generally compared to historical fire severity.

The wildfire modeling utilized the 99th percentile monthly weather conditions from ca. 1964–2007 during the June-September fire season derived from four RAWS stations (listed above). Starting with a fixed southwest wind of 74 km/h, which is the 99.9th percentile daily wind speed during the primary fire season (June–September, ca. 1964–2007), we used the WindNinja program to refine wind speeds based on topography. FlamMap's input variables and modeling procedure follow Platt et al. [78] and were verified for a level of realism by comparing observed fire severity from seven major fires that occurred since 2002 (see the next section; Table 2).

To place historical 99th percentile conditions into context, we first compared them to the conditions of past fires and then to predicted future climate. During the Hayman Fire (06/08/2002), temperature and relative humidity reached up to the 98th and 99th percentiles, respectively (Cheesman RAWS station). Similarly, during the High Park Fire (06/09/2012), both temperature and humidity were up to the 99th percentile (Redstone RAWS station). The RAWS stations recorded wind gusts over 56 kph during the High Park Fire and over 80 kph during the Hayman Fire. We make the assumption that the 99th percentiles are of comparable weather conditions between historical and contemporary time periods.

We then compared average monthly temperature in the study area during the June-September fire season to two climate model runs used in the IPCC Fourth Assessment Report [84–85]: (1) climate conditions in 1964–1999 from the Climate of the 20th Century (20C3M) model run and (2) climate conditions in 2064–2099 from the SRESA2 scenario. During the June–September fire season in 1964–1999 (20C3M model run), the 99th percentile

Table 3. Comparisons of historical fire severity, observed fire severity, and modeled potential fire behavior.

Historical	Observed (MTBS thematic burn severity classifications)	Modeled (FlamMap output)
Low-severity fire only	Unburned to low severity, Low severity	Surface fire, e.g. median fireline intensity <33,000 kW/m
Mixed-severity fire: evidence of varied severities from low-, moderate- to high-severity fire	Moderate severity, High severity	Crown, torch, e.g., median fireline intensity > 33,000 kW/m

monthly temperatures were 24°C, and the 50th percentile monthly temperatures were 19°C. During the June–September fire season from 2064–2099 (SRESA2 scenario), the 99th percentile temperatures are predicted to increase to 30°C, and the 50th percentile temperatures are predicted to increase to 26°C. The 99th percentile temperatures from 1964–1999 are predicted to become 39th percentile temperatures under the SRESA2 scenario. It is thus reasonable to say that the 99th percentile conditions described in this study may become average conditions in 50 or more years from now.

Verification of modeled fire behavior. To assess the realism of the FlamMap models described above, we compared modeled fireline intensity and crown fire class to observed fire severity of the seven regional fires that occurred since 2002, the year for which the LANDFIRE (fuels) data is valid (Table 2). Within the perimeter of each fire, we generated randomly placed points, and at each point calculated median fireline intensity and percentage active and passive crown fire within four MTBS classes (unburned to low, low, moderate, and high severity; Table S1; ESRI ArcGIS 10.1). There are limitations to such a comparison. In an actual fire, intensity varies minute by minute depending on factors such as wind, relative humidity, fuel, and fire suppression. In contrast, FlamMap estimates intensity for an instant in time across an entire landscape for the generic "extreme conditions" described in the paper. Thus, while we cannot comprehensively validate a fire we can check whether, overall, areas identified as severely burned by MTBS correspond to higher modeled fireline intensity and crown fire. Kruskal-Wallis tests (IBM SPSS 20) were used to determine if the distribution of fireline intensity and percentage crown fire were significantly different between MTBS classes at the $p = 0.01$ level.

Results

Across all 232 field sites, cover type was dominated by a single species (≥80% of the canopy trees >4 cm dbh) of ponderosa pine (42.2%–98 sites) and lodgepole pine (8.6%–20 sites), and co-dominated (<80% of the canopy trees a single species) by mixed-conifer types (49.2%–114 sites). Considering pure ponderosa pine together with mixed-conifer sites, 61.2% (142) of the 232 sites were dominated by ponderosa pine and the remaining sites were dominated by Douglas-fir (20.6% and 48 sites), lodgepole pine (15.5% and 36 sites), and aspen (2.6% and 6 sites). The overall accuracy of the LANDFIRE EVT classification was relatively low with 39.2% of the 232 sites misclassified: 31 pure ponderosa pine, 5 lodgepole pine and 55 co-dominated mixed-conifer sites. Thus, the proportional values of EVT cover types across the study area, and our results defined by EVT cover types (the best available cover type at the landscape scale), should be interpreted cautiously (i.e. Figure 1 pie chart). Nevertheless, field observations support the overall LANDFIRE cover type trends illustrating ponderosa pine throughout the entire study area and an increase of lodgepole pine and mixed-conifer stands in the higher elevation portions of the study area.

Site-level fire-severity

A total of 7680 tree cores and 1262 fire-scar samples were used to delineate fire dates and fire severity across all sites. Across 120 stand-level sites, there was evidence of 322 spreading fires (fires with ≥2 trees scarred per site) between 1597 and 1995 with only 71 (22%) of the spreading fire dates unique to one site; all other fire dates were recorded at two or more sites. Fire years of 1654, 1786, and 1859–60 were particularly extensive with 36%, 43% and 48% of the available recorder sites (with ≥2 fire-scarred trees) recording each fire year, respectively (Figure 2). These fires were recorded at multiple sites that extended over 9 km (1654), 7.5 km (1786) and up to 30 km (1859–60) distance away from one another (Figure 3). Other widespread fire dates recorded at four or more sites (recording on average 22% of the available recorder sites per year; ≥2 trees scarred per site) were 1624, 1712, 1793, 1809, 1813, 1842, 1850–52, 1857, 1863, 1866, 1871, 1880–81, 1886, and 1910 (Figure 2; multiple years indicate difficulty differentiating the exact fire year due to dormant season fires and/or missing rings before or after scarring).

Historical fire severity regime was delineated at individual sites using 297 fires that occurred prior to 1920 (pre-fire exclusion) and tree age data (remnant and establishment ages). Eighteen (7.8%) sites were characterized by only low-severity fires and 214 (92.2%) were characterized by mixed-severity fires. Across the montane region, the majority of trees established between 1820 and 1920 (66% of tree ages; Figure 4) primarily because of past mixed-severity fires during this period (this study; and also see [44–45,48,50–51,53,55]). Low-severity sites tended to have younger trees than mixed-severity sites (average date of tree establishment of 1907 compared to 1873, respectively; Figure 4). The low-severity fire regime was most common in pure ponderosa pine sites (12 sites; ≥80% of the canopy trees as ponderosa pine), but also occurred in mixed-conifer sites (6 sites with co-dominance of ponderosa pine and Douglas-fir). Sites with the historical low-severity regime tended to occur at lower elevations than sites with mixed-severity fires with median elevations of 2101 m and 2445 m, respectively.

Of the 214 sites with evidence of mixed-severity fire (Figure 5), 108 sites had evidence of low and moderate-severity fires (no evidence of high-severity fire) and 106 sites had evidence of high-severity fires (evidence of up to ~50 ha patches of high-severity in some stand-level sites). Interpretation of the spatial extent of historical fire severity is limited by our sampling unit size (sites ranged between ~3–232 ha in size). Nevertheless, there is strong evidence that in most years before 1915 fires were much more extensive than the size of a single site (only 15.5% of the fire years were unique to one site; 46 of 297 fire dates; Figure 2). Many sites burned at moderate or high severity during the same year with an average of 1.1 km distance between sites (and up to 7 km distance between nearest sites burning in the same year with evidence of higher-severity fire). Evidence of mixed-severity fires occurred in all dominant cover types sampled, including sites of pure (≥80% of the canopy trees) lodgepole pine (20 sites) and ponderosa pine (89 sites), as well as mixed-conifer types with dominance of aspen (6 sites), lodgepole pine (16 sites), Douglas-fir (40 sites), and ponderosa pine (43 sites). Across sites classified with mixed-severity fire regimes, moderate- and high-severity fire effects were evident in all dominant cover types including pure lodgepole pine (8 and 12 sites) and ponderosa pine (48 and 41 sites), as well as mixed-conifer types with dominance of aspen (2 and 4 sites), lodgepole pine (8 and 8 sites), Douglas-fir (21 and 19 sites), and ponderosa pine (18 and 25 sites), respectively.

Landscape-level historical fire severity

Elevation and slope steepness were the most important predictive variables delineating locations of only low-severity versus mixed-severity sites. The CART model with the best predicted occurrence of sites with historical low- and mixed-severity fire had a PCC of 80.6% (187 of 232 sites predicted correctly) and an AUC value of 0.77 for the cross-validation test. The model indicated low-severity sites tended to occur at or below 2263 m, or on slopes equal or less than 4 degrees above 2263 m (17 of 18 low-severity sites fit this model, 94.4%). Mixed-severity

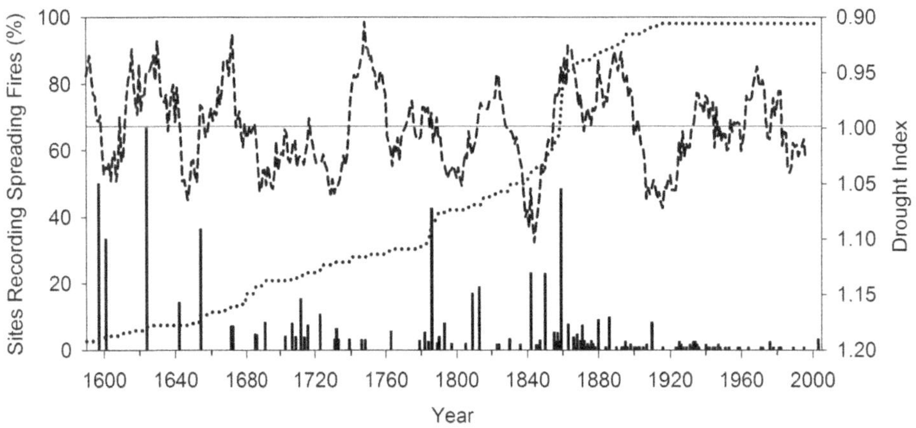

Figure 2. Percentage of sites recording spreading fires. Percentage of sites with spreading fires (minimum of 2 scarred trees per year at each stand-level site represented as histogram bars, n = 120 sites; left axis), sample depth of the percentage of sites available to record fires (minimum of 2 scarred trees available to record fires at each site; dotted line; left axis), and a 20-year smoothing average of a regional tree-ring drought index [60] with a mean value of 0.99 (dashed line with an inverted axis where smaller values are towards the top and indicate dry conditions; right axis).

sites occurred across a broad topographic gradient of the montane zone, but tended to occur on slopes greater than 4 degrees above 2263 m (170 of 214 mixed-severity sites fit this model, 79.4%). Mixed-severity sites with evidence of high-severity fire tended to occur at higher elevation, on greater slope steepness, and at greater distances from streams compared to sites with evidence of only low-severity fires. There were no strong generalizable differences between locations of mixed-severity sites with evidence of low- to high-severity fire and those without evidence of high-severity fire (only low- and moderate-severity fires). Cover type was not a significant predictor of fire severity.

The best-fit model from the CART analysis (described above) was used to map historical fire severity across the montane study area (564,413 ha). Across all cover types, 27.8% (156,198 ha) of the area was mapped with an historical regime of low-severity fire and 72.2% (406,173 ha) was mapped as mixed-severity fire (Figure 6a). Within the low-severity fire regime (27.8% of the study area) almost equal areas were classified by LANDFIRE EVT as ponderosa pine (45.4%) and mixed conifer (48.8%), intermixed with grassland/shrubland (5.7%). In comparison, areas mapped as the historical mixed-severity fire regime (72.2% of the study area) were classified with a lower percentage of ponderosa pine (16%) and grassland/shrubland (2.7%), and a higher percentage of mixed conifer (81.3%) cover types. Of areas mapped as pure ponderosa pine, 52.4% was classified as low-severity and 47.6% as mixed severity fire (Figure 6b). Of areas mapped as mixed conifer types, 18.8% was classified as low-severity and 81.2% as mixed severity fire (Figure 6c). The decrease in pure ponderosa pine dominance above 2263 m is consistent with the elevation shift in co-dominance with other species, primarily Douglas-fir, in the upper montane zone.

Comparison of historical fire severity to observed fire severity

For the nine recent fires in the study area since 2000 (Table 2; Figure 1), there was a 4% difference between the average observed (47%) and historical (51%) areas of low-severity fire severity, and a 7% difference between the total observed (39%) and historical (46%) areas of mixed-severity fire (Figure 7). The greatest differences between observed (MTBS) and expected (historical) low-severity fire was at the lowest elevations (median elevations <2155 m; Figure 7). Three fires (Picnic Rock Fire, Bobcat Fire, Crystal Fire) had lower than expected observed low-severity fire compared to the historical fire severity map (X^2 value of 6.39 for Bobcat Fire and 5.49 for Crystal Fire, $p<0.05$; the X^2 test was invalid for the Picnic Rock Fire with no pixels classified as historical mixed-severity fire). For the remaining six fires there were either no significant differences in low-severity fire (High Park Fire, Four Mile Fire, Overland Fire, Hayman Fire, X^2 value of 1.04, 2.84, 1.56, and 2.43, respectively, $p>0.05$) or higher than expected low-severity fire observed compared to the expected historical fire severity map (High Meadows Fire and Big Elk Fire, X^2 values of 5.14, and 8.25, respectively, $p<0.05$).

Verification of modeled fire behavior

Overall, we found a positive association between modeled wildfire behavior (fireline intensity and percentage crown fire) and MTBS severity class (Table S1; Figures S1 and S2). The exception was the Picnic Rock Fire, for which the modeled percentage crown fire was consistently very low across observed severity classes. This is not surprising, as the Picnic Rock Fire occurred at the lower elevation grassland ecotone with over 50% of the fire classified as unburned/low or low severity and only 21% as high severity that occurred primarily in grassland fuel types (73% of high-severity area). Overall, severely burned areas in observed fires of variable fire severities (Table 2) corresponded to higher modeled fireline intensity and proportion of crown fire (Table S1), providing confidence in the model of potential wildfire behavior and its comparison with historical fire severity.

Comparison of historical fire severity to modeled current fire behavior

Across the entire montane study area present-day potential fire behavior was 36.1% surface and 73.9% crown (1.3% torch and 62.6% crown) fire occurrence under extreme (99th percentile) weather conditions (Figure 6a). Historical fire severity for the entire study area was mapped as 27.8% low-severity only and 72.2% mixed-severity fire regimes. For the pure ponderosa pine cover type, under the extreme fire weather scenario (99th percentile), present-day fuels predicted 40.2% of the area would burn as surface fire and 59.8% as crown (3.1% torch and 56.7% crown) fire (Figure 6b). For the mixed conifer cover type, the extreme fire weather scenario and present-day fuels predicted

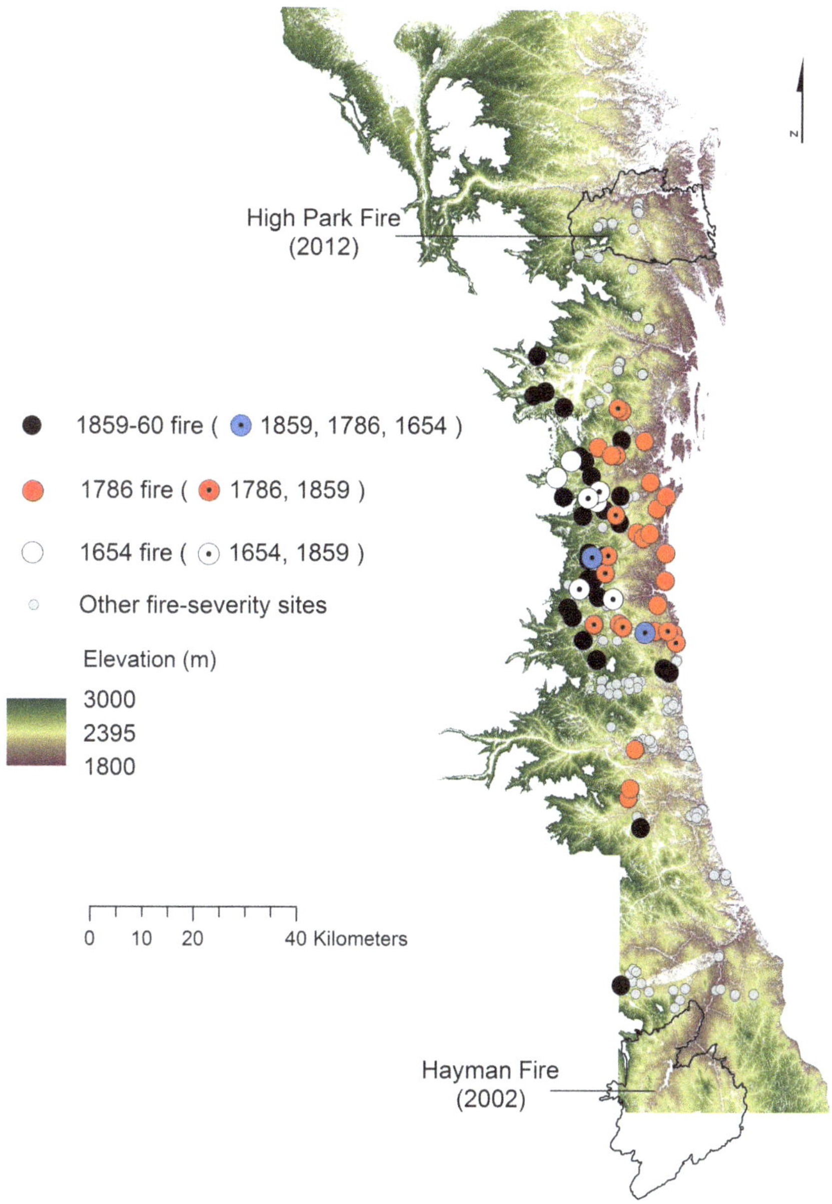

Figure 3. Historical fire-severity sites recording widespread fire dates in 1654, 1786, and 1859–60. Sites are shown across the elevation gradient of the study area. Symbols with centered dots indicate two or three of those fire years listed were recorded at that site. The two largest fires in recorded history for the region (Hayman Fire and High Park Fire) are shown for spatial comparison.

31.8% as surface and 68.2% as crown (0.7% torch and 67.5% crown) fire (Figure 6c).

Almost 12% (66,394 ha) of the entire study area showed little change in the low-severity fire regime (Figure 8 – green), based on a high spatial coincidence of historical low-severity and present-day potential for surface fire under extreme (99th percentile) weather conditions. These areas were predominantly in the lower montane zone (mean 2136 m elevation and 12 degree slope steepness), and had an average modeled fireline intensity of 14,020 kW/m. About 16% of the entire montane study area (Figure 8 – red) is now susceptible to crown fire but historically had lower-severity fire. These areas were also concentrated in the lower montane zone (mean 2177 m elevation and 16 degree slope steepness), had an average modeled fireline intensity of 43,620 kW/m, and were classified by LANDFIRE EVT as grassland or shrubland (0.5%), lodgepole pine (5.2%), ponderosa pine (43.7%), and mixed conifer (50.7%). Of areas mapped only as the historical low-severity fire regime (27.8% and 156,198 ha of the study area; Figure 8 – green and red combined), 42.5% of the area showed little change in the fire regime (Figure 8 – green), whereas 57.5% had the potential for higher-severity fire (Figure 8 – red).

Above the lower montane zone (>2263 m), the historical and modeled fire behavior under extreme conditions showed high variability from surface to crown fire, as expected in a mixed-severity fire regime. Areas above 2263 m that were mapped as historical mixed severity fire, but today have potential for surface fire and crown fire were 24.8% and 47.4% of the entire montane study area, respectively (Figure 8 – yellow and orange). Of areas mapped only with the historical mixed-severity fire regime (72.2%

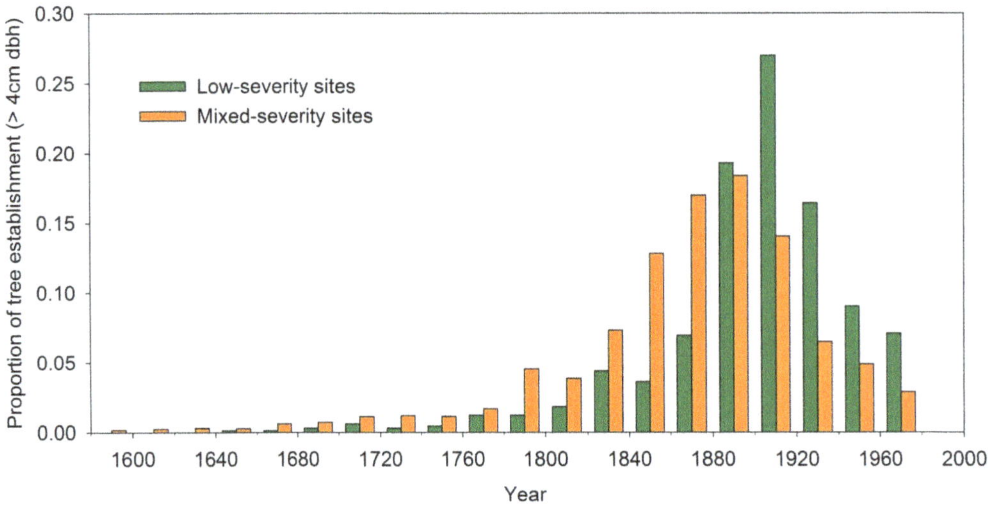

Figure 4. Proportion of tree establishment dates. Tree establishment (≥4 cm diameter) in 20-year bins for low-severity (green) and mixed-severity (orange) sites sampled at the stand-scale (120 sites). Total number of trees included are 663 and 5703, respectively.

and 406,173 ha of the study area; Figure 8 – yellow and orange), 34.4% has a low probability of crown fire today (mapped as potential for surface fire in Figure 8 (yellow); average fireline intensity of 15,861 kW/m), whereas 65.6% of the area has a high probability of crown fire activity today (mapped as potential for torch or crown fire in Figure 8 (orange); average fireline intensity of 42,290 kW/m).

Discussion

A key finding of our study of fire regime changes in the montane forest zone of the Colorado Front Range is that only 16% of the total study area recorded a shift from historical low-severity to a higher potential for crown fire today. This area of increased fire severity occurs in over half (57.5%) of the area mapped with the historical low-severity fire regime and is concentrated in the lower montane zone. A substantial portion (42.5%) of the area mapped as the historical low-severity fire regime (11.8% of the study area) shows little change in the fire regime (Figure 8), and is expected to support only surface fire even under extreme (99th percentile) weather conditions based on our modeling of present day fuels. Both areas occur primarily in the lower montane zone below 2263 m, but the areas with little change in the low-severity fire regime are on average slightly lower in elevation and slope steepness and closer to grasslands and ravine drainages than areas with higher potential for crown fire today. Observations from recent large wildfires and modeling of potential fire behavior under extreme weather conditions are consistent with the historical evidence of a varied fire regime of primarily low-severity fires at the lowest elevations to a mixed-severity fire regime at higher elevations in montane forests. Three regional fires predominantly in the lower montane forests showed the greatest differences between the observed (MTBS of recent large fires) and expected (historical fire severity) areas burned by low-severity fire. However, the differences between the observed and expected proportions of low-severity fire for the six other fires were non-significant, or showed higher than expected low-severity fire.

A decline in fire frequency over the past 100 years leading to substantial increases in stand density is supported only for the lowest elevations of forest below approximately 2200 m in the Colorado Front Range [13–14,45,51,60,86]. These areas were characterized mainly by frequent (average return intervals <30 years) low-severity fires that maintained open forests by killing mostly juvenile trees, resulting in low densities of mature trees. Greater proximity to grassland in lower elevation areas probably promoted more frequent fire due to more abundant fine, herbaceous fuels [62]. The cessation of formerly frequent fires coincides with increased stand densities broadly throughout the lower montane zone. This pattern is especially evident below 2200 m, but also occurred at some sites at higher elevations on less steep slopes most likely where montane grasslands occurred. However, overall this represents a relatively small proportion of the montane forest of the northern Colorado Front Range (27.8% of the study area is mapped with the historical low-severity fire regime).

The dominant (72.2% of the 564,413 ha study area) historical fire regime of the northern Colorado Front Range consisted of a mixed-severity regime in which stand structures were shaped primarily by moderate-severity (46.5% of sites) and high-severity (45.7% of sites) fires; only 7.8% of the sites recorded predominantly low-severity fires. At higher elevation (>2263 m), spreading fires were typically less frequent (>30 year fire intervals), and had varied fire effects (mixed-severity) that included non-stand replacing fire to canopy-replacement fire both within sites and across broad landscapes, often in relation to topographic variability. Many sites experienced intervals between successive widespread fires that would have been sufficient for conifer seedlings to reach sizes that would survive low-severity surface fires. Specifically, evidence of high-severity fire tended to occur further from grasslands and on steeper sites than those with evidence of only low-severity fire, although there is overlap between sites that show moderate- or high-severity effects particularly above the lower montane zone. Evidence of mixed-severity fires occurred in all dominant cover types sampled including sites of pure (≥80% of the canopy trees) lodgepole pine and ponderosa pine, as well as mixed-conifer types with dominance of aspen, lodgepole pine, Douglas-fir, and ponderosa pine. Consistent site- to landscape-scale evidence indicates an historical mixed-severity fire regime in which moderate- and high-severity fire effects shaped current forest age structures in the mid- and upper montane zone [13,46,51,53] (and this study).

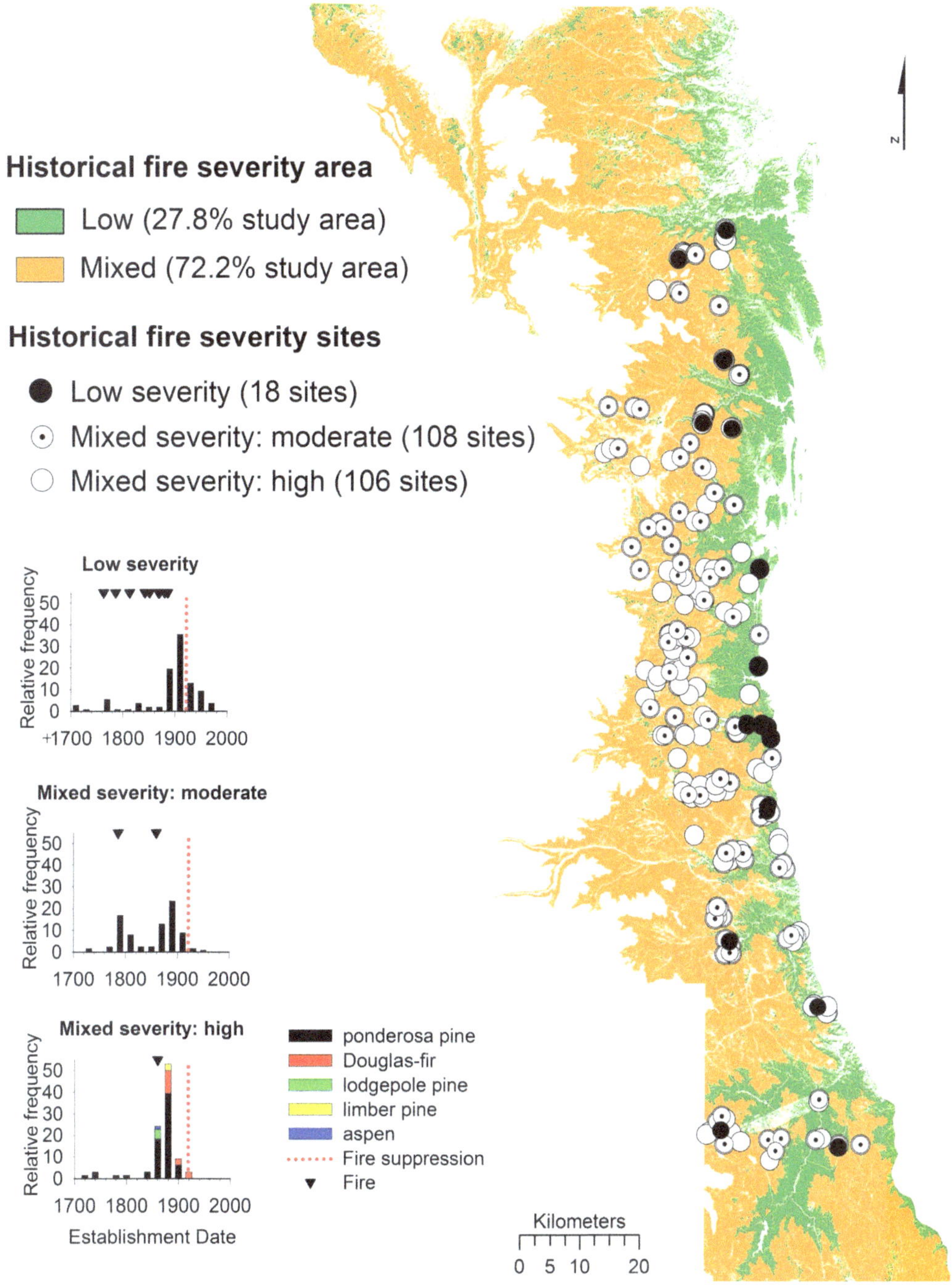

Figure 5. Distribution of 232 sites with historical (pre-1920) evidence of low-severity and mixed-severity fires. Sites with historical evidence of mixed-severity fires are differentiated with evidence of low- and moderate-severity fire (no evidence of high-severity fire; 108 sites) and low- to high-severity fire (106 sites). Areas mapped as historical low-severity fire only (27.8%) and mixed-severity fire (72.2%) are shown in green and orange, respectively. Three example graphs are also shown to illustrate the evidence (remnant trees, tree establishment and spreading fires) used to classify sites with evidence of low and mixed-severity fire effects at individual sites.

Most spreading fires recorded at our sites extended beyond a single site (88% of 322 fires with ≥2 trees scarred per site from 1597–1995). Moderate- and high-severity fires prior to the 20th century were documented at many sites, as illustrated by fires in 1654, 1786 and 1859–60 that extended at least 7–30 km between nearest sampling sites, are within the HRV for montane forests at a regional scale (Figure 3). For example, the spatial extent of sites recording fires in the same year is within the spatial scale of the largest modern wildfires on record in the study area (e.g. High Park Fire and Hayman Fire). Our ability to interpret the spatial extent of historical fire severity patches from individual fires is limited by our sampling unit size (3 ha–232 ha), subsequent fire events, and sample depth (although >20% of our sites record fires since 1700), but the evidence indicates mixed-severity fires occurred in patches up to at least 200 ha, with evidence of approximately 50 ha of high-severity in some sites. Direct

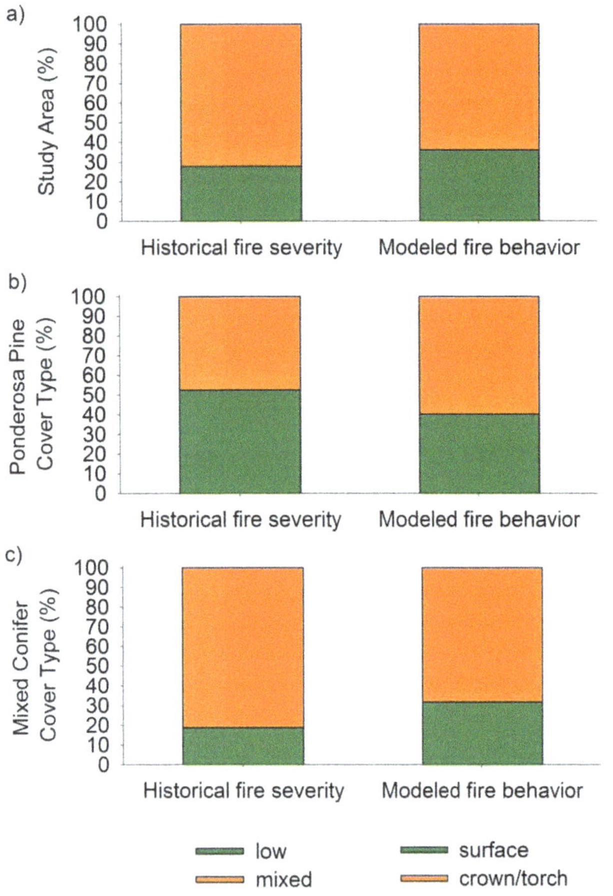

Figure 6. Percent study area and cover type classified with historical fire type and modeled fire behavior. A comparison of the percentage of the a) study area, b) ponderosa pine cover type, and c) mixed conifer cover type classified as historical fire (low-severity only or mixed-severity) and modeled current fire behavior (surface or crown/torch) under extreme weather conditions (99th percentile for 1964–2007).

comparison of patch sizes of fires is difficult given the presence of major highways and other land uses that have fragmented fuel continuity at a landscape scale. It is highly likely that variance in patch size of historical fires was great, similar to modern fire observations, rendering quantitative comparisons of rather limited value. Nevertheless, high-severity fire and high-stand densities are within the HRV for the mid- and upper montane forests (e.g. [44,51,53–54], this study), which needs to be taken into account when forest management goals consider restoring forests to pre-fire-exclusion conditions.

In the Colorado Front Range, at higher elevations (>2263 m) mixed-severity fire regimes appear to have been the predominant fire regime historically and today. The modeled potential wildfire behavior under current fuel conditions and extreme fire weather shows a mixed-severity fire regime throughout the montane forests of the Colorado Front Range. High fireline intensity and active crown fire are within the HRV of a mixed-severity fire regime particularly under extreme conditions, but would vary in extent of intensity and canopy mortality depending on existing conditions. Site-level evidence indicates high overlap in biophysical conditions that support moderate and high-severity fires (this study and [52]). These results provide a snapshot of the expectations into the near future for fire behavior in the Colorado Front Range. The advantage of comparing historical and observed fire severity with fire behavior modeling across the study area under realistic, but also extreme, weather conditions is that it provides an important present-day comparison to the past. Recent large wildfire years in the Colorado Front Range are not without precedent and similar events occurred historically (i.e. 1654, 1786, 1859–60), under similar exceptional (interannual to multi-decadal scale) climate conditions [61]. The largest fires since the year 2000 – the Hayman and High Park Fires – were both characterized by up to 98–99th percentile conditions for both temperature and relative humidity. The situation will likely be exacerbated under climate change; under the 2007 IPCC A2 scenario for the study area, the 99th percentile temperatures of the past are expected to become the 39th percentile temperatures by 2064–2099 (also see [87]). Thus, the range of weather conditions during local to more widespread fire years in the last 300+ years likely represents a similar range of potential fire-climate conditions presently, and informs expectations of potential fire behavior that may occur under severe, yet increasingly more common weather conditions.

We recognize that each of our spatial datasets (mapping of historical, observed, and potential fire) have their own set of limitations. For example, distance to grassland was not a predictor in the best-fit model used to map the historical fire severity landscape. However, our prior research indicates areas adjacent to grasslands experience more frequent fires than sites farther from grasslands, owing to the proximity to prevalent fine fuels that increase the sensitivity of fire activity to interannual climate variability [62]. Thus, the historical model likely under-estimates the amount of low-severity fire in some areas at higher elevation within or adjacent to grasslands where we have limited or no sample sites (e.g. high elevation grassland areas in the northern portion of the study area - Figures 1 and 8). This could explain why some of the study area (Figure 8) shows a shift from historical mixed-severity fire to potential low-severity fire in the contemporary forest even under the extreme weather scenario. The observed fire behavior dataset (MTBR Thematic Burn Severity) has its own limitations. To classify fire severity into 'high' 'moderate', and 'low' severity classes, image analysts determine thresholds in dNBR values based on visual interpretation of imagery, field plot data, and expert knowledge. Though dNBR is sensitive to initial vegetation conditions [88], alternatives such as RdNBR are less parsimonious and in practice may not capture burn severity with greater accuracy than dNBR [89]. Also, while the fire behavior modeling from FlamMap operates on a fixed set of fuels and weather conditions, the historical severity model is based on the fire regime over a long period of time during which fires of different severity have the opportunity to burn. Because of these limitations, we report proportional area of landscape-level changes rather than finer-scale interpretations, and emphasize broad trends evident from multiple lines of inquiry. Additionally, while our model was based on the most recently available fuels data from LANDFIRE, characterizing vegetation changes across the landscape due to wildfire and management through 2010, mortality from recent insect outbreaks, for example, are not reflected. Inferring the effect of fuel and fire behavior changes in our model due to tree mortality from insects is challenging as such fuel changes are: 1) highly dependent on the timing and severity of the outbreak, 2) not characterized well in standard fuel models, and 3) not validated for fire behavior as few empirical studies have examined how the timing and extent of insect outbreaks in montane forests affects observed fire behavior.

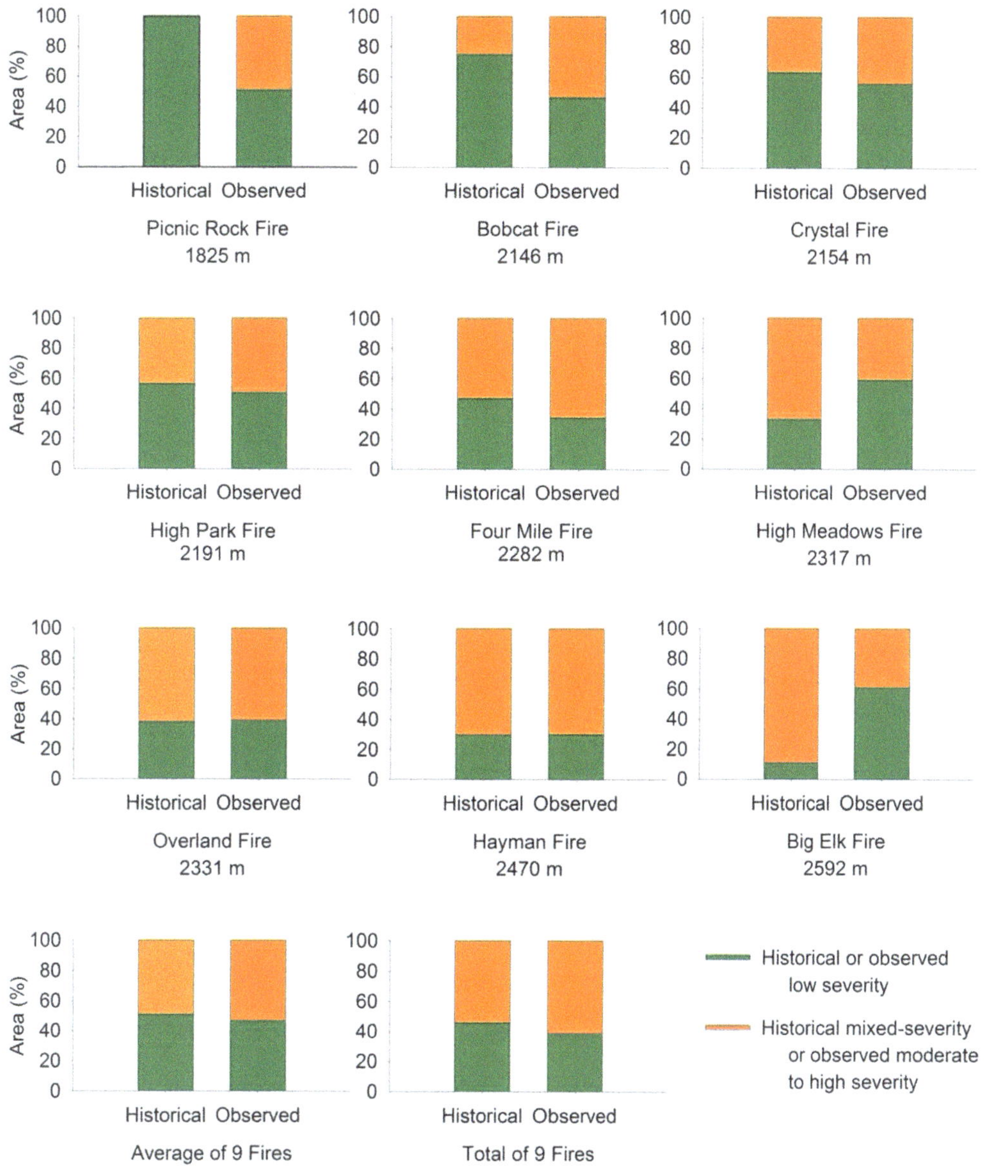

Figure 7. Comparisons of the historical and observed fire severities for nine recent fires (2000–2012). The comparison shows the proportion of low- (unburned/low and low) and mixed- (moderate and high) severity fire within the perimeter of nine recent fires (2000–2012), and the average and total (all pixels) proportions of fire severities across all nine fires in the study area. The median elevation for each fire is given under the fire name.

Implications for fuels management and ecological restoration

A clear delineation of the spatial extent of past fire regime types is a major concern for ecosystem managers in the context of wildfire risk and ecological restoration [90–92]. Evidence from previous studies in ponderosa pine and other montane forests in the Colorado Front Range have shown that the historical fire regime was variable over time and space, represented by a mixed-severity regime [44,46–48,50–54]. Goals of ecological restoration and wildland fire hazard mitigation are both compatible with management practices, like prescribed fire and thinning to reduce fuels, below approximately 2200 m in our study area, which experienced the greatest increase in fire severity, and likely fuels, since fire exclusion [93]. Disturbance from grazing and logging as well as periods of favorable climate probably also contributed to increased tree establishment in the late 19th and early 20th century, but seedling survival clearly depended on the long fire-free periods. Even at low elevations, however, some sites had historical fire regimes dominated by infrequent rather than frequent fires (e.g. steep and north-facing slopes [86]). These infrequent fires, inferred to be high-severity fires from age structure data, killed high percentages of trees within a fire perimeter, and promoted the establishment of naturally dense forest patches [44,51,86]. This suggests that some of the areas (e.g. north-facing and steep slopes) with potential for crown fire today may support characteristic fire behavior and are not necessarily out of the HRV even in the lower montane zone.

Mixed-severity fire regimes where spreading fires occurred at lower frequencies (>30 year fire intervals) are less clearly candidates for thinning than are low-severity fire regimes and a

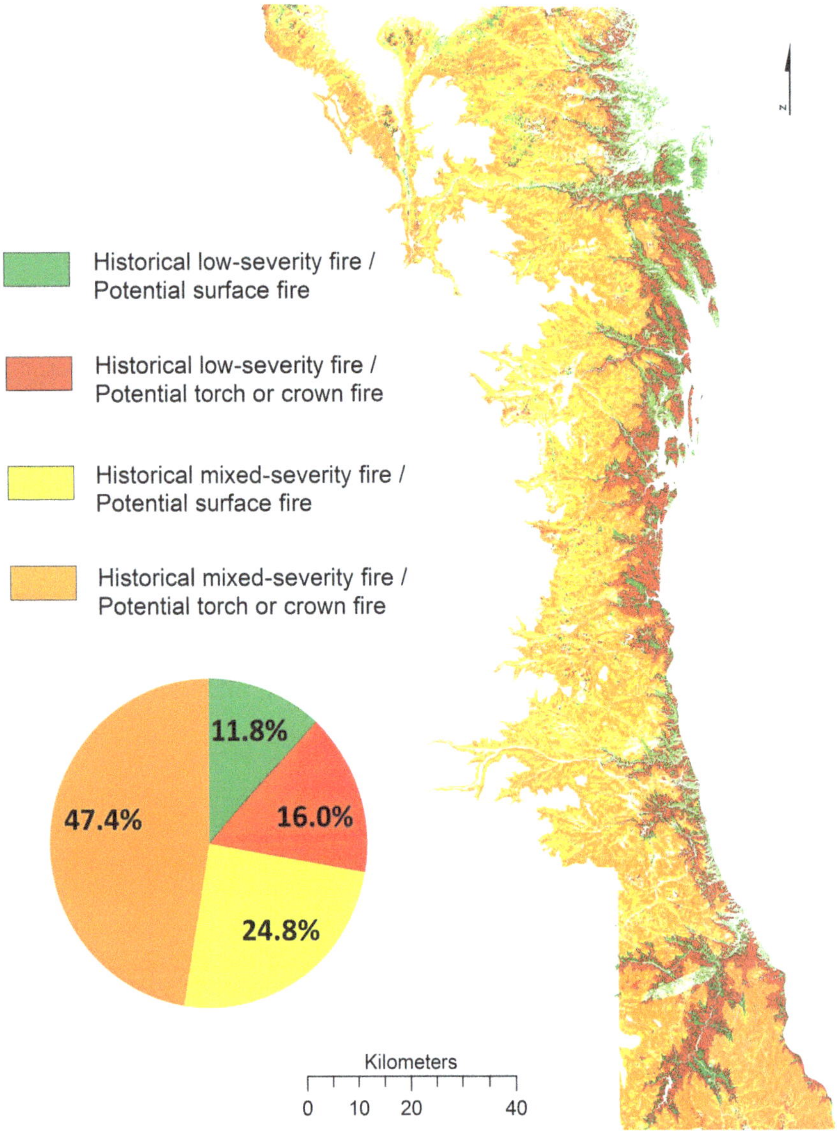

Figure 8. Historical fire severity overlaid with a model of fire potential under extreme (99th percentile weather). The areas and proportion of the study area mapped as historical low-severity fire with current potential for surface fire (green), historical low-severity fire with current potential for torch or crown fire (red), historical mixed-severity with current potential for surface fire (yellow), and historical mixed-severity with current potential for torch or crown fire (orange). Current potential fire behavior is modeled under extreme (99th percentile) weather (1964–2007).

cautious approach to restoration efforts has been recommended [15,40]. In areas naturally characterized by lower frequencies of moderate- to high-severity fires, fuel reduction through prescribed fire and thinning will likely not achieve both ecological restoration and fire hazard mitigation goals. Restoration thinning treatments will not return the fire regime to one of low severity across the Front Range montane zone, which was historically predisposed to periodic fires of varying severities, and are of questionable effectiveness in preventing severe wildfires [2]. Where extreme fire behavior appears within HRV, high-severity fire may largely be explained by extreme weather conditions (for example, high winds and low humidity during severe drought) rather than quantity of woody fuels [38]. This is illustrated by the 2002 Hayman Fire in Colorado in which more than 24,000 ha burned at high severity throughout ponderosa pine and mixed-conifer forests of variable stand structures in a single day [94]. The fire-weather conditions presented here (99th percentile for the late 20th century and early 21st century) represent only the projected average expected conditions by the mid-21st century, which is a pressing issue for existing and future WUI development [95] and fire management [2]. Learning from recent experiences where wildfire damage has been high in the WUI [2], along with considering the costs of suppressing future wildland fires (i.e. reducing forest resiliency, unsustainable federal costs to society [1]) and the sociopolitical expectations of wildland fire management are critical for managing future fire risks and forest integrity.

For the ponderosa pine cover type, an increasing number of studies in the Pacific Northwest [15,96–100] have also documented fire regimes of variable severity. These studies and ours illustrate the importance of collecting evidence on the severity of past fires (i.e. remnant tree and cohort ages, growth releases, tree mortality) because fire severity cannot be reliably interpreted from fire interval data alone, which has often the approach applied to the study of fire history in ponderosa pine forests. Although current implementation of national programs (Collaborative Forest Landscape Restoration Program) and databases (LANDFIRE) recognize variation in fire regimes within the same cover type across different geographical regions, fire regimes of a cover type within a region can still be considered relatively uniform. This is problematic in the Front Range and potentially elsewhere where the historical fire regime varied from low-severity at the lowest elevations to mixed-severity over most of the montane zone regardless of forest type. Additionally, the accuracy of the current cover type (EVT) layer for the region is low. Thus, the spatial and proportional reliability of the LANDFIRE classifications of fire regimes and vegetation do not accurately reflect the historical or current landscape, and therefore will not be highly effective in prioritizing locations for fire management and forest restoration objectives. Further corroboration of these issues are also documented at a national scale (see p. 19 in [92]) "The accuracy of various aspects of the LANDFIRE data is questionable, even when used at intended scale…Without accurate data, many assumptions and actions based on this data will be compromised. There is a need for more realistic and accurate depiction of where wildland fire hazard/risk actually occurs across the country, which can be used to base decisions upon."

Our study, along with prior studies, shows that an assumption of fire regime uniformity is not valid for the ponderosa pine and other montane cover types in the Colorado Front Range and elsewhere (i.e. [15]). These findings challenge ecologists and managers to re-consider the degree of variation in fire regimes within broadly distributed forest types. Thus, management efforts to create large areas of open woodlands in the higher elevation areas of the montane zone of the Colorado Front Range would not be consistent with historical fire regimes and stand structures.

Supporting Information

Figure S1 MTBS fire severity classification (observed) and median fire line intensity (modeled). The comparison was used as a measure of verification of the modeled fire behavior.

Figure S2 MTBS fire severity class (observed) and percentage of crown fire (modeled). The comparison was used as a measure of verification of the modeled fire behavior.

Table S1 Summary table of biophysical factors for each observed fire and potential fire model in analysis.

Table S2 List of plot-level sites. The fire severity classification for each site was documented from forest structure and fire scars in the field using an existing protocol [52].

Acknowledgments

We thank student assistants from Humboldt State University (Alicia Iverson, Kelly Muth, Madelinn Schriver) and Gettysburg College (Erin Palmer) for help with compiling datasets. We also thank Cameron Naficy, Ben Bond-Lamberty, and three anonymous reviewers for insightful comments that improved the manuscript.

Author Contributions

Conceived and designed the experiments: RLS RVP TTV TLS. Performed the experiments: RLS RVP TTV TLS MHG. Analyzed the data: RLS RVP. Wrote the paper: RLS RVP TTV TLS MHG.

References

1. Gorte R (2013) The rising cost of wildfire protection. Headwaters Economic Research Report. http://headwaterseconomics.org/wildfire/fire-cost-background. Accessed 2014 August 23.
2. Calkin DE, Cohen JD, Finney MA, Thompson MP (2013) How risk management can prevent future wildfire disasters in the wildland-urban interface. Proceedings of the National Academy of Science 756–751, doi: 10.1073/pnas.1315088111.
3. Mackun P, Wilson S (2011) Population distribution and change: 2000 to 2010. U.S. Department of Commerce Economics and Statistics Administration. U.S. Census Bureau. http://www.census.gov/prod/cen2010/briefs/c2010br-01.pdf. Accessed 2014 Aug 23.
4. Williams J (2013) Exploring the onset of high-impact mega-fires through a forest land management prism. Forest Ecology and Management 294: 4–10.
5. Stephens SL, Agee JK, Fulé PZ, North MP, Romme WH, et al. (2013) Managing forests and fire in changing climates. Science 342 (6154): 41–42, doi 10.1126/science.12402994.
6. Dillon GK, Holden ZA, Morgan P, Crimmins MA, Heyerdahl EK, et al. (2011) Both topography and climate affected forest and woodland burn severity in two regions of the western US, 1984–2006. Ecosphere 2: Article 130. http://www.esajournals.org/doi/pdf/10.1890/ES11-00271.1.
7. Odion DC, Hanson CT (2008) Fire severity in the Sierra Nevada revisited: conclusions robust to further analysis. Ecosystems 11: 12–15.
8. Miller JD, Safford HD, Crimmins M, Thode AE (2009) Quantitative evidence for increasing forest fire severity in the Sierra Nevada and southern Cascade Mountains, California and Nevada, USA. Ecosystems 12: 16–32.
9. Miller JD, Skinner CN, Safford HD, Knapp EE, Ramirez CM (2012) Trends and causes of severity, size, and number of fires in northwestern California, USA. Ecological Applications 22: 184–203.
10. Miller JD, Collins BM, Lutz JA, Stephens SL, van Wagtendonk JW, et al. (2012) Differences in wildfires among ecoregions and land management agencies in the Sierra Nevada region, California, USA. Ecosphere 3: Article 80.
11. Miller JD, Safford HD (2012) Trends in wildfire severity: 1984 to 2010 in the Sierra Nevada, Modoc Plateau, and southern Cascades, California, USA. Fire Ecology 8: 41–57.
12. Hanson CT, Odion DC (2014) Is fire severity increasing in the Sierra Nevada mountains, California, USA? International Journal of Wildland Fire. 23: 1–8.
13. Williams MA, Baker WL (2012) Comparison of the Higher-Severity Fire Regime in Historical (A.D. 1800s) and Modern (A.D. 1984–2009) Montane Forests Across 624,156 ha of the Colorado Front Range. Ecosystems 15: 832–847.
14. Fulé PZ, Swetnam TW, Brown PM, Falk DA, Peterson DL, et al. (2013) Unsupported inferences of high-severity fire in historical dry forests of the western United States: response to Williams and Baker. Global Ecology and Biogeography. doi: 10.1111/geb.12136.
15. Odion DC, Hanson CT, Arsenault A, Baker WL, DellaSalla DA, et al. (2014) Examining historical and current mixed-severity fire regimes in ponderosa pine and mixed-conifer forests of western North America. PLoS ONE. doi: 10.1371/journal.pone.0087852.
16. Morgan P, Aplet GH, Haufler JB, Humphries HC, Moore MM, et al. (1993) Historical Range Of Variability: A Useful Tool for Evaluating Ecosystem Change. In: Sampson RN, Adams DL, editors. Assessing Forest Ecosystem Health in the Inland West. Sun Valley: Haworth Press. pp. 87–109.
17. Keane RE, Hessburg PF, Landres PB, Swanson FJ (2009) The use of historical range of variability (HRV) in landscape management. Forest Ecology and Management 258: 1025–1037.
18. Hayward GD, Veblen TT, Suring LH, Davis B (2012) Challenges in the application of historical range of variation to conservation and land management. *In* Historical Environmental Variation in Conservation and Natural Resource Management. *Eds.* Wiens JA, Hayward GD, Safford HD, Giffen CM. John Wiley & Sons, Ltd.
19. Millar CI, Stephenson NL, Stephens SL (2007) Climate change and forests of the future: managing in the face of uncertainty. Ecological Applications 17: 2145–2151.
20. Wiens JA, Hayward GD, Safford HD, Giffen C (2012) Historical environmental variation in conservation and natural resource management. John Wiley & Sons.
21. Moritz MA, Hurteau MD, Suding KN, D'Antonio CM (2013) Bounded ranges of variation as a framework for future conservation and fire management. Annals of the New York Academy of Sciences 1286: 92–107.
22. DeLeo GA, Levin S (1997) The Multifaceted Aspects of Ecosystem Integrity. Conservation Ecology 1: 3.
23. Parrish JD, Braun DP, Unnasch RS. (2003) Are we conserving what we say we are? Measuring ecological integrity within protected areas. Bioscience 53: 851–860.

24. Healthy Forests Restoration Act. HR 1904 (2003) http://www.forestsandrangelands.gov/resources/overview/hfra-implementation12-2004.shtml. Accessed 2014 August 23.
25. Peterson DL, Millar CI, Joyce LA, Furniss MJ, Halofsky JE, et al. (2011) Responding to climate change in National Forests: A guidebook for developing adaptation options. USDA Forest Service General Technical Report PNW-GTR-855.
26. Wildland Fire Executive Council (2014) The National Strategy: The Final Phase in the Development of the National Cohesive Wildland Fire Management Strategy. http://www.forestsandrangelands.gov/strategy/documents/strategy/CSPhaseIIINationalStrategyApr2014.pdf. Accessed 2014 August 23.
27. Brown JK, Oberheu RD, Johnston CM (1982) Handbook for inventorying surface fuels and biomass in the Interior West. USDA Forest Service General Technical Report. INT-129.
28. Covington WW, Moore M (1994) Southwestern Ponderosa Forest Structure: changes since Euro-American settlement. Journal of Forestry 92: 39–47.
29. Covington WW (2000) Helping western forests heal. Nature 408: 135–136.
30. Agee JK (2002) The Fallacy of Passive Management Managing for Firesafe Forest Reserves. Conservation in Practice 3: 18–26.
31. Allen CD, Savage M, Falk DA, Suckling KF, Swetnam TW, et al. (2002) Ecological restoration of southwestern ponderosa pine ecosystems: a broad perspective. Ecological Applications 12: 1418–1433.
32. Agee JK, Skinner CN (2005) Basic principles of forest fuel reduction treatments. Forest Ecology and Management 211: 83–96.
33. Stephens SL, Ruth LW (2005) Federal forest-fire policy in the United States. Ecological Applications 15: 532–542.
34. Roos CI, Swetnam TW (2012) A 1416-year reconstruction of annual, multidecadal, and centennial variability in area burned for ponderosa pine forests of the southern Colorado Plateau region, Southwest USA. The Holocene 22: 281–290.
35. United States Forest Service (2002) National Database of State and Local Wildfire Hazard Mitigation Programs, http://www.wildfireprograms.usda.gov. Accessed 2014 August 23.
36. Mell WE, Manzello SL, Maranghides A, Butry DT, Rehm RG (2010) The wildland–urban interface fire problem – current approaches and research needs. International Journal of Wildland Fire 19: 238–251.
37. Veblen TT (2003) An introduction to key issues in fire regime research for fuels management and ecological restoration. In: Omi P, Joyce L, Eds. Fire, fuel treatments and ecological restoration: conference proceedings 2002, 16–18 April, Fort Collins, CO. USDA, Forest Service, RMRS-P-29.
38. Schoennagel T, Veblen TT, Romme WH (2004) The Interaction of Fire, Fuels, and Climate across Rocky Mountain Forests. BioScience 54: 661–676.
39. Nagel TA, Taylor AH (2005) Fire and persistence of montane chaparral in mixed conifer forest landscapes in the northern Sierra Nevada, Lake Tahoe Basin, California, USA. Journal of the Torrey Botanical Society 132: 442–457.
40. Baker WL, Veblen TT, Sherriff RS (2007) Fire, fuels and restoration of ponderosa pine–Douglas fir forests in the Rocky Mountains, USA. Journal of Biogeography 34: 251–269.
41. Klenner W, Walton R, Arsenault A, Kremsater L (2008) Dry forests in the Southern Interior of British Columbia: Historic disturbances and implications for restoration and management. Forest Ecology and Management 256: 1711–1722.
42. Hutto RL (2008) The ecological importance of severe wildfires: some like it hot. Ecological Applications 18: 1827–1834.
43. Colombaroli D, Gavin DG (2010) Highly episodic fire and erosion regime over the past 2,000 y in the Siskiyou Mountains, Oregon. Proceedings of the National Academy of Sciences 107: 18909–18915.
44. Veblen TT, Lorenz DC (1986) Anthropogenic disturbance and recovery patterns in montane forests, Colorado Front Range. Physical Geography 7: 1–24.
45. Veblen TT, Lorenz DC (1991) The Colorado Front Range: A Century of Ecological Change. Salt Lake: University of Utah Press. 186 p.
46. Mast JN, Veblen TT, Linhart YB (1998) Disturbance and climatic influences on age structure of ponderosa pine at the pine/grassland ecotone, Colorado Front Range. Journal of Biogeography 25: 743–755.
47. Brown P, Kaufmann M, Shepperd W (1999) Long-term, landscape patterns of past fire events in a montane ponderosa pine forest of central Colorado. Landscape Ecology 14: 513–532.
48. Kaufmann MR, Regan CM, Brown PM (2000) Heterogeneity in ponderosa pine/Douglas-fir forests: age and size structure in unlogged and logged landscapes of central Colorado. Canadian Journal of Forest Research 30: 698–711.
49. Huckaby LS, Kaufmann MR, Stoker JM, Fornwalt PJ (2001) Landscape Patterns of Montane Forest Age Structure Relative to Fire History at Cheesman Lake in the Colorado Front Range. U.S.D.A. Forest Service Proceedings RMRS-P-22.
50. Ehle DS, Baker WL (2003) Disturbance and stand dynamics in ponderosa pine forests in Rocky Mountain National Park, USA. Ecological Monographs 73: 543–566.
51. Sherriff RL, Veblen TT (2006) Ecological effects of changes in fire regimes in Pinus ponderosa ecosystems in the Colorado Front Range. Journal of Vegetation Science 17: 705–718.
52. Sherriff RL, Veblen TT (2007) A spatially-explicit reconstruction of historical fire occurrence in the ponderosa pine zone of the Colorado Front Range. Ecosystems 10: 311–23.
53. Schoennagel T, Sherriff RL, Veblen TT (2011) Fire history and tree recruitment in the Colorado Front Range upper montane zone: implications for forest restoration. Ecological Applications 21: 2210–2222.
54. Williams MA, Baker WL (2012) Spatially extensive reconstructions show variable-severity fire and heterogeneous structure in historical western United States dry forests. Global Ecology and Biogeography 21: 1042–1052.
55. Romme WH, Veblen TT, Kaufmann MR, Sherriff RL, Regan CM (2003) Ecological effects of the Hayman fire, Part I: Historical (pre-1860) and current (1860–2002) fire regimes. In: Graham RT, editor. Hayman Fire Case Study. Fort Collins CO, Rocky Mountain Research Station: USDA Forest Service, General Technical Report RMRS-GTR-114. pp. 181–195.
56. Kaufmann MR, Veblen TT, Romme WH (2006) Historical fire regimes in ponderosa pine forests of the Colorado Front Range, and recommendations for ecological restoration and fuels management. Front Range Fuels Treatment Partnership Roundtable, findings of the Ecology Workgroup. www.frftp.org/roundtable/pipo.pdf.
57. Colorado Climate Center (2013) http://climate.atmos.colostate.edu/. Accessed 2014 August 23.
58. Graham RT (2003) Hayman Fire Case Study. Gen. Tech. Rep. RMRS-GTR-114. Ogden, UT: U.S. Department of Agriculture, Forest Service, Rocky Mountain Research Station. 396 p.
59. National Fire Protection Association Codes and Standards (1989) www.nfpa.org. Accessed 2014 August 23.
60. Veblen TT, Kitzberger T, Donnegan J (2000) Climatic and Human Influence on Fire Regimes in Ponderosa Pine Forests in the Colorado Front Range. Ecological Applications 10: 1178–1195.
61. Sherriff RL, Veblen TT (2008) Variability in fire–climate relationships in ponderosa pine forests in the Colorado Front Range. International Journal of Wildland Fire 17: 50–59.
62. Gartner MH, Veblen TT, Sherriff RL, Schoennagel TL (2012) Proximity to grasslands influences fire frequency and sensitivity to climate variability in ponderosa pine forests of the Colorado Front Range. International Journal of Wildland Fire 21: 562–571.
63. Duncan (1989) An evaluation of errors in tree age estimates based on increment cores in Kahikatea (Dacrycarpus dacrydioides). New Zealand Natural Sciences Journal 16: 31–37.
64. Barrett SW, Arno SE (1998) Increment-borer methods for determining history in coniferous forests. USDA Forest Service, General Technical Report INT-244.
65. Agee JK (1993) Fire Ecology of the Pacific Northwest Forest. Island Press, Washington DC. pp. 493.
66. Hessburg P, Salter RB, James K (2007) Re-examining fire severity relations in pre-management era mixed conifer forests: inferences from landscape patterns of forest structure. Landscape Ecology 22: 5–24.
67. Wilson JP, Gallant JC (2000) Terrain analysis: principles and applications. New York: Wiley.
68. Jenson SK, Domingue JO (1988) Extracting topographic structure from digital elevation data for geographic information system analysis. Photogramm Eng Remote Sens 54: 1593–1600.
69. Breiman L, Friedman JH, Olshen RA, Stone CJ (1984) Classification and Regression Trees. Monterrey: Wadsworth and Brooks/Cole. 358 p.
70. Breiman L (2001) Random Forests. Machine Learning 45: 5–32.
71. Moisen GG, Frescino TS (2002) Comparing five modelling techniques for predicting forest characteristics. Ecological Modelling 157: 209–225.
72. Qian SS (2010) Environmental and Ecological Statistics. Boca Raton: RC Press.
73. Eidenshink J, Schwind B, Brewer K, Zhu Z, Quayle B, et al. (2007) A project for monitoring trends in burn severity. Fire Ecology 3: 3–21.
74. Scott JH, Burgan RE (2005) Standard fire behavior fuel models: A comprehensive set for use with Rothermel's surface fire spread model. In: Fort Collins, CO: USDA Forest Service Rocky Mountain Research Station, General Technical Report RMRS-GTR-153. 72 p.
75. LANDFIRE (2010) U.S. Department of Agriculture, Forest Service, U.S. Department of Interior. http://www.landfire.gov/index.php. Accessed 2014 August 23.
76. Krasnow K, Schoennagel T, Veblen TT (2009) Forest fuel mapping and evaluation of LANDFIRE fuel maps in Boulder County, Colorado, USA. Forest Ecology and Management 257: 1603–1612.
77. Finney MA (2006) An overview of FlamMap fire modeling capabilities. In: Fuels management-how to measure success: conference proceedings. 2006 March 28–30; Portland, Oregon. Proceedings RMRS-P-41. Fort Collins, CO: U.S. Department of Agriculture, Forest Service, Rocky Mountain Research Station: 213–220.
78. Platt RV, Schoennagel T, Veblen TT, Sherriff RL (2011) Modeling wildfire potential in residential parcels: A case study of the north-central Colorado Front Range. Landscape and Urban Planning 102: 117–126.
79. Rothermel RC (1972) A mathematical model for predicting fire spread in wildland fuels. Ogden, UT: USDA Forest Service Intermountain Forest and Range Experiment Station, General technical report INT-115.
80. Van Wagner CE (1977) Conditions for the start and spread of crown fire. Canadian Journal of Forest Research 7: 23–34.

81. Rothermel RC (1991) Predicting behavior and size of crown fires in the Northern Rocky Mountains. Ogden, UT: USDA Forest Service Intermountain Forest and Range Experiment Station, Research paper INT-438. 46 p.
82. Nelson RM (2000) Prediction of diurnal change in 10-hour fuel moisture content. Canadian Journal of Forest Research 31: 1071–1087.
83. Scott JH, Reinhardt ED (2001) Assessing crown fire potential by linking models of surface and crown fire behavior. In: Fort Collins CO: U.S. Department of Agriculture, Forest Service, Rocky Mountain Research Station, Research Paper RMRS-RP-29. 59 p.
84. Intergovernmental Panel on Climate Change (2007) https://www.ipcc.ch/publications_and_data/publications_and_data_reports.shtml. Accessed 2014 August 23.
85. NCAR Science and Research Data for IPCC AR4 Data Guide (2012) http://eos-earthdata.sr.unh.edu/data/dataGuides/ci_science_dg.pdf. Accessed 2014 August 23.
86. Sherriff RL (2004) The historic range of variability of ponderosa pine in the northern Colorado Front Range: past fire types and fire effects. Ph.D. Dissertation. University of Colorado, Boulder, Colorado.
87. Rocca ME, Brown PM, MacDonald LH, Carrico CM (2014) Climate change impacts on fire regimes and key ecosystem services in Rocky Mountain forests. Forest Ecology and Management 327: 290–305.
88. Miller JD, Thode AE (2007) Quantifying burn severity in a heterogeneous landscape with a relative version of the delta Normalized Burn Ratio (dNBR). Remote Sensing of Environment 109: 66–80.
89. Soverel NO, Perrakis DDB, Coops NC (2010) Estimating burn severity from Landsat dNBR and RdNBR indices across western Canada. Remote Sensing of Environment 114: 1896–1909.
90. Front Range Fuels Treatment Partnership (2006) Living with Fire: Protecting Communities and Restoring Forests – Findings and Recommendations of the Front Range Fuels Treatment Partnership Roundtable. http://frftp.org/roundtable/report.pdf. Accessed 2014 August 23.
91. Schoennagel T, Nelson CR (2011) Restoration relevance of recent National Fire Plan treatments in forests of the western United States. Frontiers in Ecology and the Environment 9: 271–277.
92. The National Cohesive Wildland Fire Management Strategy and Risk Analysis – Phase III Report (2013) http://www.forestsandrangelands.gov/. Accessed 2014 August 23.
93. Platt RV, Veblen TT, Sherriff RL (2006) Are Wildfire Mitigation and Restoration of Historic Forest Structure Compatible? A Spatial Modeling Assessment. Annals of the Association of American Geographers 96: 455–470.
94. Finney MA, Bartlette R, Bradshaw L, Close K, Collins BM, et al. (2003) Fire behavior, fuel treatments, and fire suppression on the Hayman Fire. *In* Hayman Fire case study. Ed. Graham RT. USDA Forest Service, Rocky Mountain Research Station, General Technical Report RMRS-GTR-114, pp. 33–179.
95. Theobald DM, Romme WH (2007) Expansion of the US wildland-urban interface. Landscape and Urban Planning 83: 340–354.
96. Arno SF, Scott JH, Hartwell MG (1995) Age-class structure of old growth ponderosa pine/Douglas-fir stands and its relationship to fire history. USDA Forest Service, Research Paper INT-481.
97. Wright CS, Agee JK (2004) Fire and vegetation history in the eastern Cascade mountains. Ecological Applications 14: 443–459.
98. Everett RL, Schellhaas R, Keenum D, Spurbeck D, Ohlson P (2000) Fire history in the ponderosa pine/Douglas-fir forests on the east slope of Washington Cascades. Forest Ecology and Management 129: 207–225.
99. Hessburg PF, Salter RB, James KM (2007) Re-examining fire severity relations in pre-management era mixed conifer forests: inferences from landscape patterns of forest structure. Landscape Ecology 22: 5–24.
100. Perry DA, Hessburg PF, Skinner CN, Spies TA, Stephens SL, et al. (2011) The ecology of mixed severity fire regimes in Washington, Oregon, and northern California. Forest Ecology and Management 262: 703–717.

11

Predicted Effects of Gypsy Moth Defoliation and Climate Change on Forest Carbon Dynamics in the New Jersey Pine Barrens

Alec M. Kretchun[1]*, Robert M. Scheller[1], Melissa S. Lucash[1], Kenneth L. Clark[2], John Hom[3], Steve Van Tuyl[3]

1 Portland State University, Department of Environmental Science and Management, Portland, Oregon, United States of America, **2** Silas Little Experimental Forest, USDA Forest Service, New Lisbon, New Jersey, United States of America, **3** United States Department of Agriculture Forest Service, Newtown Square, Pennsylvania, United States of America

Abstract

Disturbance regimes within temperate forests can significantly impact carbon cycling. Additionally, projected climate change in combination with multiple, interacting disturbance effects may disrupt the capacity of forests to act as carbon sinks at large spatial and temporal scales. We used a spatially explicit forest succession and disturbance model, LANDIS-II, to model the effects of climate change, gypsy moth (*Lymantria dispar* L.) defoliation, and wildfire on the C dynamics of the forests of the New Jersey Pine Barrens over the next century. Climate scenarios were simulated using current climate conditions (baseline), as well as a high emissions scenario (HadCM3 A2 emissions scenario). Our results suggest that long-term changes in C cycling will be driven more by climate change than by fire or gypsy moths over the next century. We also found that simulated disturbances will affect species composition more than tree growth or C sequestration rates at the landscape level. Projected changes in tree species biomass indicate a potential increase in oaks with climate change and gypsy moth defoliation over the course of the 100-year simulation, exacerbating current successional trends towards increased oak abundance. Our research suggests that defoliation under climate change may play a critical role in increasing the variability of tree growth rates and in determining landscape species composition over the next 100 years.

Editor: Michael L. Fine, Virginia Commonwealth Univ, United States of America

Funding: This work was supported by the National Fire Plan, Forest Health Monitoring (NE-F-08-01), and the Eastern Landfire Prototype. The funders had no role in study design, data collection and analysis, decision to publish, or preparation of the manuscript.

Competing Interests: The authors have declared that no competing interests exist.

* Email: AlecKretchun@Gmail.com

Introduction

Globally, forests are responsible for 2.4±04 Pg of carbon (C) storage, an amount equivalent to ~30% of annual anthropogenic carbon emissions [1]. The temperate forests of the United States in particular are estimated to have increased their carbon uptake by up to 33% from the 1990s to the 2000s [1]. There is some uncertainty as to whether forest growth can continue to offset global CO_2 emissions in the coming century [2]. Broadly speaking, forests in the United States are still recovering from post-settlement (1700-1935) carbon losses, and this recovery is expected to continue, though disturbance and future climate will likely be huge determinants [3].

Disturbance regimes within forests can have significant impacts on carbon storage and cycling. 'Discrete' disturbances are usually short-lived, often severe events that can have significant impacts on C fluxes (e.g. annual growth) and reservoirs (e.g. woody plant tissue)[4]. Wildfire and harvesting are two such disturbances – though they often last no more than days or weeks, their impacts can be substantial [5]. In contrast, gypsy moth outbreaks are seasonal events which can persist for several years across a region [6]. Their direct effects, such as seasonal loss of leaf area or increased moisture stress, impact ecosystem function and structure, particularly in forests of long-lived trees [7–9].

Climate change can be thought of as a unique, long-term disturbance in itself, or alternatively, as a driver of forest change. Climate change may alter forest structure and function in many ways, causing range shifts, influencing growth patterns [10,11], and ultimately altering successional trends in forests. However, it remains uncertain whether and to what degree disturbance regimes will mitigate or exacerbate the effects of climate change. In many cases, a degree of synergism between climate change and disturbance regimes is expected. For example, warming temperatures may allow insects to expand into previously climatically-restricted geographic areas, which could change tree species composition and carbon sequestration potential [8,12]. In other cases, the outcomes of one disturbance might partially offset the other. Increased growth or a longer growing season caused by warming temperatures could help mitigate some carbon loss from increased disturbance under a changing climate, though such mitigation might be lessened with the possibility of increases in ecosystem respiration [11,13].

Because of the complexity of the interactions, understanding what processes are ultimately responsible for driving forest change is challenging. Furthermore, systems can be driven by emergent behaviors of a combination of processes, which may not become evident unless examined at decadal or centurial time scales [14]. Landscape models can help address both these issues [15,16]. Landscape models allow individual processes to be isolated such that interactions between succession, disturbance, and climate can be decomposed to reveal how each may influence the others. These influences can then be examined at spatial and temporal scales that may not accessible by other means.

The New Jersey Pine Barrens (NJPB) is a system characterized by diverse disturbance regimes and is emblematic of risks to carbon storage in temperate forests. Previous research has suggested that the forests of the NJPB are a net C sink [7,17–19], however the interactive effects of climate change and gypsy moth defoliation on C dynamics are unexplored. The area is well instrumented for analysis of forest change and carbon cycling; eddy flux data from three representative sites quantified the effects of insect defoliation and fire on NEE and evapotranspiration [7,20].

In this study, our objectives were to investigate the landscape-scale effects of gypsy moth defoliation on the forests of the NJPB, and to understand how this disturbance interacts with wildfire and climate change. We used the LANDIS-II modeling framework to project and analyze C dynamics and species composition changes over the next 100 years. Comparison of simulations incorporating wildfire, climate change, and gypsy moth defoliation provide insight into how C dynamics may be affected by these disturbances and how major tree species may respond across the landscape. This research effort builds upon previous research in the area, including empirical data analysis [7] and simulation modeling [17,21,22].

Methods

Study Area

The New Jersey Pine Barrens (NJPB) is the largest contiguous forest along the Northeastern coastal plain (Fig. 1). The terrain of the NJPB is relatively homogeneous, with plains, low-angle slopes, and wetlands. The maximum elevation is 63 meters. Temperatures are cool, ranging from monthly mean temperatures of 0.3°C in January and 23.8°C in June (New Jersey State Climatologist, 1930-2009). Mean annual precipitation is 1123±182 mm. The soils are sandy, nutrient-poor, acidic, and have low cation exchange capacity [23].

From pre-settlement to ~1940, fire played a substantial ecological role. Following the abandonment of industrial forestry in the region, a period of intense wildfires occurred, a consequence of the young age and density of the resultant stands. Fire suppression and increasing fragmentation greatly reduced the presence of wildfire on the landscape starting around 1940 [24]. Though many legacy stands remain from this period, land use change and fire suppression has greatly reduced the occurrence of large wildfires, causing a gradual shift towards a more oak- (*Quercus* sp.) dominated system [21]. Though fire occurrence has been reduced, it continues to shape forest structure.

The forests of the New Jersey Pine Barrens (NJPB) are differentiated into three primary forest types largely based on differences in soil type and disturbance history. Upland forests (58% by land area) are dominated by pitch pine (*Pinus rigida* Mill.) and several species of oak, including chestnut oak (*Quercus prinus* L.), white oak (*Q. alba* L.) and black oak (*Q. velutina*) [25]. The understory is comprised of moderate to dense shrub cover of genera *Quercus, Gallussacia, and Vaccinuim* [26]. Soils in the uplands forests are generally sandy, acidic, and nutrient-deficient [7]. Wetland forests (38% of forested land) are dominated by Atlantic white cedar (*Chamaecyparis thyoides*), moisture sensitive oaks, shortleaf (*P. echinata* Mill.) and pitch pines, and some non-oak hardwoods (e.g., blackgum, *Nyssa sylvatica*) [25]. Wetland forest soils have higher water holding capacity and this forest type is primarily associated with the lowland streams and rivers that extend from the coast and reticulate the area. The smallest forest type (<5% by land area) is the xeric pine plains. These forests are almost exclusively 'pygmy' pitch pine, their low productivity a consequence of the deep, well-drained, nutrient poor soils.

Gypsy moths spread into New Jersey in 1966 after their original introduction near Boston, MA in the late 1860s. Although a single year's defoliation is often not sufficient to cause substantial (>20%) mortality of overstory oaks, outbreaks typically last for several years; this cumulative defoliation is apparently the main cause of moth-attributed mortality to affected trees. Outbreaks in more xeric forests typically occur with a 5–10 year periodicity, and can last from one to three years in mixed oak-pine forests such as those in the NJPB [27]. Moth damage in the area has peaked four times since establishment: 1972, 1981,1990, and 2007 [20].

Modeling Framework

To evaluate landscape disturbances, tree species dynamics, and C fluxes, we used the landscape forest succession and disturbance model LANDIS-II [16,28] to model the ~400,000 ha of southern New Jersey that falls within the Pinelands Management Area, which is managed jointly by the Pinelands Commission and the New Jersey Department of Environmental Protection. LANDIS-II simulates disturbance, management, and succession in a framework that emphasizes spatial interactions across the landscape and among processes (e.g., climate change, insects, fire, harvesting, succession, and seed dispersal) projected over decades or centuries (Fig. 2).

Rather than representing establishment and growth of individual trees, LANDIS-II bins trees into 'cohorts' – a grouping that organizes trees by their species and age range (e.g. pitch pine aged 11-20 years) while tracking their above ground biomass. Aggregation of trees into species-age cohorts is well suited for large study areas, as modeling individual trees would increase computation time significantly. Life history traits such as longevity and sexual maturity are required for each modeled species. At any given site, the successional dynamics of tree species are determined by their unique life history traits and their ability to establish and compete in the community in which they are found. The landscape itself is represented as a grid of interconnected cells, across which disturbances spread and dispersal occurs.

The NJPB landscape was classified into 7 different ecoregions: three upland types, three lowland types, and the xeric pine plains. Upland, lowland, and plains categorizations were determined using landcover and forest maps; the upland and lowland categories were further stratified by soil water holding capacity using the NRCS SSURGO dataset [17]. Climate and soils were assumed to be homogeneous within each ecoregion, though climate is relatively consistent among ecoregions as well [17].

With the exclusion of non-forested land use types, the total area modeled was 236,381 ha. Cell resolution was 1 ha and simulations were run for 100 years (2000–2100) at annual time steps to capture the univoltine biology of gypsy moths. Each scenario was replicated five times, the maximum allowable given our computational limits.

Carbon and Nitrogen Simulation

The Century Succession (v3) extension for LANDIS-II ('Century extension') was used to simulate both above and belowground C and N dynamics and the regenerative processes of our 14 trees

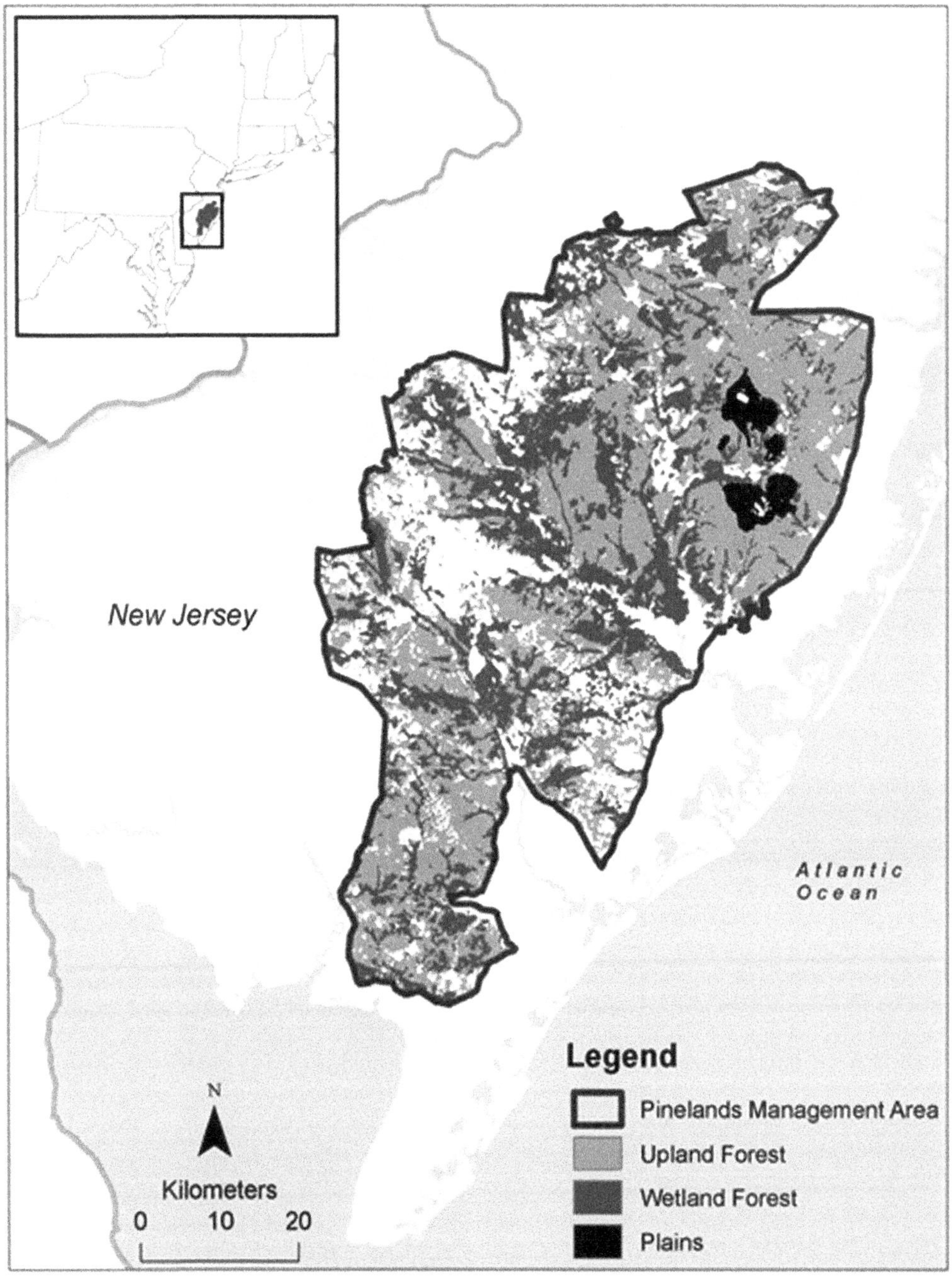

Figure 1. The New Jersey Pine Barrens (NJPB) and associated forest types. The Pinelands Management Area is outlined in black.

species [17,29–31]. Cohort biomass within the Century extension is partitioned into above- and belowground components of wood, leaves, fine roots and coarse roots. Aboveground net primary productivity (ANPP), the accumulation of aboveground biomass by tree cohorts, is determined by cohort age, longevity, and inter- and intraspecific competition for resources. Species-specific inputs determine allocation between the coarse wood and fine leaf pools. Belowground productivity is defined as a fixed percentage of aboveground productivity, with separate ratios for coarse and fine material. The Century extension tracks cohort growth at a monthly time step.

In addition to tracking cohort reproduction, growth, and mortality, the Century extension simulates changes in dead biomass (i.e., detritus) and soil organic carbon (SOC). Within the Century extension, soils are divided into fast, slow, and passive organic matter pools [30,31]. Soil respiration is a product of climate, edaphic properties, and structural and chemical properties of detrital inputs [29]. The difference between ANPP and soil respiration is calculated by the Century extension to allow the calculation of C sequestration rates (NEE). Probability of establishment (P_{est}), which describes the probability that a new cohort of a given species will become established given sufficient light and propagule presence [28], is calculated within the Century extension as a function of climate and soils.

Model calibration/validation for two critical ecosystem carbon fluxes (ANPP, NEE) against empirical data was done by matching

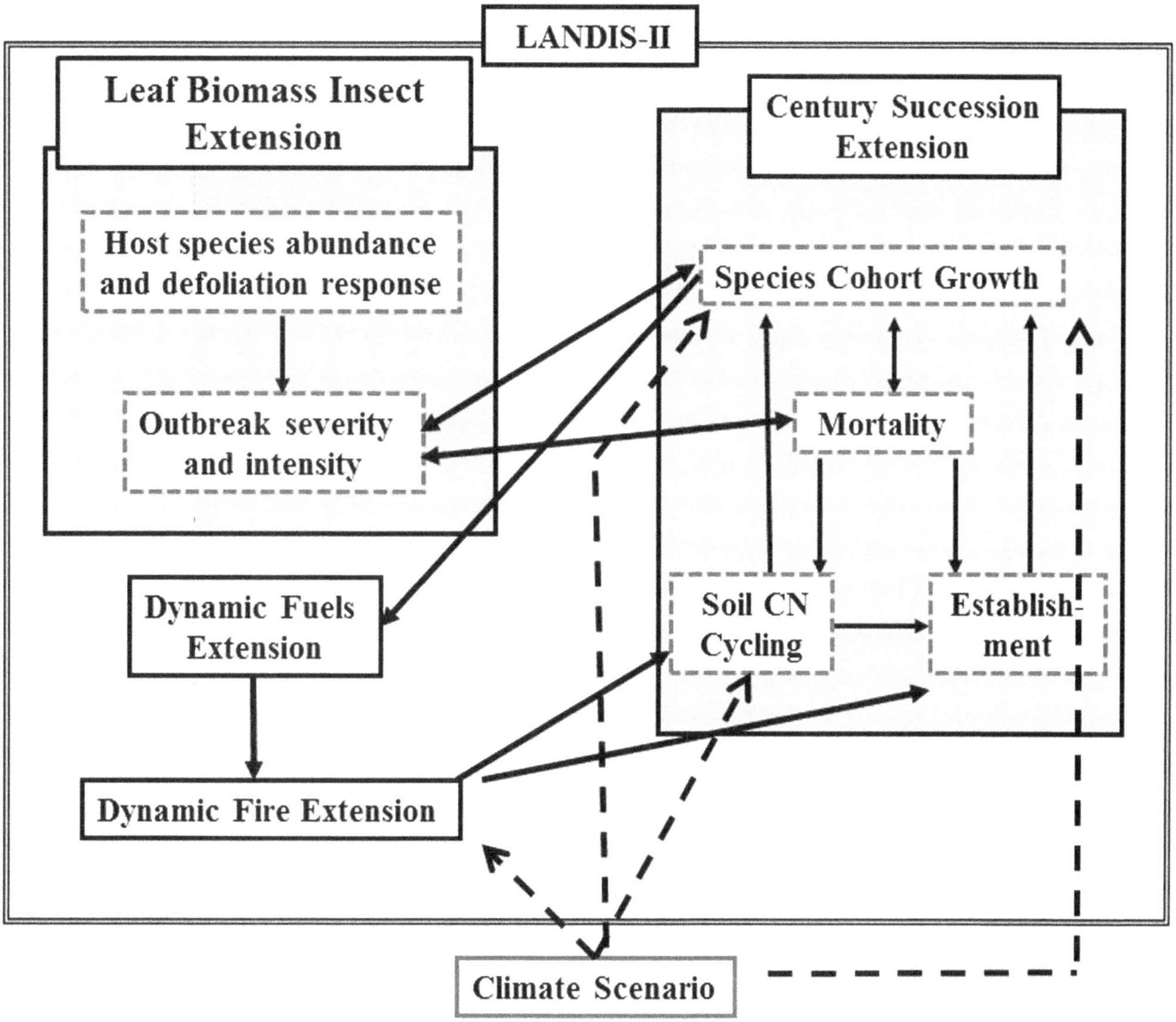

Figure 2. Conceptual diagram of LANDIS-II model. Arrows indicate interactions between model core, model extensions and climate scenarios.

literature values of ANPP and fitting simulated NEE values against local flux tower sites. Simulated ANPP values (450 g C m^2) fell within the range of literature values for the region (~440 g C m^2) [18]. Simulated NEE was calibrated against monthly data from 2005–2007 using two eddy flux tower sites from the area, with a third used for validation. These towers were located within areas of three representative forest types in the region: oak-pine, pine-oak, and pitch pine-scrub oak [7,20]. Model NEE estimates fit empirical flux tower data with a significant correlation of $R^2>0.6$, meeting calibration criteria previously established [17]. A more detailed description of the Century extension parameterization process, along with all input parameters used, can be found elsewhere [21,22].

Wildfire Simulation

The Dynamic Fire system extension and Dynamic Leaf Biomass Fuels extension (known collectively as the Dynamic Fuels and Fire system, DFFS) were used to simulate fuels and fire across the landscape [32]. Wildfire ignitions are stochastically generated. The fire then spreads based upon a fire weather index (FWI) derived from climate inputs and fuel type. The Dynamic Leaf Biomass Fuels extension uses cohort characteristics (species, age-ranges, aboveground biomass) to determine fuel types for each individual cell. These fuel classifications were previously described [17]. FWI and fuel type classifications combine to determine fire intensity, which along with species-specific fire tolerances, determines cohort mortality (younger cohorts are more vulnerable) and the amount of wood and litter consumed. Model calibrations were made for current conditions; the simulation of climate change impacts on fires is an emergent property of mechanistic simulation of altered fire weather patterns [17]. Previous simulation results using LANDIS-II in the area resulted in reasonable ignition density, fire size, and acreage burned [17].

Climate Change Data

Hadley Community Model 3 (HadCM3) climate projections were acquired from the Intergovernmental Panel on Climate Change (IPCC) data center (http://www.ipcc-data.org/). Baseline, or 'current', climate data was created from 30 year (1961–1990) monthly averages from the US EPA Center for Exposure

Assessment Modeling (www.epa.gov/ceampubl). The A2 or 'high emissions' scenario was used to simulate climate change [17]. Due to ocean influences on climate, a low emission scenario (B1) did not exhibit substantial deviation from current climate (data not shown). Under the A2 scenario, temperatures steadily rose over the 100-year simulation, projecting an increase of 7.6°C in maximum summer temperatures and 5.1°C in minimum winter temperatures. Total precipitation increased slightly and became much more variable over time. The seasonality of precipitation also shifted, peaking in August at year 2000 and November at year 2100.

Climate scenarios were consistent across extensions, though each extension requires different temporal scaling of the climate stream. For the Century extension, which operated at a 5-year time step, average monthly means and standard deviations for temperature maxima, minima and precipitation were calculated from the HadCM3 A2 high emissions climate projection. The DFFS extension required daily weather and in the case of climate change scenarios, this daily weather must match the projected changes in temperature and precipitation from the HadCM3 model. Downscaling of the HadCM3 monthly climate projections was conducted using a stochastic weather generator, LARS-WG to generate daily maximum and minimum temperatures and precipitation based on the HadCM3 A2 scenario [22]. This is done by combining a reference set of daily temperature and precipitation values (collected at Atlantic City, NJ) with the HadCM3 monthly projections within LARS-WG.

Gypsy Moth Simulation, Parameterization, and Calibration

The Leaf Biomass Insect (v1.0) extension was used to simulate landscape level effects on forest growth and mortality due to outbreaks of gypsy moths, a defoliating insect [33]. This extension stochastically introduces defoliation events uniquely parameterized for individual insect species. Temporal aspects of outbreaks are defined by specifying the duration and time between individual outbreaks. Outbreak spatial characteristics (patch size, within-patch variability, among-patch disturbance spread) are also simulated. At the cell level, outbreak spread is dependent upon adjacent defoliation levels and tree species composition.

Defoliation within the extension is defined as the amount of foliage removed from the canopy and transferred to the forest floor as unconsumed leaf matter and insect frass, resulting in reduced ANPP and possible mortality. Chemical composition of both removed foliage and insect frass is distinguished from senescent leaves within the internal nutrient cycling dynamics of the Century extension. For example, unconsumed litter that falls to the forest floor typically has much higher N content and decomposes more rapidly than litter deposited later in the year. Also, gypsy moth frass is also more readily decomposed than leaf litter. If an outbreak persists over many years, the decline in ANPP and likelihood of tree mortality increase. These two tree responses (ANPP reduction and mortality) are unique to each tree species, and operate based on 'cumulative defoliation' which is defined as the sum of defoliation over all the years an outbreak persists. This cumulative defoliation is then translated into percentage of cohorts removed by user defined relationships. Species growth reduction and mortality curves for gypsy moth were previously determined for all functional groups of trees within the NJPB using empirical defoliation data from mixed oak/conifer stands in Maryland [33]. These functions (growth reduction and mortality) are defined by a slope and intercept. Growth reductions were negative linear relationships for most tree species, while mortality curves are defined by exponential relationships (see File S1). Actual defoliation is determined by neighborhood defoliation intensity distributions and tree species susceptibility values. Previously developed neighborhood defoliation intensity distributions were implemented in this project.

The Leaf Biomass Insect extension was calibrated using a combination of empirical data and previous research. Defoliation data in the form of aerial flyover surveys was acquired from the US Forest Service Forest Health Monitoring program site [34]. Maps and associated attribute tables contain several aspects of forest condition, one of which is observed pest and disease damage. Where present, this damage was categorized by forest composition type, damage type (e.g. defoliation, dieback, etc.), severity (categorical), and likely agent. These maps, drawn by hand with geospatial reference aids, were digitized into files that contain all documented information on forest condition for southern New Jersey.

Defoliation patches were determined by selecting all damage attributed to gypsy moths; two categorical severity classes were present: low and high. High severity damage was classified as > 75% reduction in deciduous canopy foliage, following NJ state protocols. For this project, only areas of high severity were chosen to calibrate initial outbreak patches and the damage agent classification.

Temporal characteristics of gypsy moth outbreaks as well as host species preferences were parameterized from local empirical data. Outbreak patch size distributions were determined with the R v.2.13 statistical software package [35]. Growth reduction and mortality response curve parameters were taken from previous work [33]. Total defoliated sites in the initial year of simulated outbreaks were matched with total defoliation recorded in the flyover maps for the year 2007, the first year of the most recent outbreak in the area. Because of the binary nature of the empirical defoliation data, matching total defoliation area in the initial year of actual outbreaks with total defoliation area in the initial year of simulated outbreaks, was the most practical method for calibrating total area defoliated. Calibration involved iteratively changing the 'Initial area calibrator' parameter until mean simulated landscape defoliation fell within mean observed landscape outbreak area. See File S1 (Insect Extension Parameters) for complete list of input parameters.

Results

Calibration of the Leaf Insect Biomass extension based on total defoliated area was generally successful. Total empirical defoliation across the landscape for 2007 was 63,886 ha. Average defoliated sites (1 ha) for the initial year of simulated outbreaks across simulation replicates was 64,663±6,326 ha. Patches were clustered in both empirical and modeled outbreaks. Patches sizes followed a Gaussian distribution ($\lambda = 156.86$, $k = .57$), with small patches (<20ha) much more frequent than large ones (>200ha) in both empirical and modeled outbreaks. The largest empirically observed contiguous patch (year 2007) in the study area was 22,861ha, while the smallest was 1ha. Maximum modeled contiguous patches were 20,148±1460ha, Minimum outbreak patch size was restricted to 1ha, as this was the model resolution.

Simulated fire and gypsy moth outbreaks were spatially disparate. Over the course of a single model replicate, fire occurred on an average of 0.08% of sites undergoing an active gypsy moth outbreak. This trend was consistent across both current climate and the high emissions climate scenario. Differences in fire severity in the presence of gypsy moth defoliation when present were small and showed no consistent pattern across ecoregions. Finally, carbon volatilized by wildfire was not significantly different between ecoregions or gypsy moth scenarios.

Simulated ANPP sharply declined during years of defoliation under a high emissions climate scenario (A2) and simulated wildfire (Fig. 3). These declines were most apparent in the upland and wetland forests. However reductions were temporary and growth rebounded quickly, causing a short-term spike in ANPP. Ultimately, gypsy moths did not significantly alter the magnitude but greatly increased variability of ANPP in the uplands and wetlands by the end of the simulation. In the plains, gypsy moths marginally increased ANPP beginning in simulation year 30 under climate change.

Carbon sequestration rates exhibited similar trends to ANPP, showing no major changes in landscape-level NEE due to gypsy moths over the course of the simulation (data not shown). During modeled outbreaks, individual sites became temporary C sources, but the landscape as a whole remained a net C sink throughout the simulation period.

Gypsy moths did not drastically shift change the temporal pattern of most major C pools, but did affect the magnitude of total carbon. Under climate change, gypsy moth presence reduced total C in the upland and wetland ecoregions (Fig. 4). In the plains, total C was slightly higher in the presence of gypsy moths, though this increase was within the bounds of our 5-replicate variability.

In addition to examining the effects of defoliation under simulated wildfire and climate change (Figures 3 and 4), we also looked at the relative magnitude of each individual disturbance and their interactive effects on total C (sum of all above- and belowground C pools). When comparing the effects of different disturbances on total ecosystem carbon, climate change had the largest effect on total C, followed by gypsy moth presence (Fig. 5).

Functional tree species types were differentially affected by gypsy moth defoliation and the effects varied by ecoregion. Under high emissions climate, the pines decreased in biomass in the upland and wetland ecoregions with gypsy moth defoliation (Fig. 6a). Oak biomass significantly increased with gypsy moth defoliation in the upland and wetland ecoregions under baseline climate at simulation year 2100 (Fig. 6b). Under climate change, this increase in oaks was seen in the uplands, but was not present in the wetlands or plains (Fig. 6b).

"Probability of establishment (P_{est}) for both oak and pine functional groups fell throughout the 100 year simulations (Fig 6c and 6d). P_{est} in the oak group fell at a slower rate than that of the pines, resulting in significantly higher establishment probability by the middle of the 100 year simulations. Table 1 compares the average difference in aerially-weighted biomass at simulation year 100 for four scenario pairs: two functional tree types under two climate scenarios with and without gypsy moth defoliation. The oak functional group showed statistically significant increases with gypsy moth defoliation regardless of climate scenario. The pine

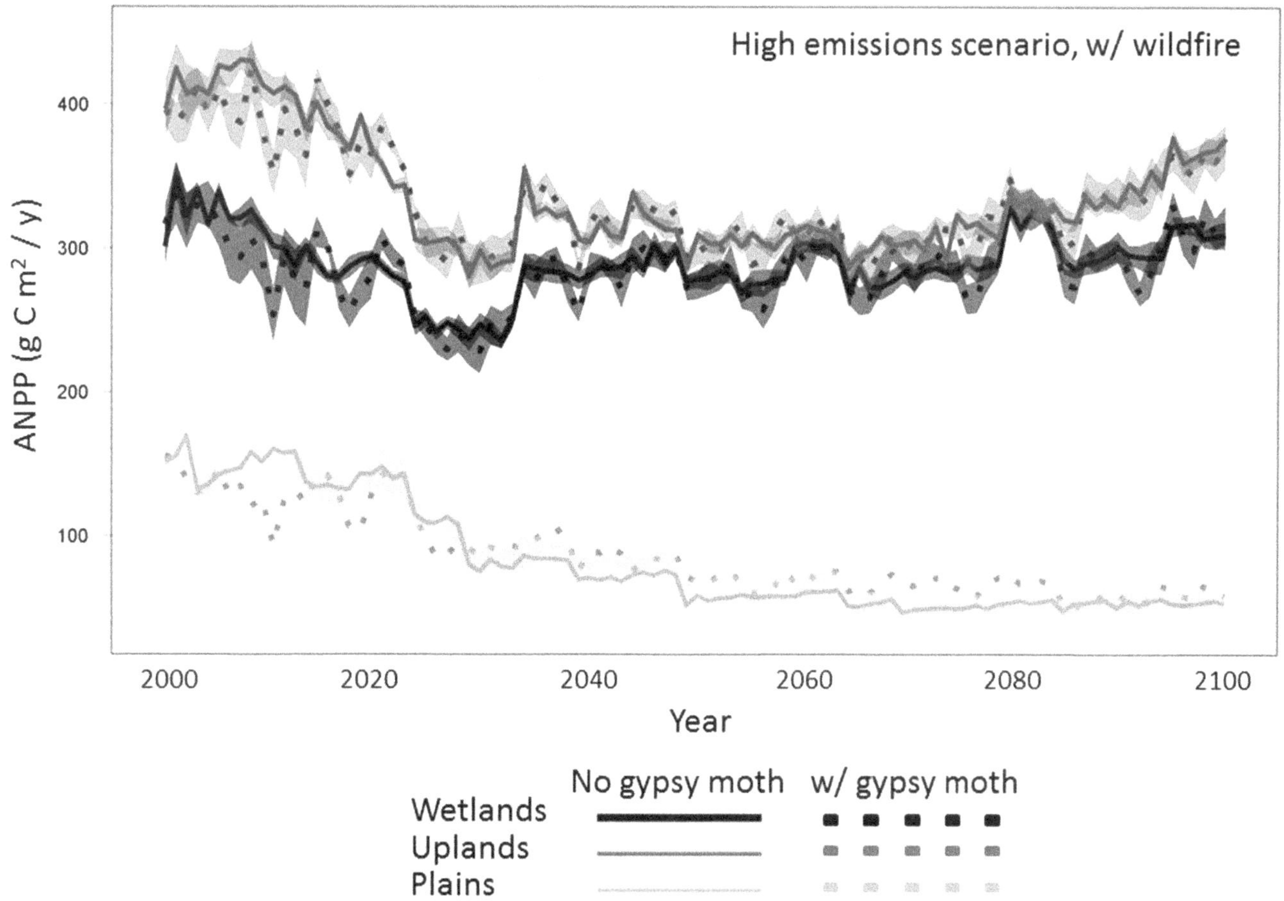

Figure 3. Aboveground net primary production (ANPP) in the three modeled ecoregion types under A2 (high emissions) climate scenario. Scenarios which included gypsy moth defoliation are seen in the dotted lines. Ribbons surrounding each line represent the standard error, based on five model replicates.

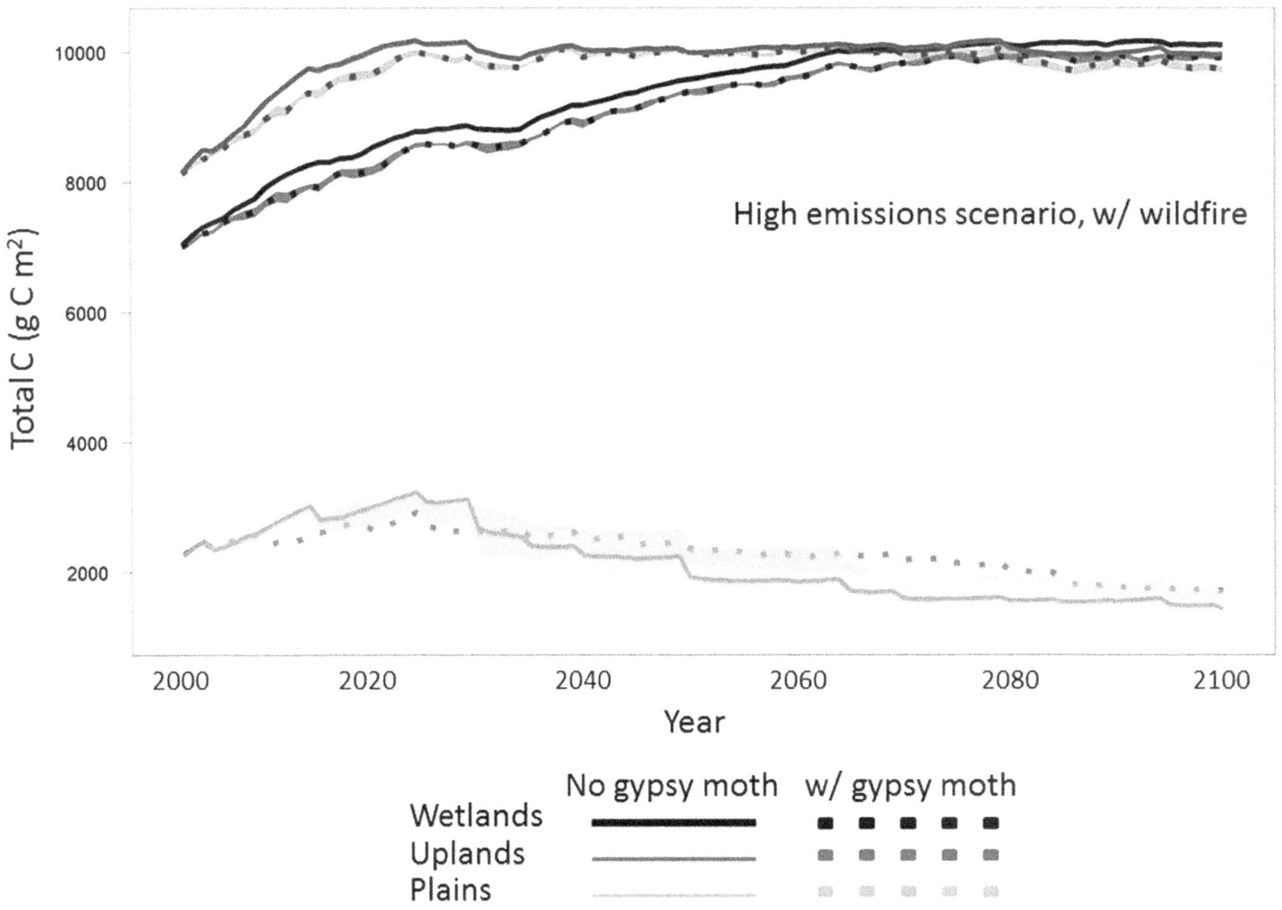

Figure 4. Total carbon under A2 (high emissions) climate scenario for all modeled ecoregion types. Reductions in the uplands and wetlands are a consequence of gypsy moth defoliation (dotted lines). Total carbon inthe plains, represented in light grey, did not significantly change over the 100-year scenario.

group showed decreases under both climate scenarios with gypsy moth defoliation; however the decrease was much greater and proved statistically significant under high emissions climate.

Discussion

Our research objective was to understand the interactive effects of gypsy moth and climate change under a fire regime on the forests of the NJPB with respect to landscape C pools and fluxes, as well as forest species composition and distribution. The spatially explicit and dynamic nature of our simulations allowed for analysis of these effects both across the entire landscape and at the one hectare scale. Our results suggest that long-term patterns in total C will be more affected by climate change than by fire or gypsy moth defoliation over the next century. When simulated gypsy moth outbreaks occurred, defoliation had a large effect on C sequestration rates (i.e. NEE) at the local scale, but not at the landscape scale. At the landscape level, gypsy moth had minor effects on ANPP and NEE, although substantially altered species composition in favor of oaks to the detriment of pines.

One of the primary effects of gypsy moth defoliation is an inhibition of tree growth via a decline in leaf biomass [7,9,36]. Such inhibition is species-specific in its magnitude, but can be cumulatively substantial at the landscape scale. This growth reduction was captured in our simulations, which generated short-term declines in ANPP during gypsy moth outbreaks in the uplands and wetlands (Fig. 3). Despite this reduction, ANPP rebounded to previous levels within 1–2 years, suggesting that defoliation of gypsy moths may have little impact on growth rates at the landscape-level. Though few studies have looked at ANPP specifically in response to defoliation, long-term data from tree rings in the region suggest that radial growth in affected trees is reduced by the previous year's defoliation [37]. Likewise, projected outbreaks of spruce budworm (*Choristoneura spp*) in Eastern Canadian forests also result in reduced net primary productivity, consequently reducing carbon sequestration [38].

The transient nature of growth reduction emphasizes the need for careful consideration of temporal and spatial scale when simulating forest disturbances. At the landscape level, defoliated sites represented an average 27% of the modeled study extent during individual major gypsy moth outbreaks. This was similar to estimates of the most recent major defoliation event occurring in 2005–2008 (20.1% of upland forests damaged) [7], though matching the complexity and heterogeneity of tree damage within outbreak patches is difficult due to the categorical nature of the aerial survey maps used to calibrate gypsy moth patch size. Defoliation in our scenarios caused small, short-term changes in ANPP during an outbreak, but the overall magnitude and trajectory of productivity were not significantly altered over the long term at the landscape scale (Fig. 3). Additionally, during modeled outbreaks many individual sites became temporary C sources; however the potential of these forests across the landscape

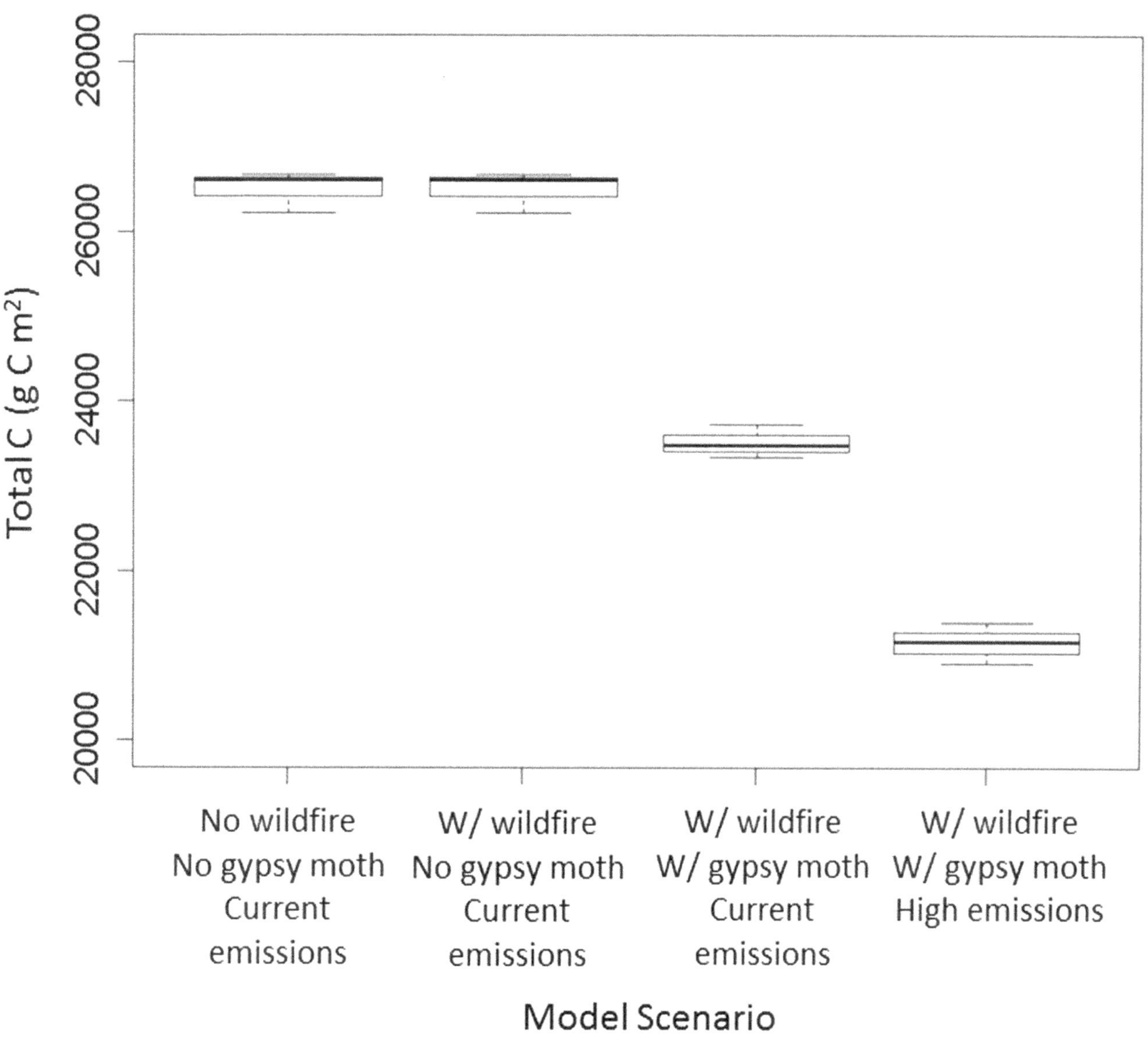

Figure 5. Total ecosystem carbon in four different model scenarios at simulation year 100. Scenarios 1–3 use baseline (current) climate, while the fourth uses the A2 (high emissions) scenario. 'w/GM' implies inclusion of gypsy moth outbreaks in the scenario. Confidence intervals represent standard errors based on 5 model replicates. Total carbon, was lowest when all three disturbances were operating on the landscape. Simulated wildfire had little impact on total C.

to act as a C sink was not wholly compromised. Medvigy et al [39] also noted the discrepancy between regionally averaged defoliation effects and effects within defoliation patches.

Total carbon shows modest decreases with gypsy moth defoliation, as leaf biomass represents a relatively small proportion of the carbon stored in forests (Fig. 4) [7]. Though a spin-up phase is incorporated into the model, disturbances are not included in this period and thus initial decreases in ANPP and modeled carbon stocks (Figs. 3, 4, and 5) are likely due to the 'introduction' of gypsy moth at year 2010 of our simulations. This may account for the large initial decline in ANPP, though reductions continue throughout the simulated century.

Following defoliation events, tree species that are less susceptible to gypsy moth defoliation (e.g. pines) may have a competitive advantage over preferentially selected host species (e.g. oaks), perhaps shifting stand composition towards the former over long periods of time [40,41]. Our results instead projected a perpetuation of the current compositional trend towards more oak domination in the NJPB [21,42]. During large outbreaks, gypsy moths are relatively indiscriminate in their host choice, and though oaks are generally more preferred hosts, many oak species are more resilient or resistant to defoliation-caused mortality than conifers [43]. Typically, only latter instar gypsy moth larvae will defoliate conifers and only when gypsy moth densities are particularly high [37]. Our simulations of mortality in response to defoliation made conifers more susceptible to mortality during outbreaks [33]. Similarly, Lovett et al [43] observed differing mortality rates in mixed oak/conifer stands were responsible for propagation of oaks after gypsy moth outbreaks.

There is some uncertainty to the long-term successional trends of defoliated forests, particularly when focusing on forests of mixed oak-conifer stands [43]. Such uncertainty is compounded in our study, which considers gypsy moth impacts, but also projected climate change influences as well as fire. Although prescribed

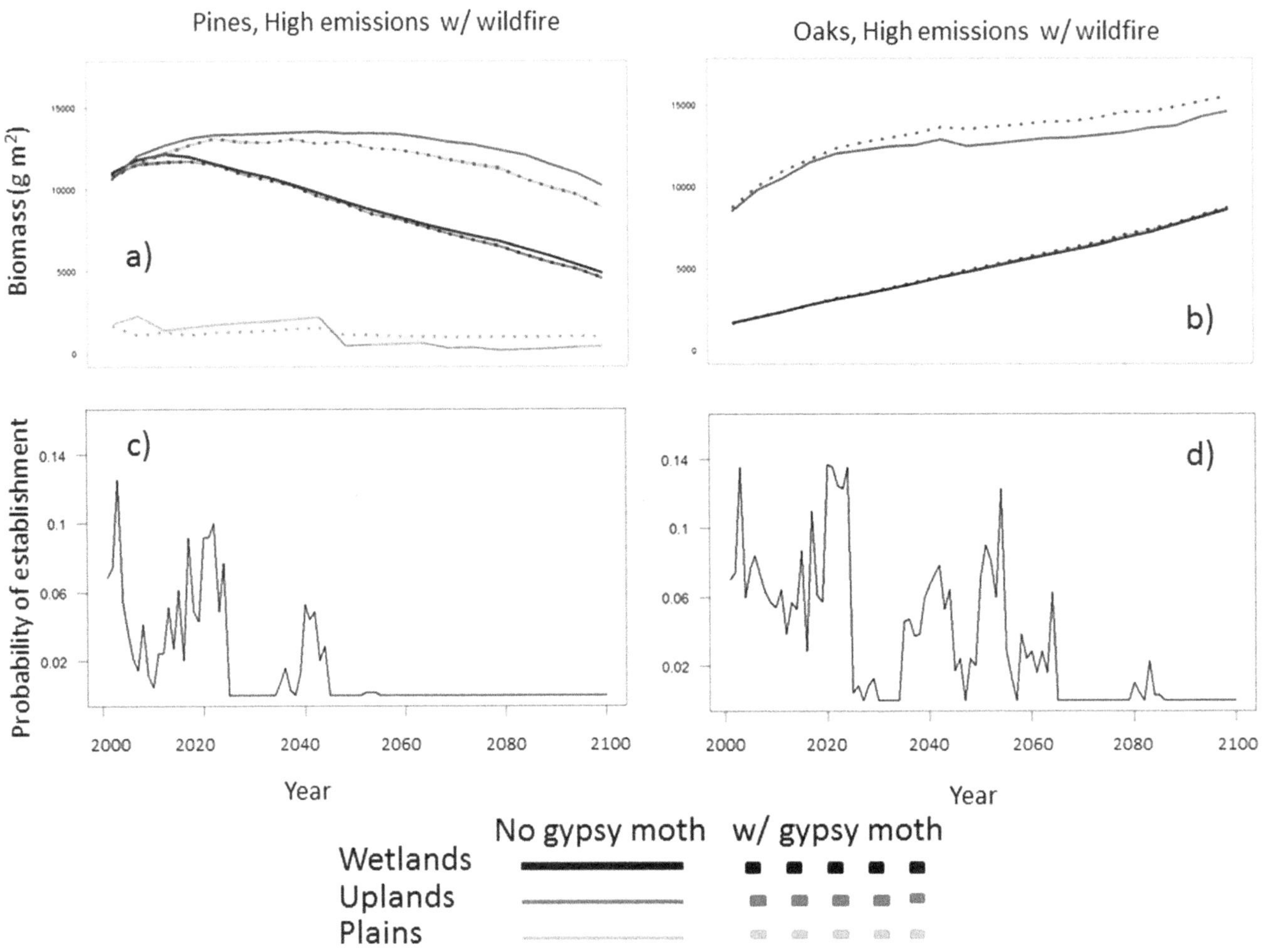

Figure 6. Aboveground biomass (g m^{-2}) for 'Oak' and 'Pine' functional groups over the course of the 100-year simulations. Plots *a* and *b* are comparisons of these groups under high emissions (A2) climate; Dotted lines represent scenarios with gypsy moth defoliation present. Ribbons surrounding each line are standard errors, based on five model replicates. Plots *c* and *d* show landscape probability of establishment for pine and oak functional groups under high emissions climate.

burns and wildfire do occur on the landscape [17,36], they more frequently occur in the 'sand plains' where gypsy moth outbreaks are not prevalent. As such the interaction between these two disturbances was limited. Field measurements from affected oak-dominated stands in the region have shown increased growth increments in pines and a reduction in live oak biomass immediately following an outbreak, as well as release of understory shrubs [37]. The discrepancy between field observations and model results is due to 1) differences in mortality rates between oaks and pines and the resulting long-term effects of regular outbreaks on species composition, and 2) declining pine establishment caused by changing climate over one hundred years. Though establishment in both pitch pine and white oak fell throughout the course of the climate change scenario, the decline in establishment of pitch pine was more severe [22] (Fig 6). Therefore at the scale of our study, higher pine mortality rates during outbreaks, along with

Table 1. Difference in biomass of functional species groups w/gypsy moth defoliation.

Functional group	Scenario comparison	Biomass difference (g C m^2)
Oaks	Baseline emissions, baseline emissions w/GM	+283.69*
	High emissions, high emissions w/GM	+191.33*
Pines	Baseline emissions, baseline emissions w/GM	−76.6
	High emissions, high emissions w/GM	−232.70*

Legend/footnotes: Difference in area-weighted biomass of two major functional groups for 4 scenario comparisons at simulation year 100 across the landscape. GM represents inclusion of gypsy moth in the scenario. Positive values indicate higher values in the second listed scenario, negative larger in the first scenario. * denotes significance at $p = 0.05$ as determined by pairwise Tukey's HSD.

projected declines in pine establishment, were stronger than the differences in host preference. Although some previous studies have concluded the opposite (e.g.[44]), work by Foster [33], which we used to parameterize the gypsy moth outbreaks in this study, found that pine mortality exceeded oak mortality in mixed oak-conifer stands. These effects exacerbate the current trend towards more oak-domination, a consequence primarily of succession and fire exclusion [21,42] and strongly suggest that long-term climate effects are additive with shorter-term gypsy moth outbreaks (Table 1).

For all modeled processes we are limited, as always, by the availability of data used for parameterization. Many parameters used in the succession and disturbance extensions, though derived from empirical data, are themselves uncertain. Data used in establishing growth declines and mortality rates in response to gypsy moth were empirically derived, though the study sites were primarily in Maryland. Soil carbon estimates used for this study are based on limited sampling extrapolated across a broad scale, a weakness previously noted in Scheller et al. [17]. Confidence in many species' life history attributes and disturbance parameters is high, primarily because the NJPB has been an extensively studied system. However, uncertainty in the GCM climate projections used is extremely high, a fact underscored by the sheer number of available projections. Also, the direct impacts of temperature on gypsy moth development and dispersal are not explicitly modeled within this study. Instead, the primary interaction between gypsy moth and climate is expressed through host distribution and abundance, which is considered a determining factor of outbreak severity [45].

Though the cumulative uncertainty in parameters is not trivial, we believe that our results can be used to reliably infer the magnitude and direction of relationships among processes and offer insight into the interactions of global change processes that would be impossible with any other approach.

Though interactions among many important processes were investigated, some model uncertainty comes from the necessary exclusion of several others, namely continuing fragmentation [42] and the recently introduced southern pine beetle (*Dendroctonus frontalis* Zimmermann). Southern pine beetle infestation risk is expected to increase with climate change [46], and with the high pine mortality rates associated with these infestations, this newly introduced forest pest will likely have significant impacts on the NJPB. Atmospheric influences on tree growth and development related to climate change, including possible CO_2 fertilization and changing O_3 levels, were not included in our climate change projections [19]. Previous research suggests that these two greenhouse gases affect tree growth in opposing ways, though these effects are not reflected in our model.

In summary, our results suggest that gypsy moth defoliation may not affect landscape-level ANPP or C sequestration rates, but rather accelerate the shift from an oak-pine/pine-oak system to an oak-dominated one. These simulations incorporate a number of interacting forest processes (e.g. soil C and nutrient cycling, species composition), each uniquely sensitive to both disturbance and climate which are key to forecasting C dynamics at a landscape scale. The considerable variation between model replicates is due to the stochastic nature of simulated wildfire and modeled gypsy moth outbreak events. This variation draws attention to the uncertainty inherent in ecological forecasting [17,29].

Acknowledgments

We would like to thank the valuable input from three anonymous reviewers.

Author Contributions

Conceived and designed the experiments: AMK RMS MSL. Performed the experiments: AMK. Analyzed the data: AMK RMS MSL KLC. Wrote the paper: AMK. Coordinated data acquisition: JH. Gathered and summarized parameterization data: AMK. Created fuel types and conducted cross-tabulation with FBFM fuel types: SVT.

References

1. Pan Y, Birdsey RA, Fang J, Houghton R, Kauppi PE, et al. (2011) A large and persistent carbon sink in the world's forests. Science 333: 988–993. Available: http://www.ncbi.nlm.nih.gov/pubmed/21764754. Accessed 7 November 2013.
2. Canadell JG, Raupach MR (2008) Managing forests for climate change mitigation. Science (80-) 320: 1456–1457.
3. McKinley DC, Ryan MG, Birdsey RA, Giardina CP, Harmon ME, et al. (2011) A synthesis of current knowledge on forests and carbon storage in the United States. Ecol Appl 21: 1902–1924. Available: http://dx.doi.org/10.1890/10-0697.
4. Hicke JA, Allen CD, Desai AR, Dietze MC, Hall RJ, et al. (2012) Effects of biotic disturbances on forest carbon cycling in the United States and Canada. Glob Chang Biol 18: 7–34. Available: http://doi.wiley.com/10.1111/j.1365-2486.2011.02543.x. Accessed 12 November 2013.
5. Campbell J, Donato D, Azuma D, Law B (2007) Pyrogenic carbon emission from a large wildfire in Oregon, United States. J Geophys Res 112: G04014. Available: http://doi.wiley.com/10.1029/2007JG000451. Accessed 12 November 2013.
6. Liebhold A, Elkinton J, Williams D, Muzika R-M (2000) What causes outbreaks of the gypsy moth in North America? Popul Ecol 42: 257–266. Available: http://link.springer.com/10.1007/PL00012004.
7. Clark KL, Skowronski N, Hom J (2010) Invasive insects impact forest carbon dynamics. Glob Chang Biol 16: 88–101. Available: http://doi.wiley.com/10.1111/j.1365-2486.2009.01983.x. Accessed 12 November 2013.
8. Kurz WA, Dymond CC, Stinson G, Rampley GJ, Neilson ET, et al. (2008) Mountain pine beetle and forest carbon feedback to climate change. Nature 452: 987–990. Available: http://www.ncbi.nlm.nih.gov/pubmed/18432244. Accessed 7 November 2013.
9. Fajvan MA, Wood JM (1996) Stand structure and development after gypsy moth defoliation in the Appalachian Plateau. For Ecol Manage 89: 79–88.
10. Iverson LR, Prasad AM (2002) Potential redistribution of tree species habitat under five climate change scenarios in the eastern US. For Ecol Manage 155: 205–222. Available: http://linkinghub.elsevier.com/retrieve/pii/S0378112700559X.
11. White A, Carnell MGR, Friend AD (1999) Climate change impacts on ecosystems and the terrestrial carbon sink: a new assessment. Glob Environ Chang 9: S21–S30.
12. Ungerer MJ, Ayres MP, Lombardero MJ (1999) Climate and the northern distribution limits of *Dendroctonus frontalis* Zimmermann (Coleoptera: Scolytidae). J Biogeogr 26: 1133–1145. Available: http://doi.wiley.com/10.1046/j.1365-2699.1999.00363.x.
13. Davidson E, Janssens I (2006) Temperature sensitivity of soil carbon decomposition and feedbacks to climate change. Nature 440: 165–173. Available: http://www.ncbi.nlm.nih.gov/pubmed/16525463. Accessed 6 November 2013.
14. Gustafson EJ, Shvidenko AZ, Sturtevant BR, Scheller RM (2010) Predicting global change effects on forest biomass and composition in south-central Siberia. Ecol Appl 20: 700–715. Available: http://www.ncbi.nlm.nih.gov/pubmed/20437957.
15. He HS (2008) Forest landscape models: Definitions, characterization, and classification. For Ecol Manage 254: 484–498. Available: http://linkinghub.elsevier.com/retrieve/pii/S0378112707006093. Accessed 12 November 2013.
16. Scheller RM, Domingo JB, Sturtevant BR, Williams JS, Rudy A, et al. (2007) Design, development, and application of LANDIS-II, a spatial landscape simulation model with flexible temporal and spatial resolution. Ecol Modell 201: 409–419. Available: http://linkinghub.elsevier.com/retrieve/pii/S0304380006004893. Accessed 12 November 2013.
17. Scheller RM, Van Tuyl S, Clark KL, Hom J, La Puma I (2011) Carbon Sequestration in the New Jersey Pine Barrens Under Different Scenarios of Fire Management. Ecosystems 14: 987–1004. Available: http://link.springer.com/10.1007/s10021-011-9462-6. Accessed 12 November 2013.

18. Pan Y, Birdsey R, Hom J, Mccullough K, Clark K (2006) Improved Estimates of Net Primary Productivity from Modis Satellite Data at Regional and Local Scales. Ecol Appl 16: 125–132.
19. Pan Y, Birdsey R, Hom J, McCullough K (2009) Separating effects of changes in atmospheric composition, climate and land-use on carbon sequestration of U.S. Mid-Atlantic temperate forests. For Ecol Manage 259: 151–164. Available: http://linkinghub.elsevier.com/retrieve/pii/S0378112709007117. Accessed 12 November 2013.
20. Energy UD of Oak Ridge National Laboratory Amerflux Network. Available: http://ameriflux.lbl.gov/SitePages/Home.aspx. Accessed December 12, 2010.
21. Scheller RM, Van Tuyl S, Clark K, Hayden NG, Hom J, et al. (2008) Simulation of forest change in the New Jersey Pine Barrens under current and pre-colonial conditions. For Ecol Manage 255: 1489–1500. Available: http://linkinghub.elsevier.com/retrieve/pii/S0378112707008717. Accessed 12 November 2013.
22. Scheller RM, Kretchun AM, Van Tuyl S, Clark KL, Lucash MS, et al. (2012) Divergent carbon dynamics under climate change in forests with diverse soils, tree species, and land use histories. Ecosphere 3: 1–16. doi:10.1890/ES12-00241.1
23. Tedrow JCF (1986) Soils of New Jersey. New Jersey.
24. Forman RTT, Boerner RE (1981) Fire Frequency and the Pine Barrens of New Jersey. Bull Torrey Bot Soc 108: 34–50.
25. Lathrop RG Jr, Kaplan MB (2004) New Jersey, land use/land cover update: 2000/2001. New Jersey Dep Environ Prot, New Brunswick, New Jersey, USA.
26. McCormick J, Jones L (1973) The Pine Barrens: vegetation and geography. New Jersey State Museum Research R.
27. Johnson DM, Liebhold AM, Bjørnstad ON (2006) Geographical variation in the periodicity of gypsy moth outbreaks. Ecography (Cop) 29: 367–374.
28. Mladenoff DJ (2004) LANDIS and forest landscape models. Ecol Modell 180: 7–19. Available: http://linkinghub.elsevier.com/retrieve/pii/S0304380004003461. Accessed 7 November 2013.
29. Scheller RM, Hua D, Bolstad PV, Birdsey R, Mladenoff DJ (2011) The effects of forest harvest intensity in combination with wind disturbance on carbon dynamics in Lake States Mesic Forests. Ecol Modell 222: 144–153. Available: http://linkinghub.elsevier.com/retrieve/pii/S0304380010004801. Accessed 12 November 2013.
30. Parton WJ, Ojima DS, Cole CV, Schimel DS (1994) A general model for soil organic matter dynamics: Sensitivity of litter chemistry, texture, and management. Quant Model soil Form Process SSSA Spec Publ 39: 147–167.
31. Parton WJ, Anderson DW, Cole CV, Stewart JWB (1983) Simulation of soil organic matter formation and mineralization in semiarid agroecosystems. Athens, Georg 436 Univ Georg Coll Agric Exp Station: 533–550.
32. Sturtevant BR, Scheller RM, Miranda BR, Shinneman D, Syphard A (2009) Simulating dynamic and mixed-severity fire regimes: A process-based fire extension for LANDIS-II. Ecol Modell 220: 3380–3393. Available: http://linkinghub.elsevier.com/retrieve/pii/S030438000900533X. Accessed 12 November 2013.
33. Foster J (2011) Foret insect defoliation patterns and carbon dynamics: linking remote sensing with simulation modeling.
34. Service UF Forest Health and Protect Aerial Results Viewer. Available: http://na.fs.fed.us/fhp/ta/av/. Accessed November 10, 2010.
35. R: A language and environment for statistical computing (2010). Available: http://www.r-project.org.
36. Naidoo R, Lechowicz MJ (2001) Effects on gypsy moth on radial growth of deciduous trees. For Sci 47: 338–348.
37. Muzika RM, Liebhold AM (1999) Changes in radial increment of host and nonhost tree species with gypsy moth defoliation. Can J For Res 29: 1365–1373. Available: http://www.nrc.ca/cgi-bin/cisti/journals/rp/rp2_abst_e?cjfr_x99-098_29_ns_nf_cjfr29-99.
38. Dymond CC, Neilson ET, Stinson G, Porter K, MacLean D, et al. (2010) Future sSpruce bBudworm oOutbreak mMay cCreate a cCarbon sSource in eEastern Canadian fForests. Ecosystems 13: 917–931. Available: http://link.springer.com/10.1007/s10021-010-9364-z. Accessed 12 November 2013.
39. Medvigy D, Clark KL, Skowronski NS, Schäfer KVR (2012) Simulated impacts of insect defoliation on forest carbon dynamics. Environ Res Lett 7: 045703. Available: http://stacks.iop.org/1748-9326/7/i=4/a=045703?key=crossref.47e93f5a8744e2e7bf6670a9aeae4096. Accessed 12 November 2013.
40. Campbell RW, Garlo AS (1982) Gypsy moth in New Jersey pine-oak. J For 80: 89–90.
41. Davidson CB, Gottschalk KW, Johnson JE (1999) Tree mortality following defoliation by the European gypsy moth (*Lymantria dispar* L.) in the United States: A review. For Sci 45.
42. La Puma IP (2012) Fire in the plains: A landscape perspective of human-induced ecological change in the pinelands of New Jersey.
43. Lovett GM, Canham CD, Arthur M, Weathers KC, Fitzhugh RD (2006) Forest ecosystem rResponses to eExotic pPests and pPathogens in eEastern North America. Bioscience 56: 395. Available: http://www.jstor.org/stable/4099670.
44. Muzika RM, Liebhold AM (2001) Effects of gypsy Moth Defoliation in Oak-Pine Forests in the Northeastern United States. Procceedings: Integrated Management and Dynamics of Forest defoliating insects USDA Fores: 117–123.
45. Williams DW, Liebhold AM (1995) Herbivorous insects and global change: Ppotential changes in the spatial distribution of forest defoliator outbreaks. J Biogeogr 22: 665–671.
46. Gan J (2004) Risk and damage of southern pine beetle outbreaks under global climate change. For Ecol Manage 191: 61–71. Available: http://linkinghub.elsevier.com/retrieve/pii/S0378112703005152. Accessed 12 November 2013.

12

A High-Resolution Chronology of Rapid Forest Transitions following Polynesian Arrival in New Zealand

David B. McWethy[1]*, Janet M. Wilmshurst[2,4], Cathy Whitlock[1,3], Jamie R. Wood[2], Matt S. McGlone[2]

1 Department of Earth Sciences, Montana State University, Bozeman, Montana, United States of America, **2** Landcare Research, Lincoln, New Zealand, **3** Institute on Ecosystems, Montana State University, Bozeman, Montana, United States of America, **4** School of Environment, University of Auckland, Auckland, New Zealand

Abstract

Human-caused forest transitions are documented worldwide, especially during periods when land use by dense agriculturally-based populations intensified. However, the rate at which prehistoric human activities led to permanent deforestation is poorly resolved. In the South Island, New Zealand, the arrival of Polynesians c. 750 years ago resulted in dramatic forest loss and conversion of nearly half of native forests to open vegetation. This transformation, termed the Initial Burning Period, is documented in pollen and charcoal records, but its speed has been poorly constrained. High-resolution chronologies developed with a series of AMS radiocarbon dates from two lake sediment cores suggest the shift from forest to shrubland occurred within decades rather than centuries at drier sites. We examine two sites representing extreme examples of the magnitude of human impacts: a drier site that was inherently more vulnerable to human-set fires and a wetter, less burnable site. The astonishing rate of deforestation at the hands of small transient populations resulted from the intrinsic vulnerability of the native flora to fire and from positive feedbacks in post-fire vegetation recovery that increased landscape flammability. Spatially targeting burning in highly-flammable seral vegetation in forests rarely experiencing fire was sufficient to create an alternate fire-prone stable state. The New Zealand example illustrates how seemingly stable forest ecosystems can experience rapid and permanent conversions. Forest loss in New Zealand is among the fastest ecological transitions documented in the Holocene; yet equally rapid transitions can be expected in present-day regions wherever positive feedbacks support alternate fire-inhibiting, fire-prone stable states.

Editor: Christopher Carcaillet, Ecole Pratique des Hautes Etudes, France

Funding: This research was supported in part by National Science Foundation Grants OISE-0966472 (DBM, CW), and BCS-1024413 (DBM, CW), New Zealand Foundation for Research Science and Technology funding to JMW, JRW and MSM, and Royal Society of New Zealand Marsden funding to MSM, JMW, and CW. The funders had no role in study design, data collection and analysis, decision to publish, or preparation of the manuscript.

Competing Interests: The authors have declared that no competing interests exist.

* Email: dmcwethy@montana.edu

Introduction

The composition and structure of forests worldwide are being altered by intensification and expanding land-use, changing climatic conditions and increasing incidence of large fires [1,2]. Sudden transitions between forest, savanna and grassland biomes can be linked to feedbacks between land-use practices, climate change and disturbances such as fire [3]. Research suggests that the extent of forest cover reinforces conditions that maintain or inhibit fire (e.g., high forest cover often promotes greater fuel moisture), and consequently, the stability of forested, savanna or grassland vegetation [4]. Recent deforestation has been shown to initiate positive feedbacks replacing forested landscapes that rarely burned with fire-prone shrubland, savanna and grasslands [5]. Human-mediated forest transitions are not a new phenomenon, however. Humans have effected forest transitions for millennia through their use of fire, and widespread rapid transitions in the past are often linked to expanding human settlements and cultivation [6,7].

New Zealand provides a prime example of a relatively recent and major prehistoric forest loss. Here, the arrival of humans c. 750 years ago [8] instigated a period of deliberate burning, termed the Initial Burning Period (IBP), that led to the loss of nearly 50% of New Zealand's forests [9,10] and conversion of podocarp dominated forests to open shrubland vegetation. The conversion of forests in the South Island, presents a striking paradox, that extremely small populations (by any estimate – human mtDNA sequences suggest an initial founding population of between 50–100 females and a low population growth rate of c. 1% [11]) were able to instigate widespread and permanent forest transitions. Paleoenvironmental records from New Zealand suggest that ecosystem feedbacks during the initial burning period accelerated the rapid rate of forest demise and help explain this paradox [5,9]. In this paper, we examine the rate at which relatively stable forest biomes in New Zealand were converted to open shrublands in the 13th century, and how forest transitions following early human-set fires in the South Island of New Zealand compares with other regions in the world currently experiencing deforestation.

During the IBP, nearly all pollen and charcoal records derived from the drier parts of New Zealand show a distinct decline in pollen from native podocarp tree taxa (and to a lesser extent beech forest trees), an increase in pollen from seral taxa particularly bracken fern (*Pteridium esculentum*) and grasses (Poaceae), and an increase in charcoal (e.g., [9,12,13]. Previous results from radiocarbon dated high-resolution microscopic charcoal analyses from a network of sites across the central South Island showed a distinct increase in charcoal accumulation rates during the IBP within 200 years following Polynesian (Māori) settlement, although the increase is diachronous among sites [9]. At all but the most remote and wet sites, paleofire reconstructions suggest that repeated fires (2 or more) were responsible for the watershed conversion of native forest to a fern/scrub or grassland. [9,10,14]. However, the age-depth chronologies described in [9,14] were based on a small number of radiocarbon dates selected prior to and after the IBP, and consequently, the rate of forest loss remains poorly constrained.

To precisely determine the rate at which prehistoric forest transitions occurred following initial human arrival in the South Island, New Zealand, we developed high-resolution reconstructions of vegetation (pollen) and fire (macroscopic charcoal) from radiocarbon dated lake-sediment records from two small, closed-basin lakes, Lake Kirkpatrick and Dukes Tarn (Fig. 1). We specifically choose two sites that represent different vulnerabilities to human-set fires: one drier lowland site and a second, wetter high-elevation site. The chronology of these reconstructions and rates-of-change calculations are based on a stratigraphic series of Accelerator Mass Spectrometry (AMS) radiocarbon dates from the sediment cores (Table S1).

Results

Chronology model results were highly convergent, suggesting forest transitions occurred within decades of the first human-set fires (Figs. S1–S5). Significant (>15% decline in native taxa) forest loss occurred within 17 yrs (SD = 7) at Lake Kirkpatrick and 48 yrs (SD = 19) at Dukes Tarn (Fig. 2). At Lake Kirkpatrick, the IBP was punctuated by two fire episodes at c. AD 1367 and AD 1391 (Fig. 3, Bchron chronology). Following the initial increase in charcoal accumulation rates (CHAR; particles cm^{-2} yr^{-1}), fire episodes occurred periodically (every 50–100 yrs) until c. AD 1600 when fire activity decreased. A second increase in fire activity coincided with European arrival ca. AD 1800. Vegetation assemblages changed dramatically following the first fires associated with human arrival. Native trees (e.g., *Nothofagus menziesii*, Nothofagus spp., *Podocarpus* spp. and *Prumnopitys* spp.) declined from 99 to 47% of the total terrestrial pollen percentage from 1331 to 1391 yr AD, whereas disturbance related-taxa (e.g. Poaceae and *Pteridium*) increased from up to 25 and 27%, respectively, over the same period (see Fig. S6 for additional pollen information). Pollen of native trees increased slightly to 64% at c. 1642 yr AD at the expense of grass pollen. Tree pollen percentages then declined, recovered again at c. 1792 yr AD before declining to <30% in the late 20th century. Pollen of exotic taxa (Pinaceae, *Rumex* spp. and *Taraxacum* spp.) reached 2% by the mid 19th century and increased to >10% at c. 1947 yr AD.

At Dukes Tarn, charcoal data register initial fire events at c. AD 1337, and 1387, and the pollen data indicate that forests recovered within decades even though fire events occurred periodically (once every 50–100 yrs) from the IBP until the last century (Fig. 3). Pollen of some native trees (*Nothofagus fuscospora* type) collectively declined from 97 to 83% whereas disturbance-related taxa, notably Poaceae and *Pteridium*, increased from 1 to 9% and 0 to 7% respectively during the interval from c. AD 1355 to 1414 (see Fig. S7 and [9] for additional pollen information). In contrast to Lake Kirkpatrick, pollen of native trees increased to 94% by c. AD 1502 at the expense of Poaceae, then fluctuated between 85–94%, and accounted for 97% at c. 1694 yr AD (back to pre-deforestation levels). Tree percentages decreased to 83% in the last century. Non-native plant taxa (Pinaceae, *Rumex* spp. and *Taraxacum* spp.) account for >1% of the pollen at c. 1928 yr AD and 5% at c. 1980 yr AD before declining to <1% in recent decades. Fires that were recorded during the same time period at both sites (Fig. 4) likely represent a pattern of burning associated with humans moving through the landscape to enhance resource conditions and facilitate travel to nearby greenstone quarries [15] and could have been facilitated by extreme fire weather (e.g., hot, dry, windy conditions). Compared to Lake Kirkpatrick where forest transitions led to open vegetation that persists to the present, native vegetation at Duke's Tarn experienced relatively minor modification and showed partial recovery within several decades of the first IBP fires and full recovery 350 years after the IBP. Post IBP fires at Duke's Tarn seem to have had less of a lasting impact on the long-term structure and composition of vegetation.

Statistical analyses of rates of change reflect the rapid transition from native trees to grasses and bracken, evident in changes in pollen percentages (Fig. 3). Sørensen's index quantifying dissimilarity in pollen taxa prior to and during the IBP shows sharp increases in the dissimilarity in vegetation composition and dominance, with transitions in vegetation taxa peaking during the IBP at both Lake Kirkpatrick and Dukes Tarn. The Sørensen's index (dissimilarity between each sample and the previous sample) was greatest at c. 1363 yr AD (10^{-2}) at Lake Kirkpatrick and at c. 1375 AD (10^{-2}) at Dukes Tarn, coinciding with the first IBP fires at both sites. Dissimilarity between pre-human vegetation assemblages and post-IBP vegetation taxa increased from present to the pre-IBP period at both sites but was much more pronounced at Lake Kirkpatrick where dramatic vegetation transitions from forest to open shrublands persisted until present. Changes in vegetation assemblages at Dukes Tarn are characterized by an initial forest loss, a late recovery of native trees, followed by an increase in grasses and the presence of introduced taxa associated with European colonization.

Discussion

Human-caused forest transitions are documented worldwide, especially during periods when land use by dense agriculturally-based populations intensified [6,16,17]. However, the rate at which prehistoric human activities led to a persistent biome shift is commonly poorly resolved [18]. In the South Island of New Zealand, the arrival of Polynesians resulted in dramatic forest loss, with conversion of nearly half of native forests to open vegetation by a small number of sparsely distributed populations [10]. Paleoenvironmental records suggest that this conversion occurred within decades to centuries of human arrival but the speed at which site conversion took place once burning commenced has been unclear until now. Chronological models (Bchron, Oxcal and MCAgeDepth) that explicitly estimate joint uncertainty of all radiocarbon calibration dates suggest that the forest loss during IBP was rapid, occurring in 10–24 years at the drier site, Lake Kirkpatrick (median = 17) and 29–67 years at the wetter site, Dukes Tarn (median = 48). The speed at which vegetation conversion occurred is likely even more rapid than estimates based on age-depth chronologies alone because they incorporate joint uncertainty originating from both AMS radiocarbon dating and radiocarbon calibration distributions. Despite inherent limitations in calculating the rate of past vegetation change from lake-sediment records, forest loss in New Zealand was rapid by any measure. These rates of deforestation parallel those made for the eastern North Island where dry, lowland podocarp forests

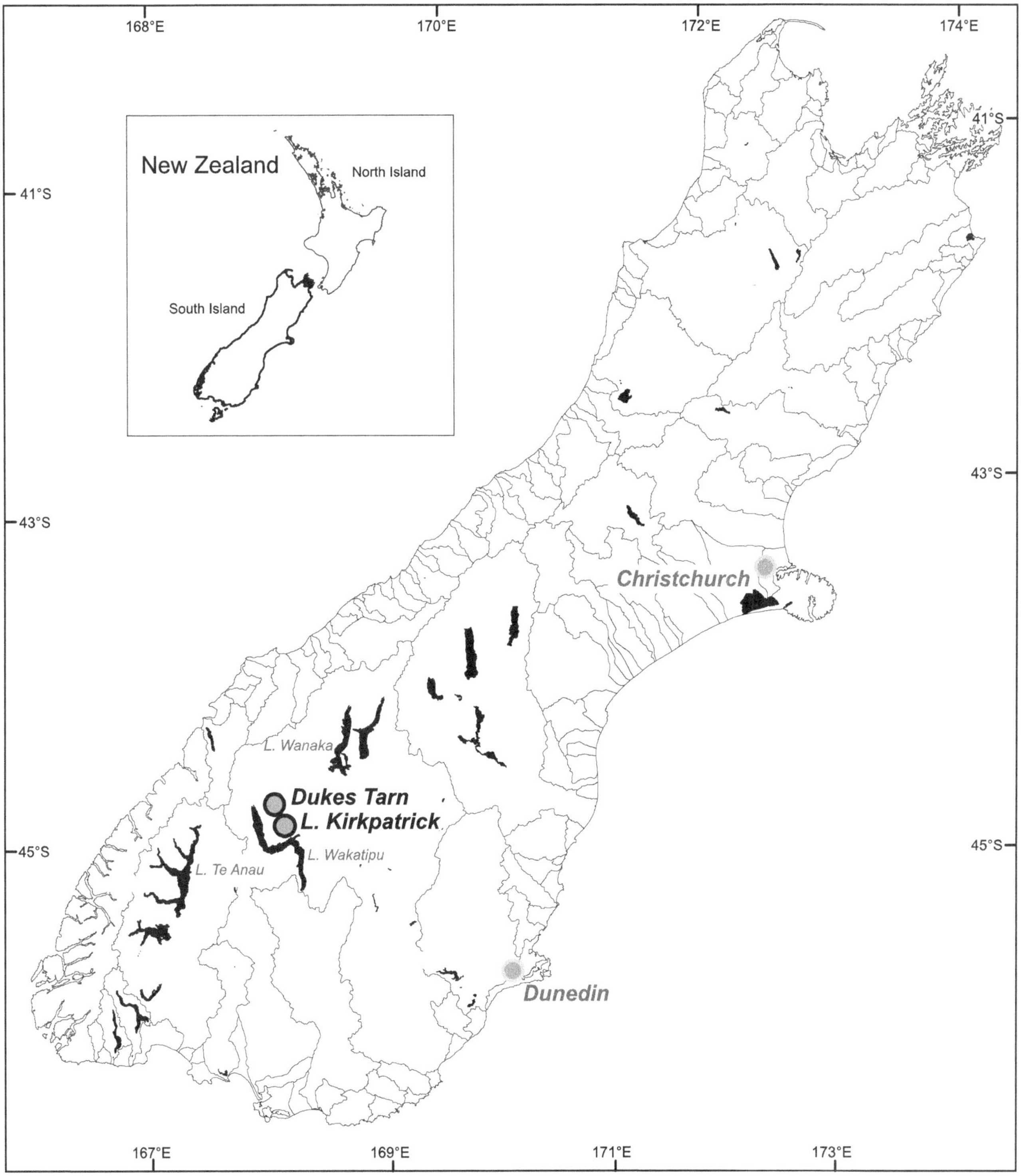

Figure 1. Location of study sites, South Island, New Zealand.

transitioned to bracken fern shrubland within decades of the first human-set fires whereas loss of more mesic, upland forests took a century or longer [19].

For most of the South Island, fire impacts on forested ecosystems were ecologically insignificant throughout the Holocene, in that changes in the structure and composition of vegetation were poorly related to the occurrence of fire [20]. The widespread and mostly one-way shift from forest to open shrubland that occurred with the arrival of Polynesians, and then following European arrival, was unprecedented in the Holocene, even after extreme disturbance events (e.g., large volcanic eruptions). This was the case for Lake Kirkpatrick and many other sites throughout similarly dry forests of eastern New Zealand including the eastern North Island [21]. The higher-elevation site,

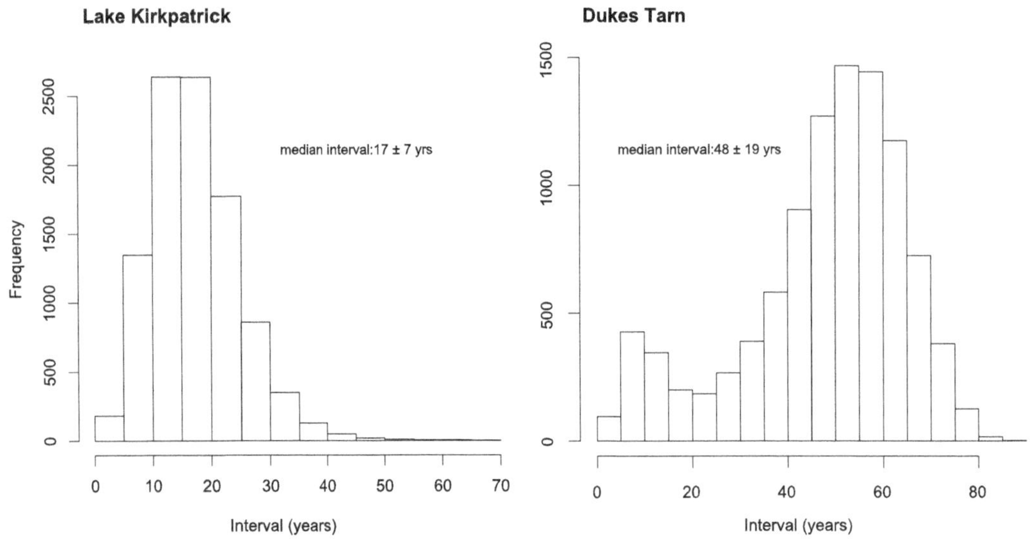

Figure 2. Bchron posterior frequency distribution of the interval (yrs) associated with forest transitions at Lake Kirkpatrick and Dukes Tarn, South Island, New Zealand. The interval associated with forest transitions was identified using two criteria: 1) visible change in core lithology, and 2) a rapid decline in pollen percentages of native forest taxa and a coincident increase in pollen percentages of open vegetation taxa.

Dukes Tarn, experienced equally rapid vegetation change, but declines in forest taxa at this site were less pronounced. This is most likely because this site is wetter and because the deeply-incised forested drainage basin south of the tarn has provided an ongoing refuge from fire, contributing significant wind-transported pollen to the tarn sediments. Differences in the extent to which fire activity and vegetation changed following human arrival at Lake Kirkpatrick and Dukes Tarn may be explained by variation in the intensity of human activity at these two sites; the most parsimonious explanation, however, centers on differences in rainfall and inferred fuel moisture at these two sites and is supported by previous results comparing a network of sites from the South Island ([9]). Paleofire records at even wetter sites (>1600 mm/yr) near or west of the Southern Alps of the South Island indicate only limited fire activity and minor associated vegetation change [9,22]. Similarly, along the western coast of the North Island moist conditions limited the extent to which Māori were able to burn forests [23]. Here, Māori gardens were established by burning small patches of forests but the scale of forest clearing was insignificant compared to deforestation that occurred in drier regions of the North and South Island sites [23]. What is surprising is the extent and rapid rate of forest clearance that occurred in the South Island where prehistoric human populations were at their lowest compared to the North Island, and cultivation was limited to a few coastal sites north of latitude 44° South.

Dynamic Global Vegetation Models that estimate rates of prehistoric deforestation at a regional to global scale suggest that the extent and pattern of preindustrial deforestation in most regions was strongly linked to population densities, inherent landscape productivity and lifestyles and technological innovation [16,17]. Estimates of population density and reconstructions of environmental change in these models are derived from archeological and paleoenvironmental records, and in most regions, rates of deforestation are poorly resolved. Our high-resolution chronology of environmental change from the South Island, New Zealand indicates rapid forest transitions occurred during the IBP during a time of exceedingly low population numbers and limited use of technological innovation (i.e., intensive cultivation). Why then were New Zealand's forests particularly vulnerable to fire? The lowland evergreen conifer/angiosperm forests were characterized as being both fire sensitive and ignition limited [20]. Low- to mid-elevation forests experiencing moderate to low levels of annual rainfall (300–1600 mm/yr) persisted for millennia simply because ignitions were rare and fuels were typically wet [20,24]. It was only when human-set fires expanded the proportion of flammable early-seral vegetation on the landscape that widespread forest transitions were possible [5]. Positive feedbacks between the fire sensitivity of the vegetation, introduction of human-set ignitions, and changes in landscape flammability with the spread of early-seral vegetation offer an explanation for these rapid transitions. Perry et al. [5,22] provide empirical support for these feedbacks, showing that dry, low- to mid-elevation evergreen conifer/angiosperm forests in the South Island New Zealand were vulnerable to fire, and that targeted burning created positive feedbacks where burning of forest resulted in more flammable seral vegetation which was more vulnerable to ignitions. The model estimated that conversion of the forest to shrubland could have occurred in 50–200 years if fires were targeted both spatially (in highly-flammable early-seral vegetation) and temporally (every few decades to maintain early-seral vegetation).

Forest transitions occurring at present in a number of regions around the world bring new attention to conditions and feedbacks that drive rapid transitions between seemingly stable states [25–28]. Analyzing global tree cover and fire activity data, Staver et al. [4] show that mesic landscapes (>2500 mm/y rainfall) with >60% tree cover are resistant to changes in fire activity and biome shifts, whereas areas with intermediate levels of rainfall (1000–2500 mm/y) and <50% tree cover are vulnerable to rapid transitions to savanna. Hirota et al. [29] suggest that landscapes with intermediate levels of tree cover are extremely rare because the interaction between climate, edaphic conditions and fire activity interact to attract vegetation to relatively stable states of higher (forest) and lower (savanna) levels of tree cover. A number of mechanisms create positive feedbacks that trigger and maintain these multiple stable states. Landscapes with high canopy cover of mature forests are resilient to increased fire activity because they maintain fuel moisture levels that suppress fire and often lack

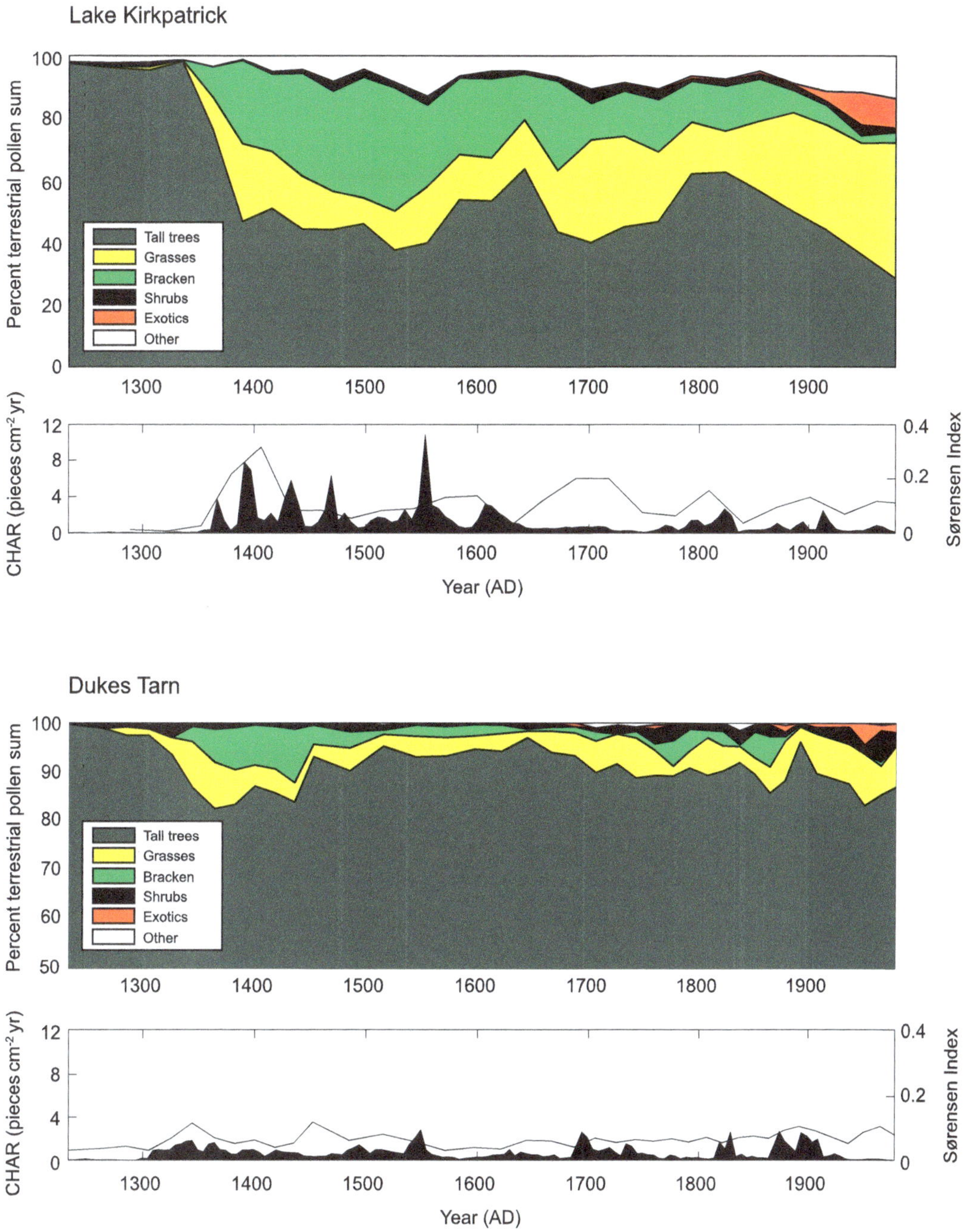

Figure 3. Statistical analysis of vegetation change and rate of vegetation change (based on Sørensen's similarity index) and changes in fire activity (based on CHAR, charcoal particles cm^{-2} yr^{-1}, and fire event determination), Lake Kirkpatrick and Dukes Tarn. Colored panels show change in percent of total terrestrial pollen percentages for native trees and disturbance-associated taxa (e.g., Poaceae, *Pteridium*) and non-native taxa (Pinaceae, *Rumex*, *Taraxacum*-type) introduced by Europeans for Lake Kirkpatrick (top) and Dukes Tarn (bottom). Black and white panels show charcoal accumulation rates and Sørensen's distance between each pollen sample and the next oldest pollen sample.

continuous understory fuels for carrying fire [3]. Dry landscapes reinforce open conditions by supporting low fuel moisture and species with plant functional traits that promote fire.

A consequence of these attracting states is the potential for human activities to reduce forest density and/or canopy cover past a tipping point where they become increasingly vulnerable to rapid transitions to more fire-prone states [30]. This has been shown to be particularly true for forests where flammability decreases with forest age. Kitzberger et al. [3] demonstrate that even small increases in the frequency of ignitions and/or climate variability in these forests can result in dramatic shifts in fire regimes at large scales. The New Zealand example suggests that targeted ignitions alone where sufficient to initiate feedbacks promoting high landscape flammability and increased fire activity.

Today, the vulnerability of temperate forests to rapid transitions is evident in a number of regions. For example, alterations to fuel and microclimate conditions following logging activities in evergreen forests of the northwestern US increase post-fire

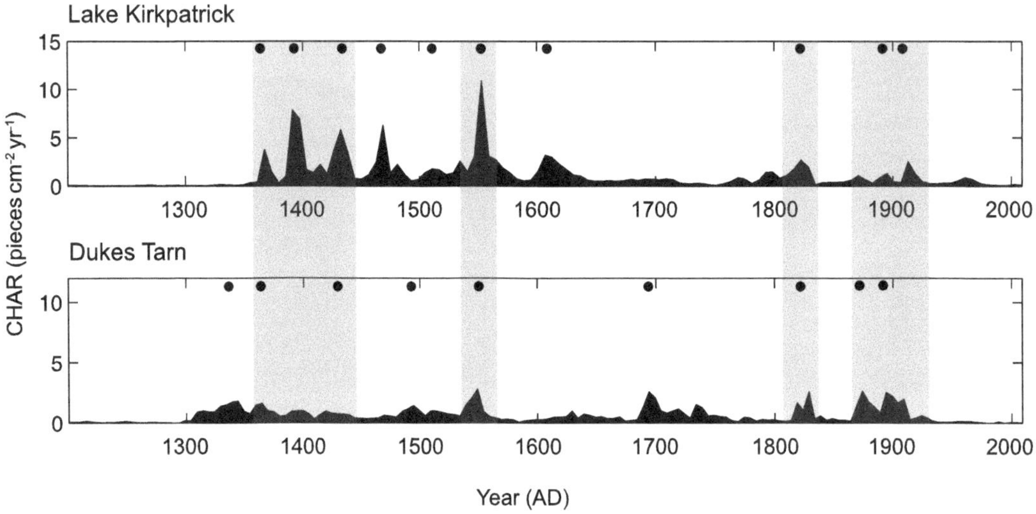

Figure 4. Variation in CHAR (charcoal particles cm^{-2} yr^{-1}), and peak identification for Lake Kirkpatrick and Dukes Tarn. Black circles identify peaks in CHAR that are statistically significant from background variation in CHAR and are thus likely to be local fires (within 1–3 kms of lake). Shaded areas highlight time periods when fire events were recorded at both lakes.

flammability, ultimately increasing fire severity such that forest recovery is delayed or non-existent [31,32]. Lindenmayer et al. [27] describe a similar feedback in mountain ash forests in Victoria, Australia where logging activity results in an abundance of fine fuels, slash and more flammable early-seral vegetation. While fire occurred in these forests in the past, it was typically infrequent, and eventually followed by conditions that promoted robust recruitment of mountain ash seedlings. In recent decades, the increase in fine fuels, slash and persistently drier microclimatic conditions following logging activity and subsequent increased fire activity interact to create these new fire-prone 'landscape traps' [27]. Likewise, Cochrane et al. [33] demonstrate that previously burned forests in Amazonia are highly vulnerable to subsequent fires that further promote more open, highly-flammable vegetation. In southern South America, the combination of increased anthropogenic fires and herbivory of native seedlings by introduced species is rapidly facilitating the replacement of mesic *Nothofagus* forests with fire-prone shrubs [34]. On Great Barrier Island North Island, New Zealand, Perry et al. [35] found that conversion of mesic broadleaf forests to shrublands initiated positive feedbacks in which an increase in fire activity not only reinforced the spread and persistence of highly flammable early-seral woody vegetation but also promoted flammable invasive species and degraded soil conditions, especially on drier aspects. Lacking fire, these forests would likely recover but Perry et al. [35] observe that, along with a continued loss of topsoil, seed trees and dispersal agents (e.g. avian frugivores) are becoming increasingly rare such that forest recovery becomes ever more improbable. These recent forest transitions demonstrate the strong influence of human-initiated positive feedbacks that increase landscape flammability and highlight the inherent vulnerability of seemingly stable temperate forests to rapid biome and fire regime shifts. Widespread forest transitions are predicted for the future in regions with similar age-flammability and low-ignition characteristics [36].

Conclusions

The timing of an abrupt change from forest to open shrubland following human-set fires at two sites in the South Island of New Zealand is well-constrained in sediment cores using high-resolution chronology models, suggesting widespread deforestation occurred within decades of the first anthropogenic fires in the most vulnerable drier regions. The fire history of New Zealand is an example of how a seemingly stable ecosystem can pass natural tipping points and experience persistent transformations. The rate at which anthropogenic forest loss occurred was rapid, taking place within decades following the initial arrival of a small founding population. The rapidity of the response was partly facilitated by positive feedbacks created by the introduction of a new and frequent ignition source by humans. In addition, the dominant forest taxa lacked natural adaptations to fire (e.g., serotiny, resprouting) which made them highly vulnerable to increased ignitions [24]. The newly created fire-prone landscapes persisted even when fire activity decreased after the IBP. These shrublands will likely be maintained for centuries if not millennia unless fire is removed from the system for long periods of time (>100 years). As with temperate forests experiencing similar transitions today, simply removing fire from New Zealand landscapes could eventually lead to forest recovery. However, recent examples suggest that a confluence of factors that degrade biophysical conditions, impede seedling regeneration, remove source populations and dispersal agents and introduce fire-adapted weedy species will continue to hinder native forest recovery in New Zealand and elsewhere. Even in the absence of fire, management of landscapes to promote open vegetation, agriculture, and exotic forestry plantations promises to reinforce conditions that prevent regeneration of native species. The rapid rate of forest transitions witnessed in New Zealand 750 years ago portends potentially rapid transitions in biomass-rich regions experiencing similar land-use-disturbance interactions and feedbacks and highlights challenges to native forest recovery.

Materials and Methods

Study Sites

The study regions lie in the southeastern hill country and southern alpine region of the South Island of New Zealand, an area that ranges between 250 and 1900 m elevation (Fig. 1). The climate is cool year-round and moderately wet with most of the precipitation occurring in winter. There is a strong steep east/west gradient both in precipitation and elevation. During the Holocene the vegetation of the region was dominated by closed-canopy broadleaf evergreen forests [20]. Natural fires occurred in areas of forest in the dry southeastern region of New Zealand but were characterized by long intervals between fires [37]. The original forest extent and composition are inferred from pollen records, isolated surviving forest stands and remnant wood [38,39]. Permission to conduct fieldwork at the two study sites was granted by the New Zealand Department of Conservation (DOC).

Lake Kirkpatrick (lat. 45.03°S, long. 168.57°E; 567 masl). The Lake Wakatipu catchment is a mid-elevation (570 masl) site receiving moderate levels of rainfall (1077 mm/yr) and was part of a trade route for highly prized greenstone that the indigenous population (Māori) used and traded [15]. Lake Kirkpatrick lies within the Lake Wakatipu catchment. The small lake (~3.0 ha) is surrounded by pastureland with browntop (*Agrostis capillaris*), Poa spp. and other introduced grasses, clover (*Trifolium* spp.), *Hypericum* spp.; pine plantations (*Pinus radiata*); and small areas of native silver beech forest (*Nothofagus menziesii*).

Dukes Tarn (lat. 44.96°S, long. 168.49°E; 825 masl). Dukes Tarn is approximately 9.5 km northwest of Lake Kirkpatrick at 830 masl and also located within the Lake Wakatipu catchment. Dukes Tarn (~1.5 ha) receives more annual rainfall (1340 mm/yr) than Lake Kirkpatrick and vegetation surrounding the lake consists of patches of black beech (*Nothofagus solandri*), small trees and shrubs (e.g., *Coprosma* spp. *Phyllocladus*, *Dracophyllum longifolium*, *Hebe* spp., *Gaultheria* spp.), native tussock (*Chionochloa* spp.) and introduced pasture grasses (*Festuca* spp.). Deeply-incised drainages supporting native forest remnants border Dukes Tarn to the south whereas open vegetation surrounds the tarn on rolling terrain to the north.

Lake Sediment Cores

Lake sediment cores were obtained from Lake Kirkpatrick and Dukes Tarn using a polycarbonate tube (Klien core). Cores were split at University of Minnesota's LacCore facility for charcoal and pollen analysis. A 195 cm sediment core was retrieved from Lake Kirkpatrick in 2009 and a 173 cm sediment core was retrieved from Dukes Tarn in 2008. Core lithology for Lake Kirkpatrick consisted of light green gyttja 0–27 cm (depth below surface), gray silty clay 27–29 cm, dark green gyttja 29–105 cm, light gray silty clay 105–115 cm, and light green gyttja 115–195 cm. Core lithology for Dukes Tarn consisted of light green gyttja 0–42 cm, silty clay 42–43 cm, light green gyttja 43–128 cm, dark silty clay with charcoal 128–136 cm, and light green gyttja 138–173 cm.

Reconstruction of Vegetation and Fire Histories

Pollen analysis followed the preparation methods of Moore et al. [40]. Pollen counts exceeded 250 terrestrial pollen grains and bracken (*Pteridium*) spores. Percentage calculations were based on this terrestrial sum (excluding other ground ferns and tree fern spores as they tend to be over-represented). We followed the protocol of McGlone and Wilmshurst [10] to detect initial human impact on the vegetation and grouped pollen taxa into tall forest taxa (predominantly *Fuscospora* spp. and podocarps); disturbance-related taxa including ferns (*Pteridium esculentum*), *Coriaria* spp. and grasses (Poaceae); and introduced taxa, including Pinaceae and *Rumex* spp. Remaining taxa, mostly herbaceous species, were labeled "other". These groupings were used to identify the transition from closed forest to open shrubland and grassland.

High-resolution charcoal analysis followed methods of Whitlock and Larsen [41]. Charcoal particles greater (>125 μm in diameter) were examined to reconstruct local fire events based on changes in charcoal accumulation rates (CHAR; particles cm^{-2} yr^{-1}). Decomposition of the charcoal time series employed a Gaussian mixture model to identify the mean and variance of the background CHAR (BCHAR) distribution [42]. The 99th percentile of this distribution is defined as the threshold value separating peak fire events from "noise". Significant charcoal peaks, determined by CHAR values greater than a locally-defined threshold value, identified specific fire events. Because each 1-cm vertical interval in the lakes encompasses 1–10 years at both sites, a charcoal peak is assumed to represent one or more fires occurring within the time span of a charcoal peak. The statistical significance of each peak was evaluated by comparing the original charcoal counts against the values in samples occurring 35 years before the peak. If the maximum count of a peak had a >5% chance of coming from the same Poisson-distributed population as the minimum charcoal count within the preceding 35 years, then a "peak" was not identified. Fire size and/or intensity were inferred from the magnitude of individual charcoal peaks (particles cm^{-2} yr^{-1}) [43].

Chronology Models

We obtained 22 radiocarbon dates at two sites to develop high-resolution chronologies (Table S1). Radiocarbon dates were obtained from macrofossils found during the interval of the first IBP fires where changes in charcoal accumulation rates and pollen percentages indicate dynamic changes in fire activity and vegetation. The chronology of these reconstructions and rates-of-change calculations are based on a stratigraphic series of AMS radiocarbon dates from the sediment cores. We incorporated this suite of radiocarbon dates into three readily accessible chronology models designed to estimate lake-sediment age-depth chronologies and chronological uncertainty: Bchron [44], Oxcal [45] and McAgeDepth [43]. Model outputs were highly convergent for all three models (Figs. S1–S5). We highlight results from Bchron because the chronology model has an environment event analysis function which explicitly estimates the duration of the interval at which significant environmental events occurred (Fig. 2).

Accelerator mass spectrometry (AMS) ^{14}C dates obtained on twig charcoal and terrestrial plant macrofossils, provided the chronology for precisely estimating rates of past environmental change. Thirteen radiocarbon dates were obtained from Lake Kirkpatrick and nine from Dukes Tarn. Eleven radiocarbon dates were obtained from across 20 centimeters of sediment where charcoal and pollen percentages identify the IBP at Lake Kirkpatrick and six radiocarbon dates were obtained across the comparable 10-centimeter interval at Dukes Tarn. We used this set of radiocarbon dates in a suite of age-depth chronology models to estimate the duration at which forest transitions occurred.

To derive age-depth chronologies we used three statistical age-depth chronology models: Bchron [44], Oxcal [45] and McAgeDepth [43] which all address two primary types of uncertainty: uncertainty in radiocarbon dates and their calibration, and variability in the sedimentation process itself. These models share a number of features that are desirable for developing a continuous age-depth chronology for sediment cores including, monotonicity (deeper sediments are likely older), variable sedimentation (rates are assumed to be variable), probabilistic interpolation based on uncertainty associated with calibrated probability distributions (posterior samples or chronologies are based on joint estimates of uncertainty of all dates), and increased uncertainty at depths far from radiocarbon-dated material [46]. Each model uses a Markov-Chain Monte-Carlo algorithm to estimate model parameters, which provides a distribution of the chronologies that are most likely given all radiocarbon-dated probability distributions. Most importantly, they explicitly account for and estimate joint uncertainty of all of the radiocarbon dates and consider the probability density functions of calibrated age ranges in chronology interpolation. All high-resolution chronology models achieved convergence and identified no outliers at Dukes Tarn, but two outliers at Lake Kirkpatrick (macrofossil dated material at depths 110 and 111 cm), which were subsequently left out of modeled chronologies. The radiocarbon material at these two depths were identified as having low mass during AMS analyses (KCCAMS lab), which likely increased their age uncertainty and inaccuracy.

Rate of vegetation change

We used the Sørensen's similarity statistic [47] to assess the rate of vegetation change between pollen samples. Sørensen's index, similar to the Bray-Curtis dissimilarity [48] is a similarity index that measures compositional similarity between two sites, or in this case, compositional similarity in pollen taxa at two depths where pollen percentages are measured. Sørensen's distance statistic is calculated between each pollen sample and the preceding pollen sample providing a statistical measure of vegetation change over time.

Supporting Information

Figure S1 Comparison of chronology model estimates (circles and bars represent median and 95% Confidence Intervals) for the interval associated with forest transitions that followed the first human-set fires, Lake Kirkpatrick and Dukes Tarn, South Island, New Zealand. Median (and 95% confidence estimate ranges) for the interval associated with forest transitions for Lake Kirkpatrick were 17 yrs (3–13, Bchron), 27 yrs (7–47, MCAgeDepth), and 31 yrs (16–46, Oxcal). Median and 95% confidence estimate ranges (in parentheses) for the interval associated with forest transitions for Dukes Tarn were 47 yrs (35–59, Bchron), 44 yrs (24–64, MCAgeDepth), and 25 yrs (14–36, Oxcal).

Figure S2 MCAgeDepth age depth model for Lake Kirkpatrick with sedimentation rates and sample resolution. Chronologies developed using a weighted cubic smoothing spline which takes into account the number and uncertainty of age estimates. Gray shading represent 95% confidence intervals reflect the combined uncertainty of all age estimates in a model and are derived from 1000 bootstrapped chronologies.

Figure S3 MCAgeDepth age depth model for Dukes Tarn with sedimentation rates and sample resolution. Chronologies developed using a weighted cubic smoothing spline which takes into account the number and uncertainty of age estimates. Gray shading represent 95% confidence intervals reflect the combined uncertainty of all age estimates in a model and are derived from 1000 bootstrapped chronologies.

Figure S4 Oxcal age-depth model for Lake Kirkpatrick using the P_Sequence deposition model which assumes deposition to be random (*k* parameter set at 1 – which assumes a postulated event spacing of approximately 1 cm). Light grey distributions represent likelihoods for single calibrated dates and in darker grey shading represents the marginal posterior distributions which take into account the depth model; the depth model curves show envelopes for the 95% and 68% highest probability density (HPD) ranges.

Figure S5 Oxcal age-depth model for Dukes Tarn using the P_Sequence deposition model which assumes deposition to be random (*k* parameter set at 1 – which assumes a postulated event spacing of approximately 1 cm). Light grey distributions represent likelihoods for single calibrated dates and in darker grey shading represents the marginal posterior distributions which take into account the depth model; the depth model curves show envelopes for the 95% and 68% highest probability density (HPD) ranges.

Figure S6 Pollen diagram for Lake Kirkpatrick, South Island, New Zealand.

Figure S7 Pollen diagram for Dukes Tarn, South Island, New Zealand.

Table S1 AMS radiocarbon dating determined age information for Lake Kirkpatrick and Dukes Tarn, New Zealand.

Acknowledgments

We thank Mairie Fromont for conducting pollen analyses for Dukes Tarn, John Southon (University of California-Irvine, Keck-Carbon Cycle AMS Facility) for radiocarbon dates and Philip Higuera for help with Figure 3 and the application of Sørensen's similarity index to pollen data.

Author Contributions

Conceived and designed the experiments: DBM CW JMW. Performed the experiments: DBM CW JMW JRW. Analyzed the data: DBM CW JMW JRW. Contributed reagents/materials/analysis tools: DBM CW JMW JRW. Wrote the paper: DBM CW JMW JRW MSM.

References

1. Nepstad DC, Stickler CM, Soares B, Merry F (2008) Interactions among Amazon land use, forests and climate: prospects for a near-term forest tipping point. Philosophical Transactions of the Royal Society B-Biological Sciences 363: 1737–1746.
2. Mayer AL, Khalyani AH (2011) Grass Trumps Trees with Fire. Science 334: 188–189.
3. Kitzberger T, Araoz E, Gowda J, Mermoz M, Morales J (2011) Decreases in fire spread probability with forest age promotes alternative community states, reduced resilience to climate variability and large fire regime shifts. Ecosystems: 1–16.
4. Staver AC, Archibald S, Levin SA (2011) The global extent and determinants of savanna and forest as alternative biome states. Science 334: 230–232.
5. Perry GLW, Wilmshurst JM, McGlone MS, McWethy DB, Whitlock C (2012) Explaining fire-driven landscape transformation during the Initial Burning Period of New Zealand's prehistory. Global Change Biology 18: 1609–1621.
6. Pongratz J, Reick C, Raddatz T, Claussen M (2008) A reconstruction of global agricultural areas and land cover for the last millennium. Global Biogeochem Cycles 22: GB3018.
7. Kaplan JO, Krumhardt KM, Ellis EC, Ruddiman WF, Lemmen C, et al. (2011) Holocene carbon emissions as a result of anthropogenic land cover change. The Holocene 21: 775–791.
8. Wilmshurst JM, Anderson AJ, Higham TFG, Worthy TH (2008) Dating the late prehistoric dispersal of Polynesians to New Zealand using the commensal Pacific rat. Proceedings of the National Academy of Sciences 105: 7676–7680.
9. McWethy DB, Whitlock C, Wilmshurst JM, McGlone MS, Fromont M, et al. (2010) Rapid landscape transformation in South Island, New Zealand, following initial Polynesian settlement. Proceedings of the National Academy of Sciences 107: 21343–21348.
10. McGlone MS, Wilmshurst JM (1999) Dating initial Maori environmental impact in New Zealand. Quaternary International 59: 5–16.

11. Murray-McIntosh RP, Scrimshaw BJ, Hatfield PJ, Penny D (1998) Testing migration patterns and estimating founding population size in Polynesia by using human mtDNA sequences. Proceedings of the National Academy of Sciences of the United States of America 95: 9047–9052.
12. Newnham R, Lowe DJ, McGlone MS, Wilmshurst JM, Higham TFG (1998) The Kaharoa Tephra as a critical datum for earliest human impact in northern New Zealand. Journal of Archaeological Science 25: 533–544.
13. Horrocks M, Deng Y, Ogden J, Alloway B, Nichol S, et al. (2001) High spatial resolution of pollen and charcoal in relation to the c. 600 yr BP Kaharoa Tephra: implications for Polynesian settlement of Great Barrier Island, northern New Zealand. Journal of Archaeological Science 28: 153–168.
14. McWethy DB, Whitlock C, Wilmshurst JM, McGlone MS, Li X (2009) Rapid deforestation of South Island, New Zealand by early Polynesian fires. The Holocene 19: 883–897.
15. CINZAS (2008) Central Index of New Zealand Archaeological Sites. nzarchaeologyorg.
16. Kaplan JO, Krumhardt KM, Zimmermann N (2009) The prehistoric and preindustrial deforestation of Europe. Quaternary Science Reviews 28: 3016–3034.
17. Pfeiffer M, Spessa A, Kaplan JO (2013) A model for global biomass burning in preindustrial time: LPJ-LMfire (v1.0). Geosci Model Dev\ 6\: 643\-685\.
18. Williams M (2000) Dark ages and dark areas: global deforestation in the deep past. Journal of Historical Geography 26: 28–46.
19. Wilmshurst JM, McGlone MS, Partridge TR (1997) A late Holocene history of natural disturbance in lowland podocarp/hardwood forest, Hawke's Bay, New Zealand. New Zealand Journal of Botany 35: 79–96.
20. Ogden J, Basher L, McGlone M (1998) Fire, forest regeneration and links with early human habitation: Evidence from New Zealand. Annals of Botany 81: 687–696.
21. Wilmshurst JM (1997) The impact of human settlement on vegetation and soil stability in Hawke's Bay, New Zealand. New Zealand Journal of Botany 35: 97–111.
22. Perry GLW, Wilmshurst JM, McGlone MS, Napier A (2012) Reconstructing spatial vulnerability to forest loss by fire in pre-historic New Zealand. Global Ecology and Biogeography: no-no.
23. Wilmshurst JM, Higham TFG, Allen H, Johns D, Phillips C (2004) Early Maori settlement impacts in northern coastal Taranaki, New Zealand. New Zealand Journal of Ecology 28: 167–179.
24. Perry G, Wilmshurst J, McGlone M (2014) The ecology and long-term history of fire in New Zealand's ecosystems. New Zealand Journal of Ecology 38.
25. Kitzberger T, Brown PM, Heyerdahl EK, Swetnam TW, Veblen TT (2007) Contingent Pacific-Atlantic ocean influence on multi-century wildfire synchrony over western North America. Proceedings of the National Academy of Sciences 104: 543–548.
26. Field RD, Van der Werf G, Shen SSP (2009) Human amplification of drought-induced biomass burning in Indonesia since 1960. Nature Geoscience 2: 185–188.
27. Lindenmayer DB, Hobbs RJ, Likens GE, Krebs CJ, Banks SC (2011) Newly discovered landscape traps produce regime shifts in wet forests. Proceedings of the National Academy of Sciences 108: 15887–15891.
28. Odion DC, Moritz MA, DellaSala DA (2010) Alternative community states maintained by fire in the Klamath Mountains, USA. Journal Of Ecology 98: 96–105.
29. Hirota M, Holmgren M, Van Nes EH, Scheffer M (2011) Global Resilience of Tropical Forest and Savanna to Critical Transitions. Science 334: 232–235.
30. McWethy DB, Higuera PE, Whitlock C, Veblen TT, Bowman DMJS, et al. (2013) A conceptual framework for predicting temperate ecosystem sensitivity to human impacts on fire regimes. Global Ecology and Biogeography 22: 900–912.
31. Thompson JR, Spies TA, Ganio LM (2007) Reburn severity in managed and unmanaged vegetation in a large wildfire. Proceedings of the National Academy of Sciences 104: 10743–10748.
32. Odion DC, Frost EJ, Strittholt JR, Jiang H, Dellasala DA, et al. (2004) Patterns of fire severity and forest conditions in the western Klamath Mountains, California. Conservation Biology 18: 927–936.
33. Cochrane MA, Alencar A, Schulze MD, Souza CM Jr., Nepstad DC, et al. (1999) Positive Feedbacks in the Fire Dynamic of Closed Canopy Tropical Forests. Science 284: 1832–1835.
34. Veblen TT, Holz A, Paritsis J, Raffaele E, Kitzberger T, et al. (2011) Adapting to global environmental change in Patagonia: What role for disturbance ecology? Austral Ecology: no-no.
35. Perry GLW, Ogden J, Enright NJ, Davy LV (2010) Vegetation patterns and trajectories in disturbed landscapes, Great Barrier Island, northern New Zealand. New Zealand Journal of Ecology: 311–323.
36. Moritz MA, Parisien M-A, Batllori E, Krawchuk MA, Van Dorn J, et al. (2012) Climate change and disruptions to global fire activity. Ecosphere 3: art49.
37. McGlone MS, Wilmshurst JM, Leach HM (2005) An ecological and historical review of bracken (Pteridium esculentum) in New Zealand, and its cultural significance. New Zealand Journal of Ecology 29: 165–184.
38. McGlone MS (1995) Lateglacial landscape and vegetation change and the Younger Dryas climatic oscillation in New Zealand. Quaternary Science Reviews 14: 867–881.
39. Wardle P (2001) Distribution of native forest in the upper Clutha district, Otago, New Zealand. New Zealand Journal of Botany 39: 435–446.
40. Moore PD, Webb JA, Collinson ME (1991) Pollen Analysis. Oxford, UK: Blackwell Scientific.
41. Whitlock C, Larsen C. (2001) Charcoal as a Fire Proxy. In: Smol JP, Birks, H.J.B, Last, W M., editor.Tracking Environmental Change Using Lake Sediments: Volume 3 Terrestrial, Algal, and Siliceous indicators.Dordrecht: Kluwer Academic Publishers. pp. 75–97.
42. Higuera PE, Peters ME, Brubaker LB, Gavin DG (2007) Understanding the origin and analysis of sediment-charcoal records with a simulation model. Quaternary Science Reviews 26: 1790–1809.
43. Higuera PE, Brubaker LB, Anderson PM, Hu FS, Brown TA (2009) Vegetation mediated the impacts of postglacial climate change on fire regimes in the south-central Brooks Range, Alaska. Ecological Monographs 79: 201–219.
44. Haslett J, Parnell A (2008) A simple monotone process with application to radiocarbon-dated depth chronologies. Journal of the Royal Statistical Society: Series C (Applied Statistics) 57: 399–418.
45. Bronk Ramsey C (2008) Deposition models for chronological records. Quaternary Science reviews 27: 42–60.
46. Parnell AC, Buck CE, Doan TK (2011) A review of statistical chronology models for high-resolution, proxy-based Holocene palaeoenvironmental reconstruction. Quaternary Science Reviews 30: 2948–2960.
47. Sørensen T (1948) A method of establishing groups of equal amplitude in plant sociology based on similarity of species and its application to analyses of the vegetation on Danish commons. Kongelige Danske Videnskabernes Selskab 5: 1–34.
48. Bray JR, Curtis JT (1957) An ordination of upland forest communities of southern Wisconsin. Ecological Monographs 27: 325–349.

13

Effects of Climate Variability and Accelerated Forest Thinning on Watershed-Scale Runoff in Southwestern USA Ponderosa Pine Forests

Marcos D. Robles[1]*, Robert M. Marshall[1], Frances O'Donnell[2], Edward B. Smith[3], Jeanmarie A. Haney[1], David F. Gori[4]

1 The Nature Conservancy Center for Science and Public Policy, Tucson, Arizona, United States of America, **2** Northern Arizona University, Flagstaff, Arizona, United States of America, **3** The Nature Conservancy in California, Sacramento, California, United States of America, **4** The Nature Conservancy in New Mexico, Santa Fe, New Mexico, United States of America

Abstract

The recent mortality of up to 20% of forests and woodlands in the southwestern United States, along with declining stream flows and projected future water shortages, heightens the need to understand how management practices can enhance forest resilience and functioning under unprecedented scales of drought and wildfire. To address this challenge, a combination of mechanical thinning and fire treatments are planned for 238,000 hectares (588,000 acres) of ponderosa pine (*Pinus ponderosa*) forests across central Arizona, USA. Mechanical thinning can increase runoff at fine scales, as well as reduce fire risk and tree water stress during drought, but the effects of this practice have not been studied at scales commensurate with recent forest disturbances or under a highly variable climate. Modifying a historical runoff model, we constructed scenarios to estimate increases in runoff from thinning ponderosa pine at the landscape and watershed scales based on driving variables: pace, extent and intensity of forest treatments and variability in winter precipitation. We found that runoff on thinned forests was about 20% greater than unthinned forests, regardless of whether treatments occurred in a drought or pluvial period. The magnitude of this increase is similar to observed declines in snowpack for the region, suggesting that accelerated thinning may lessen runoff losses due to warming effects. Gains in runoff were temporary (six years after treatment) and modest when compared to mean annual runoff from the study watersheds (0–3%). Nonetheless gains observed during drought periods could play a role in augmenting river flows on a seasonal basis, improving conditions for water-dependent natural resources, as well as benefit water supplies for downstream communities. Results of this study and others suggest that accelerated forest thinning at large scales could improve the water balance and resilience of forests and sustain the ecosystem services they provide.

Editor: Julia A. Jones, Oregon State University, United States of America

Funding: The authors have no support or funding to report.

Competing Interests: The authors have declared that no competing interests exist.

* Email: mrobles@tnc.org

Introduction

As we transition into a warmer century, managing forests in the southwestern United States for resilience is more urgent and necessary than ever. Forest conditions across the region have declined significantly. In ponderosa pine (*Pinus ponderosa*) forests of central Arizona, stand densities range from 2 to 44 times greater than during pre-settlement conditions and total basal areas range from 2 to 4 times greater [1–3]. Severe wildfires and drought have caused high tree mortality rates across 14–18% of forests and woodlands in Arizona and New Mexico [4] and these results do not include mortality from two high-severity wildfires in 2011, the Wallow and the Las Conchas wildfires, which burned 217,000 and 63,500 ha (535,000 and 157,000 acres) in Arizona and New Mexico, respectively. Persistent drought conditions have also adversely impacted water-dependent habitats and species within and downstream of forest uplands, as evidenced by heightened mortality of a riparian tree species in central Arizona [5] and declines in number of species and densities of native fish in the Gila River in southwestern New Mexico [6], [7]. Researchers suggest that warming is amplifying the size and severity of wildfire- and drought-mortality events and decreasing snow pack levels [8–11]. Suggested mechanisms involve changes in the water cycle: increased evapotranspiration losses, extended water-stress periods, earlier snowmelt, and lengthened fire seasons. These changes and trends suggest a widespread drying of forests and an increasing risk of uncharacteristic fire and competition-induced water stress and mortality.

At the same time, many communities in the western U.S. are facing critical water shortages and river flows in some basins have diminished or been lost altogether due to unsustainable water management practices [12–15]. Recent studies at the scale of both the Colorado River basin and the state of Arizona concluded that water supplies for many communities will be inadequate to meet future projected demands [16],[17]. Moreover, elevated temperatures are predicted to directly impact river flows. A recent study estimated that for every one-degree Celsius rise in temperature, Colorado river flows will decline by 3–10% [18].

In this study, we explored the ways in which a new era of forest management, including accelerating the pace and scale of forest thinning, could be used to address declining conditions in forests in central and northern Arizona. At about 6,000 hectares per year (15,000 acres/year), the current pace of mechanical thinning in these forests does not begin to match the magnitude of recent forest disturbances. However, a new congressionally funded program called the Four Forest Restoration Initiative (hereafter "4FRI") will accelerate the use of mechanical thinning and prescribed burns across four national forests, treating 238,000 ha (588,000 acres) in the first analysis area over the next 10 years. The objective of the 4FRI project is "...to re-establish forest structure, pattern, and composition, which would lead to increases in forest resiliency and function," including reductions in fire risk and improved watershed function [19]. Mechanical thinning alone or in combination with prescribed fires reduced fire risk, increased runoff, and improved tree water-stress at the plot scale [20–23]. The 4FRI project provided the opportunity to evaluate the effectiveness of forest thinning to improve forest resilience and ecosystem functioning at larger scales.

The objectives of this study were to: (1) investigate how variation in climate and the pace and extent of thinning of ponderosa pine forests affects runoff in the Salt and Verde watersheds of central Arizona; and (2) explore the management implications of using accelerated thinning to improve forest resilience in a time of declining forest and water conditions. We chose the Salt-Verde watersheds because forests from these watersheds provide water for 1.5 million people in Phoenix, which is the 6th largest city in the United States. We focused on ponderosa pine because this forest type produces 50% of the runoff in these watersheds even though it accounts for only 20% of the area [24]. We modified a forest runoff model developed from historical paired watershed experiments in the Beaver Creek sub-watershed within our study area [20] and ran the model in multiple scenarios to estimate additional annual runoff from mechanical thinning. The scenarios simulated forest treatment at two scales (a) landscape-scale thinning planned within the first analysis area of the 4FRI project and (b) watershed-scale thinning in forests across the Salt-Verde watersheds (Figure 1). Scenarios were designed to account for variability in winter precipitation, as well as the pace and extent of forest treatments. We estimated runoff in periods with below-average precipitation, herein referred to as droughts, and in periods with above-average precipitation, herein referred to as pluvials, using a model of 20th century precipitation in ponderosa pine forests in the region [25]. We believe this is the first attempt to estimate the influence of mechanical thinning on runoff over multi-year broad-scale restoration projects that accounts for the effects of climate variability.

Materials and Methods

The subsequent sections provide detailed information on the study area, the development of the modified runoff model, the method we used to simulate variability in winter precipitation, the 4FRI runoff scenarios, and the Salt-Verde runoff scenarios. All statistical analyses and graphics were completed using the statistical software SigmaPlot 11 (Systat software Inc, San Jose, CA), except for regression model fitting which was performed using Matlab 2013 (Mathworks, Natick, MA). Methods and results are reported in SI units, but for accessibility to forest and water managers, we reported English units in parentheses within the text and included all study figures and tables in English units as File S1. Area and water volume amounts were rounded off to three significant figures.

Study Area

Ponderosa pine forests occupy approximately 0.68 million ha (1.68 million acres) of the 3.46 million ha (8.56 million acres) within the Salt-Verde watersheds in central Arizona (Figure 1). A portion of these forests, all within the Verde watershed, will be mechanically thinned in the 4FRI project in the next 10 years. Within the Salt-Verde watersheds, ponderosa pine forests grow at mid-elevations, from 1,800 to 2,600 m (5,900–8,600 feet) [26], [27], on soils derived from basalt (59%) and sedimentary rock (41%) [28]. Mean annual precipitation for ponderosa forests in the southwestern United States ranges from 510 to 760 mm (20 to 30 inches) and is highly variable [29]. Sixty percent of annual precipitation falls primarily as snow in the winter and spring months from October to April, and the melting of the resulting snowpack accounts for 80 to 90% of annual stream flow [29].

Forest Treatment Runoff Model

For the purposes of this study, we were interested in estimating the additional runoff that becomes available due to forest thinning on an annual basis, not absolute values of watershed runoff. We define "additional runoff due to thinning" as the portion of precipitation that appears as surface water at the sub-watershed outlet and that is directly attributable to mechanical thinning treatments. This additional runoff can be considered as "in-place" or "in-situ" as we did not attempt to model runoff accumulation, routing, groundwater outflows or inflows, or channel losses in downstream watersheds.

We modified the Baker-Kovner logistical regression equation from the historical Beaver Creek sub-watershed experiments to calculate annual runoff in our study [20]. We selected this model because the Beaver Creek sub-watershed is actually within our study area (Figure 1), and because the model was based upon measured empirical data. These experiments and others like it demonstrated that reductions in forest basal area of 30–100% in ponderosa forests on basalt-derived soils increased runoff between 15–40% for up to six years after the initial treatment [29]. Runoff increases were negligible for basal area reductions below 30%, six years after treatments, and for years with winter precipitation below 230 mm (9 inches) [20], [29]. The original model predicted annual watershed runoff based on three independent variables: winter precipitation (Oct-Apr), forest basal area, and a solar insolation index, and explained a fair amount of variability in total watershed runoff [20].

To derive increases in runoff directly associated with thinning using the paired watershed method, researchers in the Beaver Creek experiments first established a relationship between stream flows in the control and treated watersheds before treatments [30]. This relationship was used to predict what would have been baseline flows in the treated watershed. The difference between measured flows in the treated watershed and this predicted value were attributed to thinning effects. We compared data calculated in this fashion in the Beaver Creek experiments [30], [31] and one additional nearby site, Castle Creek [32], to output generated by the Baker-Kovner model. The output from the model was calculated as the difference in post-treatment and pre-treatment watershed runoff using post- and pre-treatment basal areas respectively and holding all other variables constant. In this comparison, we found the fit of the model output to the portion of runoff due to thinning to be much poorer ($r^2 = 0.43$, Figure 2a)

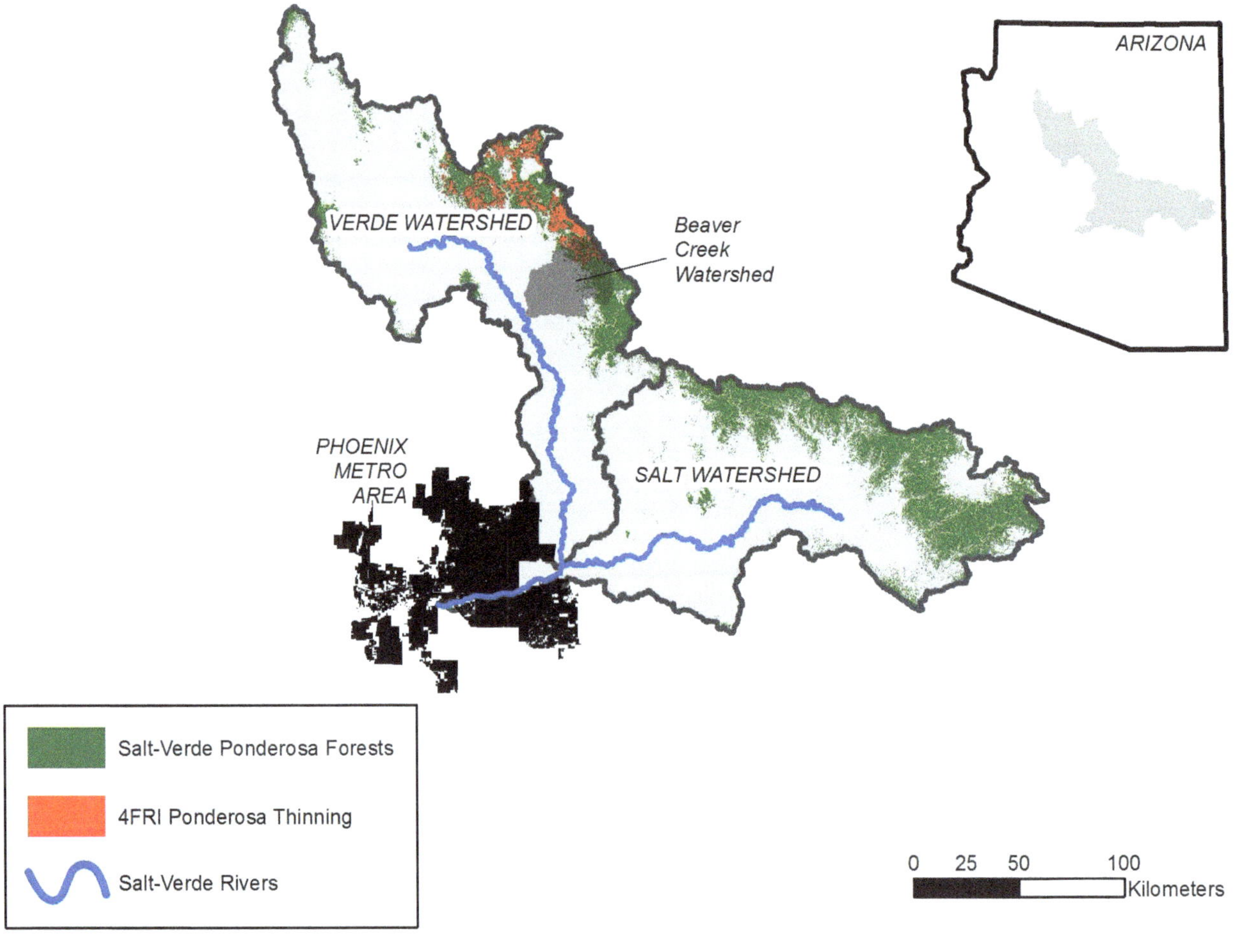

Figure 1. Map of Study Area. Map showing ponderosa pine forests in Salt-Verde watersheds in central Arizona, including those forests that are slated for mechanical thinning within the 4FRI project. Runoff from snowmelt in these forests is primary source of flow to Salt-Verde rivers which in turn are major sources of water for communities in the Phoenix Metro Area. Study used runoff model developed from experimental studies conducted in Beaver Creek watershed [20]. *Inset*: Location of study area in Arizona.

than model fit to total watershed runoff ($r^2 = 0.69$). We concluded that the original regression model was relatively insensitive to the direct effects of forest treatments, including time since treatment, on runoff.

We used a stepwise regression procedure to develop a revised model, beginning with an initial model that included the independent variables (winter precipitation and basal area) that were found to be significant in historical experiments [20]. Input data for all independent variables used to develop the model were derived from empirical data collected during the Beaver Creek and Castle Creek historical experiments and archived in databases or published reports [30–32]. We added independent variables and used an F-test to test the null hypothesis that the coefficient of each term was zero at the 95% significance level. To ensure that the model was not over-fitted, we repeated the stepwise regression procedure with 70% of the original data, selected randomly. We determined that the structure of the model did not change and provided a good fit to the remaining 30% of the data. We examined the residuals from the model and found no discernible patterns of dispersion or bias. Including terms for slope, percent of watershed area with a south-facing aspect, and mean, minimum, and maximum winter temperatures did not significantly improve the model.

The final model included three terms: winter precipitation, an interaction term between winter precipitation and years since treatment, and an interaction term between winter precipitation and basal area. It can be expressed as:

$$R_{treatment} = -28.464 + 0.148 * P - 0.015 * P * Y - 0.092 * P * [\exp(-BA_1/10.33) - \exp(-BA_2/10.33)]$$

$r^2 = 0.67$

where

$R_{treatment}$ = Increase in Annual Runoff attributed to Forest Thinning in mm

P = Total Winter Precipitation (Oct-Apr) in mm

Y = Years since Treatment

BA_1 = Basal Area before Treatment in m^2/ha

BA_2 = Basal Area after Treatment in m^2/ha

Figure 2b shows that the revised model explained more of the variability in predicting additional runoff associated with forest treatments ($r^2 = 0.67$) than did the original model. The model predicted runoff increases in a linear fashion with winter precipitation and basal area reductions, and decreases linearly

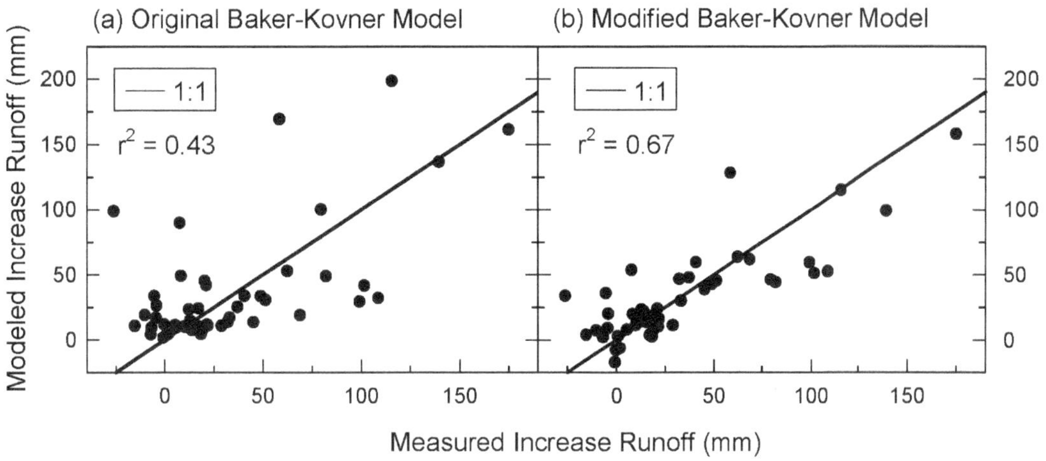

Figure 2. Comparison of models to observed runoff. Fit of (a) original [20] and (b) modified Baker-Kovner regression model output to increases in runoff associated with forests treatments in central Arizona, from Beaver Creek [30],[31] and Castle Creek [32] watersheds.

with years since treatment (Figure 3). To compare the original and modified Baker-Kovner models in English Units, see File S2.

We did use the original Baker-Kovner regression equation in this study to calculate total watershed runoff for forests in their current condition so that we could estimate the percentage increase in runoff associated with treatments. For these calculations, we used the average value for the solar insolation index, r = 0.7, reported from the Beaver Creek experiments [20]. In a sensitivity analysis, we found that calculating absolute runoff based on the low (r = 0.66) and high values (r = 0.74) of this index changed the estimated increase from treatments by only 1–2 percentage points.

Winter Precipitation

In order to account for inter-annual and decadal variability of winter precipitation, we used the PRISM model [25] to extract pluvials and droughts from the 20th century and inserted levels of winter precipitation from these periods into our runoff scenarios. We first extracted total winter precipitation, summed across October to April, for every year from 1900–2012 from the PRISM modeled dataset [25] across 2 scales: (a) ponderosa pine forests within the Verde watershed and (b) these forests within the Salt-Verde watersheds. We selected these geographies to be most representative of conditions for the 4FRI and Salt-Verde runoff scenarios, respectively. From 1900–2012, mean winter precipitation in Verde ponderosa pine forests was 394 mm (15.5 inches) with a range of 99–815 mm (3.9–32.1 inches) (Figure 4). For Salt-Verde forests, the mean was 368 mm (14.5 inches) and the range was 89–747 mm (3.5–29.4 inches) for this same period of time.

To test the accuracy of these modeled data, we compared values from the Verde ponderosa pine PRISM modeled data to measurements of total winter precipitation recorded during the original Beaver Creek experiments [31]. Given that we found the distribution of the data to be non-normal, we ran the non-parametric Spearman's rank order correlation. We found a strong association between modeled and observed data (Rho = 0.9577, $p<0.001$) although the model underestimated measured winter precipitation over the time period by approximately 25 mm (1 inch) and consistently under-estimated precipitation in wet winters (Figure 4 inset).

We calculated mean winter precipitation across 15-, 25-, and 35-year periods from this modeled data and used this information to select droughts and pluvials. For each scenario, we moved the start and end years that thinning was simulated to occur. For example, time periods for our 15-year scenarios included 1944–1958, 1945–1959, and 1946–1960 (examples illustrated as shaded areas in Figure 4). Using this method, we found two instances of pluvials – early 20th century and late 1970s to late-1990s – and two droughts – during the 1950s and the current drought.

Estimating runoff from 4FRI project

We used the modified regression model to calculate additional annual runoff from individual forests stands that will be thinned in the first analysis area of the 4FRI project, established an annual thinning schedule, and then summed increases in runoff from individual stands across treatment years. We ran a total of 26 scenarios where in each scenario we inserted a different 15-year winter precipitation sequence.

We obtained a Geographic Information System (GIS) dataset of alternative C in the draft environmental impact statement for the first analysis area of the 4FRI project [19]. For each stand to be mechanically thinned, this dataset contained estimates of current basal areas and desired post-treatment basal areas. Forest prescriptions for "group-selection" stands called for two post-treatment basal areas, one for "open" areas that would be evenly thinned and another for "group" areas where thinning would be minimal to enhance wildlife habitats by retaining greater tree densities. For these stands, we calculated a post-treatment basal area that was an average of these two basal areas, weighted by the proportional area of the open and group areas. We selected 4,064 ponderosa pine stands within the Verde watershed slated for mechanical thinning, where the prescription will reduce basal areas by at least 30%. They ranged in size from 0.4 to 217 ha (1 to 536 acres), for a total of 61,900 ha (153,000 acres). The prescriptions within these stands called for reductions of basal area between 30% and 70% with a mean reduction of 48% (Figure 5).

Based on consultation with US Forest Service staff, we found that it was reasonable to assume that an equal number of hectares will be mechanically thinned every year across the 10-year treatment period. So for the purposes of our 4FRI scenarios, we assigned each of these stands to one of ten cohorts until the area of each cohort equaled 1/10th of total area or approximately 6,190 ha (15,300 acres). Figure 6a illustrates the treatment schedule for

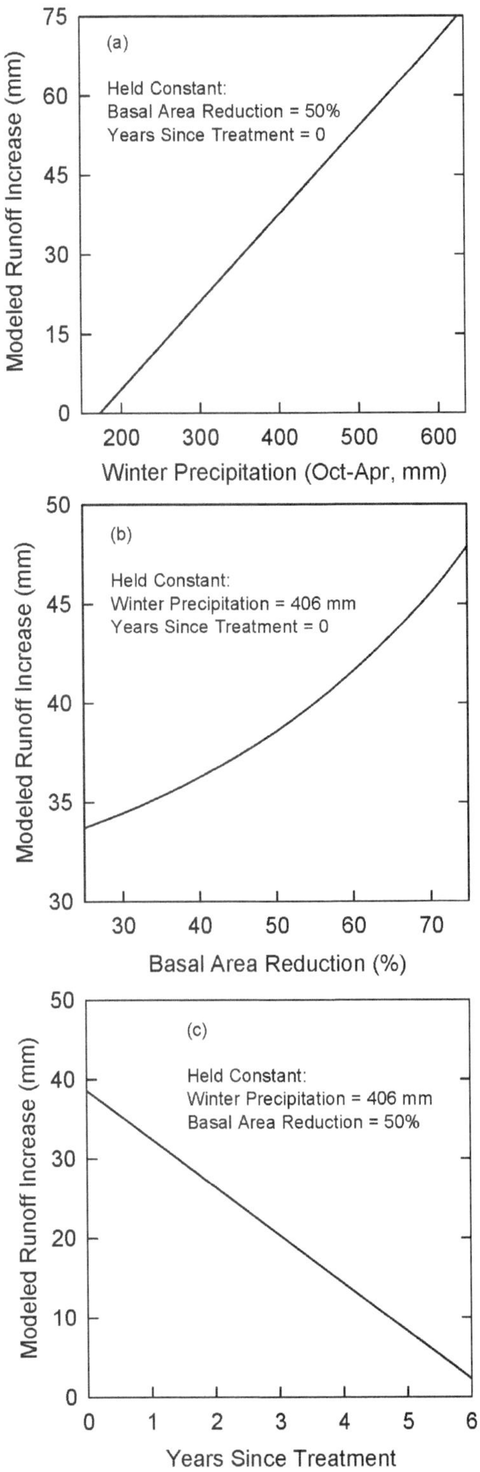

Figure 3. Influence of independent variables on runoff model. Relationships between model output to values for independent variables, including (a) winter precipitation, (b) percent basal area reduction, and (c) years since treatment. In all cases, other independent variables are held constant in order to view relationship of independent variable plotted on X-axis to model output.

the 15-year 4FRI scenarios reported in this study. Cohorts were treated consecutively in the first 10 years and it was assumed that cohorts influenced runoff for 6 years. So, for example, stands treated in cohort 1, contributed to additional runoff in scenario years 1-6 and stands treated in cohort 10 contributed from years 10–15.

We calculated additional runoff associated with thinning and total watershed runoff by inserting values of the independent variables into the revised and original Baker-Kovner regression models, respectively. Pre-treatment and post-treatment basal area values were derived from the 4FRI stand data. Years since treatment ranged from y = 0 to 5. As described previously, winter precipitation values were droughts and pluvials drawn from the PRISM model for ponderosa forests in the Verde watershed. We set runoff increase to zero in years when winter precipitation was less than 230 mm (9 inches). We transformed the unit of runoff from mm to volume (million m^3) using the area of each stand. We estimated annual increases in runoff at the landscape scale by summing stand level amounts for each scenario year, and calculated summary statistics (mean, median, max, cumulative) that allowed for a comparison of scenarios.

Estimating runoff from Salt-Verde watersheds

Unlike the first analysis area of the 4FRI project, there was not a planned forest restoration project across the larger geography of the Salt-Verde watersheds. We developed a range of estimates for the extent, pace, and intensity of forest thinning that could be conducted over this larger geography, grouped these estimates into runoff scenarios, and ran the scenarios using the revised and original regression models to estimate additional runoff from treatments and total watershed runoff. To estimate the potential areal extent of thinning, we subtracted from the total forested areas those land uses that are typically considered unsuitable for mechanical thinning. We bracketed this initial area estimate with lower values to account for the lack of comprehensive spatial information. Assuming that the pace of treatments at this scale could take longer, we constructed 15-, 25-, and 35-year runoff scenarios. Finally, we assumed that the intensity of forest thinning at this scale would be similar to the range of basal area reductions planned for the first analysis are of the 4FRI project. The remaining parts of this section describe these steps in greater detail.

Based on a methodology developed in a previous study that estimated wood supply on four National Forests in northern Arizona [33], we subtracted from the total area of ponderosa pine in the watersheds – 0.68 million ha (1.68 million acres) – those areas that are typically excluded from mechanical thinning projects due to steepness, restrictions in land management, previous treatments, or considerations of soil, habitat, or wildlife conditions (Table 1). We adopted 6 of the 7 "exclusion" criteria from that study and added one additional factor. We removed one criterion – Northern goshawk nest areas –because the mean basal area reduction for these areas in the 4FRI project was greater than 30% [19] suggesting that thinning in these areas could result in additional runoff based on the results of the Beaver Creek experiments [20]. We added one criterion – high severity burn patches – because a nearby study on the recovery of ponderosa pine forests after wildfire demonstrated that mature forests suitable for thinning would not likely develop in areas that had burned with high severity within the timeframe of this study [34]. While two of the three 4FRI prescriptions for Mexican Spotted Owl Protected Activity Centers (MSO PAC) would result in negligible changes in basal area, one would allow for basal area reductions greater than 30% [19]. However, we were not able to differentiate among these prescriptions from the data we obtained, so we opted for a conservative estimate for this factor by excluding MSO PAC areas from our analysis. Compiling the publicly available geospatial data

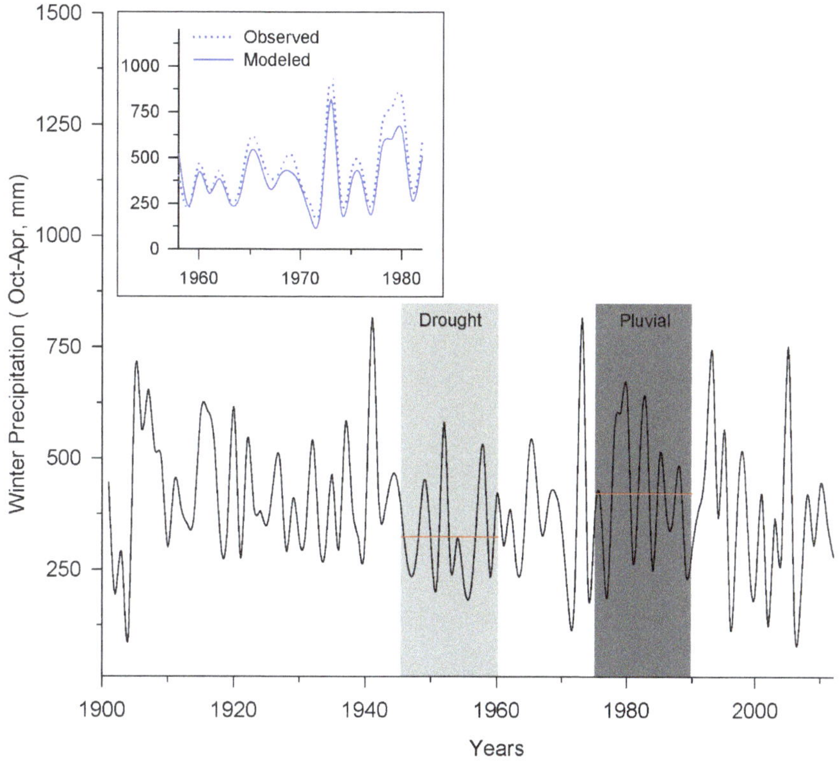

Figure 4. Variability in winter precipitation in ponderosa pine forests. Estimates of historical winter precipitation from 1900–2012 in ponderosa pine forests within Verde watershed from PRISM model [25]. Shaded areas are examples of 15-year droughts and pluvials that were used in study scenarios; horizontal red lines represent mean winter precipitation within these shaded areas. *Inset*: Comparison of measured winter precipitation observed during the historical Beaver Creek watershed experiments [20] from 1958–1982 versus modeled winter precipitation data shown in main figure.

for these factors [26], [27], [33], [35–38], we found that 401,000 ha (992,000 acres), or 59% of ponderosa pine forests, were available for mechanical thinning after accounting for these exclusion areas. The percentage of forests excluded in this estimate – 41% – compared favorably with the average of 37% of forests that were excluded in dozens of restoration projects evaluated in the northern Arizona [33]. We bracketed this high estimate with a moderate estimate – 45% or 301,000 ha (743,000 acres) – and a low estimate – 30% or 204,000 ha (505,000 acres) – because quantitative geo-spatial data on some exclusion areas was missing or incomplete. As illustrated graphically for the moderate estimate in Figure 6b-d, we divided these areas into 10-, 20-, and 30-year treatment cohorts assuming that equal areas would be thinned every year.

We scaled up the forest condition information contained in the 4FRI stand data using a proportion metric, assuming that these stands were representative of stand-level forest conditions across the Salt-Verde watersheds. For example, 4FRI stands that comprised 1,500-ha in a 15,000-ha 4FRI treatment cohort were estimated to be 3,000-ha in a 30,000-ha Salt-Verde cohort. If the thinning prescription for these stands called for a reduction in basal area from 33 to 17 m^2/ha, then the Salt-Verde scenarios calculated additional runoff from 3,000-ha that are thinned from 33 to 17 m^2/ha. After scaling up the area of each stand to the watershed scale, the procedure for estimating runoff in the Salt-Verde scenarios was the same as described for the 4FRI scenarios. We ran 147 scenarios in total, 67 fifteen-year, 52 twenty-five year, and 28 thirty-five year scenarios.

Results

Four Forest Restoration Initiative Runoff Scenarios

Mechanical thinning of ponderosa pine forests in the first analysis area of the 4FRI project – 6,190 ha/year for ten years totaling 61,900 ha (15,300 ac/year; 153,000 acres total) – increased mean annual runoff from 3.13 million m^3 (2,540 acre-feet) in a simulated drought to 7.27 million m^3 (5,890 acre-feet) in a pluvial. Differences in winter precipitation over the 15-year simulation periods significantly influenced runoff gains. A difference in mean winter precipitation of only 130 mm (5 inches), from 330 mm (13 inches) in drought scenario to 460 mm (18 inches) in a pluvial scenario, resulted in a doubling of the annual increase in runoff from treatments (Figure 7).

Inter-annual variability in winter precipitation was also important. Figure 8 illustrates year to year increases in runoff for those scenarios that resulted in the lowest and highest gains in runoff. Years with high winter precipitation played a disproportionate role in additional runoff in both droughts and pluvials. In the majority of scenarios ran (15 out of 26), winter precipitation from only 5 of the 15 years accounted for at least 75% of the increased runoff. Cumulative increases in runoff across 15-year periods resulted in a total increase from 54.3 to 111 million m^3 (44,000–89,800 acre-feet) (Figure 9). Runoff from thinned forests was approximately 20% greater than unthinned forests (as estimated using original Baker-Kovner regression model) in both droughts and pluvials (data not shown).

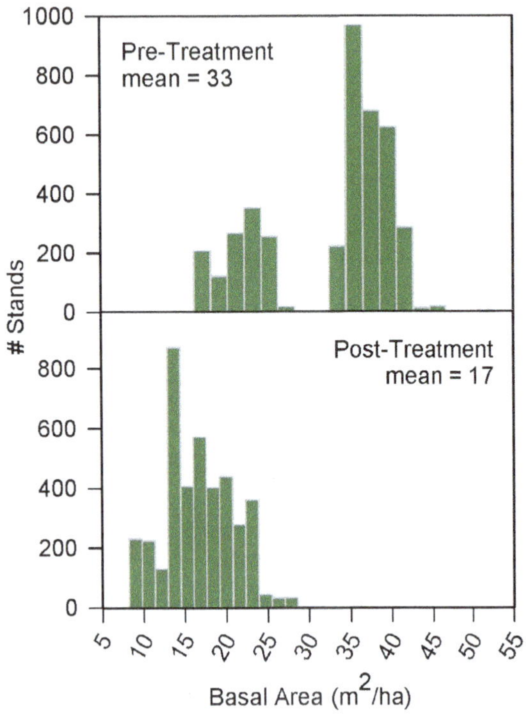

Figure 5. Ponderosa pine basal area reductions in 4FRI project. Histograms showing (top) pre-thinning and (bottom) desired post-thinning basal areas (in m^2/ha) of ponderosa pine stands in the first analysis area of the 4FRI project (excluding stands where basal area reduction $<=30\%$).

Salt-Verde Runoff Scenarios

Depending on winter precipitation and the forest treatment schedule, mean annual increases in runoff from thinning of ponderosa forests across the Salt-Verde watersheds ranged from 4.76 to 15.0 million m^3 (3,860–12,200 acre-feet) over a 35-year treatment period, 6.18 to 23.4 million m^3 (5,010 to 19,000 acre-feet) over 25 years, and 9.23 to 42.8 million m^3 (7,480 to 34,700 acre-feet) over 15 years (Table 2). Similar to the 4FRI scenarios, additional runoff in the Salt-Verde watersheds was 1.6–2.3 times greater in pluvials than in droughts. Regardless of whether the scenarios occurred in a drought or pluvial, cumulative runoff gains in thinned forests were 20-26% greater than unthinned forests. Cumulative gains ranged from 167 to 525 million m^3 (135,000–426,000 acre-feet) in the 35-year scenarios, 154 to 585 million m^3 (125,000–474,000 acre-feet) in 25-year scenarios, and 138 to 643 million m^3 (112,000–521,000 acre-feet) in 15-year scenarios. In both droughts and pluvials, runoff increased in a positive linear fashion with increases in the pace and the extent of forest thinning (Figure 10). See File S1 to see study figures and tables in English units.

Discussion

This study demonstrated that the pace, extent, and intensity of forest thinning that is planned under the first analysis area of 4FRI and at larger scales could measurably increase runoff in ponderosa pine forests in central Arizona. Modeled runoff from thinned forests was approximately 20% greater than runoff from unthinned forests which is within the range of 10–40% increases in runoff demonstrated in the Beaver Creek experiments [20], [29]. The Salt-Verde runoff scenarios showed that levels of

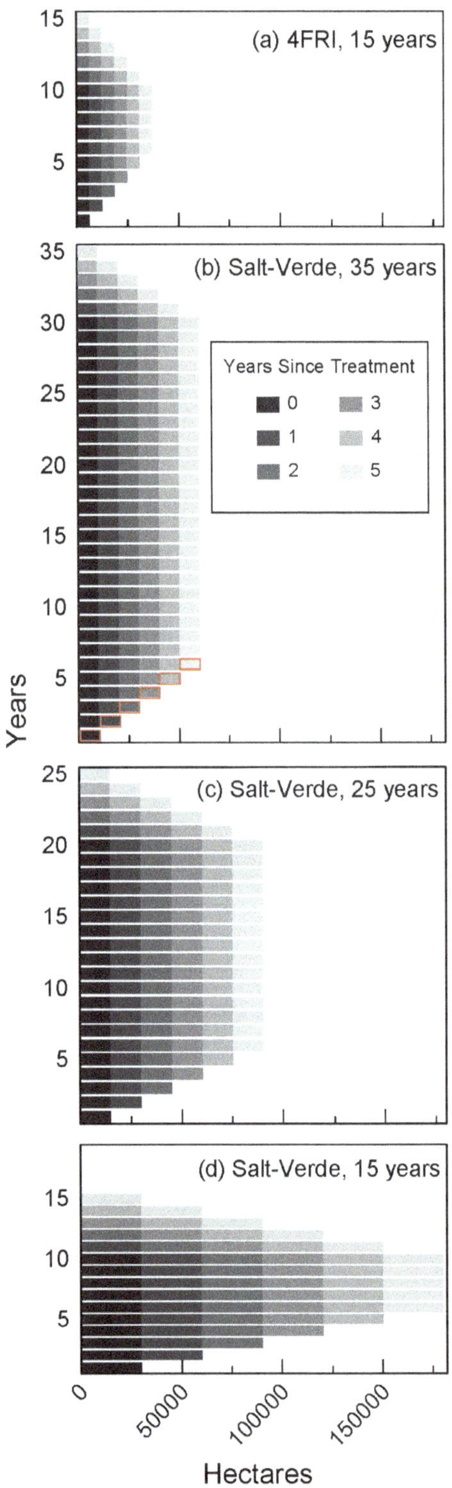

Figure 6. Forest treatment schedules for study scenarios. Graphical depiction of mechanical thinning treatment schedules for (a) 15-year 4FRI scenarios and (b-d) 35-year, 25-year, and 15-year Salt-Verde moderate thinning scenarios (total thinned area was 301,000 ha or 743,000 acres). Scenarios assumed consecutive treatments for 10-, 20-, and 30-year treatment periods shown as black bars in the bottom left portion of each of the figures. Bars outlined in red in (b) show the contribution of one cohort of stands through six years in the scenario.

Table 1. Estimate of hectares of ponderosa pine forests available to mechanical thinning in Salt-Verde watersheds.

Category	**Hectares**	**%**	*Sources*
Salt-Verde Watersheds Ponderosa Pine Forests	*681,000*	-	[27]
Exclusions			
Steep Slopes greater than or equal to 40%	68,000	10%	[26]
Hi-Severity Burn Patches within Wallow and Rodeo-Chedeski Fires	43,700	6%	[35]
Specially Designated Areas including Wilderness Areas, Research Natural Areas, State Wildlife Management Areas[a]	21,800	3%	[36],[37]
Streamside Management Zones within 100 feet perennial reaches in national forests and 200 feet in tribal areas	8,050	1%	[38]
Mexican Spotted Owl Protected Activity Centers where owls have been found to be nesting[b]	57,500	8%	[33]
Erodible Soils where thinning is unsuitable due to erosion risk or rocky conditions[c]	93,100	14%	[33],[37]
Completed Treatments that have been thinned in last 10 years[d]	35,200	5%	[37]
Sub-Total Exclusions (accounting for overlap between layers)	*280,000*	*41%*	
Sub-Total Forest Available Mechanical Thinning	*401,000*	*59%*	

[a] Estimate derived from USDA Forest Service Southwestern Region Wilderness Status layer and Forest Level Special Interest Management Areas layers [37].
[b] Data available for Coconino, Apache-Sitgreaves, Kaibab, and Tonto National Forests only. Added qualitative estimate for this exclusion factor for tribal lands and Prescott National Forest assuming land excluded in these land units would be proportional to lands excluded for this factor on the four national forests above.
[c] List of excluded soils for Coconino, Apache-Sitgreaves, Kaibab, and Tonto National Forests based on Hampton et al. [33]. This list of excluded soils was augmented based on consultation with USDA Forest Service southwestern region soil scientists. Geospatial data for these soils derived from USDA Forest Service Forest Level Terrestrial Ecological Units layers [37]. Added qualitative estimate for this exclusion factor for tribal lands and Prescott National Forest where geospatial data was unavailable, assuming land excluded in these land units would be proportional to lands excluded for this factor on the four national forests above.
[d] Geospatial data for this factor derived from USDA Forest Service Forest Level Activities Layer joined to FACTs table for Coconino, Apache-Sitgreaves, Kaibab, Tonto, and Prescott National Forests [37]. Added qualitative estimate for this exclusion factor for tribal lands based on consultation with Bureau of Indian Affairs forestry personnel.
Summary of hectares and percentages of ponderosa pine forests within Salt-Verde watersheds in relation to potential for thinning treatments, including total hectares, those areas typically excluded from mechanical thinning, and the remaining hectares available for mechanical thinning. Estimates rounded off to three significant digits.

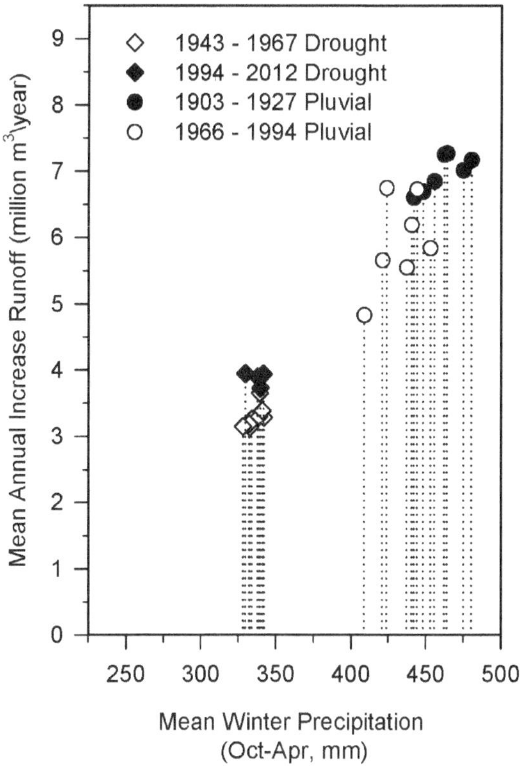

Figure 7. Increases in mean annual runoff from thinning in 4FRI project. Results from 26 scenarios with varying levels of winter precipitation showing increases in mean annual runoff associated with mechanical thinning of ponderosa pine forests in the first analysis area of the 4FRI project. In order to compare scenarios, only increases in *mean* annual runoff are shown. Annual variability in runoff for two of these scenarios is shown in Figure 8.

additional runoff increased proportionally with increases in the pace and extent of thinning. Runoff gains occurred in droughts and pluvials, a surprising outcome that underscored the importance of inter-annual and decadal variability in precipitation. These increases in runoff would likely improve conditions for water-dependent natural resources, such as cienegas, riparian areas, and aquatic habitats, which are vulnerable to low flows that are experienced seasonally, especially in summer months, and also during droughts [5–7]. Runoff gains also could provide incidental benefits to the water supply of downstream users, but the increases were more modest when compared to total runoff from the Salt-Verde watersheds. They would comprise a 0–3% increase from mean annual surface flows from the Salt-Verde rivers of 1.39 billion m^3 (1.13 million acre-feet) and approximately a 1-9% increase of Salt-Verde flows supplied to municipal users in the Phoenix Metro Area on an annual basis [39].

Consistent with the historical experiments, thinning effects on runoff were temporary in this study. Additional runoff would cease six years after the multi-year thinning schedules in our scenarios. Beaver Creek investigators speculated that thinning effects were short-lived because of subsequent regrowth of understory species [29]. We note that prescribed fires are planned as subsequent treatments for all stands that are initially thinned within the first analysis area of 4FRI [19]. These maintenance treatments should provide additional information as to how effective subsequent actions are in terms of removing understory vegetation and sustaining runoff gains.

Our study adds to previous research on the potential for increases in runoff associated with thinning of ponderosa pine forests in the Southwest. Using a simulation model, Brown and Fogel [40] found that thinning approximately 9,300 ha (23,000 acres) of ponderosa pine per year over a 10-year period produced increases in runoff that ranged from 0.28 to 11.8 million m^3/year (230–9,600 acre-feet/year). This range overlapped with but was

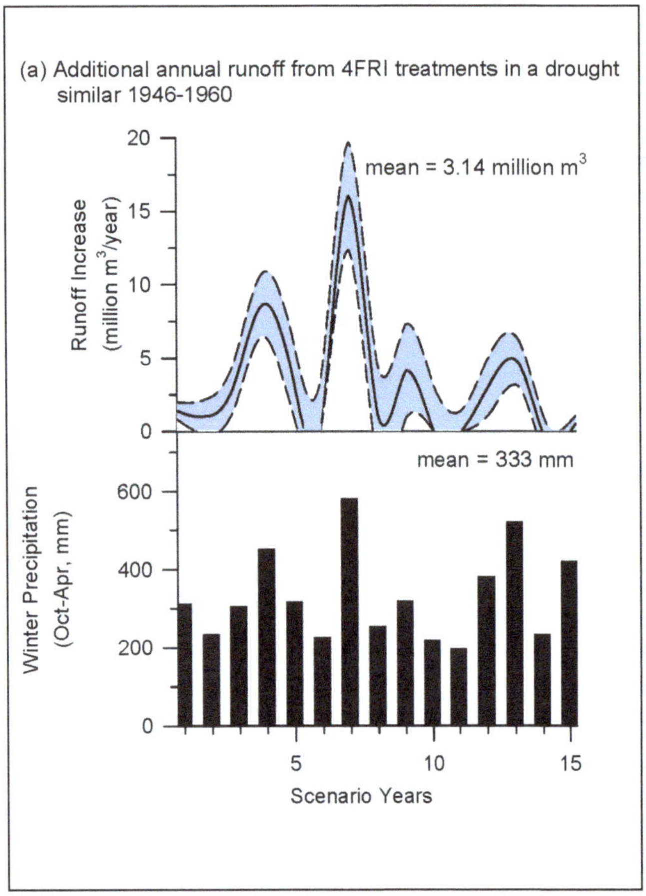

Figure 8. Year to year variability in runoff increases from thinning in 4FRI project. Modeled increases in annual runoff associated with mechanical thinning of ponderosa pine forests in the first analysis area of the 4FRI project during the (a) drought and (b) pluvial that produced lowest and highest gains in runoff respectively. Top panes show increases in annual runoff in million m^3/year. Solid black lines are output values from regression model; dotted lines and blue areas represent 90% confidence intervals. Bottom panes show corresponding winter precipitation values (Oct-Apr, mm) used as one of the independent variables to calculate runoff.

lower than the 6.18 to 11.9 million m^3/year (5,010–9,660 acre-feet/year) increases in runoff for the two Salt-Verde runoff scenarios with annual thinning of approximately 10,000 ha (25,000 acres) (Table 2). One potential reason for this difference was that Brown and Fogel [40] modeled the treatment-runoff relationship to be constant over time, 18 mm (0.71 inches), and did not evaluate variability in precipitation but instead assumed average winter precipitation of 396 mm (15.6 inches). Using similar values for winter precipitation and basal area reduction, our model simulated the runoff response to vary depending on years since treatment, from a high of 38 mm (1.5 inches) in the year immediately following treatments to a low approaching zero mm after six years (Figure 3c).

Study Limitations

In this study, we revised a statistical runoff model from the historical Beaver Creek watershed experiments to explore how variations in climate and forest thinning would affect runoff. To the extent possible, we limited the application of the runoff model to the range of climatic and environmental conditions under which the original data were collected and model developed. However, our modeling approach was constrained by several uncertainties that will require further investigation. The model was limited by the scope of forest practices conducted in the Beaver Creek watershed experiments [20], [30–32]. These experiments used thinning techniques such as strip-thinning and patch clearing that had a similar range of basal area reductions but a different spatial configuration from contemporary thinning prescriptions. They did not measure the effects of maintenance treatments on runoff. They measured "in-place" runoff at sub-watershed outlets and did not measure other aspects of the water budget, such as evapotranspiration, soil moisture, surface routing, and groundwater recharge. The revised model represented statistical relationships between annual runoff and winter precipitation, forest basal area reductions and time since treatment, but these relationships were not necessarily related to physical processes or functions that control water balance [20]. We summarized how important factors that are not directly modeled may increase or decrease the range or runoff due to thinning that is documented in this study in File S3.

Of all the factors where uncertainties in forest hydrology and runoff remain, perhaps the most significant were the effects of soil types, long-term climate variability, and climate change. In terms of soil types, the revised regression model in this study was derived from experimental thinning on basalt-derived soils that support 59% of ponderosa pine forests in the Salt-Verde watersheds. The runoff effects of thinning on the remaining 41% of forests on sedimentary-derived soils are unknown. Two studies demonstrated that runoff from untreated ponderosa forests on sedimentary-derived soils can be

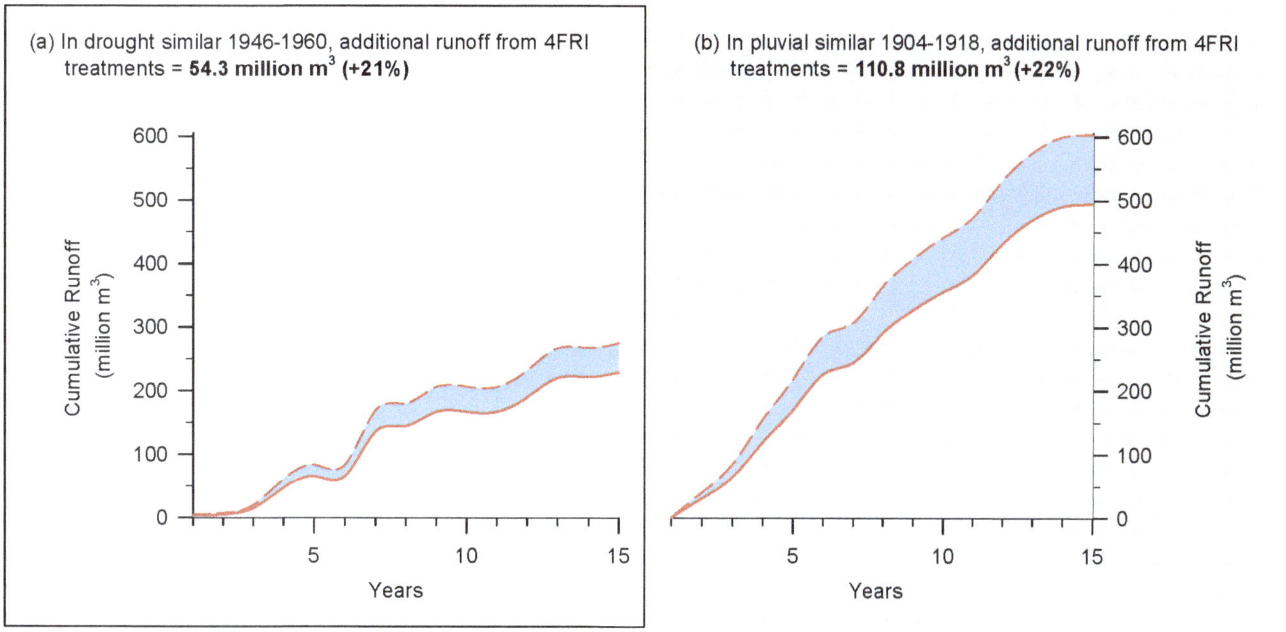

Figure 9. Cumulative runoff increases from thinning in 4FRI project. Estimates of cumulative increases in runoff (million m^3) from planned mechanical thinning of ponderosa pine forests in the first analysis area of the 4FRI project under (a) drought and (b) pluvial conditions. Solid red lines are estimates of cumulative runoff under current forest conditions using original Baker-Kovner regression model [20]. Dotted red lines represent increases in cumulative runoff associated with 4FRI treatments using modified Baker-Kovner regression model. Difference between these two values, shown with blue shading, is additional runoff from forest thinning treatments. Estimated increases in runoff ceased after 15 years.

substantially less than runoff from volcanic soils [41], [42]. Several properties of sedimentary-derived soils relative to basalt-derived soils may lead to greater water permeability and infiltration, including lower clay content, extensive fracturing, and deeper soil horizons [30], [43], [44]. Depending on a number of factors, increased infiltration may lead to increased recharge, which we would expect to result in increased discharge to surface water at some location downstream.

A nearby paired-watershed study in mixed conifer forests on sedimentary-derived soils found that overstory removal of 30-80% led to a slightly higher range of increased runoff, 30–110%, that persisted for more years than the Beaver Creek experiments [45],[46]. Direct comparisons with ponderosa pine forests are not possible because this study was conducted within a higher-elevation mixed-conifer forest that had higher initial basal areas and higher mean annual precipitation.

Relative to climate, this study's scenarios captured the high end of variability associated with pluvials but probably not the low end of variability associated with droughts. Specifically, the early 20th century pluvial may have been the wettest pluvial that has occurred in the last 12 centuries [47], whereas the two occurrences of 20th century drought were eclipsed in terms of duration and severity by several multi-decade "mega-droughts" [48–51]. Output from global circulation models indicates that climate variability will continue to be an important characteristic of the region in the future [52], but that climate change may increase the risk of extreme climatic events such as multi-decade droughts and extreme winter precipitation [53], [54]. Some global circulation models also project that mean winter precipitation in the Southwest will decline by up to 10% [52], but it may take many years to detect effects on stream flows because of precipitation variability [55]. The net effect of changes in precipitation on forest condition and hydrology still needs to be resolved and will require more investigation.

Unlike precipitation, the effects of anthropogenic warming on forest hydrology are clearer. Studies have found that warmer temperatures in recent decades help explain a downward trend in snowpack in the western United States, even after patterns of natural climate variability have been considered [9], [56]. In our revision of the historical runoff model, we attempted to add several parameters of temperature – seasonal means, maximums, and minimums – as explanatory variables to the revised regression model but none were significant. This suggests that the temporal and spatial range of winter temperatures that occurred during the years when the watershed experiments were conducted was not sufficient to add explanatory value to predicting runoff. In a separate trend analysis of winter temperatures (Oct-Apr), we found that temperatures in the months of March and April in the last 25 years, 1988–2012, were significantly warmer by 2 and 1 degrees Celsius (3.5 and 2 degrees Fahrenheit), respectively, than temperatures for the same months in the 25 years when the experiments were conducted, 1958–1982 (see File S4 for more detailed information). The magnitude of observed declines in snowpack in the Southwest, in the range of 20%, is similar to the increases in runoff associated with thinning from this study, suggesting that accelerated thinning may at least offset or ameliorate runoff losses due to climate change.

Management Implications: Accelerated Forest Thinning to Improve Resilience

On balance, we believe there is enough evidence for managers to explore the ways in which accelerated forest thinning can restore forest resilience and functioning to ameliorate the drying of ponderosa pine forests associated with warming and past management activities. Below we summarize the management implications of this and related studies.

Table 2. Annual runoff increases from thinning in Salt-Verde watersheds.

Forest Management: Thinning[a]				Winter Precipitation Regime			Runoff: Annual Increase			Runoff: Cumulative Increase	
Scenario Name	*Total Hectares*	*Hectares per Year*	*Effective Hectares per Year*	*Type*	*Similar to 20th Century Period*	*Mean (mm/yr)*	*Mean (million m^3)*	*Median (million m^3)*	*Maximum (million m^3)*	*Total (million m^3)*	*%*
35-low	204,000	6,800	35,000	Drought	1942–1976	340	4.76	2.99	17.5	167	23%
				Pluvial	1965–1999	396	7.65	5.85	25.7	268	26%
35-mid	301,000	10,000	51,400	Drought	1942–1976	340	7.01	4.39	25.8	245	23%
				Pluvial	1965–1999	396	11.3	8.61	37.7	393	26%
35-high	401,000	13,400	68,800	Drought	1942–1976	340	9.35	5.86	34.4	327	23%
				Pluvial	1965–1999	396	15.0	11.5	50.4	525	26%
25-low	204,000	10,200	49,000	Drought	1942–1966	338	6.18	4.06	26.3	154	20%
				Pluvial	1975–1999	411	11.9	9.65	33.6	299	22%
25-mid	301,000	15,100	72,000	Drought	1942–1966	338	9.09	5.97	38.7	227	20%
				Pluvial	1975–1999	411	17.5	14.2	49.3	438	22%
25-high	401,000	20,100	96,300	Drought	1942–1966	338	12.1	7.98	51.7	303	20%
				Pluvial	1975–1999	411	23.4	19.0	66.0	585	22%
15-low	204,000	20,400	81,700	Drought	1945–1959	315	9.23	5.95	52.3	138	20%
				Pluvial	1975–1989	422	21.8	8.31	60.1	327	22%
15-mid	301,000	30,100	120,000	Drought	1945–1959	315	13.6	8.75	77.0	204	20%
				Pluvial	1975–1989	422	32.1	12.2	88.4	481	22%
15-high	401,000	40,100	161,000	Drought	1945–1959	315	18.1	11.7	103	271	20%
				Pluvial	1975–1989	422	42.8	16.3	118	643	22%

[a] 'Scenario Name' indicates number of years simulated in each scenario (15, 25, 35 years) and extent of forest thinned (low = 30% of forest thinned; mid = 45%; high = 60%). 'Total Hectares' are total number of hectares thinned in each scenario. 'Hectares per Year' are number of hectares thinned every year for 10 consecutive years (15-year scenarios), 20 consecutive years (25-year scenarios), and 30 consecutive years (35-year scenarios). 'Effective Hectares per Year' are the average number of thinned hectares across all years of the scenario that contribute to additional runoff (see figure 6).

Estimates of additional runoff associated with thinning of ponderosa pine forests across Salt-Verde watersheds accounting for variation in treatment periods, total forest extent treated, and drought and pluvial periods of winter precipitation (Oct-Apr). Estimates rounded off to three significant digits.

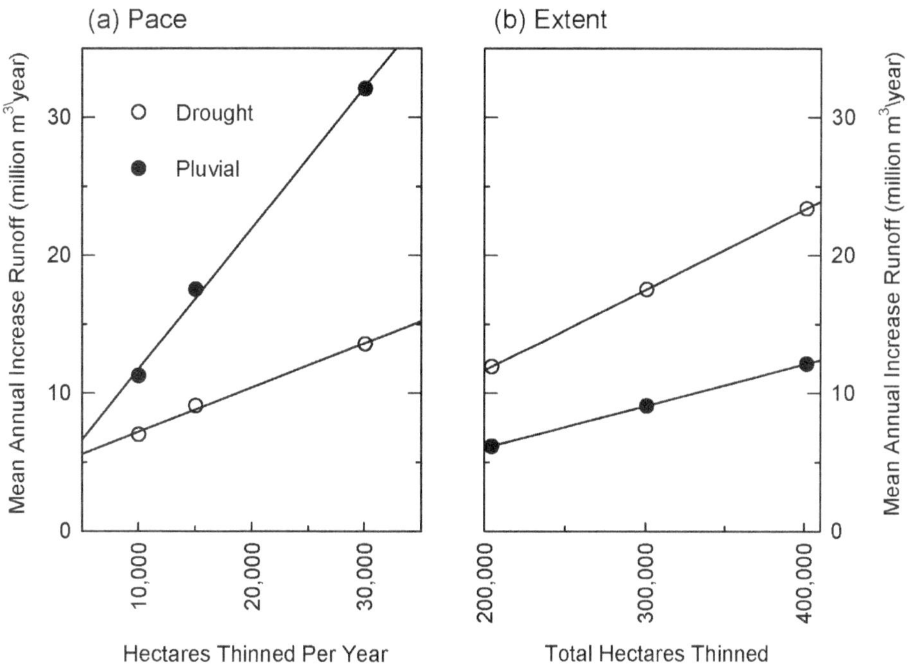

Figure 10. Scale effects of thinning on runoff in Salt-Verde watersheds. Effects of increasing (a) pace and (b) extent of thinning treatments in ponderosa pine forests in Salt-Verde watersheds on increases in mean annual runoff (million m^3/year). In (a) total area thinned is held constant at 301,000 ha (743,000 acres) (scenarios: 35mid, 25mid, 15mid) to show influence of increasing the area thinned per year. In (b) duration of thinning treatments is held constant at 25 years (scenarios: 25low, 25mid, 25high) to show influence of increasing the total area thinned across the scenario. In order to illustrate scale effects, only increases in *mean* annual runoff are shown. Statistics describing annual variability in runoff gains are shown in Table 2 and illustrated graphically for 4FRI scenario in Figure 8.

1. *Maintain & improve ecosystem function* – The current paradigm of forest management has moved away from a debate about managing forests to address one societal need, such as downstream water supply, to recognition that management strategies need to address multiple challenges [29], [57]. The 4FRI restoration project demonstrates this approach with planned benefits that include "…improved vegetation biodiversity, wildlife habitat, soil productivity, and watershed function" [19]. The impaired functioning and heightened vulnerability of southwestern forests to large-scale disturbances increases the urgency to improve ecosystem functioning, and to measure the effectiveness and trade-offs associated with these management activities using robust adaptive management and monitoring protocols. Such a focus may result in more options and flexibility in subsequent years, whereas failure to maintain functioning ecosystems may eliminate future options if soil and hydrologic processes are irreversibly altered.

Our study demonstrated that accelerated thinning can improve surface water runoff, a key ecosystem function, and provided indirect evidence of benefits for soil moisture and productivity. Another study found that average soil water content on low density ponderosa forests (250 trees/ha) was substantially higher than on high density forests (2,710 trees/ha), although these differences in forest density were not a result of mechanical thinning [58]. Researchers in the historical Beaver Creek experiments hypothesized that lower tree densities associated with thinning would reduce evapotranspiration losses, and thereby allow more water to be available for soil moisture, groundwater recharge, and surface water runoff [29]. Although we were unable to evaluate thinning effects on evapotranspiration, Dore et al. [59] found that light thinning (40% basal area reduction) in a ponderosa pine forest reduced stand-level evapotranspiration by 12% over four years, but the effect diminished over time and was not detectable in the fourth post-harvest year. This same study found that removal of all trees in a wildfire-burned site reduced annual evapotranspiration by 20%.

Other studies demonstrated that mechanical thinning can reduce the vulnerability of forests to uncharacteristic crown fires. Using a fire simulation model, Cochrane et al. [60] found that landscape-scale mechanical thinning that preceded actual wildfires could have reduced the average size of six wildfires in ponderosa pine and mixed conifer forests in California and the southwestern United States by an average of 18% (range 0.3 to 65%). A meta-analysis of 54 experimental studies showed that mechanical thinning significantly reduced the susceptibility of western ponderosa and Jeffrey pine forests to crown fires [23]. These studies describe the primary mechanism by which thinning reduces fire risk as a redistribution of fuels, thereby reducing fire spread rates and crowning behavior.

2. *Apply management tools at scale* – Managing forests at a scale that is comparable to recent forest disturbances will be another key factor for improving resilience. At 238,000 ha (588,000 acres) of ponderosa pine to be treated with mechanical thinning and prescribed fire, the 4FRI project is approaching the scale of these amplified disturbances. Our study demonstrated that potential increases in runoff associated with thinning were scale dependent: runoff increased in a linear fashion as the pace and extent of thinning increased in the Salt-Verde runoff scenarios (Figure 10). Recent efforts to reduce fire risk with landscape-scale treatments of ponderosa pine forests have been successful. For example, treatments completed before the largest wildfire in Arizona history, the Wallow Fire in 2011, effectively protected communities and towns by reducing fire severity before it reached key residential areas [61].

Further, accelerated thinning may open up opportunities to use other management tools at an accelerated pace. For example, wildland fire use has been an effective and inexpensive tool for maintaining forest resilience across large areas in remote forests and woodlands – the Gila Wilderness in New Mexico and Grand Canyon National Park in northern Arizona [62], [63] – but has been used less frequently in forests near population centers and towns. Completion of treatments in the first analysis area of the 4FRI project could have a multiplying effect on maintaining forest resilience if the thinning itself allows for increased use of wildland fire or prescribed fire over many more hectares.

3. *Leverage knowledge of climate variability* – The Beaver Creek watershed experiments and others like it demonstrated that winter precipitation was the best predictor of additional runoff from forest thinning [20], [30]. We add to this knowledge by examining how variability in winter precipitation could influence management effectiveness. Our scenarios demonstrated that gains in runoff from thinning could occur even under drought conditions. In simulated droughts, 1 out of 3 years had above-average precipitation, and these years accounted for 75% of the runoff gains in these scenarios. Even though these gains were substantially less than runoff gains during pluvials, additional runoff under drought conditions could help lessen the impact of predicted climate-driven losses in hydrologic connectivity important to native fish in the Salt and Verde watersheds [64].

This study also demonstrated that the same thinning schedule could result in a doubling in additional runoff if conducted in a pluvial typical of 20th century variability as compared to a drought. Scientists have developed decision support tools to help managers understand the likelihood that precipitation regimes will shift from wet to dry conditions or vice-versa [65], [66]. Knowledge of whether management activities are taking place in a pluvial or a drought could improve the ability of managers to predict the likelihood of success for various objectives. For example, thinning in a pluvial might meet substantive objectives to improve runoff, whereas thinning in a drought might be more significant for reducing risks of catastrophic crown fires or drought mortality.

With the notable exception of a mega-drought, the droughts and pluvials examined in this study were characteristic of climate variability in the region. An examination of hydro-climatic trends in the Salt-Verde watersheds revealed that temperatures are non-stationary, increasing significantly in recent decades, but the same is not true for winter precipitation or resulting stream flows [55]. Rather, winter precipitation has fluctuated between droughts and pluvials in these watersheds in the last century with no evidence of directional change, a pattern that has reoccurred in the southwestern United States for the last 1000 years [50]. These precipitation cycles have been linked to persistent anomalies in sea surface temperatures that vary across inter-annual and decadal time steps [67], [68]. If past patterns of precipitation variability remain stable in the near term, then it is probable that precipitation and flows in the Salt-Verde watersheds will shift into wetter conditions within the timeframe examined in this study [66].

4. *Building Knowledge to Reduce Uncertainties* – Landscape-scale restoration projects like 4FRI present the opportunity to learn about the influence of accelerated thinning on forest water budgets and resilience using modern forestry techniques and under a changing and variable climate. Stakeholders in the 4FRI project are developing a robust adaptive management and monitoring program so that progress towards objectives can be measured and timely adjustments to management made. Additionally, a paired-watershed experiment to evaluate the effects of forest treatments, including prescribed fire maintenance treatments, on watershed properties is planned within the first analysis area of 4FRI. It will monitor runoff, precipitation, evapotranspiration, snow water storage, soil moisture, groundwater recharge, and sediment yield, thereby reducing our knowledge gaps regarding the effects of forest management on the water cycle.

Conclusions

The widespread pattern of forest drying, evidenced by ecosystem level moisture deficits and larger and more severe wildfires, indicates we are entering a new era of forest management. The 4FRI project is an example of a restoration effort that is addressing this challenge in two ways: establishing objectives to improve resilience and increasing the scale at which management tools are applied. Our study demonstrated that this type of project can increase runoff by approximately 20% compared to unthinned forests, even under simulated drought conditions. If treatments at this scale are completed and repeated over the next several decades, increases in runoff could help offset the current and projected declines in snowpack and stream flow due to warming while improving the resilience of forest stands. As an incidental benefit in an era of dwindling water supplies and projected water shortages, forest thinning could play a role in augmenting river flows on a seasonal basis, improving conditions for water-dependent ecosystems, and benefitting the water supplies of downstream communities. Accelerated forest thinning to reduce water stress and wildfire risk is one of the only management options under our control, and it is probably the most critical to apply over the short term. Because the effects are temporary, it is important to frame accelerated thinning as a reset of ecosystem resilience, not a permanent cure. Rather, accelerated forest thinning creates opportunities to apply other management tools that restore forest resilience at broader scales, whereas management actions at the current pace will likely not be sufficient to address recent and projected changes in forest conditions.

Supporting Information

File S1 Methods and results figures and tables in English units.

File S2 Original and revised Baker-Kovner regression models in English units.

File S3 Additional factors that influence runoff.

File S4 Temperature trend analysis.

Acknowledgments

We thank Daniel Majka for his early contributions and input into this study. We thank Dan Huebner and Karen Koestner for providing data from the historical Beaver Creek experiments and Shaula Hedwall and Mark Nigrelli for providing data and expertise from the Four Forest Restoration Initiative and the Northern Arizona Wood Supply Study. We thank Kit MacDonald, Chris Nelson, Estella Smith, Rory Steinke, Joe Ringelberg, and Andrea Thode for providing data and expertise of forest thinning exclusions. We thank Julio Bentancourt, Michael Crimmins, and Gregg Garfin for input on climate variability and climate change, Charlie Ester for input on water management in the Salt-Verde River watersheds, and Robert Steidl for advice on statistical analyses. We thank Pat Graham, Molly Hunter, Dale Turner and Sue Sitko for critical review of this paper.

Author Contributions

Conceived and designed the experiments: MR RM FO JH. Performed the experiments: MR FO. Analyzed the data: MR FO. Contributed reagents/materials/analysis tools: MR FO JH. Wrote the paper: MR RM FO ES JH DG.

References

1. Covington WW, Moore MM (1994) Southwestern ponderosa pine forest structure: Changes since Euro-American settlement. Journal of Forestry 92(1): 39–47.
2. Fule PZ, Heinlein TA, Covington WW, Moore MM (2003) Assessing fire regimes on Grand Canyon landscapes with fire-scar and fire-record data. International Journal of Wildland Fire 12: 129–145.
3. Cocke AE, Fule PZ, Crouse JE (2005) Forest change on a steep mountain gradient after extended fire exclusion: San Francisco Peaks, Arizona, USA. Journal of Applied Ecology 42(5): 814–823.
4. Williams AP, Allen CD, Millar CI, Swetnam TW, Michaelsen J, et al. (2011) Forest responses to increasing aridity and warmth in the southwestern United States. Proceedings of the National Academy of Sciences 107(50): 21289–21294.
5. Gitlin AR, Sthultz CM, Bowker MA, Stumpf S, Paxton KL, et al. (2006) Mortality gradients within and among dominant plant populations as barometers of ecosystem change during extreme drought. Conservation Biology 20(5): 1477–1486.
6. Propst DL, Gido KB, Stefferud JA (2008) Natural flow regimes, nonnative fishes, and native fish persistence in arid-land river systems. Ecological Applications 18(5): 1236–1252.
7. Stefferud JA, Gido KB, Propst DL (2011) Spatially variable response of native fish assemblages to discharge, predators and habitat characteristics in an arid-land river. Freshwater Biology 56: 1403–1416.
8. Breshears DD, Cobb NS, Rich PM, Price KP, Allen CD, et al. (2005) Regional vegetation die-off in response to global-change-type drought. Proceedings of the National Academy of Sciences 102(42): 15144–15148.
9. Mote PW (2006) Climate-driven variability and trends in mountain snowpack in western North America. Journal of Climate 19: 6209–6220.
10. Westerling AL, Hidalgo HG, Cayan DR, Swetnam TW (2006) Warming and earlier spring increase Western US forest wildfire activity. Science 313: 940–943.
11. van Mantgem PJ, Stephenson NL, Byrne JC, Daniels LD, Franklin JF, et al. (2009) Widespread increase of tree mortality rates in the Western United States. Science 323: 521–524.
12. Thomas BE, Pool DR (2006) Trends in Streamflow of the San Pedro River, Southeastern Arizona, and Regional Trends in Precipitation and Streamflow in Southeastern Arizona and Southwestern New Mexico. U.S. Geological Survey Professional Paper 1712. Reston, VA: U.S. Geological Survey. 79 p.
13. Timmons D, Springer A (2006) Prescott AMA groundwater flow model update report. Available: http://www.azwater.gov/AzDWR/Hydrology/Modeling/documents/Prescott_Model_Update_Report.pdf. Accessed 6 May 2014.
14. Webb RH, Turner RM, Leake SA. (2007) The Ribbon of Green: Change in Riparian Condition in the Southwestern United States. Tucson, AZ: The University of Arizona Press. 480 p.
15. Marshall RM, Robles MD, Majka DR, Haney JA (2010) Sustainable Water Management in the Southwestern United States: Reality or Rhetoric? PLoS ONE 5(7): e11687. doi:10.1371/journal.pone.0011687.
16. Water Resources Development Commission (2011) Water Resources Development Commission Final Report: Volume II Committee Reports. Phoenix, AZ: Water Resources Development Commission. 685 p.
17. United States Department of Interior Bureau of Reclamation (2012) Colorado River Basin Water Supply and Demand Study: Study Report. Washington DC: U.S. Department of Interior, Bureau of Reclamation. 95p.
18. Vano JA, Udall B, Cayan DR, Overpeck JT, Brekke LD, et al. (2014) Understanding uncertainties in future Colorado River streamflow. Bulletin of the American Meteorological Society 95(1): 59–78.
19. United States Department of Agriculture Forest Service (2013) Draft environmental impact statement for the Four-Forest Restoration Initiative. Flagstaff, AZ: USDA Forest Service, Southwestern Region, Coconino and Kaibab National Forests. 744 p.
20. Brown HE, Baker MB, Rogers JJ, Clary WP, Kovner JL, et al. (1974) Opportunities for increasing water yields and other multiple uses on ponderosa pine forest lands. USDA Forest Service Research Paper RM-129. Fort Collins, CO: U.S. Department of Agriculture, Forest Service, Rocky Mountain Forest and Range Experiment Station. 36 p.
21. Zausen GL, Kolb TE, Bailey JD, Wagner MR (2005) Long-term impacts of stand management on ponderosa pine physiology and bark beetle abundance in northern Arizona: A replicated landscape study. Forest Ecology and Management 218: 291–305.
22. Stephens SL, Moghaddas JJ, Edminster C, Fiedler CE, Haase S, et al. (2009) Fire treatment effects on vegetation structure, fuels, and potential fire severity in western US forests. Ecological Applications 19(2): 305–320.
23. Fule PZ, Crouse JE, Roccaforte JP, Kalies EL (2012) Do thinning and/or burning treatments in western USA ponderosa pine or Jeffrey pine-dominated forests help restore natural fire behavior? Forest Ecology and Management 269: 68–81.
24. Barr GW (1956) Recovering rainfall. Tucson, AZ: Department of Agricultural Economics, University of Arizona. 33 p.
25. Daly C, Halbleib M, Smith JI, Gibson WP, Doggett MK, et al. (2008) Physiographically-sensitive mapping of climatological temperature and precipitation across the conterminous United States. International Journal of Climatology 28(15): 2031–2064.
26. Gesch D, Oimoen M, Greenlee S, Nelson C, Steuck M, et al. (2002) The National Elevation Dataset. Photogrammetric Engineering and Remote Sensing 68(1): 5–11.
27. US Geological Survey, Gap Analysis Program (2010) National Land Cover, Version 1. US Geological Survey. Available: http://gapanalysis.usgs.gov/gaplandcover/viewer/. Accessed 2012 Dec 23.
28. Arizona Geological Survey (2011) Geologic Map of Arizona. Available: http://www.azgs.az.gov/services_azgeomap.shtml. Accessed 2013 Apr 9.
29. Baker MB (2003) Hydrology. In: P . Friederici, editor. Ecological Restoration of Southwestern Ponderosa Pine Forests. Washington DC: Ecological Restoration Institute, Northern Arizona University, Island Press. pp. 161–174.
30. Baker MB (1986) Effects of ponderosa pine treatments on water yield in Arizona. Water Resources Research 22(1): 67–73.
31. Neary D (2011) Managing Arid and Semi-Arid Watersheds: Beaver Creek Watershed. U.S. Department of Agriculture, Forest Service, Rocky Mountain Research Station. Available: http://www.rmrs.nau.edu/watersheds/highlands/watersheds/beavercreek/. Accessed 2012 Oct 29.
32. Rich LR (1972) Managing a ponderosa pine forest to increase water yield. Water Resources Research 8(2): 422–428.
33. Hampton HM, Sesnie SE, Bailey JD, Snider GB (2011) Estimating regional wood supply based on stakeholder consensus for forest restoration in Northern Arizona. Journal of Forestry: 15–26.
34. Haire SL, McGarigal K (2010) Effects of landscape patterns of fire severity on regenerating ponderosa pine forests (***Pinus ponderosa***) in New Mexico and Arizona, USA. Landscape Ecology 25: 1055–1069.
35. Eidenshink J, Schwind B, Brewer K, Zhu Z, Quayle B, et al. (2007) A project for monitoring trends in burn severity. Journal of Association of Fire Ecology 3(1): 3–21.
36. Conservation Biology Institute (2010) Protected Areas Database of the United States, Version 2. Available: http://consbio.org/products/projects/pad-us-cbi-edition. Accessed 2013 Apr 9.
37. United States Department of Agriculture Forest Service (2013) GIS datasets for the Southwestern Region, United States Department of Agriculture, Forest Service. Available: http://www.fs.usda.gov/detail/r3/landmanagement/gis/?cid=stelprdb5202474. Accessed 2013 Apr 9.
38. Turner DS, List MD. 2007. Habitat mapping and conservation analysis to identify critical streams for Arizona's native fish. Aquatic Conservation: Marine and Freshwater Systems 17: 737–748.
39. Arizona Department of Water Resources (2010) Arizona Water Atlas Volume 8: Active Management Areas Water Atlas. Available: http://www.azwater.gov/azdwr/StatewidePlanning/WaterAtlas/ActiveManagementAreas/default.htm. Accessed 2014 May 5.
40. Brown TC, Fogel MM (1987) Use of streamflow increases from vegetation management in the Verde River Basin. Water Resources Bulletin 23(6): 1149–1160.
41. Campbell RE, Baker MB, Ffolliott PF, Larson FR, Avery CC (1977) Wildfire effects on a ponderosa pine ecosystem: An Arizona case study. USDA Forest Service Research Paper RM-191. Fort Collins, CO: U.S. Department of Agriculture, Forest Service, Rocky Mountain Forest and Range Experiment Station. 12 p.
42. Ffolliott PF, Baker MB (1977) Characteristics of Arizona ponderosa pine stands on sandstone soils. USDA Forest Service General Technical Report RM-44. Fort Collins, CO: U.S. Department of Agriculture, Forest Service, Rocky Mountain Forest and Range Experiment Station. 7 p.
43. Pearson GA (1931) Forest types in the Southwest as determined by climate and soil. Technical Bulletin No. 247. Washington DC: U.S. Department of Agriculture, Forest Service, Southwestern Forest and Range Experiment Station. 144 p.
44. Bosch JM, Hewlett JD (1982) A review of catchment experiments to determine the effect of vegetation changes on water yield and evapotranspiration. Journal of Hydrology 55: 3–23.
45. Rich LR, Gottfried GJ (1976) Water yields resulting from treatments on the Workman Creek experimental watersheds in Central Arizona. Water Resources Research 12(5): 1053–1060.
46. Hibbert AR, Gottfried GJ (1987) Streamflow responses to forest treatments on two Arizona mixed conifer watersheds. In: Troendle CA, Kaufmann MR, Hamre RH, Winokur RP editors. Management of subalpine forest: Building on 50 years of research. USDA Forest Service General Technical Report RM-149. Silver Creek, CO: U.S. Department of Agriculture, Forest Service, Rocky Mountain Forest and Range Experiment Station. pp. 189–194.

47. Woodhouse CA, Kunkel KE, Easterling DR, Cook ER (2005) The twentieth-century pluvial in the western United States. Geophysical Research Letters 32: L07701, doi:10.1029/2005GL022413.
48. Woodhouse CA, Overpeck JT (1998) 2000 years of drought variability in the central United States. Bulletin of the American Metereological Society 79(12): 2693–2714.
49. Stahle DW, Cook ER, Cleaveland MK, Therrell D, Meko DM, et al. (2000) Tree-ring data document 16th century megadrought over North America. Eos 81(12): 121–125.
50. Ni F, Cavazos T, Hughes MK, Comrie AC, Funkhouser G (2002) Cool-season precipitation in the southwestern USA since AD 1000: Comparison of linear and nonlinear techniques for reconstruction. International Journal of Climatology 22: 1645–1662.
51. Cook ER, Seager R, Cane MA, Stahle DW (2007) North American drought: Reconstructions, causes, and consequences. Earth-Science Reviews 81: 93–134.
52. Garfin G, Jardine A, Merideth R, Black M, LeRoy S, editors (2013) Assessment of Climate Change in the Southwest United States: A Report Prepared for the National Climate Assessment. A report by the Southwest Climate Alliance. Washington DC: Island Press. 531 p.
53. Ault TR, Cole JE, Overpeck JT, Pederson GT, Meko DM (2014) Assessing the risk of persistent drought using climate model simulations and paleoclimate data. Journal of Climate e-View doi:http://dx.doi.org/10.1175/JCLI-D-12-00282.1.
54. Dominguez F, Rivera E, Lettenmaier DP, Castro CL (2012). Changes in winter precipitation extremes for the western United States under a warmer climate as simulated by regional climate models. Geophysical Research Letters (39) L05803, doi:10.1029/2011GL050762.
55. Murphy KW, Ellis AW (2014) An assessment of the stationarity of climate and stream flow in watersheds of the Colorado River Basin. Journal of Hydrology 509: 454–473.
56. Pierce DW, Barnett TP, Hidalgo HG, Das T, Bonfils C, et al. (2008) Attribution of declining Western U.S. snowpack to human effects. Journal of Climate 21: 6425–6444.
57. Grant GE, Tague CL, Allen CD (2013) Watering the forest for the trees: An emerging priority for managing water in forest landscapes. Frontiers in Ecology and the Environment 11: 314–321.
58. Zou CB, Breshears DD, Newman BD, Wilcox BP, Gard MO, et al. (2008) Soil water dynamics under low-versus high-ponderosa pine tree density: Ecohydrological functioning and restoration implications. Ecohydrology 1: 309–315.
59. Dore S, Montes-Helu M, Hart SC, Hungate BA, Koch GW, et al. (2012) Recovery of ponderosa pine ecosystem carbon and water fluxes from thinning and stand-replacing fire. Global Change Biology 18: 3171–3185.
60. Cochrane MA, Moran CJ, Wimberly MC, Baer AD, Finney MA, et al. (2012) Estimation of wildfire size and risk changes due to fuel treatments. International Journal of Wildland Fire 21(4): 357–267.
61. Kennedy MC, Johnson MC (2014) Fuel treatment prescriptions alter spatial patterns of fire severity around the wildland-urban interface during the Wallow Fire, Arizona, USA. Forest Ecology and Management 318: 122–132.
62. Fule PZ, Laughlin DC (2007) Wildland fire effects on forest structure over an altitudinal gradient, Grand Canyon National Park, USA. Journal of Applied Ecology 44(1): 136–146.
63. Hunter ME, Iniguez JM, Lentile LB (2011) Short- and long-term effects on fuels, forest structure, and wildfire potential from prescribed fire and resource benefit fire in southwestern forests, USA. Fire Ecology 7(3): 108–121.
64. Jaeger KL, Olden JD, Pelland NA (2014) Climate change poised to threaten hydrologic connectivity and endemic fishes in dryland streams. Proceedings of the National Academy of Sciences. doi: 10.1073/pnas.1320890111.
65. Enfield DB, Cid-Serrano L (2006) Projecting the risk of future climate shifts. International Journal of Climatology 26: 885–895.
66. Gangopadhyay S, McCabe GJ (2010) Predicting regime shifts in flow of the Colorado River. Geophysical Research Letters 37: L20706, doi:10.1029/2010GL044513.
67. Sheppard PR, Comrie AC, Packin GD, Angersbach K, Hughes MK (2002) The climate of the Southwest. Climate Research 21: 219–238.
68. Mantua NJ, Hare SR (2002) The Pacific Decadal Oscillation. Journal of Oceanography 58: 35–44.

14

Does Fire Influence the Landscape-Scale Distribution of an Invasive Mesopredator?

Catherine J. Payne[1], Euan G. Ritchie[1], Luke T. Kelly[2], Dale G. Nimmo[1]*

1 Centre for Integrative Ecology, School of Life and Environmental Sciences, Deakin University, Melbourne, Victoria, Australia, **2** Australian Research Council Centre of Excellence for Environmental Decisions, School of Botany, University of Melbourne, Melbourne, Victoria, Australia

Abstract

Predation and fire shape the structure and function of ecosystems globally. However, studies exploring interactions between these two processes are rare, especially at large spatial scales. This knowledge gap is significant not only for ecological theory, but also in an applied context, because it limits the ability of landscape managers to predict the outcomes of manipulating fire and predators. We examined the influence of fire on the occurrence of an introduced and widespread mesopredator, the red fox (*Vulpes vulpes*), in semi-arid Australia. We used two extensive and complimentary datasets collected at two spatial scales. At the landscape-scale, we surveyed red foxes using sand-plots within 28 study landscapes – which incorporated variation in the diversity and proportional extent of fire-age classes – located across a 104 000 km^2 study area. At the site-scale, we surveyed red foxes using camera traps at 108 sites stratified along a century-long post-fire chronosequence (0–105 years) within a 6630 km^2 study area. Red foxes were widespread both at the landscape and site-scale. Fire did not influence fox distribution at either spatial scale, nor did other environmental variables that we measured. Our results show that red foxes exploit a broad range of environmental conditions within semi-arid Australia. The presence of red foxes throughout much of the landscape is likely to have significant implications for native fauna, particularly in recently burnt habitats where reduced cover may increase prey species' predation risk.

Editor: R. Mark Brigham, University of Regina, Canada

Funding: Funding for the site-scale study was provided by the Victorian government's Department of Environment and Primary Industries, under the Mallee Hawkeye project (Contract Number 313764). Funding and logistical support for the landscape-scale study was provided by Land and Water Australia, the Mallee Catchment Management Authority, Parks Victoria, Department Sustainability and Environment Victoria, Department Environment and Heritage SA, Lower Murray-Darling Catchment Management Authority, Department Environment and Climate Change NSW, Australian Wildlife Conservancy and Birds Australia. The funders had no role in study design, data collection and analysis, decision to publish, or preparation of the manuscript.

Competing Interests: The authors have declared that no competing interests exist.

* Email: dale@deakin.edu.au

Introduction

Predators shape ecosystems worldwide [1]. They can exert top-down regulation of lower trophic levels [2] and induce trophic cascades which flow through entire ecosystems [3]. Predators introduced to areas outside of their native range can have a particularly strong effect on native species [4], and have caused population declines and extinctions in a range of ecosystems [5]. Many invasive predators are 'mesopredators': smaller predator species that increase in abundance or activity following the removal of apex predators [6]. For example, in Australia, persecution of the native apex predator, the dingo (*Canis dingo*), has led to increases in the density or activity of invasive mesopredators (e.g. the red fox [*Vulpes vulpes*]) throughout large portions of the continent [3].

Fire is another globally significant process that affects environments worldwide [7]. Fire influences ecosystems via bottom-up control by altering the availability of key resources for biota. Fire incinerates plant matter, altering vegetation structure [8,9], which in turn affects the distribution and abundance of animals [10].

Invasive mesopredators and fire share an important characteristic from a conservation perspective: both can be manipulated through management interventions. Invasive mesopredators are managed using lethal control and exclusion fencing, and fire using suppression or prescribed burning. However, management of mesopredators and fire usually occurs in isolation, without consideration of the potential effects of fire *on* mesopredators [11]. It is important to rapidly address this significant knowledge gap because some fire regimes may exacerbate the effects of invasive mesopredators by simplifying vegetation and amplifying predation risk [12,13]. For example, interactions between fire regimes and invasive mesopredators have been hypothesised as a cause of lower survival of reptile species in recently-burned areas [14], and a contributor to the collapse of small mammal communities in northern Australia [15].

The red fox is one of the world's most widely distributed mesopredators. It is common in both the northern and southern hemispheres. Foxes, and a second introduced mesopredator, the feral cat (*Felis catus*), are widely regarded as the primary cause of extinctions and declines of Australia's marsupial fauna [5]. Evidence for the negative impact of foxes has been demonstrated through predator-control experiments that have shown that prey species increase in both range and activity when foxes are removed [16,17]. Further evidence comes from dietary studies showing foxes eat a wide range of native mammal, reptile, bird, and invertebrate prey [18–20].

Despite indications that foxes may inhibit the recovery of native species following fire [12,21], whether foxes are themselves influenced by fire remains poorly known. This knowledge gap limits the ability of land managers to consider the effects of fire management on red foxes, which could have negative ramifications for native biodiversity. While foxes are widely considered as habitat generalists, they do display local variability in occurrence related to habitat or landscape structure [22]. For example, in some regions, foxes prefer heterogeneous landscapes [22], as they are able to use multiple landscape elements on a daily or seasonal basis [23,24]. Fire management in many regions seeks to maximise landscape heterogeneity by creating mosaics of fire ages (i.e. 'patch mosaic burning'; [25]). Does such management inadvertently favour invasive mesopredators?

The few studies that have explored the topic have focused on relatively short temporal scales (<30 years and often <10 years post fire) or small spatial scales (but see [26]). However, in some ecosystems, post-fire vegetation recovery continues for a century or more after fire [27]. Consequently, animal species respond to fire over similarly long time-frames [28]. The effects of fire can also occur across multiple spatial scales [29]; while time since fire may affect a species' occurrence at any *point* in the landscape, the area and composition of fire-ages within a 'whole' landscape can play a critical role in affecting species' landscape-level distributions [30]. This is likely to be especially true for large, mobile species, such as the red fox.

In addition to the effects that fire may have on species' occurrence, other environmental factors may be locally important. With regard to foxes, this includes climate [26], the distribution of vegetation types [31], and the distance to roads [24] and agricultural land [22]. Foxes rely on free standing water for drinking, particularly when temperatures are high (>30°C), as is common in many semi-arid environments. Hence, as annual rainfall decreases (aridity intensifies) permanent water may be reduced in its availability and limit fox occurrence. Foxes are often thought of as edge specialists [22]. They often prefer to hunt in open areas such as resource-rich agricultural fields or structurally simple vegetation types adjacent to more complex vegetation which provides cover during the day [22,32]. Their ability to hunt may be further enhanced where roads create easy access and increased visibility in otherwise structurally complex habitats [24,33].

Here, we examine what drives the occurrence (reporting rate) of red foxes in semi-arid Australia at multiple spatial scales, with a particular emphasis on the role of fire. We conducted two large-scale natural experiments. First, we explored landscape-scale patterns of fox occurrence in relation to the properties of fire mosaics; namely, the amount and diversity of fire age-classes within each of 28 study landscapes (each 12.6 km^2). Second, we explored site-scale patterns of fox occurrence in relation to fire history at 108 sites stratified along a century-long post-fire chronosequence. In both cases, we also quantified the influence of other environmental variables such as vegetation type and distance to agricultural land. Our aims were: 1) to determine the drivers of fox distribution in semi-arid Australia; and 2) to understand the specific role of fire in influencing fox occurrence at large scales relevant to fire and mesopredator management.

Materials and Methods

Study region

This study was undertaken in the Murray Mallee region of south-eastern Australia (Fig. 1). The climate in the region is semi-arid, with mean annual rainfall of 200–350 mm and average daily maximum temperatures are 30–33°C in summer and 15–18°C in winter (Australian Bureau of Meteorology; http://www.bom.gov.au). The vegetation is predominantly 'tree mallee' characterised by an overstorey of *Eucalyptus* species (<5–8 m) with a multi-stemmed growth form [34]. Two vegetation types are common throughout region [35]. 'Triodia Mallee' has a canopy of *Eucalyptus dumosa* and *E. socialis* with an understorey of *Triodia scariosa* and mixed shrubs, and occurs mainly on sandier soils typical of dunes. 'Chenopod Mallee' has a canopy of *E. oleosa* and *E. gracilis* with an open understorey of chenopod species, and occurs on heavier soils typical of swales.

Mallee vegetation is fire-prone with large fires (i.e. > 100,000 ha) occurring somewhere in the region on a bidecadal basis [36], although individual sites can go long periods without fire (i.e. >100 years; [37]). Fire is actively managed in the region through prescribed burning and suppression for both asset protection and conservation objectives [36]. Most wildfires are ignited by lightning strikes and are stand-replacing, essentially resetting vegetation succession to 'year-zero' (Fig. 2; [8]).

Site selection

We refer to two datasets in this study derived from two different natural experiments that differed in both their spatial grain and extent. We refer to these as 'landscape-scale' and 'site-scale' datasets throughout, in reference to the spatial scale of the response and predictor variables (i.e. the spatial grain) of the respective datasets.

Landscape-scale data. The landscape-scale dataset consists of 28 study landscapes, each with a 4 km diameter circle (12.6 km^2; Fig. 1), distributed throughout a 104, 000 km^2 study area. These landscapes were selected as part of a broad-scale natural experiment: the Mallee Fire and Biodiversity Project. Study landscapes were selected to allow a comparison of the effects of different approaches to patch mosaic burning on biodiversity, with a particular emphasis on the role of the area and diversity of fire-ages ('pyrodiversity', see [25]). Thus, landscapes were stratified according to number and spatial extent of fire-age classes within the landscape [29]. The fire history of the region was mapped using the ENVI package [38] and then converted to shape files for use in ArcMap version 9.2 [39]. Only fires that occurred post-1971 were mapped due to limited availability of Landsat imagery prior to this time (see [36]).

Site-scale data. We collected site-scale data within a subset of the 28 study landscapes located within the region's largest national park; Murray Sunset National Park (6630 km^2; Fig. 1). Ten sites were established within each of 10 of the original study landscapes. Sites were distributed to incorporate a range of fire-age classes (range = 7–105 years), as well as capturing geographic and topographic variation. We established an additional landscape, containing 12 sites, following an experimental burn during the study (fire age = 0 years), resulting in 11 landscapes containing 112 sites. We omitted four sites to comply with ethics permits due to their close proximity to active nesting sites of the endangered malleefowl (*Leipoa ocellata*). This resulted in a total of 108 sites being surveyed. All sites were a minimum of 200 m apart and typically >100 m from the edge of a fire-age class.

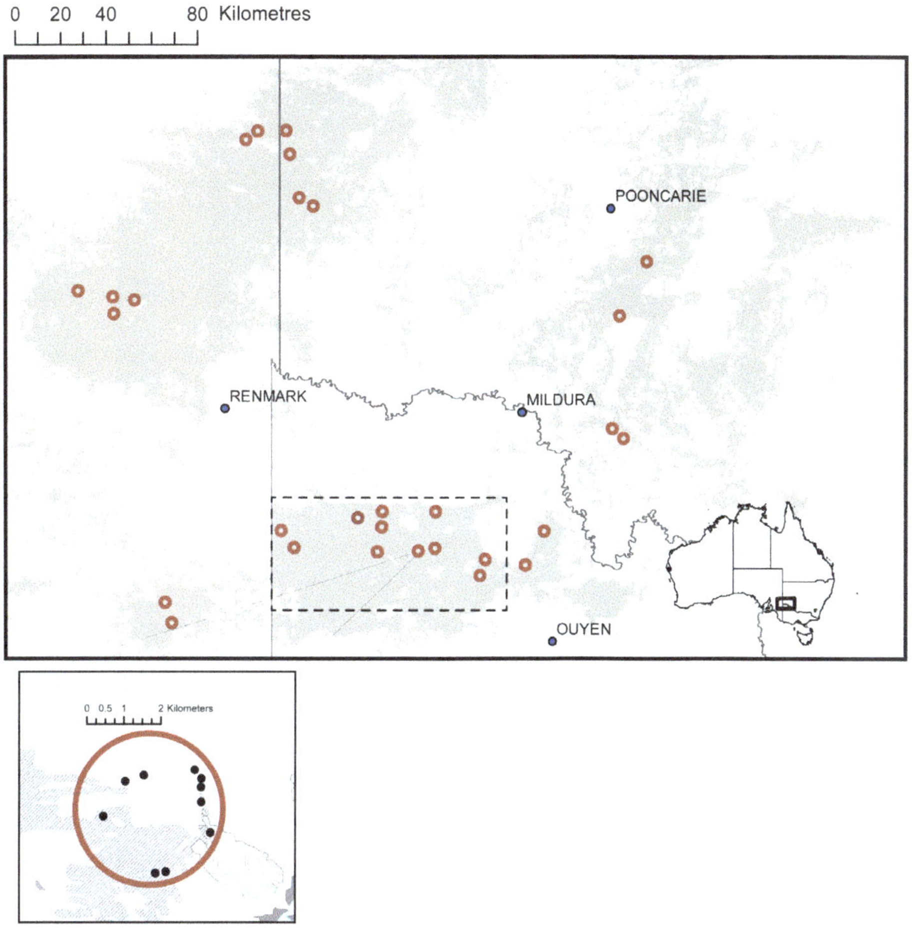

Figure 1. Map of study area showing all landscapes (circles) considered in this study (grey shading indicates mallee vegetation; majority of white areas indicates agricultural land used for grazing and cereal crops). The dashed box shows the spatial extent of the site-scale study. An inset shows an example of a study landscape including the position of 10 sites within where site-scale data were collected. Within the inset, different hatching represents different fire ages.

Predator surveys

Landscape-scale data. We surveyed large mammalian predators using track surveys from three sand-plots within each study landscape (n = 84 in total). Each sand-plot was a 100 m×2 m area smoothed out by dragging a weight along an unsealed vehicle track. The locations of sand-plots within landscapes were chosen to incorporate variation in the topography (dunes and swales) within each landscape. Sand-plots were typically >500 m apart. We checked each sand-plot for tracks once per day by walking along the transect and identifying tracks to species level for three consecutive days in spring (October–November 2007), and again in summer (January–March 2008), resulting in six survey nights for each sand-plot and thus 18 survey nights per landscape. Following checking, sand-plots were smoothed over in preparation for the following day. If the sand-plot was heavily disrupted on one day (due to weather or vehicle disturbance), it was surveyed for an additional day.

Site-scale data. We used camera traps (Passive ScoutGuard 550; ScoutGuard IR Cameras, Australia) to survey for mammalian predators at the site-scale during April–July 2012. We installed one camera per site and deployed each for a minimum of 15 nights. We attached cameras to a post at a height of 0.5 m and positioned them facing southward. Vegetation was removed within the immediate area of the camera to reduce false triggering. A 15 second video was taken each time the camera motion sensor was triggered. To attract predators to the front of the camera from the local vicinity, we placed a scent lure of tuna oil soaked into chemical wadding inside a bait holder made from PVC piping with steel mesh at one end. We positioned the lure 3 m from the base of the camera post, and secured it to the ground with a peg.

Predictor variables

Landscape-scale data. Six predictor variables were chosen to represent the properties of the study landscapes (Table 1). Three of these variables represent the fire history of the landscape: (1) the extent of recently burnt vegetation in the landscape (<10 years since fire; 'recently burned'); (2) the extent of long unburnt vegetation (unburned since 1972; 'long unburned'); and (3) the diversity of fire-ages within a landscape ('fire diversity'). Fire diversity was calculated as the Shannon-Wiener diversity index of the proportional cover of fire age-classes within each landscape.

Three predictor variables were chosen to describe properties of the study landscapes other than fire history. We used a measure of mean solar radiation ('solar radiation') as a surrogate for aridity across the region. The solar radiation variable represents the total amount of solar energy falling on a horizontal space per day (MJ/m^2). We derived these values from a gridded data set (5 km resolution) extending over 18 years (1990–2008; Australian

Figure 2. Examples of mallee vegetation with differing fire histories. (a) A recently burned site; (b) A long unburned site.

Bureau of Meteorology http://www.bom.gov.au, 2009). Solar radiation was the mean of the 18 yearly averages of the grids that overlaid each landscape. Solar radiation is negatively correlated with annual rainfall and positively correlated with temperature. We used the proportional extent of *Triodia* mallee vegetation ('Triodia Mallee') within the landscape to capture differences in vegetation types. The extent of mallee vegetation in the study area was mapped in previous work (see [35]). Finally, we used the distance from the centre point of each landscape to the closest area of contiguous non-mallee vegetation ('distance to agricultural land'), to capture the context of landscapes with respect to landscape modification. The area surrounding each reserve is comprised almost entirely of grazing land and grain crops. We calculated distance to agricultural land using ArcGIS [39].

Site-scale data. Eight predictor variables were chosen at the site level (Table 1). The fire history of sites was represented by the time since the last fire ('time since fire'; range: 0–105 years). This was determined using two methods. Recent fire history (since 1972) was calculated using the fire history maps (see [36]). Fire-ages for sites burnt prior to the availability of satellite imagery (i.e. before 1972) were estimated using regression models of the relationship between stem diameter and tree age, and then using stem diameter to estimate the age of trees in areas where fire history was unknown (see [37] for detailed methods). This extended the time since fire axis from 0–32 years to 0–105 years.

Vegetation type was considered as a categorical variable with two levels: Triodia Mallee or Chenopod Mallee ('vegetation type'). We again considered the effects of landscape modification by including the distance of sites to both the border of the National Park (1.72–21.28 km; 'distance to edge') and dirt roads (range: 28–1044 m; 'distance to road'). We used park boundary as a proxy for an edge habitat because the park forms abrupt boundaries with cleared agricultural land and other non-mallee vegetation. We calculated distance variables using ArcGIS [39]. Aridity (solar radiation) was not considered at this scale as the data were collected from a single reserve.

Four additional predictor variables were included to describe vegetation structure at the sites. We established vegetation transects in representative areas 15 m from each camera location.

Table 1. Predictor variables included in models using the landscape-scale and the site-scale datasets.

Dataset	Predictor variable	Description
Landscape-scale data	Recently burned	Extent of landscape burned within 10 years of surveys
	Long unburned	Extent of landscape not burned since 1972 (>35 years since fire)
	Fire diversity	Shannon-Wiener diversity index of the extent of three fire age classes (0–10 years, 11–35 years and >35 years)
	Solar radiation	Long-term average monthly gridded solar exposure (MJ/m^2) from 1990–2008 for each landscape
	Triodia Mallee	Extent of landscape comprised of vegetation type in which *Triodia scariosa* typically occurs
	Distance to agricultural land	Distance from the centre of each landscape to contiguous non-mallee vegetation (m)
Site-scale data	*Time since fire*	Amount of time since a site last experienced fire (years)
	Bare ground cover	Cover of bare ground present
	Triodia cover	Cover of *Triodia scariosa* <1 m
	Eucalypt cover	Cover of eucalypt shrubs <1 m
	Shrub cover	Cover of non-eucalypt shrubs <1 m
	Vegetation type	Broad vegetation classification (Triodia Mallee or Chenopod Mallee)
	Distance to edge	Distance from each site to the nearest park boundary (m)
	Distance to road	Distance from each site to the nearest road (m)

We recorded substrate type and vegetation structure at 1 m intervals along a 50 m transect using a 2 m structure pole (2 cm diameter) held vertically above the ground. The four variables considered in the analysis represent the cover of open, bare ground ('bare ground cover'), spinifex ('*Triodia* cover'), eucalypt shrubs (defined as Eucalypt trees <3 m in height 'eucalypt cover') and non-eucalypt shrubs ('shrub cover'). Bare ground was included because it gives an approximation of the 'openness' of the vegetation at the ground level. The cover of spinifex, eucalypt shrubs, and non-eucalypt shrubs were included as they form the majority of the ground and understorey structural complexity, and are known to drive fauna in the region [34,40].

Response variables

For both datasets, the response variable was the 'reporting rate' of foxes. At the landscape-scale, we defined reporting rate as the number of nights that a fox was recorded as 'present' and 'absent', respectively, at a sand pad over the 18 nights of sampling per landscape (i.e. three sand-plots surveyed for six nights in each landscape). Likewise, at the site-scale, reporting rate is the number of nights that foxes were and were not detected at the site, respectively, over the course of sampling (i.e. 15 nights).

Statistical analysis

We used generalised linear mixed models (GLMMs) with the Laplace approximation [41] to examine the relationship between response and predictor variables at both landscape and site-scales. In landscape-scale models, we included 'reserve' as a random effect to account for spatial clustering of landscapes in conservation reserves (Fig. 1). Similarly, in the site-scale models, we included 'landscape' as a random effect to account for potential spatial correlation due to the clustering of sites into landscapes. Because we were studying the reporting rate of red foxes, a proportion, we modelled the response variable (at both scales) using a binomial distribution of errors and a logit link function.

For the landscape-scale dataset, we developed a set of candidate models that included all combinations of the six landscape-scale predictor variables. At the site-level, we developed two separate sets of models. As fire affects the variables used to describe vegetation structure (e.g. *Triodia* cover, bare ground cover; [8]), including both fire and vegetation structure variables in the same model could result in unreliable parameter estimates due to colinearity between predictor variables [42]. Thus, one model set (model set 1) included time since fire, vegetation type, distance to edge and distance to road, and a second model set (model set 2) included the vegetation structure variables (bare ground cover, *Triodia* cover, eucalypt cover and shrub cover). All combinations of predictors within the two sets of models were considered, meaning all eight site-level variables were in the same number of models overall. All variables included within a model set had low levels of colinearity (i.e. r <0.5). We tested both datasets for overdisperson using Pearson's residuals [43], and found no evidence of overdispersion.

We compared each set of candidate models using Akaike's Information Criterion corrected for small sample sizes (AICc; [44]). To compare the level of support for each model relative to the most parsimonious model, we calculated the difference (Δ_i) between the AIC_c value of the best model (lowest AIC_c value) and the AIC_c value of each candidate model [44]. We considered models with $\Delta_i<2$ to have substantial support [44]. We also calculated the Akaike weight (w_i) for each model. By summing these weights to calculate predictor weights ($\sum w_i$) for each variable, we were able to explore the influence of individual predictor variables at both the landscape and site level.

When there was no clear 'best model' (i.e. the most parsimonious model was not strongly weighted [$w_i<0.9$]), we used model averaging to determine the direction and magnitude of the effect of each predictor variable [44]. We considered a variable as important when the associated 95% confidence interval of the averaged estimate did not overlap with zero. We performed all statistical analyses in R version 2.15.1 [45] using the lme4 package [41] and the MuMIn package [46].

Ethics statement

The landscape-scale data were collected with approval from animal ethics committees at La Trobe University (approval number AEC06/07[L]V2) and Deakin University (approval number A41/2006), and permits from the Department of Sustainability and Environment, Victoria (permit 10003791), the Department of Environment and Heritage, South Australia (permit 13/2006), and the National Parks and Wildlife Service, NSW (license number S12030). The site-scale data were collected in accordance with the regulations of the Deakin University Animal Ethics Committee (approval number B10-2012) and in accordance with Department of Sustainability and Environment, Victoria (approval number 10006279).

Results

At the landscape-scale, we recorded fox tracks in 24 of 28 (86%) study landscapes. We detected foxes on 3.32±0.49 (mean ± standard error) of 18 nights per landscape over the total sampling period. Other large-bodied, mammalian predators were uncommon: we detected cats at only 7 of 28 landscapes (25%). At the site-scale, we observed foxes at 62 of 102 (61%) sites (six cameras failed to reach the full 15 day survey period due to fault and were excluded from further analysis i.e. n = 102) and found the species to be widely distributed across the study area. We did not detect any cats at the site-scale over the 15 night sampling period.

At the landscape-scale, all models were a poor fit for the data and explained <6.5% of the variation in the data (% deviance explained). At the site-scale, all models explained <3.5% of the variation in the data. For both datasets, model selection indicated there was a similar level of support for several models ($\Delta_i<2$; Table 2), including the intercept-only model (i.e. only an intercept terms, no predictor variables), which received substantial support at both scales. As no single model was supported as being clearly best (i.e. $w_i>0.9$; Table 2), we employed multi-model inference using model averaging to estimate the size, direction and uncertainty of parameter effects for fox explaining reporting rate in both datasets.

The model-averaged coefficients for each predictor variable, in both datasets, were small and uncertain. The 95% confidence intervals of all predictor variables overlapped with zero (Fig. 3). The $\sum w_i$ for all predictor variables was low: <0.5 and <0.6 for the landscape- and site-scale datasets respectively.

Graphical exploration of the data further highlights that fox activity was not strongly linked to key predictor variables (Fig. 4). In summary, the data shows that neither fire, nor any other predictor variable measured, affected the reporting rate of foxes at either the landscape- or site-scale.

Discussion

Introduced mesopredators and fire are two processes that shape ecosystems around the world [4,7]. Here, we have shown that a widespread and ecologically devastating mesopredator, the red fox [5], is largely unaffected by fire and is an extreme habitat

Table 2. Model selection results for red fox reporting rate for landscape-scale and sits-scale datasets.

Candidate model	df	LogLik	AIC_c	Δ_i	w_i	%Dev
Landscape-scale dataset						
Null model (intercept only)	2	−32.37	69.2	0.00	0.14	0.00
Distance to agricultural land	3	−31.21	69.4	0.21	0.12	3.57
Distance to agricultural land + Triodia Mallee	4	−30.53	70.8	1.58	0.06	5.68
Triodia Mallee	3	−31.99	71.0	1.76	0.06	1.17
Fire diversity	3	−32.10	71.2	1.99	0.05	0.82
Site-scale dataset						
Bare ground cover	3	−65.90	138.0	0.00	0.13	2.16
Bare ground cover + *Triodia* cover	4	−65.07	138.6	0.51	0.10	3.39
Triodia cover	3	−66.26	138.8	0.73	0.09	1.62
Null model (intercept only)	2	−67.36	138.8	0.79	0.09	0.00
Bare ground cover + eucalypt cover	4	−65.67	139.8	1.71	0.06	2.50

Models are shown for which Δ_i<2.0.

generalist in semi-arid Australia. This result was confirmed using two large, complementary datasets, collected at different times and characterised by differing spatial scales and sampling strategies.

Fire and the red fox

Our findings show that fire does not exert a strong influence on the distribution of the red fox in semi-arid mallee ecosystems. Despite conducting two intensive natural experiments across a broad geographic region, we did not detect a relationship between the reporting rate of foxes and fire history at either the landscape- or site-scale. At the landscape-scale, the red fox was recorded equally often in landscapes dominated by recently burned or long unburned vegetation, and in landscapes with a single fire age-class as those with a diversity of fire ages. At the site-scale, the red fox has a similar reporting rate in recently burned sites as in sites unburned for over a century. The post-fire preferences of the red fox are thus extremely broad, both spatially and temporally (also see [12,47]).

Fire causes significant changes to vegetation structure over century-long time frames in mallee ecosystems [8]. In doing so, fire affects the distribution of a large range of fauna species [34]. Indeed, work conducted within the same study landscapes has shown the large and long-term effects fire has on birds, reptiles, and small mammals [28–30]. The lack of a response to fire by foxes is therefore not typical of native fauna in the region. It also suggests that foxes are not restricted to areas with particular soil or vegetation attributes for denning. This is consistent with foxes not being affected by any of the vegetation attributes measured (e.g. *Triodia* cover, shrub cover etc.).

A related way that fire could influence foxes is by altering the distribution of prey resources. As mentioned above, the distribution of many prey species are significantly affected by fire in the study region (e.g. birds, mammals, reptiles). Thus, foxes occupy a range of post-fire ages despite the strong influence of fire on the type and abundance of prey available. Red foxes have a broad and generalist diet [48], being able to consume a wide range of prey including both vertebrates and invertebrates, and even vegetation [19,20]. Furthermore, foxes are capable of prey switching to capitalize on the most abundant prey source available [18,49], thereby reducing their reliance on any particular prey item. This flexibility in their diet is likely to be a key component of their life history that allows them to occur within such a broad range of post-fire conditions.

One objection to our findings at the site-scale may be that the local site is not a relevant spatial scale to characterize the effects of fire, as foxes are a relatively large and mobile species. Given the large estimated home ranges of foxes in other parts of arid Australia (e.g. 8–33 km^2; [31]), foxes may select broader areas (i.e. kms^2) that capture their resource requirements across entire landscapes, and this might include a large area of a particular fire-age, or multiple fire ages. Such use of multiple habitat types by foxes has been demonstrated in other systems [23,24]. Our landscape-scale study characterized land mosaics at a large scale relevant to the home range of foxes (12.6 km^2), and still failed to detect any relationship between fox activity and fire history. Therefore, our results suggest that the lack of relationships between fox reporting rate and fire history does not stem from spatial scaling issues. Instead, foxes are resilient towards the effects of fire at multiple temporal and spatial scales.

Climate and distance to modified land

In addition to fire, we examined other variables that could influence the distribution of the red fox. Here, we again found red foxes to be flexible to a broad range of ecological conditions. Foxes displayed no response to an aridity gradient across the study region. This lack of response to aridity is unsurprising, as the geographic range of the red fox spans the northern hemisphere and much of Australia, suggesting the species is capable of coping with a range of climatic conditions.

Despite foxes occupying a broad climatic niche in space, fluctuations in populations do occur in response to extreme weather events. For example, fox populations in arid areas rise rapidly following high rainfall events, in response to increased prey availability [50]. Our site-scale study was carried out during a year of record high rainfall (Australian Bureau of Meteorology, Ouyen Station). Considered in isolation, this may suggest that the wide distribution of the fox was partly due to a productivity-related increase in food resources (predominantly populations of native and introduced rodents; [40]). However, the landscape-scale data were collected near the end of a severe, decade-long drought. Foxes were widely distributed across the region despite the drought. This indicates that, in semi-arid Australia, foxes can be

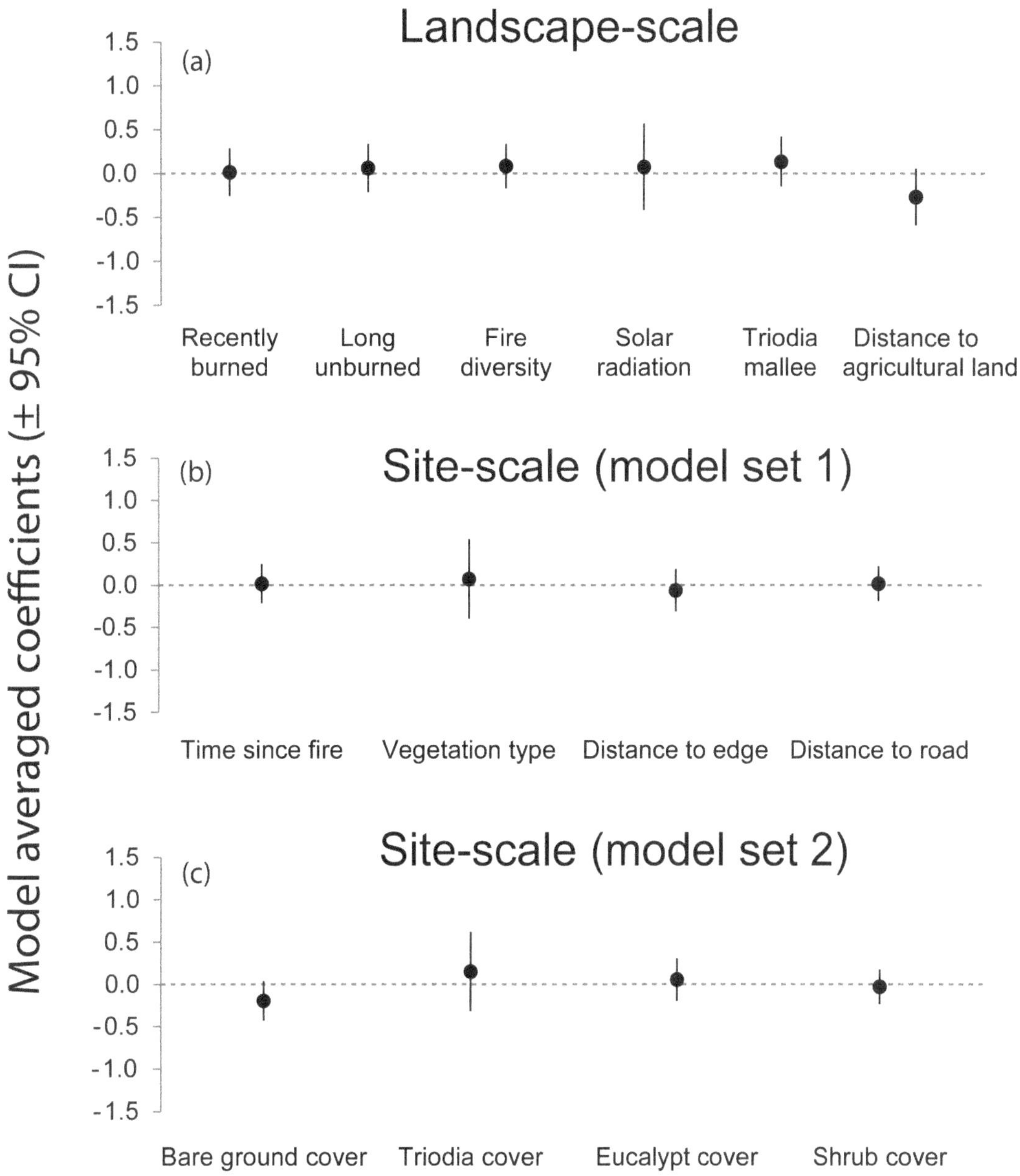

Figure 3. Model-averaged regression coefficients and 95% confidence intervals of models describing the reporting rate of foxes at both the landscape-scale (a) and site-scale (b and c).

widespread during a broad range of climatic conditions and despite fluctuations in their prey populations which accompany climactic extremes [40].

Some studies have found that foxes are positively associated with edges between fragments and modified land (e.g. agricultural land) [22,32]. Our results indicate that foxes do not show a preference for edge habitats in mallee ecosystems, despite our sites and landscapes capturing a broad gradient of distances to agricultural land, from <2 km to >30 km. Edge habitats may be more important for foxes in highly fragmented landscapes, where they occur with small remnant patches of wooded vegetation which provide the only available cover [22]. While the mallee region has been subject to large amounts of land clearing, there are still relatively large intact areas of native vegetation. Edges may be less important in this region because the interior mallee vegetation provides sufficient shelter and prey. Nevertheless, it is also possible that edge effects occur closer to the agricultural boundary than we sampled (i.e. <2 km).

The use of roads and tracks by foxes is also well documented [51,52]. Foxes have been found to be more abundant along roadsides [33]. In the mallee system, however, we found similar reporting rates at varying distances (28–1044 m) from roads, indicating foxes use areas well away from roads equally as often as sites close to roads. One hypothesis for the use of roads by foxes is that they provide 'runways' which facilitate movement and allow access to foraging areas that would be otherwise difficult to reach [51,52]. In contrast to environments with a dense understory, mallee vegetation is relatively open, and is unlikely to limit the movement of foxes to roadsides. This may explain the lack of preference for sites near roads in the current study.

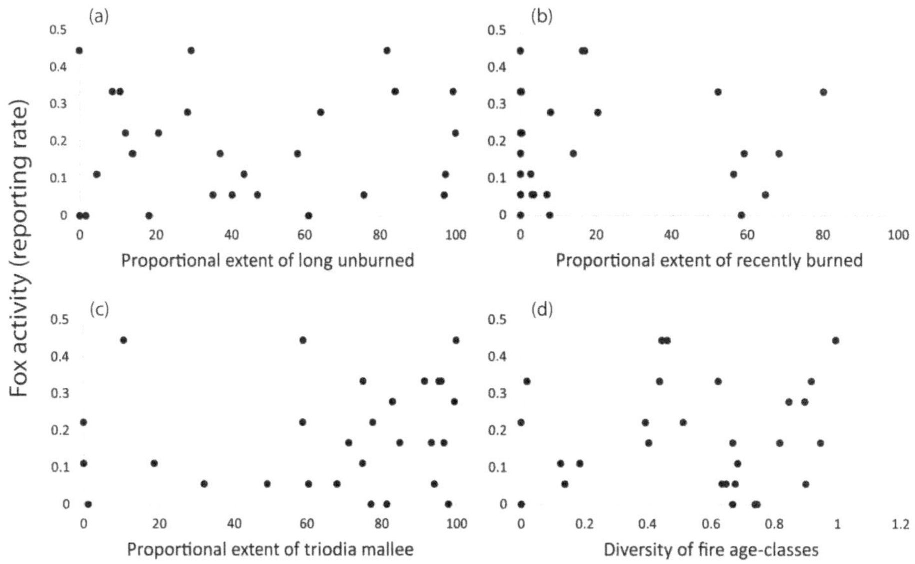

Figure 4. Relationships between the reporting rate of the red fox and the properties of fire mosaics. Circles are raw data points.

Implications

Fire is used as a conservation tool in Australia and around the world [25]. This study suggests it is unlikely that any particular approach to fire management will alter the reporting rate of the red fox in the semi-arid mallee systems of Australia. However, the presence of foxes in recently burned sites and landscapes is a concern. Predation by invasive mesopredators has been hypothesized as a cause of low post-fire survival in reptiles [14] and mammals [12,13], due to the reduced cover available in burnt habitats. Although we found no effect of fire history on red fox occurrence, it is possible that predation pressure differs across fire ages due to increased predation risk in recently burned areas. Thus, assessing predation pressure directly across a range of post-fire ages is an important area for further research.

The loss of apex predators can cause smaller predators to increase in abundance, expand their range, and change their temporal activity; this is known as 'mesopredator release' [2,6]. Red foxes have been shown to select particular habitats which may allow them to avoid dominant predators (e.g. coyotes; [53]). As such, one further explanation for the lack of obvious habitat selection by foxes in this system may be the lack of regulating predators. In other Australian systems, the presence of the dingo, Australia's largest terrestrial apex predator, has been shown to affect fox distributions [3]. Dingoes are largely extinct from the study area but were once common, and as such there is no direct regulation of the abundance or distribution of foxes via biotic interactions. Thus, one potential way to control red foxes in mallee communities is by reinstating dingoes as the apex predator. As this is likely to be a controversial idea owing to the proximity of mallee vegetation to agricultural land and livestock, trialing reintroductions in a controlled and experimental way would be an important first step towards a proof of concept, and a potential solution to this complex conservation issue.

Acknowledgments

Thanks to all members of the Mallee Fire and Biodiversity Team, particularly Andrew Bennett, Mike Clarke, Lisa Farnsworth and Lauren Brown, and to the many volunteers, agency staff, and land owners who assisted with logistics and field work.

Author Contributions

Conceived and designed the experiments: DGN CJP EGR LTK. Performed the experiments: DGN CJP LTK. Analyzed the data: DGN CJP. Contributed to the writing of the manuscript: CJP DGN LTK EGR.

References

1. Estes JA, Terborgh J, Brashares JS, Power ME, Berger J, et al. (2011) Trophic downgrading of planet Earth. Science 333: 301–306.
2. Ritchie EG, Johnson CN (2009) Predator interactions, mesopredator release and biodiversity conservation. Ecology Letters 12: 982–998.
3. Letnic M, Ritchie EG, Dickman CR (2012) Top predators as biodiversity regulators: the dingo *Canis lupus dingo* as a case study. Biological Reviews 87: 390–413.
4. Salo P, Korpimäki E, Banks PB, Nordström M, Dickman CR (2007) Alien predators are more dangerous than native predators to prey populations. Proceedings of the Royal Society B: Biological Sciences 274: 1237–1243.
5. Johnson C (2006) Australia's Mammal Extinctions: a 50000 year history. New York: Cambridge University Press.
6. Crooks KR, Soule ME (1999) Mesopredator release and avifaunal extinctions in a fragmented system. Nature 400: 563–566.
7. Bowman DMJS, Balch JK, Artaxo P, Bond WJ, Carlson JM, et al. (2009) Fire in the Earth system. Science 324: 481–484.
8. Haslem A, Kelly LT, Nimmo DG, Watson SJ, Kenny SA, et al. (2011) Habitat or fuel? Implications of long-term, post-fire dynamics for the development of key resources for fauna and fire. Journal of Applied Ecology 48: 247–256.
9. Smit IPJ, Asner GP, Govender N, Kennedy-Bowdoin T, Knapp DE, et al. (2010) Effects of fire on woody vegetation structure in African savanna. Ecological Applications 20: 1865–1875.
10. Letnic M, Tamayo B, Dickman CR (2005) The responses of mammals to La Nina (El Nino Southern Oscillation) - associated rainfall, predation and wildfire in central Australia. Journal of Mammalogy 86: 689–703.
11. Driscoll DA, Lindenmayer DB, Bennett AF, Bode M, Bradstock RA, et al. (2010) Fire management for biodiversity conservation: Key research questions and our capacity to answer them. Biological Conservation 143: 1928–1939.
12. Arthur AD, Catling PC, Reid A (2012) Relative influence of habitat structure, species interactions and rainfall on the post-fire population dynamics of ground-dwelling vertebrates. Austral Ecology 37: 958–970.

13. Sutherland EF, Dickman CR (1999) Mechanisms of recovery after fire by rodents in the Australian environment: a review. Wildlife Research 26: 405–419.
14. Smith AL, Bull CM, Driscoll DA (2012) Post-fire succession affects abundance and survival but not detectability in a knob-tailed gecko. Biological Conservation 145: 139–147.
15. Woinarski JCZ, Legge S, Fitzsimons JA, Traill BJ, Burbidge AA, et al. (2011) The disappearing mammal fauna of northern Australia: context, cause, and response. Conservation Letters 4: 192–201.
16. Kinnear JE, Onus ML, Bromilow RN (1988) Fox control and rock-wallaby population dynamics. Wildlife Research 15: 435–450.
17. Risbey DA, Calver MC, Short J, Bradley JS, Wright IW (2000) The impact of cats and foxes on the small vertebrate fauna of Heirisson Prong, Western Australia. II. A field experiment. Wildlife Research 27: 223–235.
18. Catling PC (1988) Similarities and contrasts in the diets of foxes, *Vulpes vulpes*, and cats, *Felis catus*, relative to fluctuating prey populations and drought. Wildlife Research 15: 307.
19. Glen AS, Fay AR, Dickman CR (2006) Diets of sympatric red foxes *Vulpes vulpes* and wild dogs *Canis lupus* in the Northern Rivers Region, New South Wales. Australian Mammalogy 28: 101–104.
20. Risbey DA, Calver MC, Short J (1999) The impact of cats and foxes on the small vertebrate fauna of Heirisson Prong, Western Australia. I. Exploring potential impact using diet analysis. Wildlife Research 26: 621–630.
21. Letnic M and Dickman CR (2005) The responses of small mammals to patches regenerating after fire and rainfall in the Simpson Desert, central Australia. Austral Ecology 30: 24–39.
22. Graham CA, Maron M, McAlpine CA (2012) Influence of landscape structure on invasive predators: feral cats and red foxes in the brigalow landscapes, Queensland, Australia. Wildlife Research 39: 661–676.
23. Lucherini M, Lovari S, Crema G (1995) Habitat use and ranging behaviour of the red fox (Vulpes vulpes) in a Mediterranean rural area: is shelter availability a key factor? Journal of Zoology 237: 577–591.
24. Meek PD, Saunders G (2000) Home range and movement of foxes (*Vulpes vulpes*) in coastal New South Wales, Australia. Wildlife Research 27: 663–668.
25. Parr CL, Andersen AN (2006) Patch mosaic burning for biodiversity conservation: a critique of the pyrodiversity paradigm. Conservation Biology 20: 1610–1619.
26. Southgate R, Paltridge R, Masters P, Ostendorf B (2007) Modelling introduced predator and herbivore distribution in the Tanami Desert, Australia. Journal of Arid Environments 68: 438–464.
27. Gosper CR, Yates CJ, Prober SM (2013) Floristic diversity in fire-sensitive eucalypt woodlands shows a 'U'-shaped relationship with time since fire. Journal of Applied Ecology 50: 1187–1196.
28. Watson SJ, Taylor RS, Nimmo DG, Kelly LT, Haslem A, et al. (2012) Effects of time since fire on birds: How informative are generalized fire response curves for conservation management? Ecological Applications 22: 685–696.
29. Nimmo DG, Kelly LT, Spence-Bailey LM, Watson SJ, Taylor RS, et al. (2013) Fire mosaics and reptile conservation in a fire-prone region. Conservation Biology 27: 345–353.
30. Kelly LT, Nimmo DG, Spence-Bailey LM, Taylor RS, Watson SJ, et al. (2012) Managing fire mosaics for small mammal conservation: a landscape perspective. Journal of Applied Ecology 49: 412–421.
31. Moseby KE, Stott J, Crisp H (2009) Movement patterns of feral predators in an arid environment – implications for control through poison baiting. Wildlife Research 36: 422–435.
32. Catling P, Burt R (1995) Why are red foxes absent from some eucalypt forests in eastern New South Wales? Wildlife Research 22: 535–545.
33. Towerton AL, Penman TD, Kavanagh RP, Dickman CR (2011) Detecting pest and prey responses to fox control across the landscape using remote cameras. Wildlife Research 38: 208–220.
34. Bradstock RA, Cohn JS (2002) Fire regimes and biodiversity in semi-arid mallee ecosystems. In: R. A Bradstock, J. E Williams and M. A Gill, editors. Flammable Australia: The fire regimes and biodiversity of a continent. Cambridge: Cambridge University Press. 238–258.
35. Haslem A, Callister KE, Avitabile SC, Griffioen PA, Kelly LT, et al. (2010) A framework for mapping vegetation over broad spatial extents: A technique to aid land management across jurisdictional boundaries. Landscape and Urban Planning 97: 296–305.
36. Avitabile SC, Callister KE, Kelly LT, Haslem A, Fraser L, et al. (2013) Systematic fire mapping is critical for fire ecology, planning and management: A case study in the semi-arid Murray Mallee, south-eastern Australia. Landscape and Urban Planning 117: 81–91.
37. Clarke MF, Avitabile SC, Brown L, Callister KE, Haslem A, et al. (2010) Ageing mallee eucalypt vegetation after fire: insights for successional trajectories in semi-arid mallee ecosystems. Australian Journal of Botany 58: 363–372.
38. ITT (2005) ENVI. Version 4.2. Boulder, Colorado: ITT Industries.
39. Environmental Systems Research Institute (2007) Arc View. Version 9.2. Redlands, California: ESRI.
40. Kelly LT, Dayman R, Nimmo DG, Clarke MF, Bennett AF (2013) Spatial and temporal drivers of small mammal distributions in a semi-arid environment: The role of rainfall, vegetation and life-history. Austral Ecology 38: 786–797.
41. Bates D, Maechler M, Bolker B (2012) lme4: Linear mixed-effects models using S4 classes. R package (Version 0.999999-0). Available: http://CRAN.R-project.org/package=lme4.
42. Quinn GP, Keough MJ (2002) Experimental design and data analysis for biologists. New York: Cambridge University Press.
43. Zuur AF, Ieno EN, Walker N, Saveliev AA, Smith GM (2009) Mixed effects models and extensions in ecology with R. New York: Springer.
44. Burnham KP, Anderson DR (2002) Model selection and multi-model inference. New York: Springer.
45. R Development Core Team (2012) R: A language and environment for statistical computing. Vienna, Austria. Available: http://www.R-project.org/: R Foundation for Statistical Computing.
46. Bartoń K (2012) MuMIn: Multi-model inference. R package (Version 1.7.11). Available: http://CRAN.Rproject.org/package=MuMIn.
47. Catling PC, Coops N, Burt RJ (2001) The distribution and abundance of ground-dwelling mammals in relation to time since wildfire and vegetation structure in south-eastern Australia. Wildlife Research 28: 555–565.
48. White JG, Gubiani R, Smallman N, Snell K, Morton A (2006) Home range, habitat selection and diet of foxes (*Vulpes vulpes*) in a semi-urban riparian environment. Wildlife Research 33: 175–180.
49. Leckie FM, Thirgood SJ, May R, Redpath SM (1998) Variation in the diet of red foxes on Scottish moorland in relation to prey abundance. Ecography 21: 599–604.
50. Pavey CR, Eldridge SR, Heywood M (2008) Population dynamics and prey selection of native and introduced predators during a rodent outbreak in arid Australia. Journal of Mammalogy 89: 674–683.
51. Carter A, Luck GW, McDonald SP (2012) Ecology of the red fox (Vulpes vulpes) in an agricultural landscape. 2. Home range and movements. Australian Mammalogy 34: 175–187.
52. Frey SN, Conover MR (2006) Habitat use by meso-predators in a corridor environment. The Journal of Wildlife Management 70: 1111–1118.
53. Gosselink TE, Deelen TRV, Warner RE, Joselyn MG (2003) Temporal habitat partitioning and spatial use of coyotes and red foxes in east-central Illinois. The Journal of Wildlife Management 67: 90–103.

15

Landscape Management of Fire and Grazing Regimes Alters the Fine-Scale Habitat Utilisation by Feral Cats

Hugh W. McGregor[1,2]*, Sarah Legge[3], Menna E. Jones[1], Christopher N. Johnson[1]

1 School of Biological Sciences, University of Tasmania, Hobart, Tasmania, Australia, **2** Australian Wildlife Conservancy, Derby, Western Australia, Australia, **3** Australian Wildlife Conservancy, Mornington Wildlife Sanctuary, Derby, Western Australia, Australia

Abstract

Intensification of fires and grazing by large herbivores has caused population declines in small vertebrates in many ecosystems worldwide. Impacts are rarely direct, and usually appear driven via indirect pathways, such as changes to predator-prey dynamics. Fire events and grazing may improve habitat and/or hunting success for the predators of small mammals, however, such impacts have not been documented. To test for such an interaction, we investigated fine-scale habitat selection by feral cats in relation to fire, grazing and small-mammal abundance. Our study was conducted in north-western Australia, where small mammal populations are sensitive to changes in fire and grazing management. We deployed GPS collars on 32 cats in landscapes with contrasting fire and grazing treatments. Fine-scale habitat selection was determined using discrete choice modelling of cat movements. We found that cats selected areas with open grass cover, including heavily-grazed areas. They strongly selected for areas recently burnt by intense fires, but only in habitats that typically support high abundance of small mammals. Intense fires and grazing by introduced herbivores created conditions that are favoured by cats, probably because their hunting success is improved. This mechanism could explain why, in northern Australia, impacts of feral cats on small mammals might have increased. Our results suggest the impact of feral cats could be reduced in most ecosystems by maximising grass cover, minimising the incidence of intense fires, and reducing grazing by large herbivores.

Editor: Paul Adam, University of New South Wales, Australia

Funding: The research was funded by the Australian Research Council, the Western Australian Department of Parks and Wildlife (Kimberley Science and Conservation Strategy), and supporters of the Australian Wildlife Conservancy. The funders had no role in study design, data collection and analysis, decision to publish, or preparation of the manuscript.

Competing Interests: The authors have declared that no competing interests exist.

* Email: hugh.mcgregor@australianwildlife.org

Introduction

Predator-prey relationships are strongly influenced by the structure and quality of habitat, principally its vegetation [1,2,3]. Variability in vegetation structure may be used by predators to increase hunting success. For example, lions use dense vegetation to hide their approach from prey [4]. Also, it may be used by prey to help them evade predators, such as elk using woodlands as a refuge from wolves [5]. Changes in habitat structure may therefore shift the relationships between predators and prey [6,7]. Such changes can determine the extent to which some prey are threatened with extinction by heavy predation [2,8].

One of most pervasive impacts on vegetation structure arises from changes to fire and grazing regimes. While drivers of such changes vary immensely, the impacts on fauna communities display some general trends. For example, small mammal populations are especially sensitive, and the vast majority of studies detecting declines in populations in response to either intense fire events or intense grazing [9,10,11,12,13], unless they occur in ecosystems with few predators [14,15]. The underlying mechanisms of these declines remain elusive, but are likely to be indirect rather than through direct effects such as being burnt by the fires or trampled by cattle [16]. Instead, such disturbances may improve habitat for predators in ways that increases their impacts on prey [9,17,18], although no field data are available to confirm this.

Many small mammal species are declining in the savannas of northern Australia, and several are threatened with extinction [18,19]. Declines have been greatest in areas subject to intense fires [20,21] and recent experimental evidence also supports an association of grazing by introduced herbivores (cattle, horses, donkeys, buffalo) with the magnitude of small-mammal decline [22]. Both fire and grazing regimes in northern Australia have intensified substantially over recent decades in ways that could contribute to the contemporary native mammal decline. These changes to fire and grazing have generally made grass communities less complex and more open [23].

Predation by feral cats *Felis catus* may also be contributing to the declines. This is suggested by three lines of evidence. First, the declining species fall within the preferred prey-size range of cats [18,24]. Second, mammal populations in complex rocky habitats have been less affected than those in more productive woodlands and savanna, suggesting a predation effect [25,26]. Finally, populations of declining mammal species are more stable in the absence of cats, on island or in large enclosures [27,28]. However, there is a temporal mismatch between the arrival of cats in northern Australia up to 170 years ago [29], and mammal declines observed in the last 20 years [19].

The apparent mismatch in timing of the early arrival of cats and recent mammal declines could be explained by the hypothesis that cat predation has its largest impact when it interacts with fire and grazing regimes established in the more recent past. Reduction of structural complexity of vegetation and increased openness due to fire and grazing might increase the exposure of small mammals to predators, making prey easier to detect and capture [15]. Small mammals are cats' preferred prey [24,30]. If cats preferentially use the open and relatively simple habitats created by fire and grazing, the result could be a higher predation impacts on small mammals. This has been suggested as a possible mechanism for these declines [19,31], however, until now there has been no evidence.

If cats do favour the conditions created by fire and grazing, this should be revealed by the patterns of movement of individual cats in heterogeneous landscapes with variable effects of fire and grazing. We tested this hypothesis using intensive GPS tracking of a large sample of individual cats; both within and outside of a large 40 300 hectare area that has been destocked of all introduced herbivores (cattle, horses, and donkeys) [22], and spanning contrasting fire patterns (mild control fires or intensive wildfires). A dynamic habitat map was created so that fire and vegetation attributes at any location or point in time could be determined. This was used to generate a parsimonious model of fine-scale habitat selection by cats. We predicted that feral cats would select for open grass cover to improve hunting success. If so, we hypothesise that cats would increase their use of habitats that have been recently burnt or intensely grazed, and that this relationship would be stronger in areas of high small-mammal abundance.

Methods

Study area

Our study area encompassed three large properties in the central Kimberley of north-western Australia (**17°01′S, 126°01′E**, see Fig. 1). One property is managed for commercial cattle production (Glenroy, 1455 km^2) and two are ex-pastoral leases managed for conservation by the Australian Wildlife Conservatory (Mornington and Marion Downs Sanctuaries, 3225 km^2 and 2676 km^2 respectively). Habitats are mostly savanna woodlands with a perennial grass layer, dissected by riparian vegetation along the edges of creeks. The region has a tropical monsoon climate with three broad seasons: the wet (December – March), early dry (April – July) and late dry (August – November). Fire is managed on all three properties to promote biodiversity values. This fire management aims to reduce the incidence of extensive, high-intensity, uncontrolled fires in the late dry season using strategic prescribed burning in the early dry season when fires tend to be small and of low intensity because of weather and condition of the grass layer. In addition, when uncontrolled late dry season fires occur, they are suppressed where possible. All large introduced herbivores (cattle, horses and donkeys) have been removed from a 40,300 ha fenced section of Mornington since 2005 [22], - hereafter referred to as the 'destocked' zone. The only other large mammalian predator in the study area is the dingo *Canis dingo*. Whilst they are controlled elsewhere in Australia, they are not persecuted in the study area, and occurred at a density of ~0.2 per km^2 [32].

Cat capture and tracking

Feral cats were captured between September 2010 and June 2013, using either large wire cage traps, leg-hold traps (soft-jaw, size #1.5) or by spotlighting and netting with the assistance of dogs trained to locate and bail cats up trees. If a cat was either bailed up a tree or required examination of possible injury, it was sedated with Zolotil at a rate of 0.5 cc/kg via intramuscular injection. Cats were fitted with GPS collars (Telemetry Solutions Quantum 4000 enhanced). Cats weighing between 2 and 3.3 kg were fitted with a 70 g collar, and those weighing more than 3.3 kg were fitted with a 100 g collar (<3% of body-weight). Sedated cats were released after full muscle control was regained (4–6 hours later), and non-sedated cats were released as soon as possible (2–5 minutes later). When it was necessary to replace GPS collars, the cats were recaptured using the dogs.

GPS collars were deployed on equal numbers of cats in the stocked and destocked zone, and between burnt and unburnt areas. Within burnt areas, the cats were split evenly between areas with low and high intensity fires (Table 1). The GPS units were programmed to record fixes every 15 minutes for two-day bouts, starting and finishing at 12 pm. These bouts were separated by intervals of one, two or fourteen days. All bouts were timed to commence at least 24 hrs after the cat was handled. Units were programmed to search for a satellite for 60 seconds, and to remain on for at least 5 seconds to refine the fix if there was memory from the last fix, or 15 seconds if not.

Habitat variables

Across the study area, we developed habitat maps relating to fire and grazing, along with any other variable likely to influence cat habitat selection or movement. Where necessary, the maps were updated to make them temporally dynamic, so that attributes at any given time and location could be determined. Most descriptors of habitat related to the ground layer, rather than the tree layer. Nine distinct grass communities common in the region (see Table 2) were mapped by examining colour disjunctions on aerial photos while altering light levels in Photoshop Elements v. 8, tracing the boundaries of discrete polygons onto the aerial photos, then geo-rectifying these boundaries in ArcGIS v.10. For example, spinifex *Triodia* spp. grasslands are uniquely green in dry-season aerial photos, while communities dominated by bluegrass *Dichanthium fecundum* are white. The digital map was ground-truthed at 768 plots (described later); the attribution of grass community was correct at 96% of sites.

Fire scars were initially mapped using monthly Landsat 7 remote-sensing imagery available from the US Geological Survey (2011–2013), and fire boundaries were then refined using aerial photography taken from a helicopter flying approximately 300 m above ground. For each burnt area, we assigned the date of burn and intensity (intense = 100% tree scorch and no ground cover remaining unburnt, or mild = all other fires). Relative to the date of each GPS fix, fire was considered in multiple binary variables at 30, 60, 90, 180, 360 or 600 days since fire. The Australian Wildlife Conservancy's stock-proof fence [22] separated the stocked and destocked areas (see Fig. 1).

A dynamic map was created that estimated grass cover at any given location and time since fire, based on a series of models of the response of the grass layer to fire (given grass community, stocking status). They were created from field data, where vegetation attributes were measured at 768 plots (each 10 m^2) across the study area and duration, spread equally across grass communities (see Table 1) and combinations of mild/intense fire, time since fire, and grazed and destocked areas (total of 96 plots per community). At each plot, we estimated the extent of grass

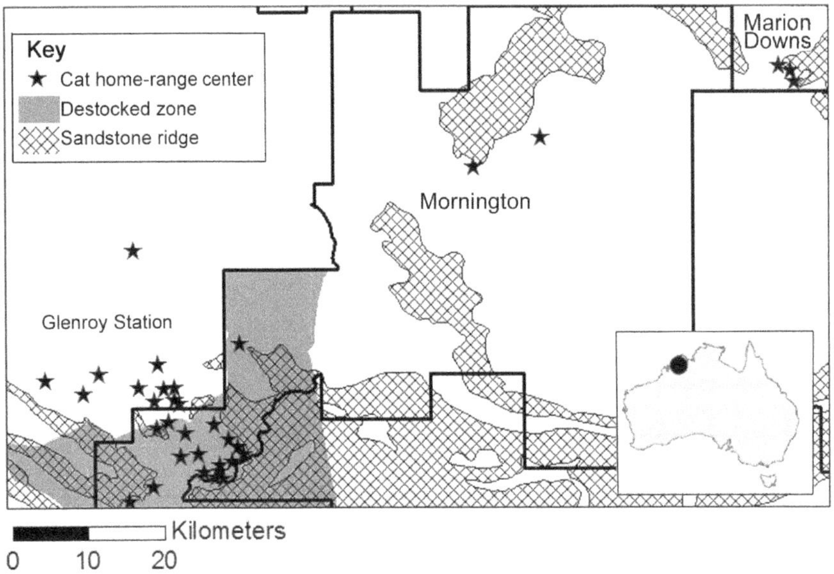

Figure 1. Map of study area in north-west Australia (see inset), including home-range centroids of feral cats used in this study. The dark grey represents the destocked zone.

cover at different heights by adapting a line-intercept method. We inserted a 100 cm pole (diameter of 1.5 cm) vertically through the grass to the ground at 50 points in a systematic grid over the plot. The number of grass intercepts was recorded in height intervals of 0–10 cm, 11–30 cm, and 31–100 cm. This was used to derive grass cover, cover of dense tussocks, and relative biomass at each plot (see Material S1 for more detail on methods). These variables were modelled against grass community, time since last fire, intensity of fire, and stocking status. Models were then used to derive values for all GPS fixes and random points used in discrete choice models (see Material S1).

The influence of grazing on grass biomass was measured as the difference in the average total number of grass intercepts per plot between grass communities in the stocked and destocked sites. We used only unburnt sites for this comparison. These averages were converted into a grazing impact rank. However, these variables would be confounded by correlation as the habitats favourable to cattle would also be favoured by feral cats (e.g. riparian areas), irrespective of impacts of cattle. Therefore, this score or rank was applied to all fixes, grazed or not, and the actual impacts of cattle grazing considered as the difference in this variable's strength between the destocked zone and outside.

Relative small-mammal abundances in different habitats were estimated from the Australian Wildlife Conservancy's annual fauna monitoring data, which is carried out across all three properties [32]. The sample at each site is based on 60 trap-nights using small mammal box traps (Elliott traps) and 12 nights of possum-sized wire traps, with 55 to 64 sites sampled each year - totalling 176 site-years between 2011–3. Small mammals were defined as those weighing between 30 g and 2000 g [18,24], and were predominantly *Rattus tunneyi* and *Pseudomys nanus* (these species comprised 87% of captures). For each year, the average rates of capture of individual small mammals were calculated for each grass community in the grazed and destocked areas (see Table 2). However, as sites were typically larger than the mapped patterns of these grass communities, results from some communities were combined (see Table 2). In the event cats altered their response to a threshold of small mammal abundance, we also included three binary variables of abundance (whether there was more than 1, 2 or 5 small mammal captures per plot) for each habitat.

We measured other spatial and temporal features likely to be important to cats to provide context for their movements. As cats may prefer hunting on edges between open and dense vegetation, we delineated a 25 m buffer either side of any linear boundary where grass cover was <25% on one side and >50% on the other (these were most often fire edges). Water features were mapped in the field, and each assigned a descriptor for the seasons that they contained water (wet season only, wet and early dry, or year-round). This was used to derive distance to known standing water at any given time throughout the study. Elevation and slope were derived from a digital elevation model of 15 m resolution. We

Table 1. The number of cats fitted with GPS collars in each of the different grazing and fire treatments. Destocked means that all introduced herbivores are excluded.

Management	No fire	Mild fire	Intense fire	Total
Stocked	8 (5♂ 3♀)	4 (3♂ 1♀)	4 (3♂ 1♀)	**16**
Destocked	8 (6♂ 2♀)	4 (4♂)	4 (4♂)	**16**
Total	**16**	**8**	**8**	**32**

Table 2. Vegetation communities mapped in the study region.

			Small mammal abundance					
		Grazing rank	Stocked			Destocked		
Grass community	Dominant grasses		2011	2012	2013	2011	2012	2013
Riparian forest[A]	*Chrysopogon fallax*	1	0.2	0.3	0.8	1.2	3.4	9.4
	Mnesithea rottboellioides							
	Various introduced grasses							
Alluvial grasslands[A]	*Crysopogon fallax*	2	0.2	0.3	0.8	1.2	3.4	9.4
	Dichanthium fecundum							
Bluegrass plains	*Aristida* spp.	3	1	1.3	4.6	0.3	2.6	2.9
	Dichanthium fecundum							
Canegrass[A]	*Mnesithea rottboellioides*	4	0.2	0.3	0.8	1.2	3.4	9.4
Mixed woodlands	*Aristida* spp.	5	0	1.3	0.6	0.2	3.8	6
	Dichanthium fecundum							
	Heteropogon contortus							
Sandseep	*Triodia* spp.	6	1.7	1.2	1.2	1.8	7.4	16.8
Hillside woodlands[B]	*Sehima nervosum*	7	0	0.6	1	0.4	3.2	4
Spinifex woodlands[B]	*Triodia* spp.	8	0	0.6	1	0.4	3.2	4
Bare ground	No or little	9	0	0	0	0.2	0.2	0.5

[A] and [B] denote grass communities where the small mammal abundance was grouped, as sites were typically larger than the mapped distributions of these communities.
Vegetation classes are ranked by their preference for grazing by domestic stock, with 9 the most impacted and 1 the least (Grazing rank). Small mammal abundance is number of small mammals captured per 100 trap nights in the stocked and destocked zone, from 2011 to 2013.

created a variable representing the spatial home-range context for each cat, by making a kernel density estimate (smooth cross-validation) for all moving fixes of each cat, and delineating kernel isopleths at 50%, 90%, 95% and 99% contours. Temporal variables included sun time [33], season (wet, early dry or late dry), number of months since the end of the wet season, and minimum nightly temperature (HOBO temperature data logger, Micro-DAQ). Finally, a cat's choice of which habitat to select in successive GPS fixes may be dependent on the type of habitat it was last in. This spatial correlation might arise when an animal selects for certain landscape features and tends to remain within them for some time, rather than making a *de novo* selection at every 15 min interval of their walk. Therefore, we fitted a binary 'carry-over' variable, which described whether the vegetation, grazing and fire values were equal to the previous fix.

Organisation of data

All GPS fixes that were likely to be erroneous, biased or to represent a stationary cat were removed from the analysis. Erroneous fixes were those representing implausible 'spikes' in movement, presumably caused by GPS error [34]. We deemed it unlikely that a cat would suddenly change course and speed, then return to the same area it was in 15 minutes ago, so spikes were defined where fixes met all the following criteria: distances from the last fix were >50 m, difference in distance from the preceding and succeeding fix was <10%, and turning angle >170° [34,35]. The HDOP values were not used to filter fixes, as a test of six GPS collars found no relationship between HDOP and distance from the GPS fix to the known location. Fixes that were potentially biased by human disruption of the behaviour of the cat were removed, being those within three hours of field VHF tracking (this was occasionally carried out in order to download data remotely from the cat GPS collar). Fast-moving fixes were also removed, as it was likely cats were moving away from something, rather than choosing habitat. For this purpose a filtering speed of greater than 2 km per hour was used, as this was where the histogram of speeds between fixes reached an asymptote [35], representing a shift in behaviour mode. A test of GPS error within the open savanna found that 95% of fixes had <5 m error (from 634 fixes on six different collars), so fixes less than 10 m from the preceding fix were classed as stationary. We considered only moving fixes, as cats may have different habitat requirements for resting versus hunting.

Data analysis

Habitat selection by cats was analysed using discrete choice modelling [36]. The range of 'available' habitats was calculated for each fix, and then we compared the option selected by the cat to the available habitats. To find the available habitats, we first constructed probability distributions of a cat's step-length and bearing over 15-minute intervals, then used these probability distributions to select five random points to sample the cat's options [36]. Each GPS fix and associated random points were attributed with the habitat variables of interest using the dynamic vegetation map.

We determined resource selection by cats by creating models with all combinations of variables, including different biologically relevant interaction terms, and comparing them within an information theory framework [37]. This produced a total of 916 models. For each interaction term a model was included with all combinations, or with only significant combination terms retained. No pairs of variables with Pearson's correlation greater than 0.5 were included in the same model. Models for habitat selection were created using standard case-control logistic regression models, and were implemented in R [38] using the 'clogit' command from the 'survival' library. Each individual cat was considered as a mixed effect in the models, using Gaussian frailty [39]. The most parsimonious models of cat habitat selection were chosen as those with an AIC score within two points of the highest-ranked model [37], and only these are presented in the results. The cat's selection is measured as an odds ratio representing the magnitude of change in the odds of selection for each unit of the predictor variable. Differences in the odds ratio are relative only to the other habitat choices immediately available to a cat.

Ethics statement

All data collection fulfilled all legal requirements in Australia, and has been approved by University of Tasmania Animal Ethics Committee (A0011661) and Western Australian Department of Parks and Wildlife Animal Ethics Committee (2010/35), with a Regulation 17 licence to research animals (SF009379). All research was conducted with permission on three pastoral leases; Mornington Wildlife Sanctuary, Marion Downs (both managed by the Australian Wildlife Conservancy, ph; +61 8 9191 7014), and Glenroy Station (ph: +61 8 9191 4703). All three leases are located around 17°01′S, 126°01′E. Field studies did not involve protected or endangered species.

Results

In total, 60 cats were captured between September 2010 and June 2013. Three cats were caught in wire cage traps (265 trap nights), 19 in leg-hold traps (940 trap nights) and 38 by spotlighting/netting with trained cat-dogs (221 hours). GPS collars were placed on 37 cats, and at least one month of GPS data was obtained from 32 cats whose locality was spread equally across stocking and fire management treatments (Table 1). There was a strong male bias in the sample of cats: males comprised 78% of all captured cats (47/60) and 78% of cats from which GPS data were obtained (25/32). Of the GPS-collared cats, four had disappeared and their fates were unknown at the end of the study, nine had died naturally, and the rest were euthanased.

From the 32 cats that provided useable GPS data, we obtained a total of 133 047 GPS fixes. Cats were moving 56% of time. Of these moving fixes, 62% were at night (between sunset and sunrise). Removal of erroneous or biased fixes and those representing high-speed movement left 38 472 choices for habitat between successive 15-minute moving fixes.

From the 916 models generated to describe habitat selection by cats, three were ranked within the candidate model set (the model with the lowest AIC score and two other models within two AIC scores of the top model). The top model carried 41% of the weight, compared to 24% and 20% for the second and third ranked models. Of these, the second and third ranked models were almost identical to the top model, but contained interaction terms that did not decrease the models AIC value. As these terms did not improve the maximum likelihood for these models, only the top model was considered further (Table 3).

The top model included negative selection for grass cover (equivalently, positive selection for open areas) and positive selection for edges (see Table 3). Both variables had a significant interaction with small-mammal density in a binary format (>2 captures per 100 trap nights), showing that cats selected particularly strongly for open areas in habitats with higher density of small mammals.

Fire was represented in the top model with fire scars up to 360 days old, fire scars <90 days old, whether the fire was intense, small-mammal abundance, and interaction terms between these

Table 3. Statistics of the top ranked model of cat habitat selection based on GPS data at 15 minute intervals from 32 individuals.

Variable	Odds ratio	robust SE	Z	Pr(>|z|)	
Grass cover with scarce small mammals[a]	−1.26	0.08	−2.88	0.004	**
Grass cover with abundant small mammals[a]	−1.85	0.1	−6.44	0.0001	***
Bare/grass edge with scarce small mammals[a]	1.2	0.07	2.81	0.005	**
Bare/grass edge with abundant small mammals[a]	1.41	0.06	6.26	<0.0001	***
Fire scar <360 days	−1.32	0.07	−4.09	<0.0001	***
Intense fire scar <360 days old	1.54	0.13	3.29	0.001	**
Intense fire scar <90 days old	−2.11	0.13	−5.67	<0.0001	***
Small mammal abundance (square-root)	3.52	0.21	6.03	<0.0001	***
Small mammal abundance, intense fire scar <360 days	−4.93	0.66	−2.4	0.0163	*
Small mammal abundance, intense fire scar <90 days	10.71	0.79	2.99	0.0028	**
Grass communities ranked on grazing impacts, in:					
- stocked areas during day	5.96	0.14	12.56	<0.0001	***
- destocked areas during day	3.52	0.25	5.07	<0.0001	***
- stocked areas over night	2.64	0.17	5.65	<0.0001	***
- destocked areas over night	2.44	0.16	5.67	<0.0001	***
Water proximity (km) by months into dry season	1.56	0.06	−7.77	<0.0001	***
- as above, by minimum nightly temperature (10°C)	1.13	0.03	3.94	<0.0001	***
Elevation (100m)	−1.93	0.28	−2.32	0.0202	*
Home range isopleth[b], adult female	3.53	0.05	27.17	<0.0001	***
Home range isopleth[b], adult male	2.28	0.1	−8.41	<0.0001	***
Home range isopleth[b], sub-adult	1.06	0.49	−0.12	0.9077	
Same habitat as last fix, if within 95% isopleth	2.06	0.02	41.4	<0.0001	***

[a]Scarce and abundant small mammals are defined as less than or greater than two individuals captured per 100 trap nights at Australian Wildlife Conservancy monitoring sites.
[b]Home range isopleth derived at 50, 90, 95 and 99% contours from kernel density estimator.
The odds ratio is the change in selection likelihood per unit of the variable.

variables. Once the odds ratios of these variables were combined, cats showed strong positive selection for areas within 90 days after an intense fire and where density of small-mammal prey was high (Fig. 2). However, in all other circumstances selection for areas burnt by intense or mild fires was negative.

Vegetation types with greater susceptibility to grazing impacts (e.g. riparian areas, Table 2) were strongly selected for in both the grazed and destocked zones. However, this relationship was significantly stronger in the stocked zone than in the destocked zone, especially during the day (see Fig. 3).

Several other variables influenced selection by cats. Cats chose to move towards water. The odds ratio of moving towards water became progressively stronger throughout the dry season (by 0.56 every month into the dry season) and on days with higher minimum temperature (by 0.13 for each 10°C increase). Cats selected against changes in elevation, with odds declining by 0.93 every 100 m. Cats tended to move into higher-use areas within their home range (determined by kernel density estimates), with an interaction with age and sex. Adult females demonstrated the strongest fidelity to home-range isopleths, while for sub-adults of either sex this was not significant ($P = 0.9$). Cats were twice as likely to select for the same habitats as the previous fix ($P < 0.001$), assuming the fix was inside a cats' home range (95% isopleth).

Discussion

Our study provides a detailed analysis of the preferences that underlie movement decisions by feral cats in a tropical savanna environment in northern Australia. We show that modifications of habitat produced by grazing and by certain types of fire have strong effects on cat movement behaviour, with the general result that fire and grazing can create habitat conditions which are

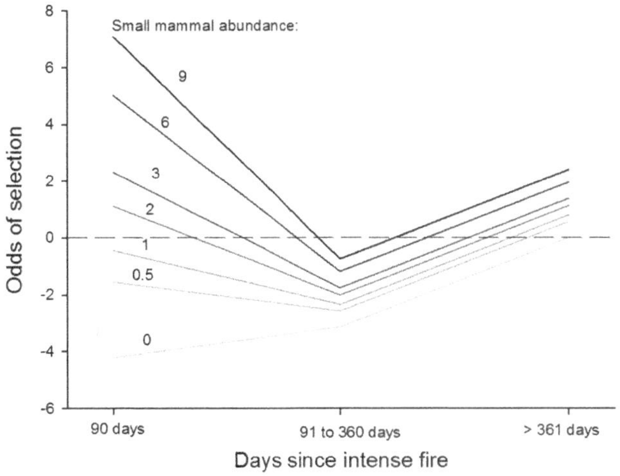

Figure 2. Change in the odds of selection ratios for different days since intense fires, at different average small mammal abundances based on capture rates per 100 trap nights (ranging from 0 to 9, lighter to darker respectively). All other variables in the model are assumed to be constant (see Table 3).

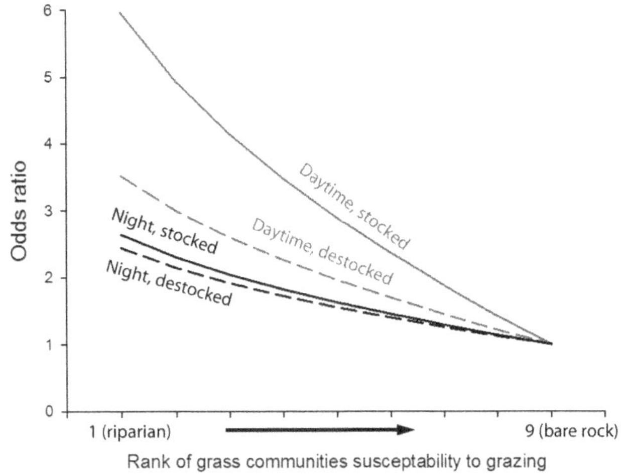

Figure 3. Odds ratios for selection of cats at night (black) and day (grey) in stocked (solid) and destocked (dashed) areas against grass communities ranked by grazing susceptibility. All other variables in model assumed to be constant (see Table 3).

strongly favoured by cats. When faced with choices about where to move, cats consistently selected sites with a more open grass layer, which had recently been subject to intense fires, and which were heavily grazed. Further, the influence of fire intensity on selection of habitat by cats was strongly affected by whether the habitats in question supported high densities of small mammals: sites in mammal-rich habitats that had recently been burnt at high intensity were especially strongly favoured.

This interaction between cat movements, fire and grazing regimes may help explain the recent declines in small mammals in northern Australia. There has been doubt that predation by cats might be driving these declines because of the mismatch in timing between the introduction of cats and small mammal declines, together with the fact that we have no evidence that small-mammal declines have coincided with an increased populations of cats [19]. As our results demonstrate, impacts of cats could have become more severe with the changes in fire and grazing regimes that began around the 1960s [40,41], even if cat density remained constant.

Selection by cats for areas with an open grass layer is a consistent finding from other studies on habitat selection by small felids [42,43], except in situations where moving into open areas leaves them exposed to larger predators [44] or where prey density is low [45]. Our cats' selection for open areas is almost certainly an expression of hunting preferences, as their selection became stronger in areas with higher small-mammal abundances. Although grass cover in itself had a large effect size (odds ratio maximum of 1.8), the effects of grazing and fire patterns on cat movements were much stronger.

Selection for fire scars was strongly dependent on fire intensity and time since fire. In general, cats avoided fire scars up to 360 days after fire. The exception was for recent scars resulting from intense fires in areas where small mammal densities were high, as this created the greatest increase in the odds ratio of any variable in this study. Fire opens up the grass layer thereby leaving prey more exposed and accessible to predators [15]. Intense fires would create pulses where prey would be easy for cats to catch - provided local abundance of prey was high. Cats did not select for recent mild fire scars, even in habitats with high small-mammal abundance. Mild fires typically leave pockets of unburnt vegetation [46], which provide protection for prey [10]. Also, mild fires are often stopped by riparian and alluvial areas [40], and such areas act as refugia for small mammals post-fire [10]. Our results suggest that cats are not able to capitalise on vulnerable small mammals after mild fires, but can after intense fires. This can explain why declines of some threatened native mammals have been so dramatic under regimes of consecutive high intensity fires [20,21].

Grazing by introduced herbivores also affected habitat choice by cats. Certain vegetation types (with more palatable grasses) were more preferred by introduced herbivores than others - typically riparian areas and bluegrass plains [47]. Cats selected these habitats even when they had been destocked, but grazing intensified this preference. Cats may have a stronger selection for these areas in the grazed zone as lower prey densities [22] require them to hunt in areas of relatively higher mammal density in order to obtain enough food. Daytime movements of cats were especially affected by grazing, suggesting that grazing creates favourable conditions for hunting of diurnal prey, such as quail [48].

Adult cat movements at small-scale 15 minute segments were dictated by their overall home range. Females had a stronger home-range fidelity than males, probably due to the importance of staying near dens that are used for rearing young [49]. Sub-adults displayed no such home range fidelity as they were probably in the process of creating and defining their home-ranges.

The top-order predator, the dingo, would also have influenced the movements of cats. Over the area and duration of the study, the dingo populations was abundant (0.2 individuals per km^2) and stable [32], and dingoes would have been a constant threat to cats [50]. A concurrent study in the same area that compared GPS movements of cats and dingoes found that although high use areas of dingoes and cats overlapped, cats were constantly avoiding the locations of individual dingoes [51]. This suggests avoidance of dingoes would have affected the timing of behavioural decisions of cats [52], but not necessarily by excluding them from certain areas [53]. Considering dingoes potential role in trophic regulation [25], further research into the relationship between dingoes, cats, and fire and grazing is warranted.

Conclusions

We demonstrate increased predator activity after intense fires and with grazing by large herbivores, which is likely to increase predation rates on small mammals. Small mammals are the preferred prey of cats, and form a substantial part of the diet of cats in the study region (55% of prey volume, from 33 stomach contents; unpublished data). Furthermore, the preference of cats for open and intensely burned areas depended on small-mammal abundance, and was reduced in habitats where abundance of small mammals was low. These patterns of habitat selection by cats correspond with declines of small mammal populations with intensifying fire and/or grazing regimes in mainland northern Australia, outside complex rocky outcrops [20,21,22,54,55,56,57]. Our work supports the hypothesis that the declines in small mammals across northern Australia are driven by cat predation facilitated by simplification of ground layer structure. However, this evidence is not direct, and other possible mechanisms including trophic alterations [25] and disease [19] may still have a role. Further research needs to measure whether mortality of prey is greater in more open areas. Regardless, the magnitude of the impacts of cats globally [58] suggests that our findings provide a general mechanism for prey decline in ecosystems with grass-dominated understoreys.

The results presented here suggest that manipulation of habitat through careful management of fire and grazing could be used to reduce impacts of cats over large landscapes. Reducing the frequency of intense fires and removing introduced herbivores is likely to be beneficial for small mammals - especially if this management is focussed on naturally mammal-rich habitats, and if it increases ground cover. Vegetation structure is pivotal in creating ideal landscapes for predators to hunt, and/or refuges for prey to hide.

Acknowledgments

Staff and volunteers from the Australian Wildlife Conservancy assisted with field work, in particular Katherine Tuft, Monica Griffith, Danni Lisle, Isabel Connell, and Lily Leahy. Anja Skroblin provided comments.

Author Contributions

Conceived and designed the experiments: HWM SL MJ CJ. Performed the experiments: HWM SL. Analyzed the data: HWM. Contributed reagents/materials/analysis tools: HWM SL CJ. Wrote the paper: HWM SL MJ CJ.

References

1. Lima SL, Dill LM (1990) Behavioural decisions made under the risk of predation: a review and prospectus. Canadian Journal of Zoology 68: 619–640.
2. Didham RK, Tylianakis JM, Gemmell NJ, Rand TA, Ewers RM (2007) Interactive effects of habitat modification and species invasion on native species decline. TRENDS in Ecology & Evolution 22: 490–496.
3. Griffiths D (1975) Prey availability and the food of predators. Ecology 56: 1209–1214.
4. Hopcraft JCC, Sinclair ARE, Packer C (2005) Planning for success: Serengeti lions seek prey accessibility rather than abundance. Journal of Animal Ecology 74: 559–566.
5. Hebblewhite M, Merrill EH, McDonald TL (2005) Spatial decomposition of predation risk using resource selection functions: an example in a wolf-elk predator-prey system. Oikos 111: 101–111.
6. Koenen MT, Leslie D, Gregory MS (1996) Habitat changes and success of artificial nests on an alkaline flat. Wilson Bulletin 108: 292–301.
7. Dickson BG, Beier P (2002) Home-range and habitat selection by adult Cougars in Southern California. Journal of Wildlife Management 66: 1235–1245.
8. Gilliam JF, Fraser DF (1987) Habitat selection under predation hazard: test of a model with foraging minnows. Ecology 68: 1856–1862.
9. Yarnell RW, Scott DM, Chimimba CT, Metcalfe DJ (2007) Untangling the roles of fire, grazing and rainfall on small mammal communities in grassland ecosystems. Oecologia 154: 387–402.
10. Banks SC, Dujardin M, McBurney L, Blair D, Barker M, et al. (2011) Starting points for small mammal population recovery after wildfire: recolonisation or residual populations? Oikos 120: 26–37.
11. Tabenia S, Ojedaa RA (2003) Assessing mammal responses to perturbations in temperate aridlands of Argentina. Journal of Arid Environments 55: 715–726.
12. Jones A (2000) Effects of cattle grazing on North American arid ecosystems: a quantitative review. Western North American Naturalist 60: 155–164.
13. Horn KJ, McMillan BR, St Clair SB (2012) Expansive fire in Mojave Desert shrubland reduces abundance and species diversity of small mammals. Journal of Arid Environments 77: 54–58.
14. Southgate R, Palmer C, Adams MA, Masters P, Triggs B, et al. (1996) Population and habitat characteristics of the Golden Bandicoot (*Isoodon auratus*) on Marchinbar Island, Northern Territory. Wildlife Research 23: 647–664.
15. Conner LM, Castleberry SB, Derrick AM (2011) Effects of mesopredators and prescribed fire on hispid cotton rat survival and cause-specific mortality. Journal of Wildlife Management 75: 938–944.
16. Driscoll DA, Lindenmayer DB, Bennett AF, Bode M, Bradstock RA, et al. (2010) Fire management for biodiversity conservation: key research questions and our capacity to answer them. Biological Conservation 143: 1928–1939.
17. Sutherland EF, Dickman CR (1999) Mechanisms of recovery after fire by rodents in the Australian environment: a review. Wildlife Research 26: 405–419.
18. Fisher DO, Johnson CN, Lawes MJ, Fritz SA, McCallum HI, et al. (2013) The current decline of tropical marsupials in Australia: is history repeating? Global Ecology and Biogeography 23: 181–190.
19. Woinarski JCZ, Legge S, Fitzsimons JA, Traill BJ, Burbidge AA, et al. (2011) The disappearing mammal fauna of northern Australia: context, cause, and response. Conservation Letters 4: 1–10.
20. Firth RSC, Brook BW, Woinarski JCZ, Fordham DA (2010) Decline and likely extinction of a northern Australian native rodent, the Brush-tailed Rabbit-rat *Conilurus penicillatus*. Biological Conservation 143: 1193–1201.
21. Pardon G, Brook BW, Griffiths AD, Braithwaite RW (2003) Determinants of survival for the northern brown bandicoot under a landscape-scale fire experiment. Journal of Animal Ecology 72: 106–115.
22. Legge S, Kennedy MS, Lloyd R, Murphy SA, Fisher A (2011) Rapid recovery of mammal fauna in the central Kimberley, northern Australia, following the removal of introduced herbivores. Austral Ecology 36: 791–799.
23. Liedloff AC, Coughenour MB, Ludwig JA, Dyer R (2001) Modelling the trade-off between fire and grazing in a tropical savanna landscape, northern Australia. Environment International 27: 173–180.
24. Dickman CR (1996) Overview of the impacts of feral cats on Australian native fauna. Sydney: University of Sydney.
25. Radford IJ, Dickman CR, Start AN, Palmer C, Carnes K, et al. (2014) Mammals of Australia's tropical savannas: A conceptual model of assemblage structure and regulatory factors in the Kimberley region. PLos ONE 9: e92341.
26. Oakwood M (2000) Reproduction and demography of the northern quoll, *Dasyurus hallucatus*, in the lowland savanna of northern Australia. Australian Journal of Zoology 48: 519–539.
27. Woinarski JCZ, Ward S, Mahney T, Bradley J, Brennan K, et al. (2011) The mammal fauna of the Sir Edward Pellew island group, Northern Territory, Australia: refuge and death-trap. Wildlife Research 38: 307–322.
28. Frank ASK, Johnson CN, Potts J, Alaric F, Lawes MJ, et al. (2014) Experimental evidence that feral cats cause local extirpation of small mammals in Australia's tropical savanna. Journal of Applied Ecology - online early.
29. Abbott I (2002) Origin and spread of the cat, *Felis catus*, on mainland Australia, with a discussion of the magnitude of its early impact on native fauna. Wildlife Research 29: 51–74.
30. Kutt AS (2012) Feral cat (Felis catus) prey size and selectivity in north-eastern Australia: implications for mammal conservation. Journal of Zoology 287: 292–300.
31. Ziembicki MR, Woinarski JCZ, Mackey B (2013) Evaluating the status of species using Indigenous knowledge: Novel evidence for major native mammal declines in northern Australia. Biological Conservation 157: 78–92.
32. Legge S, Webb T, Smith J, Tuft K, James A (2014) The ecological health of Mornington and Marion Downs Wildlife Sanctuaries, 2013 Report. Perth, WA: Australian Wildlife Conservancy.
33. Torrea I, Diazb M, Martínez-Padillac J, Bonalb RD, Viñuelae J, et al. (2007) Cattle grazing, raptor abundance and small mammal communities in Mediterranean grasslands. Basic and Applied Ecology 8: 565–575.
34. Bjørneraas K, van Moorter B, Rolandsen CM, Herfindal I (2010) Screening global positioning system location data for errors using animal movement characteristics. Journal of Wildlife Management 74: 1361–1366
35. Recio MR, Mathieu R, Virgos E, Seddon PJ (2014) Quantifying fine-scale resource selection by introduced feral cats to complement management decision-making in ecologically sensitive areas. Biological Invasions 1: 1–13.
36. Cooper AB, Millspaugh JJ (1999) The application of discrete choice models to wildlife resource selection studies. Ecology 80: 566–575.
37. Burnham KP, Anderson DR (1998) Model selection and multimodel inference: A practical information-theoretic approach: Springer Science & Business Media, Inc.
38. R Development Core Team (2008) R: a language and environment for statistical computing. Vienna: R Foundation for Statistical Computing.
39. Therneau TM, Grambsch PM, Pankratz S (2003) Penalized survival models and frailty. Journal of Computational and Graphical Statistics 12: 156–175.
40. Russell-Smith J, Yates CJ, Edwards A, Allan GE, Cook GD, et al. (2003) Contemporary fire regimes of northern Australia, 1997–2001: change since Aboriginal occupancy, challenges for sustainable management. International Journal of Wildland Fire 12: 283–297.
41. Woinarski JCZ, Mackey B, Nix HA, Traill BJ (2007) The nature of northern Australia: Natural values, ecological processes and future prospects. Canberra, ACT: ANU Press.
42. Moseby KE, Stott J, Crisp H (2009) Movement patterns of feral predators in an arid environment – implications for control through poison baiting. Wildlife Research 36: 422–435.
43. Rajaratnam R, Sunquist M, Rajaratnam L, Ambu L (2007) Diet and habitat selection of the leopard cat (*Prionailurus bengalensis borneoensis*) in an agricultural landscape in Sabah, Malaysian Borneo. Journal of Tropical Ecology 23: 209–217.
44. Genovesi P, Besa M, Toso S (1995) Ecology of a feral cat *Felis catus* population in an agricultural area of northern Italy. Wildlife Biology 1: 233–237.
45. Hall LS, Kasparian MA, Van Vuren D, Kelt DA (2009) Spatial organization and habitat use of feral cats (*Felis catus L.*) in Mediterranean California. Mammalia 64: 19–28.
46. Russell-Smith J, Ryan PG, Cheal DC (2002) Fire regimes and the conservation of sandstone heath in monsoonal northern Australia: frequency, interval, patchiness. Biological Conservation 104: 91–106.

47. Tomkins N, O'Reagain P (2007) Global positioning systems indicate landscape preferences of cattle in the subtropical savannas. The Rangeland Journal 29: 217–222.
48. Brown RL (1982) Effects of livestock grazing on Mearns Quail in southeastern Arizona. Journal of Range Management 35: 727–732.
49. Krofel M, Skrbinšek T, Kos I (2013) Use of GPS location clusters analysis to study predation, feeding, and maternal behavior of the Eurasian lynx. Ecological Research 28: 103–116.
50. Moseby KE, Neilly H, Read JL, Crisp HA (2012) Interactions between a top order predator and exotic mesopredators in the Australian rangelands. International Journal of Ecology 2012.
51. Brook LA (2013) Predator guild interactions in northern Australia: behaviour and ecology of an apex predator, the dingo *Canis lupus dingo*, and an introduced mesopredator, the feral cat *Felis catus*. Townsville, QLD: James Cook University.
52. Brook LA, Johnson CN, Ritchie EG (2012) Effects of predator control on behaviour of an apex predator and indirect consequences for mesopredator suppression. Journal of Applied Ecology 49: 1278–1286.
53. Wang Y, Fisher DO (2012) Dingoes affect activity of feral cats, but do not exclude them from the habitat of an endangered macropod. Wildlife Research 39: 611–620.
54. Woinarski JCZ, Armstrong M, Brennan K, Fisher A, Griffiths AD, et al. (2010) Monitoring indicates rapid and severe decline of native small mammals in Kakadu National Park, northern Australia. Wildlife Research 37: 116–126.
55. Kutt AS, Woinarski JCZ (2007) The effects of grazing and fire on vegetation and the vertebrate assemblage in a tropical savanna woodland in north-eastern Australia. Journal of Tropical Ecology 23: 95–106.
56. Andersen AN, Cook GD, Corbett LK, Douglas MM, Eager RW, et al. (2005) Fire frequency and biodiversity conservation in Australian tropical savannas: implications from the Kapalga fire experiment. Austral Ecology 30: 155–167.
57. Legge S, Murphy S, Heathcote J, Flaxman E, Augusteyn J, et al. (2008) The short-term effects of an extensive and high-intensity fire on vertebrates in the tropical savannas of the central Kimberley, northern Australia. Wildlife Research 35: 33–43.
58. Loss SR, Will T, Marra PP (2012) The impact of free-ranging domestic cats on wildlife of the United States. Nature Communications 4: 1396.

Fire Activity and Severity in the Western US Vary along Proxy Gradients Representing Fuel Amount and Fuel Moisture

Sean A. Parks[1]*, Marc-André Parisien[2], Carol Miller[1], Solomon Z. Dobrowski[3]

1 Aldo Leopold Wilderness Research Institute, Rocky Mountain Research Station, USDA Forest Service, Missoula, Montana, United States of America, **2** Northern Forestry Centre, Canadian Forest Service, Natural Resources Canada, Edmonton, Alberta, Canada, **3** Department of Forest Management, College of Forestry and Conservation, University of Montana, Missoula, Montana, United States of America

Abstract

Numerous theoretical and empirical studies have shown that wildfire activity (e.g., area burned) at regional to global scales may be limited at the extremes of environmental gradients such as productivity or moisture. Fire activity, however, represents only one component of the fire regime, and no studies to date have characterized *fire severity* along such gradients. Given the importance of fire severity in dictating ecological response to fire, this is a considerable knowledge gap. For the western US, we quantify relationships between climate and the fire regime by empirically describing both fire activity and severity along two climatic water balance gradients, actual evapotranspiration (AET) and water deficit (WD), that can be considered proxies for fuel amount and fuel moisture, respectively. We also concurrently summarize fire activity and severity among ecoregions, providing an empirically based description of the geographic distribution of fire regimes. Our results show that fire activity in the western US increases with fuel amount (represented by AET) but has a unimodal (i.e., humped) relationship with fuel moisture (represented by WD); fire severity increases with fuel amount and fuel moisture. The explicit links between fire regime components and physical environmental gradients suggest that multivariable statistical models can be generated to produce an empirically based fire regime map for the western US. Such models will potentially enable researchers to anticipate climate-mediated changes in fire recurrence and its impacts based on gridded spatial data representing future climate scenarios.

Editor: Matthew Germino, US Geological Survey, United States of America

Funding: This research was conducted with National Fire Plan funds provided by the Rocky Mountain Research Station, USDA Forest Service. The funders had no role in study design, data collection and analysis, decision to publish, or preparation of the manuscript.

Competing Interests: The authors have declared that no competing interests exist.

* E-mail: sean_parks@fs.fed.us

Introduction

Fire is a ubiquitous ecosystem process across the globe. The concept of the fire regime has been used to describe the role of fire in an ecosystem in terms of its spatial-temporal patterns and ecosystem impacts [1]. As such, maps of fire regimes, in terms of fire frequency and severity, have been produced [2]. These maps not only allow geographic comparisons of fire regime components [3], but are also a necessary first step for describing shifts in fire regimes resulting from factors such as climate change [4], fire suppression [5], and invasive species [6]. Though the importance of mapping fire regimes is long acknowledged [7,8], efforts to date have been largely qualitative and thus have been unable to make direct linkages to factors driving both fire activity and severity (but see [9]).

Recently, however, linkages between environmental gradients and fire activity have been identified. Several studies have shown that wildland fires across the globe tend to avoid environmental extremes of productivity and moisture [10–13]. For example, in very dry areas such as deserts, fire activity (i.e., fire occurrence and area burned) is limited by a lack of biomass [14]. Conversely, in the wettest places on Earth (e.g., rainforests), there is ample biomass but fire activity is limited because climate conditions promoting combustion are rare [15]. To date, studies of the geographic distribution of fire, or pyrogeography, have focused more on patterns of *fire activity* (e.g., [16–18]) than on explaining patterns of *fire severity* (but see [19,20]). Given that fire severity – a measure of ecosystem change – can strongly dictate the response of biological communities to fire [21–23], an understanding of its environmental controls is a prerequisite for understanding the role of fire in ecosystems. Furthermore, without an understanding of these controls, modeling and prediction of climate-mediated changes in fire regimes is tenuous.

The seminal biome classification of Whittaker [24] exemplifies the utility of representing large scale ecological features along an energy-moisture biplot (temperature and precipitation in this case). As a refinement of Whittaker's framework, Stephenson [25] developed a bivariate representation of North American's main vegetation types using actual evapotranspiration (AET) and climatic water deficit (WD; potential evapotranspiration minus AET), as measures of productivity and drought, respectively. Specifically, AET represents a measure of available moisture to plants, whereas WD is a measure of unmet atmospheric demand for water (i.e., how much water could be evaporated and transpired were it available) [25]. Though related to precipitation and temperature, AET and WD take into account the concurrent availability and demand for moisture and therefore better represent environmental controls on ecosystems. As such, these metrics have been shown to be strong predictors of plant physiognomic types [25] and tree species distributions [26].

Recent studies have shown that water balance metrics such as AET and WD are predictive of annual to decadal fire activity [27,28]. AET is strongly correlated with biomass production [29,30], and as such, it is a suitable proxy for fuel amount in many ecosystems. WD is a measure of absolute drought and should reflect the moisture status and flammability of live [31] and large-diameter dead fuels [28]. Previous research has suggested a unimodal (i.e., humped) response of fire activity to fuel amount (e.g., [11]) and fuel moisture [32]; that is, less fire occurs at the extremes of these gradients than at intermediate levels. Less clearly understood are the relationships between severity and these gradients. Some studies have suggested that fire severity depends upon amount of biomass [33] and therefore would be expected to increase with AET; the relationship between fire severity and fuel moisture, however, is unknown.

We seek to better understand how contemporary fire regimes across the western US vary along macroscale environmental gradients. Our first objective is to examine how fire activity and fire severity each vary as a function of proxy gradients for fuel amount and fuel moisture (represented by AET and WD, respectively). Our second objective is to present an empirically based characterization of fire regimes that includes both fire activity and fire severity. Specifically, we characterize fire activity and severity by ecoregion, thus quantitatively describing ecoregions within a pyrogeographic context that includes more than one fire regime component. Because anthropogenic activities can obscure fire-climate relationships [34], we focus our analysis on protected areas. This study explicitly links fire regime components with physical environmental gradients, and therefore is an important first step toward forecasting and mitigation of projected changes in fire regimes.

Methods

Data

Fire data for the 1984–2010 period were obtained from the Monitoring Trends in Burn Severity (MTBS) project [35], which has mapped fire perimeters and fire severity for all fires ≥400 ha in the western US. We define fire severity as the degree of fire-induced change to vegetation and soils as measured by the delta normalized burn ratio (dNBR), a satellite-inferred index that differences pre- and post-fire Landsat images [36]. Raw dNBR values were adjusted to account for differences due to phenology or precipitation between the pre- and post-fire images by subtracting the average dNBR of pixels outside the burn perimeter; this adjustment can be important when comparing dNBR among fires [37]. As dNBR values increase, there is generally an increase in char and scorched/blackened vegetation and a decrease in moisture content and vegetative cover [36]. The dNBR is strongly associated with field-assessed measures of fire severity over a broad range of ecosystem types [38–41]. In very limited cases, fire perimeters had no associated dNBR; this was generally the case for fires that occurred in 1984 and pre-fire Landsat TM data were unavailable.

Gridded annual AET and WD data (30 arc-second resolution; ~800 m) were obtained from Dobrowski et al. [42] and were derived from a modified Thornthwaite-Mather soil water balance model. The model was run on a monthly time step and accounts for atmospheric demand (potential evapotranspiration – calculated using the Penman-Monteith equation [43]), soil water storage, and includes the effect of temperature and radiation on snow hydrology via a snowmelt model. Annual values for AET and WD were averaged for each pixel over the 1984–2010 time period, corresponding to the fire data used in this study. Low values of AET represent areas with low potential productivity and thus limited fuel amount, whereas high values indicate the potential for high fuel amounts. Low values of WD indicate that water is less limiting (i.e., high fuel moisture), whereas high values indicate severe water limitation. Given that these water balance metrics are correlated in this study (Pearsons's $r > 0.75$), it may be difficult to disentangle the unique contribution of each term.

Analysis

The western US was partitioned into 50,000-ha (500 km^2) hexels in which fire activity and severity were summarized. Although the spatial resolution of fire-related analyses has been shown to influence outcomes [44,45], this hexel size was deemed a reasonable compromise for capturing variability among sample units while maintaining an adequate sample size for analysis, especially given the inherent spatial variability associated with the relatively short fire record we analyzed (1984–2010). Fire activity was calculated as the proportion of each hexel (excluding nonfuel, such as barren, ice/snow, etc. [3]) that burned from 1984 to 2010; this proportion was subsequently square root transformed to homogenize variance in model residuals. Fire severity was calculated by averaging the dNBR of all pixels of all fires that intersected each hexel; pixels classified as nonfuel were excluded in the calculation of the mean. AET and WD were also aggregated to the hexel scale (Fig. 1).

To characterize how fire activity varies along gradients in fuel amount and fuel moisture, we built bivariate statistical models of area burned as a function of AET and WD; similar models were produced with dNBR to characterize how fire severity varies along these gradients. The two models of area burned were generated using generalized linear models (family = quasibinomial) with a quadratic term (i.e., 2nd order polynomial) for AET and WD. dNBR was assessed as a function of AET using an ordinary least squares (ols) model, whereas dNBR was related to WD using a nonlinear exponential function. All model fits (except the dNBR vs. AET model) were evaluated using the coefficient of determination (denoted here as R^2), which is the r^2 of a linear regression between observed and predicted values (cf. [39,46]); the dNBR vs. AET model was evaluated using ols r^2. In our initial data exploration, we also evaluated dNBR as a function of AET using an exponential function, but this resulted in a decreased model strength compared to the linear fit. To reduce the effect of human infrastructure that can disrupt the spread of fires (e.g., roads) and vegetation management activities that can alter fuel structures (e.g., silviculture), we limited this analysis to hexels comprising ≥ 80% designated wilderness and national park; a total of 153 hexels met this criterion for the fire activity models. An additional criterion was added for the evaluation of fire severity to limit variability in mean dNBR values that are associated with small sample sizes (cf. [47]): at least 400 ha had to have burned within a hexel from 1984–2010. A total of 99 hexels met this additional criterion for the fire severity models.

To place our data and findings into a broader biogeographic context, we summarized the hexels (using the previously described subset based on minimum of 80% wilderness and national park) in each 'ecological section' (hereafter ecoregion) as defined by ECOMAP [48] (Fig. 1). We plotted each ecoregion along axes of AET and WD by averaging the values among hexels in each ecoregion. We similarly plotted each ecoregion along axes of area

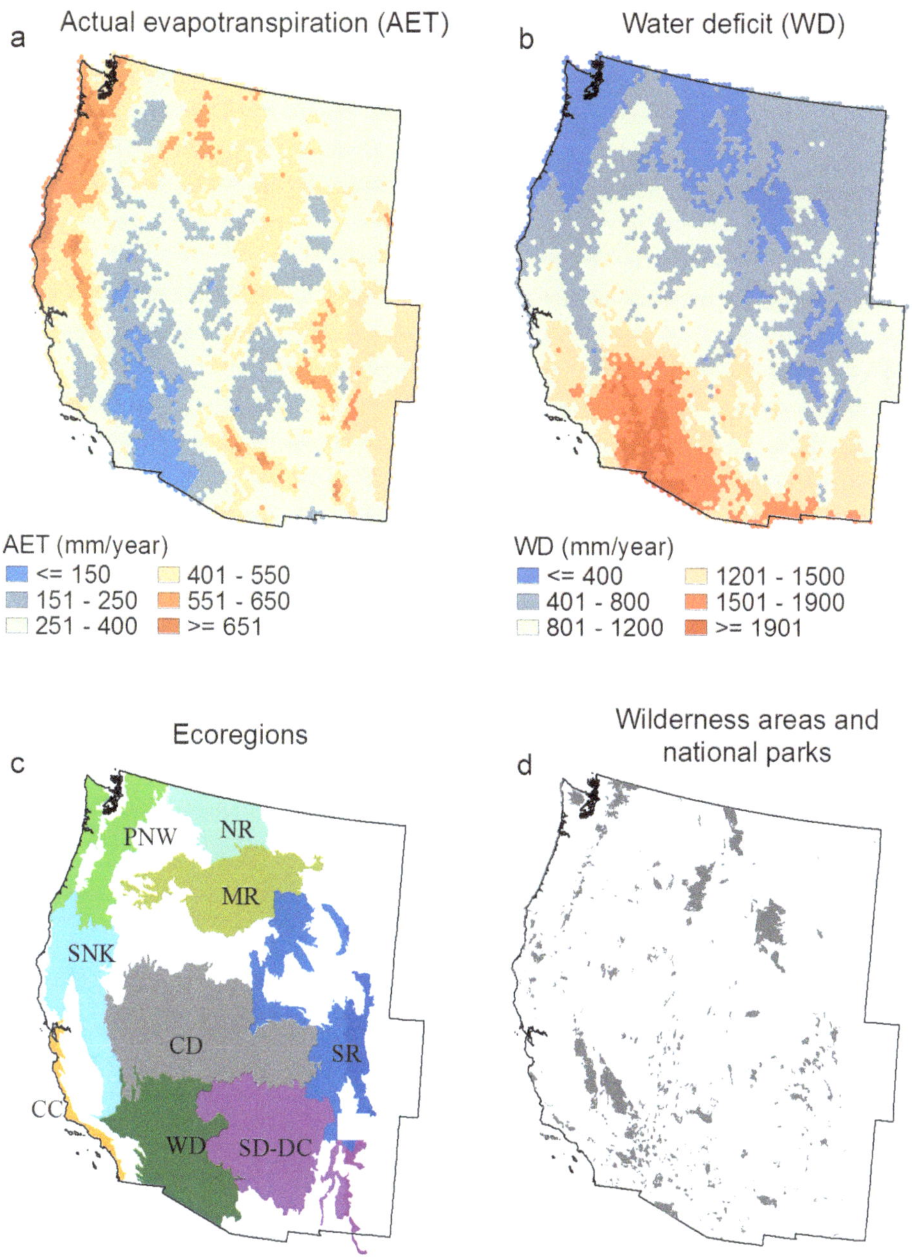

Figure 1. Mean actual evapotranspiration (AET) (a) and water deficit (WD) (b) per 50,000 ha hexel for the western US from 1984–2010. Also shown are the ecoregions (USDA Forest Service, 2007) we analyzed (c): California chaparral (CC), cold desert (CD), middle Rocky Mountains (MR), northern Rocky Mountains (NR), Pacific Northwest (PNW), semi desert – dry conifer (SD-DC), Sierra Nevada and Klamath (SNK), southern Rocky Mountains (SR), warm desert (WD). Cold desert and semi desert – dry conifer are both the result of merging two ecoregions; all ecoregions were renamed for easier interpretation. Areas in white represent ecoregions that did not contain enough fire and/or wilderness and national park area to be included in this study (see Methods). Locations of wilderness areas and national parks in the western US (d).

burned and dNBR, thereby providing a broad-scale concurrent characterization of two components of the fire regime. All ecoregions were renamed and two pairs of two ecoregions were merged to 'cold desert' and 'semi desert – dry conifer' for easier interpretation.

Although not a primary objective of our study, we were able to evaluate how changing the sampling criteria for inclusion of hexels into the models affected the strength of the relationships between the fire characteristics (area burned and dNBR) and the climatic water balance metrics (AET and WD). For each of the four relationships, we built 20 models with different minimum requirements for amount of within-hexel designated wilderness and national park, ranging from zero to 95% (5% increments).

Results

The relationships of area burned and dNBR to AET and WD are moderately strong. Area burned increases with AET ($R^2 = 0.40$), as does dNBR ($r^2 = 0.21$) (Fig. 2a, 2b). Area burned and WD have a unimodal relationship ($R^2 = 0.34$), whereas dNBR decreases with WD ($R^2 = 0.43$) (Fig. 2c, 2d).

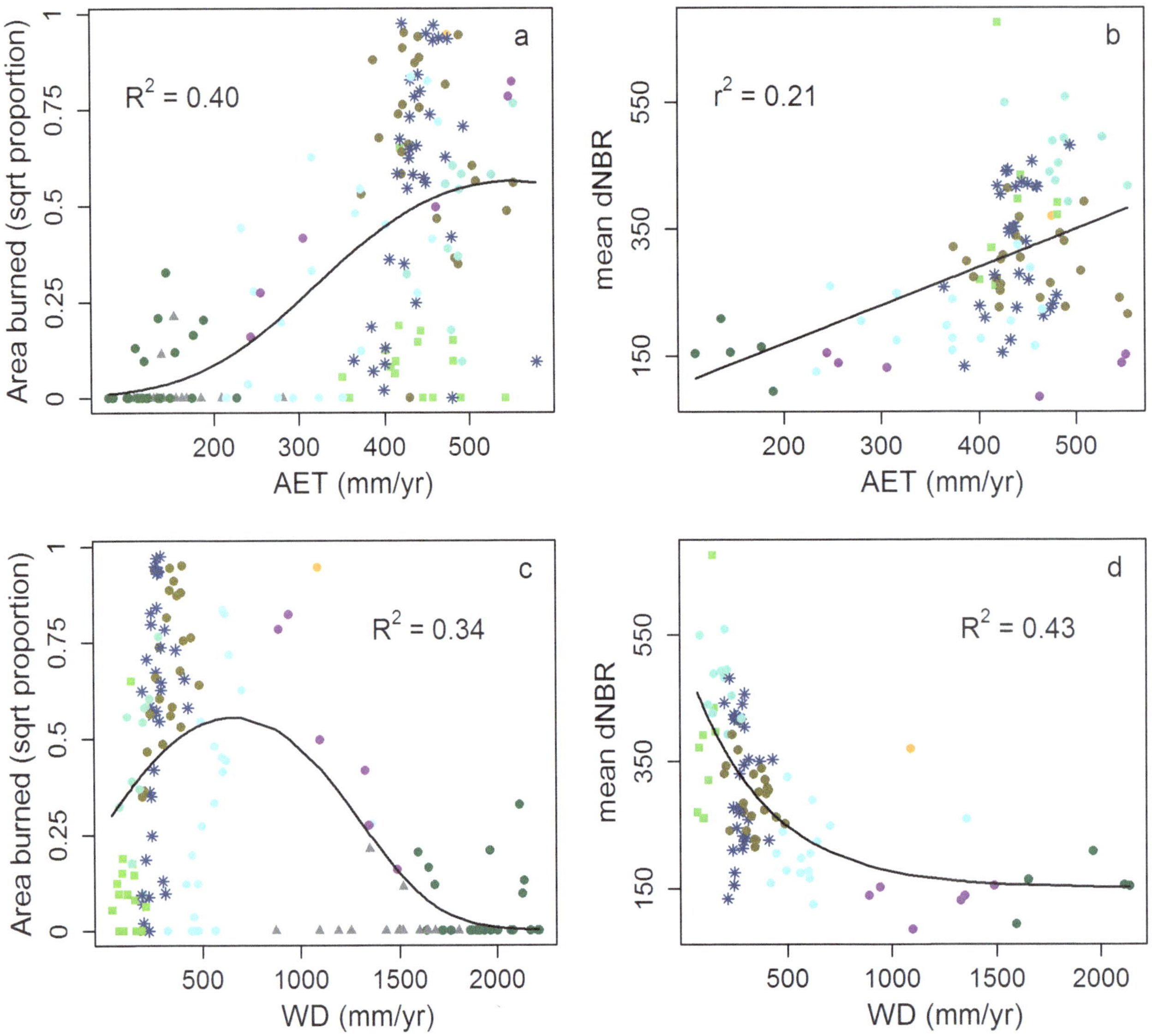

Figure 2. Area burned plotted against AET (a) and WD (b). Mean dNBR plotted against AET (c) and WD (d). Fitted lines are generated by models of each relationship (see Methods); the strength of each model is reported (R^2 or r^2). Symbol colors correspond to the ecoregion in which it is located (Fig. 1).

Broad ecoregion-level biogeographic patterns are revealed in the biplot of AET and WD (Fig. 3a); extreme differences in both AET and WD among contrasted ecoregions, such as the warm desert (WD) and Pacific Northwest (PNW), are evident. In general, ecoregions occupy distinct portions of the AET and WD bi-dimensional space, but the Pacific Northwest and all Rocky Mountain ecoregions are relatively tightly grouped (Fig. 3a). Ecoregional differences in fire regime characteristics are also apparent in the biplot of area burned and dNBR (Fig. 3b). The tight grouping of some ecoregions seen in the biplot of AET and WD (Fig. 3a) is not seen in the biplot of fire regime characteristics (Fig. 3b).

The relationships generally strengthened as we increased the requisite percent of wilderness and national park for inclusion in the models (Fig. 4). In one of the four cases (area burned vs. AET), the R^2 increased from 0.03 to 0.53 when the minimum within-hexel percent wilderness and national park increased from 0% to 95%, respectively.

Discussion

Fire activity increased with AET, the proxy we used for fuel amount; the unimodal relationship that previous theoretical and empirical studies have reported was not seen here [11–13]. This is likely because our study was limited to the western US, and therefore, we evaluated a truncated range of AET (the highest AET values in the US are located in the Southeast). Had we been able to include ecosystems with higher AET values in our analysis, we expect we would have seen a limitation on fire activity at the extreme upper values of AET. Fire activity had a unimodal relationship with fuel moisture, represented by WD, concurring with the theoretical model described by Pausas and Bradstock [32]. Consistent with the "varying constraints hypothesis" posited by Krawchuk and Moritz [11], areas in ecoregions with the highest fuel moisture (lowest WD) (i.e., Pacific Northwest) experienced relatively low fire activity because, despite being biomass rich, they are rarely dry enough to burn. At the other end

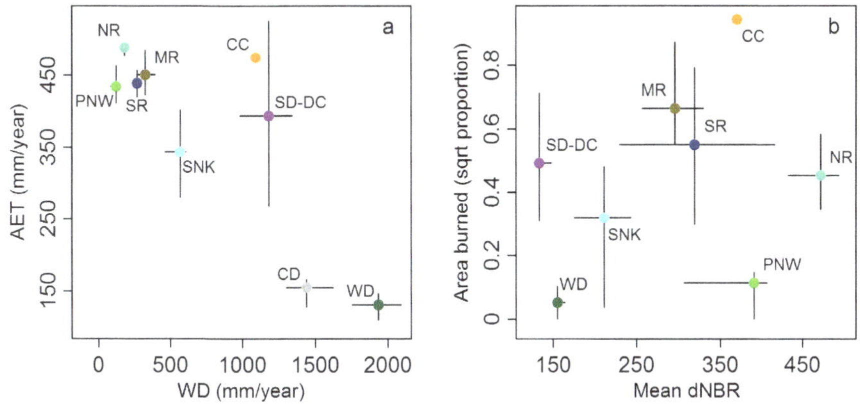

Figure 3. AET and WD averaged across hexels in each ecoregion [48] **(a).** Area burned and dNBR averaged across hexels in each ecoregion (b). Although the cold desert ecoregion did experience some fire within the hexels we analyzed (see Fig. 2), there were no dNBR data for this ecoregion (see Methods); therefore, this ecoregion is not included in the plot on the right (b). Vertical and horizontal lines represent the middle 50 percent (25th to 75th percentile) of the hexel values in each ecoregion.

of the gradient, fire activity in ecoregions with low fuel moisture (i.e., warm desert) was limited because low productivity limits the amount of biomass that can burn.

Fire severity increased with fuel amount, providing support to the idea that fire severity is limited by available biomass [33,49]. The relationship between fire severity and fuel moisture distinctly showed an exponential decrease in fire severity with increasing WD (see Fig. 2d). This is a somewhat counterintuitive and unexpected result because lower fuel moisture has been found to result in higher severity fires [20,50], as drought-stressed trees may be more susceptible to damage by fire and low fuel moisture increases the flammability of fuels. The apparent discrepancy likely involves key differences and interpretations between analyses that evaluate the influence of short-term (e.g., annual) variability (e.g., [20,50]) and those that evaluate long-term averages (our study). For example, Parisien et al. [51] found that the factors controlling fire activity were strikingly different when a statistical model was built incorporating annual variability compared to one using long-term (~30 years) averages. In our study, we evaluated fairly long-term fuel moisture (i.e., averaged over 26 years), as measured by WD, which doesn't necessarily reflect the short-term moisture conditions under which fires actually occur. Consequently, the influence of fuel moisture on fire severity seen in this study may be due to its long-term climatic influence on the establishment and perpetuation of forest types that are adapted to particular moisture conditions; those forest types in turn influence fire severity. For example, on those occasions when moist forest types experience fire (i.e., unusually dry years), they tend to burn at high severity due to their high tree density and abundance of ladder fuels [52]. Nevertheless, the aforementioned correlation between our proxies for fuel amount and moisture (AET and WD, respectively) cannot be ignored and, consequently, completely disentangling their influence is challenging.

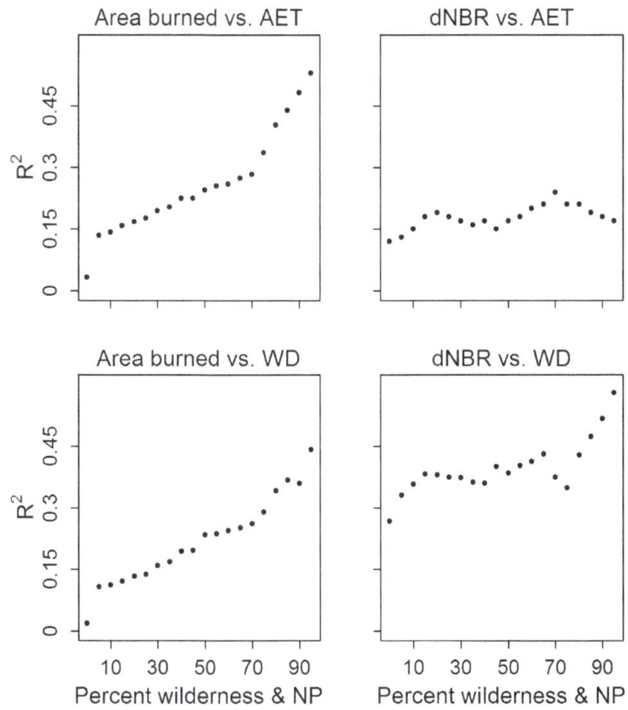

Figure 4. Strength of the relationship (R^2 for all models except the dNBR vs. AET model, which is evaluated with the r^2; see Methods) for all models as a function of increasing percentage of wilderness and national park in each hexel.

Broad biogeographic differences among ecoregions were evident when they were partitioned in terms of AET (fuel amount) and WD (fuel moisture). The concurrent characterization of fire activity and severity also clearly distinguished ecoregions. For example, the warm desert ecoregion experienced low fire activity and severity, whereas the Pacific Northwest ecoregion experienced low fire activity but high fire severity (Fig. 3b). However, ecoregions that were tightly grouped in terms of AET and WD (Pacific Northwest and all Rocky Mountain ecoregions) (Fig. 3a) did not exhibit a similar grouping of fire regime characteristics (Fig. 3b). This discrepancy could be due to difficulties associated with describing the fire regime using only AET and WD or the relatively short fire record that we analyzed, which potentially over- or under-emphasizes fire activity for certain regions (e.g., 1988 Yellowstone fires) [53]; other factors such as macroscale spatial variability in ignition sources [54] and fire suppression effectiveness could also be responsible. Ultimately, however, aggregating hexels by ecoregion may oversimplify complex fire regime dynamics and their effects on ecosystems.

Changes in fire activity and severity are certain to occur as water balance metrics shift in concert with a changing climate, perhaps in complex and unexpected ways [55]. The temporal covariance between energy and moisture availability will play a key role in determining how fuel amount and fuel moisture are geographically distributed in the future, with clear implications for pyrogeography. The complex interplay between climate and the seasonal timing of water and energy availability illustrates the need for mechanistic bioclimatic predictors that can account for these contingencies [25]. The framework developed in this study is an excellent starting point for predicting fire regime shifts under a changing climate though we acknowledge that climate-induced changes in fire regimes will likely have strong feedbacks with the dominant vegetation and ecosystem processes in any given area [56].

We focused this study on wilderness areas and national parks in the western US, where anthropogenic influences (e.g., forest management) are minimal relative to unprotected lands. One consequence of our focus on protected areas is that the fairly low number of hexels available to use in our models of area burned and fire severity represent only 2.5% and 1.6% of the land area in the western US, respectively. As such, the relationships we derived from these relatively unmanaged, natural lands may not well represent the majority of the western US; indeed, the relationships between fire and climate clearly weaken as the human footprint increases (see Fig. 4). Archibald et al. [34] noted a similar finding in the African savannas, in that annual climatic variability was strongly associated with large fire occurrence in areas of low human impact but not in human-dominated areas. Although the hexels we used admittedly represent a small proportion of the western US, the data we analyzed provide valuable insight regarding the "natural" relationships between fire and climate. Consequently, our approach can potentially be used to identify areas with disrupted fire regimes and in need of restoration treatments [57].

Although we used a single metric (mean dNBR) as a convenient measure of ecosystem change, fire severity is the result of many complex physical and ecological factors that are difficult to represent with simplistic measures such as dNBR. For example, lodgepole pine forests experience high-severity fire regimes that favor the regeneration and perpetuation of lodgepole pine on a site, because, as fire often kills most or all of the trees, it also opens their serotinous cones [58,59]. In this example, although dNBR would indicate high severity fire, the change to the ecosystem would be considered far less drastic because lodgepole pine will remain the dominant tree species. As such, our analyses do not address if fire severity is within its historical range of variability [60] for any given vegetation community. Furthermore, dNBR in some ecosystems such as semi-arid shrublands (low AET and high WD) may be relatively low even though the degree of fire-induced change may be very high due to factors such as high rates of post-fire erosion [61] and conversion of perennial woodlands to annual grasslands [62]. Consequently, the relationships reported in this study could be different were other aspects of fire severity (those that cannot be measured with satellites) evaluated. It has also been argued that *relative* metrics of burn severity (i.e., those that measure change relative to the amount of pre-fire vegetation) are more appropriate than absolute metrics (i.e., dNBR) on sites where pre-fire vegetation is low [63,64]. The reasoning for this argument is that dNBR values will be lower on sparsely vegetated sites such as deserts and shrublands, regardless of the degree of fire-induced vegetation mortality, compared to relative metrics such as the relativized delta normalized burn ratio (RdNBR) [63] or relativized burn ratio (RBR) [37]. Indeed, we explored how RBR [37] varied along gradients of AET and WD; shapes of the relationships were nearly identical to those we reported for dNBR, but the strengths of the relationships were weaker. In summary, although fire severity has ecological significance beyond what can be inferred from dNBR, we used dNBR as a convenient and standardized way to assess fire severity.

Several additional factors should be considered when interpreting our results. First, although we focused our study on wilderness and national parks, fire regimes are not entirely natural in those areas. Fire exclusion does occur, which reduces fire activity [65] and increases the potential for high severity fire in some forest types [5]. This factor may blur the relationship between fire and climate. Second, the spatial resolution of our analysis (i.e., size of hexels) does not allow us to adequately represent the substantial variability in climate and fire regime characteristics within each hexel [52]. Third, though we used AET and WD as proxies for fuel amount and fuel moisture, respectively, there may be some ecosystems where these proxies are less suitable. For example, AET may poorly represent fuel amount in ecosystems with high rates of decomposition (tropics), or those that experience frequent fire. Finally, because wilderness areas and national parks are often located at high elevations (e.g., at the tops of mountain ranges), the climatic conditions within hexels included in this study are not necessarily representative of the ecoregions as a whole [66]; for example, many of the hexels in the Pacific Northwest and Sierra Nevada and Klamath ecoregions have a relatively high proportion of unproductive alpine environments, and they hence have lower AET than the ecoregion on average.

Conclusion

A growing body of literature on pyrogeography points to relatively simple energetic controls on fire occurrence [58]: fires require fuel and that fuel needs to be dry enough to burn. To the extent that AET and WD are suitable proxies for fuel amount and fuel moisture, respectively, we found that fire activity across the western US is clearly limited by lack of available fuel and fuel moisture conditions. Furthermore, we found that fire severity is positively related to both fuel amount and fuel moisture. To our knowledge, this is the first broad-scale characterization of fire activity *and* severity along physical environmental gradients. A pyrogeographic perspective that includes both fire activity and severity provides an enhanced view of contemporary fire regimes that complements existing classifications of fire activity. As such, the concurrent characterization of fire activity and severity was effective in distinguishing contemporary fire regime properties of most ecoregions in the western US. The explicit links between fire activity and severity with physical environmental gradients provides a necessary first step for generating multivariable statistical models to produce an empirically based fire regime map for the western US. Such models will potentially enable researchers to predict the geographic distribution of future fire regime characteristics based on gridded spatial data representing future climate scenarios. Consequently, our framework should advance the study of climate-mediated impacts on fire regimes, because, although numerous studies have predicted changes in fire activity due to climate change (e.g., [67]), none have yet examined how fire severity is predicted to change.

Author Contributions

Conceived and designed the experiments: SP MAP CM SD. Performed the experiments: SP. Analyzed the data: SP. Contributed reagents/materials/analysis tools: SD. Wrote the paper: SP MAP CM SD.

References

1. Agee JK (1993) Fire Ecology of Pacific Northwest Forests. Washington, D.C.: Island Press. 493 p.
2. Schmidt KM, Menakis JP, Hardy CC, Hann WJ, Bunnell DL (2002) Development of Coarse-Scale Spatial Data for Wildland Fire and Fuel Management. USDA Forest Service, Rocky Mountain Research Station, Fort Collins, CO, GTR-RMRS-087. 42.
3. Rollins MG (2009) LANDFIRE: a nationally consistent vegetation, wildland fire, and fuel assessment. International Journal of Wildland Fire 18: 235–249.
4. Westerling AL, Hidalgo HG, Cayan DR, Swetnam TW (2006) Warming and earlier spring increase western US forest wildfire activity. Science 313: 940–943.
5. Naficy C, Sala A, Keeling EG, Graham J, DeLuca TH (2010) Interactive effects of historical logging and fire exclusion on ponderosa pine forest structure in the northern Rockies. Ecological Applications 20: 1851–1864.
6. Mack MC, D'Antonio CM (1998) Impacts of biological invasions on disturbance regimes. Trends in Ecology & Evolution 13: 195–198.
7. Morgan P, Hardy CC, Swetnam TW, Rollins MG, Long DG (2001) Mapping fire regimes across time and space: Understanding coarse and fine-scale fire patterns. International Journal of Wildland Fire 10: 329–342.
8. Murphy BP, Williamson GJ, Bowman D (2011) Fire regimes: moving from a fuzzy concept to geographic entity. New Phytologist 192: 316–318.
9. Archibald S, Lehmann CER, Gomez-Dans JL, Bradstock RA (2013) Defining pyromes and global syndromes of fire regimes. Proceedings of the National Academy of Sciences of The United States of America 110: 6442–6447.
10. Whitlock C, Higuera PE, McWethy DB, Briles CE (2010) Paleoecological perspectives on fire ecology: revisiting the fire-regime concept. The Open Ecology Journal 3: 6–23.
11. Krawchuk MA, Moritz MA (2011) Constraints on global fire activity vary across a resource gradient. Ecology 92: 121–132.
12. Pausas JG, Ribeiro E (2013) The global fire productivity relationship. Global Ecology and Biogeography 22: 728–736.
13. Ryan KC (1991) Vegetation and wildland fire - implications of global climate change. Environment International 17: 169–178.
14. Westerling AL, Gershunov A, Brown TJ, Cayan DR, Dettinger MD (2003) Climate and wildfire in the western United States. Bulletin of the American Meteorological Society 84: 595–604.
15. Littell JS, McKenzie D, Peterson DL, Westerling AL (2009) Climate and wildfire area burned in western U.S. ecoprovinces, 1916–2003. Ecological Applications 19: 1003–1021.
16. Parisien M-A, Snetsinger S, Greenberg JA, Nelson CR, Schoennagel T, et al. (2012) Spatial variability in wildfire probability across the western United States. International Journal of Wildland Fire 21: 313–327.
17. Krawchuk MA, Moritz MA, Parisien M-A, Van Dorn J, Hayhoe K (2009) Global pyrogeography: the current and future distribution of wildfire. PLoS ONE 4: e5102.
18. Guyette RP, Stambaugh MC, Dey DC, Muzika RM (2012) Predicting fire frequency with chemistry and climate. Ecosystems 15: 322–335.
19. Dillon GK, Holden ZA, Morgan P, Crimmins MA, Heyerdahl EK, et al. (2011) Both topography and climate affected forest and woodland burn severity in two regions of the western US, 1984 to 2006. Ecosphere 2: art130.
20. van Mantgem PJ, Nesmith JCB, Keifer M, Knapp EE, Flint A, et al. (2013) Climatic stress increases forest fire severity across the western United States. Ecology Letters 16: 1151–1156.
21. Turner MG, Romme WH, Gardner RH (1999) Prefire heterogeneity, fire severity, and early postfire plant reestablishment in subalpine forests of Yellowstone National Park, Wyoming. International Journal of Wildland Fire 9: 21–36.
22. Smucker KM, Hutto RL, Steele BM (2005) Changes in bird abundance after wildfire: Importance of fire severity and time since fire. Ecological Applications 15: 1535–1549.
23. McHugh CW, Kolb TE (2003) Ponderosa pine mortality following fire in northern Arizona. International Journal of Wildland Fire 12: 7–22.
24. Whittaker RH (1975) Communities and Ecosystems. New York: Macmillan.
25. Stephenson NL (1990) Climatic control of vegetation distribution: the role of the water balance. The American Naturalist 135: 649–670.
26. Lutz JA, van Wagtendonk JW, Franklin JF (2010) Climatic water deficit, tree species ranges, and climate change in Yosemite National Park. Journal of Biogeography 37: 936–950.
27. Littell JS, Gwozdz RB (2011) Climatic water balance in regional fire years in the Pacific Northwest, USA: Linking regional climate and fire and at landscape scales. In: McKenzie D, Miller C, Falk DA, editors. The Landscape Ecology of Fire: Springer Netherlands. 177–139.
28. Abatzoglou JT, Kolden CA (2013) Relationships between climate and macroscale area burned in the western United States. International Journal of Wildland Fire 22: 1003–1020.
29. Cleveland CC, Townsend AR, Schimel DS, Fisher H, Howarth RW, et al. (1999) Global patterns of terrestrial biological nitrogen (N-2) fixation in natural ecosystems. Global Biogeochemical Cycles 13: 623–645.
30. Mu Q, Heinsch FA, Zhao M, Running SW (2007) Development of a global evapotranspiration algorithm based on MODIS and global meteorology data. Remote Sensing of Environment 111: 519–536.
31. Dennison PE, Roberts DA, Thorgusen SR, Regelbrugge JC, Weise D, et al. (2003) Modeling seasonal changes in live fuel moisture and equivalent water thickness using a cumulative water balance index. Remote Sensing of Environment 88: 442–452.
32. Pausas JG, Bradstock RA (2007) Fire persistence traits of plants along a productivity and disturbance gradient in mediterranean shrublands of south-east Australia. Global Ecology and Biogeography 16: 330–340.
33. Mitchell SR, Harmon ME, O'Connell KEB (2009) Forest fuel reduction alters fire severity and long-term carbon storage in three Pacific Northwest ecosystems. Ecological Applications 19: 643–655.
34. Archibald S, Nickless A, Govender N, Scholes RJ, Lehsten V (2010) Climate and the inter-annual variability of fire in southern Africa: a meta-analysis using long-term field data and satellite-derived burnt area data. Global Ecology and Biogeography 19: 794–809.
35. Eidenshink J, Schwind B, Brewer K, Zhu ZL, Quayle B, et al. (2007) A project for monitoring trends in burn severity. Fire Ecology 3: 3–21.
36. Key CH, Benson NC (2006) Landscape assessment (LA). In: Lutes D, Keane RE, Caratti JF, Key CH, Benson NC et al., editors. FIREMON: Fire Effects Monitoring and Inventory System. Fort Collins, CO: RMRS-GTR-164. 55.
37. Parks SA, Dillon GK, Miller C (2014) A new metric for quantifying burn severity: the Relativized Burn Ratio. Remote Sensing 6: 1827–1844.
38. Wimberly MC, Reilly MJ (2007) Assessment of fire severity and species diversity in the southern Appalachians using Landsat TM and ETM plus imagery. Remote Sensing of Environment 108: 189–197.
39. Soverel NO, Perrakis DDB, Coops NC (2010) Estimating burn severity from Landsat dNBR and RdNBR indices across western Canada. Remote Sensing of Environment 114: 1896–1909.
40. Cocke AE, Fule PZ, Crouse JE (2005) Comparison of burn severity assessments using Differenced Normalized Burn Ratio and ground data. International Journal of Wildland Fire 14: 189–198.
41. Arkle RS, Pilliod DS, Welty JL (2012) Pattern and process of prescribed fires influence effectiveness at reducing wildfire severity in dry coniferous forests. Forest Ecology and Management 276: 174–184.
42. Dobrowski SZ, Abatzoglou J, Swanson AK, Greenberg JA, Mynsberge AR, et al. (2013) The climate velocity of the contiguous United States during the 20th century. Global Change Biology 19: 241–251.
43. Allen RG, Pereira LS, Raes D, Smith M (1998) Crop Evapotranspiration-Guidelines for Computing Crop Water Requirements-FAO IRRIGATION and Drainage. Food and Agriculture Organization of the United Nations, Rome, Italy.
44. Parks SA, Parisien MA, Miller C (2011) Multi-scale evaluation of the environmental controls on burn probability in a southern Sierra Nevada landscape. International Journal of Wildland Fire 20: 815–828.
45. Parisien MA, Parks SA, Krawchuk MA, Flannigan MD, Bowman LM, et al. (2011) Scale-dependent controls on the area burned in the boreal forest of Canada, 1980–2005. Ecological Applications 21: 789–805.
46. Girardin MP, Ali AA, Carcaillet C, Mudelsee M, Drobyshev I, et al. (2009) Heterogeneous response of circumboreal wildfire risk to climate change since the early 1900s. Global Change Biology 15: 2751–2769.
47. Wisz MS, Hijmans RJ, Li J, Peterson AT, Graham CH, et al. (2008) Effects of sample size on the performance of species distribution models. Diversity and Distributions 14: 763–773.
48. USDA Forest Service (2007) Ecological Subregions: Sections and Subsections for the Conterminous United States.
49. Parks SA, Miller C, Nelson CR, Holden ZA (2014) Previous fires moderate burn severity of subsequent wildland fires in two large western US wilderness areas. Ecosystems 17: 29–42.
50. Renkin RA, Despain DG (1992) Fuel moisture, forest type, and lightning-caused fire in Yellowstone National Park. Canadian Journal of Forest Research 22: 37–45.
51. Parisien MA, Parks SA, Krawchuk MA, Little JM, Flannigan MD, et al. (2014) Comparing the influence of temporally averaged versus annually varying controls on fire probabilities across the boreal forest of Canada. Ecological Applications in press.
52. Schoennagel T, Veblen TT, Romme WH (2004) The interaction of fire, fuels, and climate across rocky mountain forests. Bioscience 54: 661–676.
53. Romme WH, Boyce MS, Gresswell R, Merrill EH, Minshall GW, et al. (2011) Twenty years after the 1988 Yellowstone fires: lessons about disturbance and ecosystems. Ecosystems 14: 1196–1215.
54. McWethy DB, Higuera PE, Whitlock C, Veblen TT, Bowman D, et al. (2013) A conceptual framework for predicting temperate ecosystem sensitivity to human impacts on fire regimes. Global Ecology and Biogeography 22: 900–912.
55. Batllori E, Parisien MA, Krawchuk MA, Moritz MA (2013) Climate change-induced shifts in fire for Mediterranean ecosystems. Global Ecology and Biogeography 22: 1118–1129.
56. Turner MG (2010) Disturbance and landscape dynamics in a changing world. Ecology 91: 2833–2849.
57. Schoennagel T, Nelson CR (2010) Restoration relevance of recent National Fire Plan treatments in forests of the western United States. Frontiers in Ecology and the Environment 9: 271–277.

58. McKenzie D, Miller C, Falk DA (2011) Toward a Theory of Landscape Fire. In: McKenzie D, Miller C, Falk DA, editors. The Landscape Ecology of Fire: Springer Netherlands. 3–25.
59. Baker WL (2009) Fire ecology in Rocky Mountain Landscapes. Washington DC: Island Press. 605 p.
60. Landres PB, Morgan P, Swanson FJ (1999) Overview of the use of natural variability concepts in managing ecological systems. Ecological Applications 9: 1179–1188.
61. Wagenbrenner NS, Germino MJ, Lamb BK, Robichaud PR, Foltz RB (2013) Wind erosion from a sagebrush steppe burned by wildfire: Measurements of PM10 and total horizontal sediment flux. Aeolian Research 10: 25–36.
62. D'Antonio CM, Vitousek PM (1992) Biological invasions by exotic grasses, the grass fire cycle, and global change. Annual Review of Ecology and Systematics 23: 63–87.
63. Miller JD, Thode AE (2007) Quantifying burn severity in a heterogeneous landscape with a relative version of the delta Normalized Burn Ratio (dNBR). Remote Sensing of Environment 109: 66–80.
64. Norton J, Glenn N, Germino M, Weber K, Seefeldt S (2009) Relative suitability of indices derived from Landsat ETM+ and SPOT 5 for detecting fire severity in sagebrush steppe. International Journal of Applied Earth Observation and Geoinformation 11: 360–367.
65. Heyerdahl EK, Brubaker LB, Agee JK (2001) Spatial controls of historical fire regimes: A multiscale example from the interior west, USA. Ecology 82: 660–678.
66. Batllori E, Miller C, Parisien M-A, Parks SA, Moritz MA (2014). Is US climatic diversity well represented within the existing federal protection network? Ecological Applications in press.
67. Littell JS, Oneil EE, McKenzie D, Hicke JA, Lutz JA, et al. (2010) Forest ecosystems, disturbance, and climatic change in Washington State, USA. Climatic Change 102: 129–158.

17

The Role of Wildfire, Prescribed Fire, and Mountain Pine Beetle Infestations on the Population Dynamics of Black-Backed Woodpeckers in the Black Hills, South Dakota

Christopher T. Rota[1]*, Joshua J. Millspaugh[1], Mark A. Rumble[2], Chad P. Lehman[3], Dylan C. Kesler[1]

1 Department of Fisheries and Wildlife Sciences, University of Missouri, Columbia, Missouri, United States of America, **2** Forest and Grassland Research Laboratory, United States Forest Service Rocky Mountain Research Station, Rapid City, South Dakota, United States of America, **3** Custer State Park, South Dakota Department of Game, Fish, and Parks, Custer, South Dakota, United States of America

Abstract

Wildfire and mountain pine beetle infestations are naturally occurring disturbances in western North American forests. Black-backed woodpeckers (*Picoides arcticus*) are emblematic of the role these disturbances play in creating wildlife habitat, since they are strongly associated with recently-killed forests. However, management practices aimed at reducing the economic impact of natural disturbances can result in habitat loss for this species. Although black-backed woodpeckers occupy habitats created by wildfire, prescribed fire, and mountain pine beetle infestations, the relative value of these habitats remains unknown. We studied habitat-specific adult and juvenile survival probabilities and reproductive rates between April 2008 and August 2012 in the Black Hills, South Dakota. We estimated habitat-specific adult and juvenile survival probability with Bayesian multi-state models and habitat-specific reproductive success with Bayesian nest survival models. We calculated asymptotic population growth rates from estimated demographic rates with matrix projection models. Adult and juvenile survival and nest success were highest in habitat created by summer wildfire, intermediate in MPB infestations, and lowest in habitat created by fall prescribed fire. Mean posterior distributions of population growth rates indicated growing populations in habitat created by summer wildfire and declining populations in fall prescribed fire and mountain pine beetle infestations. Our finding that population growth rates were positive only in habitat created by summer wildfire underscores the need to maintain early post-wildfire habitat across the landscape. The lower growth rates in fall prescribed fire and MPB infestations may be attributed to differences in predator communities and food resources relative to summer wildfire.

Editor: Marco Festa-Bianchet, Université de Sherbrooke, Canada

Funding: Funding for this project was provided by the United States Forest Service Rocky Mountain Research Station agreement 07-JV-11221609-211 and the National Fire Plan agreement 10-JV-11221632-178. The South Dakota Department of Game, Fish and Parks provided funding through agreement number 08-CO-11221632-111 and the Wildlife Diversity Small Grants Program. This study was also made possible through State Wildlife Grant T-39-R-1, Study #2439, provided by the South Dakota Department of Game, Fish, and Parks. CTR was also supported by the Missouri Chapter of the Wildlife Society, the University of Missouri ORG, the James D.Chambers Memorial Scholarship, the Judy Southern Fellowship, and a TransWorld Airlines Scholarship administered by the University of Missouri. The funders had no role in study design, data collection and analysis, decision to publish, or preparation of the manuscript.

Competing Interests: A TransWorld Airlines scholarship administered by the University of Missouri was used in this study. There are no patents, products in development or marketed products to declare. This does not alter the author's adherence to all the PLOS ONE policies on sharing data and materials.

* E-mail: rotact@missouri.edu

Introduction

Western North American forests are shaped by natural disturbances. From small-scale canopy gaps to stand-replacing fires, disturbances are an important source of heterogeneity and species diversity [1]. Wildfires and mountain pine beetle (*Dentroctonus ponderosae*, hereafter MPB) infestations are disturbances of particular interest to managers because of their widespread occurrence and economic impacts. Wildfires burned an average of 2.7 million ha annually between 2001 and 2011 [2]. Many species benefit from wildfires, from xylophagous insects that reproduce in dead and dying trees [3] to various ungulates that benefit from improved forage production [4], [5]. Widespread MPB infestations occur irregularly in western forests, though eruptions impact millions of hectares and last for several years [6], [7]. Benefits of MPB infestations include a superabundant food resource (beetle larvae) that is exploited by many species [8], [9] and wildlife habitat in the form of standing dead trees [7].

Black-backed woodpeckers (*Picoides arcticus*) are emblematic of the role these natural disturbances play in creating wildlife habitat. Black-backed woodpeckers are associated with habitat created by wildfire [10–14], prescribed fire [15], and MPB infestations [16–18]. Although black-backed woodpeckers are known to occupy undisturbed forests [19–21], densities in such habitat is low relative to recently burned forests [11], [20], and woodpeckers may avoid undisturbed forest altogether if recently burned forests are in close proximity [22]. Despite the importance of recently disturbed forests to black-backed woodpeckers, wildfire and MPB infestations reduce timber value and have historically been considered undesirable. Considerable effort is thus put into preventing or mitigating the effects of these disturbances through fire suppression, post-fire salvage logging, and sanitation logging, which may contribute to habitat loss for black-backed woodpeckers [23], [24].

As a result, black-backed woodpeckers are considered a sensitive species in Region 2 of the U.S. Forest Service, a species of greatest conservation concern by the State of South Dakota [25], and have recently been petitioned for listing under the Endangered Species Act [26].

Identifying effective conservation strategies requires understanding the relative value of wildfire, prescribed fire, and MPB infestations to black-backed woodpeckers. However, considerable uncertainty exists regarding the relative value of these disturbances. Many authors consider black-backed woodpeckers a fire-dependent species [12–14], [22], [27], [28] that rely on moderate or high severity burns [29], [30]. Despite potential benefits to some wildlife species, forest managers often focus on reducing the incidence of high severity wildfires. Prescribed fire is one tool used to meet this objective [31], but prescribed fire can differ from wildfire in important ways. For example, prescribed fires often burn at low severity, while wildfires often burn at mixed or high severity. Additionally, forests are typically treated with prescribed fire during spring or fall, when fires are easier to control [32], while wildfires often burn during summer months [33]. Such differences in timing can impact post-fire arthropod communities [34], [35], potentially affecting food resources for black-backed woodpeckers. Therefore, while habitat created by prescribed fire may appear similar to habitat created by wildfire, the value of prescribed fire to black-backed woodpeckers remains unknown.

Although typically associated with post-fire habitat, black-backed woodpeckers are also attracted to MPB infestations. Black-backed woodpeckers use MPB infestations in lodgepole pine (*Pinus contorta*) forests in the Cascade Mountains of Oregon [16] and ponderosa pine (*Pinus ponderosa*) forests in the Black Hills of South Dakota [17], [18]. However, use of MPB infestations is not uniform across their range. For example, black-backed woodpeckers were rarely detected in MPB infestations in lodgepole pine forests in the northern Rocky Mountains [36] and in MPB infested lodgepole pine/Douglas fir (*Pseudotsuga menziesii*) forests in British Columbia [9]. Such apparent discrepancies cause uncertainty regarding the value of MPB infestations in various forest types to black-backed woodpeckers.

We evaluated demographic rates of black-backed woodpeckers in habitats created by summer wildfire, fall prescribed fire, and MPB infestations in the Black Hills, South Dakota. We first estimated adult and juvenile survival and reproductive rates of black-backed woodpeckers occupying each of these disturbance types. We then derived habitat-specific population growth rates as a function of demographic parameters. By evaluating habitat-specific population growth rates, we clarify the relative value of each of these disturbances for black-backed woodpeckers in the Black Hills.

Materials and Methods

Study Sites

This study occurred in the Black Hills, South Dakota at study sites representing habitat created by wildfire, prescribed fire, and MPB infestations (Table 1). Study sites were selected by first identifying forest patches that were burned by wildfire, burned by prescribed fire, or infested with MPBs. Potential study sites were then searched for signs of woodpeckers, and those study sites where woodpeckers were found were included in the study. Site selection was thus opportunistic because forest patches needed to be both disturbed and occupied by black-backed woodpeckers. Because of the opportunistic nature of site selection, the eastern-most study sites were burned by wildfire and the western-most study sites were infested with MPBs (Table 1). We do not believe such placement affected our results because the Black Hills National Forest is a maximum of 65 miles wide, and we are not aware of any east-west gradients that would influence black-backed woodpecker population dynamics. All wildfire sites burned during June or July (hereafter we use wildfire and summer wildfire synonymously; see Table 1 for the year each wildfire burned) and all prescribed fire sites burned during September or October (hereafter we use prescribed fire and fall prescribed fire synonymously; see Table 1 for the year each prescribed fire burned). All study sites were predominately monotypic stands of ponderosa pine, with occasional quaking aspen (*Populus tremuloides*), paper birch (*Betula papyrifera*), and white spruce (*Picea glauca*) [37]. All prescribed fire and wildfire study sites contained trees that burned at low, moderate, and high severity, although the proportion of trees burned at each severity category varied by study site. All MPB study sites contained trees that had been infested <1 year, 1–2 years, and >2 years. All study sites also contained live trees unaffected by wildfire or MPB infestations. Field work began in April 2008 and continued year-round through August 2011. Additional field work at prescribed fire study sites occurred from May through August 2012.

Capture and Radio-telemetry

We used very high frequency radio-transmitters (Advanced Telemetry Systems, Isanti, Minnesota, USA; Lotek Wireless Inc., Newmarket, Ontario, Canada) to collect survival data from adult and juvenile black-backed woodpeckers. We initially targeted adult woodpeckers for capture by searching study sites for signs of woodpeckers. We then used mist nets, hoop nets, and netguns [38] to capture adult woodpeckers. Once captured, we weighed all adults and attached a 3.0–3.3 g transmitter [39]. Adult black-backed woodpeckers weighed an average of 75 g (SD = 5 g), so transmitters weighed <5% of an average adult bird's mass [40]. We also marked each bird with a unique combination of colored leg bands and a uniquely numbered US Fish and Wildlife Service (USFWS) aluminum leg band. As radio-transmitter batteries failed, we attempted to recapture marked individuals and replace transmitters. We supplemented recaptured birds with unmarked birds that were captured opportunistically. Adult woodpeckers with active transmitters were relocated at least once per month, though most woodpeckers were relocated more frequently (mean number of telemetry locations per month = 6).

We captured black-backed woodpecker nestlings at the nest cavity with a 'chick-snagging' device and by accessing the nest with a hole-saw [41]. We constructed chick-snagging devices by looping fishing line through 1.27 cm plastic tubing. At approximately 3-days post-hatching, we captured nestlings with the chick-snagging device by entangling them in the looped fishing line. All nestlings captured with the chick-snagging device were fitted with a unique combination of colored leg bands and a uniquely numbered USFWS aluminum leg band but were not fitted with radio-transmitters. We only used the chick-snagging device during the 2009 breeding season. We captured nestlings during all other years by accessing the cavity with a 7.62 cm hole-saw approximately 3 days prior to fledging. All nestlings captured with the hole-saw method were weighed and fitted with a unique combination of colored leg bands and a uniquely numbered USFWS aluminum leg band. After capture with the hole saw, nestlings were placed back in the cavity and the access hole was plugged. During the 2010 breeding season, all nestlings ($n = 25$) were fitted with a 2.2 g transmitter [39]. During the 2011 breeding season, one randomly selected nestling from each nest was fitted with a 2.2 g transmitter until all available transmitters were used ($n = 6$). Nestlings weighed an average of 57 g (SD = 10 g) and transmitters weighed <5% of

Table 1. Study sites used to evaluate Demography of black-backed woodpeckers in the Black Hills, South Dakota, USA.

Site	Habitat	Coordinates	Size (Ha)[a]	Month/year disturbed[b]	Years included in study
Ricco	Wildfire	44°13′N, 103°25′W	1,602	July 2005	2008, 2010
Box Elder	Wildfire	44°9′N, 103°24′W	129	July 2007	2008, 2009
4-Mile	Wildfire	43°41′N, 103°26′W	955	June 2007	2008–2011
Bullock	Rx fire	44°0′N, 103°30′W	486	Sept. 2008	2010–2012
Bitter	Rx fire	43°58′N, 103°26′W	304	Oct. 2010	2012
Headquarters West	Rx fire	43°34′N, 103°30′W	255	Sept. 2009	2011
American Elk	Rx fire	43°61′N, 103°49′W	1376	Oct. 2010	2012
Norbeck	MPB	43°50′N, 103°30′W	>213[c]	1998	2008
Bear Mountain	MPB	43°51′N, 103°45′W	>48[c]	Before 1995	2008–2011
East Slate Creek	MPB	43°58′N, 103°44′W	>1,303[d]	Before 1995	2008–2011
Deerfield Lake	MPB	44°00′N, 103°49′W	>169[c]	Before 1995	2008
Medicine Mountain	MPB	43°52′N, 103°42′W	>1,748[d]	Before 1995	2009–2011

[a] Size of MPB infestations calculated from FHP Aerial Detection Surveys, available at <http://www.fs.usda.gov/detail/r2/forest-grasslandhealth/?cid=fsbdev3_041629> (accessed Feb. 13, 2013). This is an estimate of the minimum total area impacted by MPBs in each study site in a given year.
[b] The first year MPB infestations were detected in FHP Aerial Detection Surveys. Note there is no aerial detection data prior to 1995.
[c] Calculated from 2008 FHP Aerial Detection Survey.
[d] Calculated from 2010 FHP Aerial Detection Survey.

an average nestling's mass [40]. Fledglings with active transmitters were relocated as soon as possible after emerging from the nest and every-other week thereafter. Fledglings without transmitters were relocated by observing radio-marked parents provisioning individually-marked young.

Reproductive Success and Number of Young Fledged

We located nests by systematically searching study areas, by following birds to their nests, and opportunistically while collecting field data. We visited nests every 3–4 days during the 2009–2011 breeding seasons until either the nest failed or fledged young. Because of logistic constraints, nest visits were less regular during the 2008 and 2012 field seasons and the mean interval between nest visits was 4 days. We examined nest contents with a camera attached to a telescoping pole [42], [43]. We assumed the number of young fledged was the number of nestlings present in the nest during the last visit prior to fledging.

Ethics Statement

The majority of field work occurred on public land (Black Hills National Forest, Custer State Park, and Wind Cave National Park). We obtained permission to access Wind Cave National Park from the US National Park Service, permit number WICA-2010-SCI-0011. Two authors (MAR and CPL) were employees of the US Forest Service and Custer State Park, respectively. We therefore did not require specific permission to access the Black Hills National Forest or Custer State Park. We avoided private lands when possible, and obtained permission from landowners when necessary to access private land (e.g., when woodpeckers were located on private land).

All birds were individually marked and fitted with transmitters as quickly as possible to minimize stress. All research protocols were approved by the University of Missouri Animal Care and Use Committee, protocol numbers 4439 and 7205. Black-backed woodpeckers are protected under the Migratory Bird Act. Permission to capture and band black-backed woodpeckers was thus obtained from the US Department of the Interior, Federal Bird Banding Permit number 23574 and Federal Fish and Wildlife Permit number MB161645-0; and from the State of South Dakota Department of Game, Fish, and Parks, Free Scientific Collector's Permit License Number 10.

Estimating Adult and Juvenile Survival

We estimated adult and juvenile survival probabilities with Bayesian multi-state mark recapture models [44]. We considered a 3-state model where woodpeckers were classified as 'detected alive' (state 1), 'detected dead' (state 2), or 'undetected' (state 3). We assumed the following state-transition matrix:

$$\begin{array}{cc} & \text{stateattime } t \\ & \begin{array}{ccc} 1 & 2 & 3 \end{array} \\ \text{stateattime} t-1 \begin{array}{c} 1 \\ 2 \\ 3 \end{array} & \begin{pmatrix} d_{it}\phi_{it} & d_{it}(1-\phi_{it}) & (1-d_{it})(\phi_{it}+[1-\phi_{it}]) \\ 0 & 1 & 0 \\ d_{it}\phi_{it} & d_{it}(1-\phi_{it}) & (1-d_{it})(\phi_{it}+[1-\phi_{it}]) \end{pmatrix} \end{array}.$$

where woodpeckers transition from the state along the row to the state along the column with the associated cell probability. Detection probability, denoted d_{it}, is the probability woodpecker i is detected by the end of time step t. Survival probability, denoted φ_{it}, is the probability woodpecker i is alive at the end of time step t. We thus modeled a woodpecker's state at time t as a multinomial random variable:

$$\text{state}_t \sim \text{ltinomial}(M_{t-1},1).$$

where $\mathbf{M}_{t-1}$ is the row of the state transition matrix associated with a woodpecker's state at time t-1.

We modeled adult and juvenile detection and survival probability during each time step as a function of covariates. We assumed a 1-month time step for adult detection and survival probability and a 2-week time step for juvenile detection and survival probability. We modeled adult and juvenile detection probability as a function of whether woodpecker i had an active transmitter during time step t. Juveniles that were color-banded but did not receive a transmitter were coded as having an inactive transmitter. We modeled monthly adult survival probability as a function of sex, season (breeding, April – September; non-breeding, October – March), the habitat occupied at the end of each time step (wildfire, prescribed fire, and MPB infestation), and number of years post-fire. We modeled bi-weekly juvenile survival probability as a function of habitat, the number of years post-fire, and the number of two-week time intervals since fledging. Details of covariate models and prior distributions are described in Appendix S1. Juvenile detection histories began the last full two-week interval they remained in the nest and lasted through the end of March the following year. Thus, the first transition included the period when juveniles fledged from the nest. Note that woodpeckers originally captured as juveniles were assumed to survive as adults the 1st April post-fledging, which corresponds with the beginning of the breeding season.

Woodpeckers inhabiting established study sites were assigned the habitat category associated with that study site (e.g., woodpeckers at the 4-Mile study site were assigned to wildfire habitat; Table 1), even if woodpeckers were occasionally located in adjacent undisturbed forest (Fig. S1, Fig. S2). Habitat categories thus reflect the type of disturbed forest each woodpecker was primarily associated with, even if they occasionally occupied other forest types during a single time step. Such a coarse categorization of habitat association is appropriate given the coarse scale at which we modeled adult (1 month) and juvenile (2 weeks) survival probability. Note that woodpeckers occasionally dispersed from established study sites, sometimes moving from one disturbance type to another (e.g., moving from a wildfire to a prescribed fire). When such dispersal occurred, we modeled survival probability as a function of the habitat occupied at the end of the time step. We only observed woodpeckers dispersing to forest disturbed by wildfire, prescribed fire, or MPB infestations.

Estimating Nest Success and the Number of Young Fledged

We modeled daily nest survival probabilities with a Bayesian nest survival model [45]. We modeled daily survival probability as a function of habitat and time since fire (Appendix S1). Detection histories for all nests started the first day a nest was found. Detection histories for successful nests continued until the last date the nest was observed active [46] and detection histories for unsuccessful nests ended the first date nest failure was observed. We treated the days between the last observation of an active nest and the first observation of a failed nest as missing data and imputed the response variable.

We modeled the number of young fledged from each successful nest with a zero-truncated Poisson model. We modeled the number of young fledged per successful nest as a function of habitat (Appendix S1).

Estimating Habitat-specific Population Growth Rates and Sensitivity Analysis

We calculated habitat-specific annual population growth rates with a 2-stage female-based pre-breeding projection matrix [47]:

$$\mathbf{A}_h = \begin{bmatrix} m_h P_h^{\mathrm{juv}} & m_h P_h^{\mathrm{juv}} \\ P_h^{\mathrm{ad}} & P_h^{\mathrm{ad}} \end{bmatrix}.$$

where m_h is habitat specific fecundity (calculated as a function of nest success and the number of young fledged, Appendix S1), P_h^{juv} is the habitat-specific probability a juvenile will survive to the adult stage, and P_h^{ad} is habitat-specific annual adult female survival probability (Appendix S1). We used life stage simulation analysis (LSA) to evaluate sensitivity of population growth rates to variation in component demographic rates [48] (Appendix S1).

Results

Adult and Juvenile Survival

We censored detection histories of adult birds with inadequate data or whose mortality was possibly a result of capture. In total, we censored complete detection histories for 7 adults and censored mortality events for 4 adults that were recovered dead the first observation after a recapture. We estimated survival probabilities from detection histories of 140 adult woodpeckers, which consisted of 369, 406, and 164 cumulative months at risk of dying in habitat created by wildfire, MPB infestations, and prescribed fire, respectively. The average period between first and last detection of individual adults was 8 months (range = 1 to 40 months).

Adult survival probability was greatest in habitat created by wildfire, intermediate in habitat created by MPB infestations, and lowest in habitat created by prescribed fire (Table S1). For example, mean annual adult female survival probability was 0.75 (95% CI = [0.54, 0.91]) in 2-year post-wildfire habitat, 0.65 (95% CI = [0.45, 0.83]) in MPB infestations, and 0.50 (95% CI = [0.20, 0.79]) in 2-year post-prescribed fire habitat (Fig. 1a). Mean annual adult survival probability was nearly identical between sexes and declined slightly as time since fire increased (Fig. 1b). Finally, mean monthly survival probability was slightly greater during the breeding season relative to the non-breeding season. For example, mean monthly adult female survival probability in habitat created by MPB infestations was 0.97 (95% CI = [0.94, 0.99]) during the breeding season and 0.96 (95% CI = [0.92, 0.98]) during the non-breeding season.

We censored the detection history for 1 juvenile that died before fledging. After censoring, we estimated juvenile survival probabilities from detection histories of 72 woodpeckers, consisting of 119, 139, and 44 cumulative time steps (2-week intervals) at risk of dying in habitat created by wildfire, MPB infestations, and prescribed fire, respectively. The average period between first and last detection of individual juveniles was 10 weeks (range = 2 to 42 weeks).

Patterns of habitat-specific juvenile survival probability closely tracked patterns of habitat-specific adult survival probability. The mean probability of juvenile black-backed woodpeckers surviving to the adult stage class (defined as the probability of surviving 21 time steps [42 weeks] post-fledging, Appendix S1) was greatest in habitat created by wildfire, intermediate in habitat created by MPB infestations, lowest in habitat created by prescribed fire, and declined as the time since fire increased (Table S2). For example, the mean probability a juvenile survived to the adult stage class was 0.64 (95% CI = [0.25, 0.91]) in 2-year post-wildfire habitat, 0.35 (95% CI = [0.11, 0.62]) in MPB infestations, and 0.15 (95% CI = [0.00, 0.55]) in 2-year post-prescribed fire habitat (Fig. 2). Mean survival probability was lowest immediately after fledging, and increased as the number of time steps fledged increased (Table S2).

Nest Success and Number of Young Fledged

We suspected some nest failures were related to capture with the hoop net. We thus censored 14 capture-related nest failures, though we retained detection histories through the last day each nest was observed active. We estimated the probability of a nest successfully fledging at least one young from 95 nests: 40, 35, and 20 in habitat created by wildfire, MPB infestations and prescribed fire, respectively.

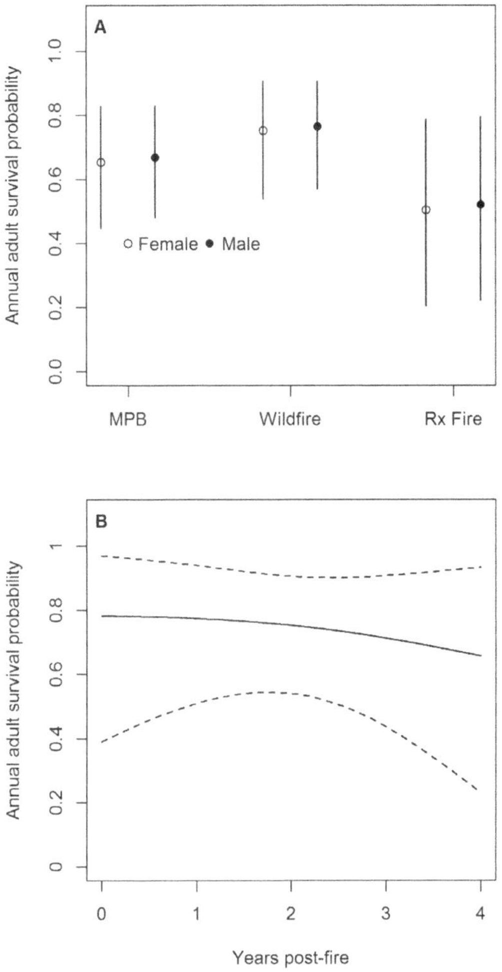

Figure 1. Estimated annual adult survival probabilities of black-backed woodpeckers in the Black Hills, South Dakota. (a) Mean posterior distribution and 95% CI of adult male and female black-backed woodpecker annual survival probabilities in habitat created by wildfire, prescribed (Rx) fire, and mountain pine beetle (MPB) infestations in the Black Hills, SD. Plots of annual survival probabilities assume habitat created by wildfire and prescribed fire are both 2 years post-fire. (b) Mean posterior distribution and 95% CI of annual survival probability as a function of time since wildfire. This figure assumes an adult female in habitat created by wildfire, but the trend is similar in habitat created by prescribed fire.

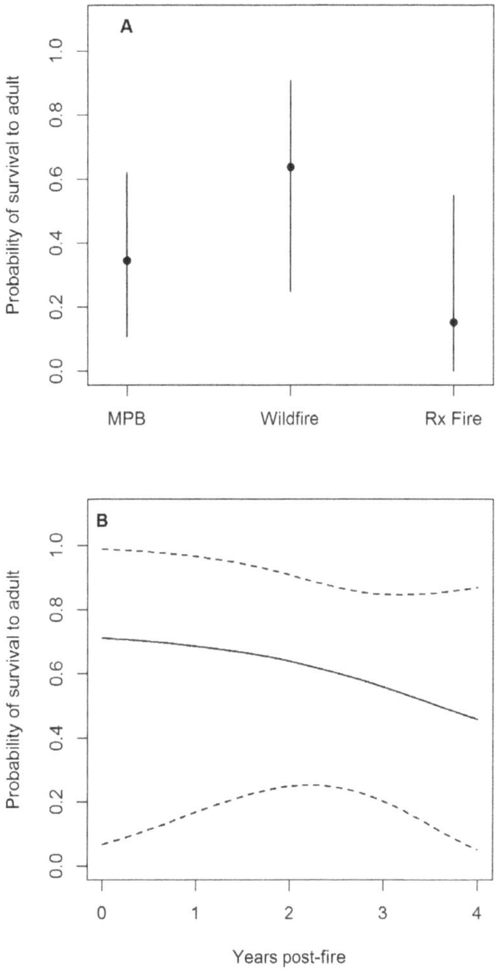

Figure 2. Estimated probabilities of juvenile black-backed woodpeckers surviving to adults in the Black Hills, South Dakota. (a) Mean posterior distribution and 95% CI of the probability juvenile black-backed woodpeckers survive to the adult stage-class in habitat created by wildfire, prescribed (Rx) fire, and mountain pine beetle (MPB) infestations in the Black Hills, SD. Plots of survival probabilities assume habitat created by wildfire and prescribed fire are both 2 years post-fire. (b) Mean posterior distribution and 95% CI of the probability a juvenile black-backed woodpecker survives to the adult stage-class as a function of time since wildfire. This figure assumes a juvenile in habitat created by wildfire, but the trend is similar in habitat created by prescribed fire.

Patterns of habitat-specific nest success followed similar patterns as habitat-specific adult and juvenile survival. The mean probability of a nest fledging at least one young was greatest in habitat created by wildfire, intermediate in habitat created by MPB infestations, lowest in habitat created by prescribed fire, and decreased as time since fire increased (Table S3). For example, the mean probability of successfully fledging at least 1 young was 0.72 (95% CI = [0.55, 0.86]) in 2-year post-wildfire habitat, 0.60 (95% CI = [0.40, 0.77]) in MPB infestations, and 0.45 (95% CI = [0.22, 0.68]) in 2-year post-prescribed fire habitat (Fig. 3).

We estimated the expected number of young fledged per successful nest from 50 successful nests: 23, 18, and 9 in habitat created by wildfire, MPB infestations, and prescribed fire, respectively. The expected number of young fledged per successful nest was greatest in habitat created by prescribed fire (mean fledged = 2.05, 95% CI = [1.44, 2.95]) and nearly identical between habitat created by wildfire (mean fledged = 1.80, 95% CI = [1.44, 2.28]) and MPB infestations (mean fledged = 1.81, 95% CI = [1.41, 2.35], Table S4).

Habitat-specific Population Growth Rates

Mean posterior distributions of population growth rates (λ) were positive only in habitat created by wildfire. For example, $\overline{\lambda} = 1.16$ in 2-year post-wildfire habitat and 84% of the posterior density of estimated population growth rate was >1 (Fig. 4b). Mean population growth rates were negative in habitat created by MPB infestations ($\overline{\lambda} = 0.84$) and 11% of the posterior density of estimated population growth rate was >1. Mean population growth rates were also negative in habitat created by prescribed fire. For example, $\overline{\lambda} = 0.57$ in 2-year post-prescribed fire habitat and <1% of the posterior density of estimated population growth rate was >1. Variation in adult and juvenile survival rates

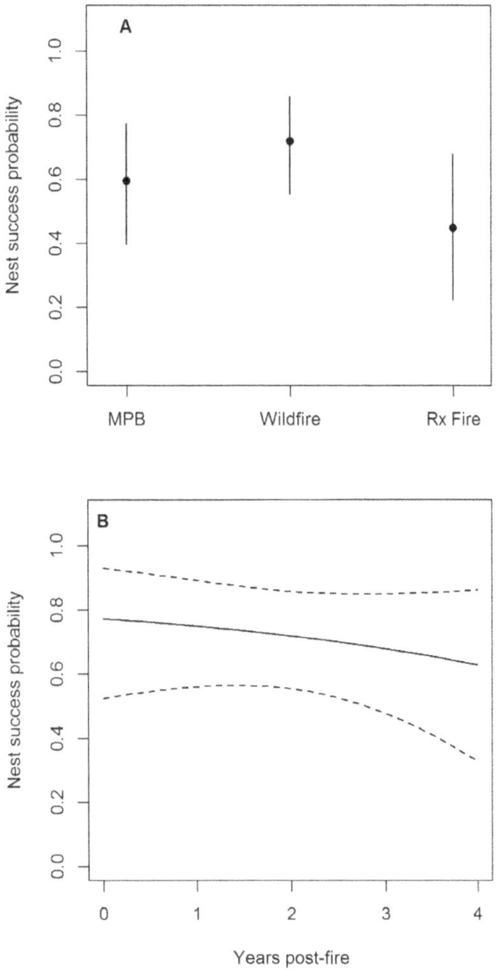

Figure 3. Estimated probabilities of black-backed woodpecker nests successfully fledging at least one young. (a) Mean posterior distribution and 95% credible intervals of the probability a black-backed woodpecker nest successfully fledged at least 1 young in habitat created by wildfire, prescribed (Rx) fire, and mountain pine beetle (MPB) infestations in the Black Hills, SD. This plot assumes nests in habitat created by wildfire and prescribed fire are both 2-years post-fire. (b) Mean posterior distribution and 95% credible intervals of the probability a nest successfully fledges at least 1 young as a function of time since fire. This figure assumes a nest in habitat created by wildfire, but the trend is similar in habitat created by prescribed fire.

consistently explained the most variation in population growth rates, while reproductive rates explained little variation in population growth rates (Table 2, Fig. S3).

We obtained reasonably distinct posterior distributions of asymptotic growth rates among habitats created by wildfire, prescribed fire, and MPB infestations despite broad and sometimes overlapping posterior distributions of individual demographic rates for two main reasons. First, annual adult survival probability, the probability a juvenile survived to an adult, and the probability of successfully fledging at least 1 young were all identically ranked across habitats. As a consequence, estimated asymptotic growth rates were highest in wildfire and lowest in prescribed fire, despite sometimes considerable uncertainty in component demographic rates. Second, asymptotic growth rates were highly sensitive to both adult and juvenile survival probabilities. As a consequence, small changes in estimated survival probabilities could mean the difference between a growing or a declining population.

Discussion

This study helps clarify the relative value of wildfire, prescribed fire, and MPB infestations to black-backed woodpeckers in the Black Hills, South Dakota. Mean population growth rates were positive only in habitat created by summer wildfire, while mean population growth rates were negative in habitats created by MPB infestations and fall prescribed fire. These findings support long-standing hypotheses that recently burned forests are population sources for black-backed woodpeckers [10]. Further, the negative population growth we observed in habitat created by fall prescribed fire indicate this management tool, under the conditions we evaluated, is not a viable substitute for summer wildfire.

Habitats created by wildfire and prescribed fire differed at our study sites in two primary ways. First, managers treated sites with prescribed fire in September and October, while wildfire sites burned during June and July. This difference in timing may affect post-fire arthropod communities. For example, some species of wood-boring beetles can detect compounds emitted from burning wood [49] and can rapidly colonize burned forest. These beetles may not be active during autumn months and may be unable to immediately colonize late-season burns (MAR, *personal observation*). Second, the prescribed fire study sites tended to be smaller in area, burn at lower severity, and thus kill fewer trees relative to the wildfire study sites, which may result in different predator communities between the two disturbance types [50].

Our sensitivity analysis provides insight into how these potential differences may affect population growth rates. In all habitats, population growth rates were most sensitive to changes in adult and juvenile survival. If the timing of prescribed fire affects wood-boring beetle abundance, this may affect the food resources available to black-backed woodpeckers, which may in turn impact survival rates. This may particularly affect juvenile survival, since fledglings rely on provisioning from adults for several weeks post-fledging. Survival rates may also be affected by potential differences in predator communities between habitats. For example, Northern Goshawks, a known predator of black-backed woodpeckers in the Black Hills [51], preferentially nest in closed canopy forest [52] and may be less abundant in severely burned forest relative to unburned forest or forest that burned at low severity.

We were surprised that mean population growth rates were negative in habitat created by MPB infestations, since woodpeckers readily occupy and successfully breed in such habitat [17], [18]. However, our finding is consistent with previous hypotheses that black-backed woodpeckers specialize on recently burned forests [30]. Furthermore, black-backed woodpeckers are much more likely to move to burned forest, relative to their availability, than to MPB infestations [51] and home range sizes in recently burned forests are much smaller than home range sizes in MPB infestations [53]. These findings may thus help explain regional differences in the use of MPB infestations. Black-backed woodpeckers are only documented using MPB infestations in isolated portions of their range, where recently burned forest may be relatively rare across the landscape and woodpeckers may face dispersal constraints (e.g., the limited extent of forested habitat in the Black Hills). In contrast, black-backed woodpeckers may rarely use MPB infestations in more contiguous portions of their range if there is a greater relative availability of recently burned forests and woodpeckers do not face such dispersal constraints.

Habitat created by MPB infestations likely harbors some value to black-backed woodpeckers, even if mean population growth rates in this habitat were negative. Indeed, 95% CIs of estimated

Table 2. Proportion of variation in black-backed woodpecker population growth rates explained by variation in component demographic rates.

	Variation Explained (r^2)[a]		
Demographic Rate	**Wildfire**	**MPB Infestation**	**Prescribed Fire**
Adult Survival	0.34	0.58	0.80
Juvenile Survival	0.48	0.31	0.17
Nest Success	0.08	0.06	0.01
No. Young Fledged	0.09	0.04	0.01

Life-stage simulation analysis was based on 300,000 random draws from the posterior distribution of each demographic rate.
[a] This table assumes 2-year post wildfire and prescribed fire.

growth rates overlapped 1, suggesting the potential for positive population growth in this habitat during some years. It is likely that mean population growth rates in MPB infestations are intermediate between early post-wildfire habitat and undisturbed forest, although additional research is needed to understand demographic rates of black-backed woodpeckers in undisturbed forests. Additionally, adult, juvenile, and nest survival all declined as burned forests aged and mean population growth rates in 4-year post-wildfire forests were similar to MPB infestations (Fig. 4d). Mountain pine beetle infestations may thus improve in relative value to black-backed woodpeckers as post-wildfire habitat ages and may help buffer population declines when recently burned forest is not available.

The differences in population growth rates between habitat created by wildfire and MPB infestations likely reflects historic disturbance patterns in the Black Hills. There were 7 discrete MPB outbreaks in the Black Hills during the 20th century, including the current infestation [6]. The extent of individual outbreaks can be larger than the total forest area burned in any given year. However, during most years, MPBs exist at 'endemic' levels, defined as <1 tree killed per 0.40 ha per year [6]. In contrast, there is a mean 16 year pre-settlement fire-return interval for ponderosa pine forests in Jewel Cave National Park, South Dakota [33]. Mean fire-return intervals reflect the average time between fires at any given location, though fires probably burned different portions of the Black Hills during most years. These fires likely burned at mixed severity, killing some trees while allowing others to live [54], [55].

We do not wish to completely discard the utility of either prescribed fire or MPB infestations as important disturbance agents for black-backed woodpeckers. Large, severe wildfires are not likely to gain widespread acceptance on public lands, and prescribed fire may be the only way to predictably introduce fire. Prescribed fire is a flexible management tool that can be applied in a variety of ways. For example, managers can vary factors including timing, severity, and extent. Further research is needed to determine the mechanisms leading to negative population growth of black-backed woodpeckers in habitat created by prescribed fire so forests can be treated in an appropriate manner. In particular, further research is needed to understand how the timing of prescribed fire may impact post-fire wood-boring beetle abundance, a primary food resource for black-backed woodpeckers [11].

Although our study evaluates demographic rates of black-backed woodpeckers over a range of disturbed forest conditions, black-backed woodpeckers are also known to use undisturbed forest across their range [19–21]. Little is known about demographic rates of black-backed woodpeckers in undisturbed forest conditions, though they are known to occur in very low densities in undisturbed forest in the Black Hills [20] and evidence suggests that black-backed woodpeckers may use undisturbed forests only when recently disturbed forest is not available [22]. Throughout our study, 1–5 year post-fire forest comprised approximately 1.0% and active MPB infestations comprised approximately 3.8% of the forested area, respectively, of the Black Hills [51]. Our dataset was comprised of 138 adult and 37 juvenile radio-marked woodpeckers, including permanent dispersal movements of 18 adult and 7 juvenile woodpeckers up to 60 kilometers away from established study sites [51], and all territories included an element of recently disturbed forest. Our failure to observe any territories without an element of wildfire, prescribed fire, or MPB infestations, particularly among dispersing woodpeckers, indicates undisturbed forests were rarely used and played only a minor role in population dynamics during the duration of our study.

Our study is the first to evaluate the demographic response of black-backed woodpeckers to a range of disturbance conditions. This and other studies are necessary to gain a better understanding of the importance of wildfire and other forest disturbances in maintaining viable black-backed woodpecker populations. Our results indicate this species is dependent on early post-wildfire habitat in the Black Hills, South Dakota, underscoring the importance of ensuring recently burned forest is present across the landscape.

Supporting Information

Figure S1 Home ranges of black-backed woodpeckers relative to the 4-Mile wildfire in Custer State Park, SD. The gray shaded polygon represents the extent of the 4-Mile wildfire, individual points represent coordinates of black-backed woodpecker locations obtained via radio-telemetry, and dashed lines represent 95% probability contours of black-backed woodpecker home ranges estimated using kernel density methods [53]. The 95% home range contours were calculated from the points of the same color. Telemetry locations represented in this figure were collected between April 1 2008 and March 31 2009. The white space outside the burn perimeter represents areas not burned by the 4-Mile wildfire.

Figure S2 Home ranges of black-backed woodpeckers relative to the East Slate Creek mountain pine beetle infestation. The green shaded polygons represent <1 year old mountain pine beetle (MPB) infestations as of autumn 2009 [56] and the red shaded polygons represent 1–2 year-old mountain pine beetle infestations as of autumn 2009 [57]. Individual points represent coordinates of black-backed woodpecker locations

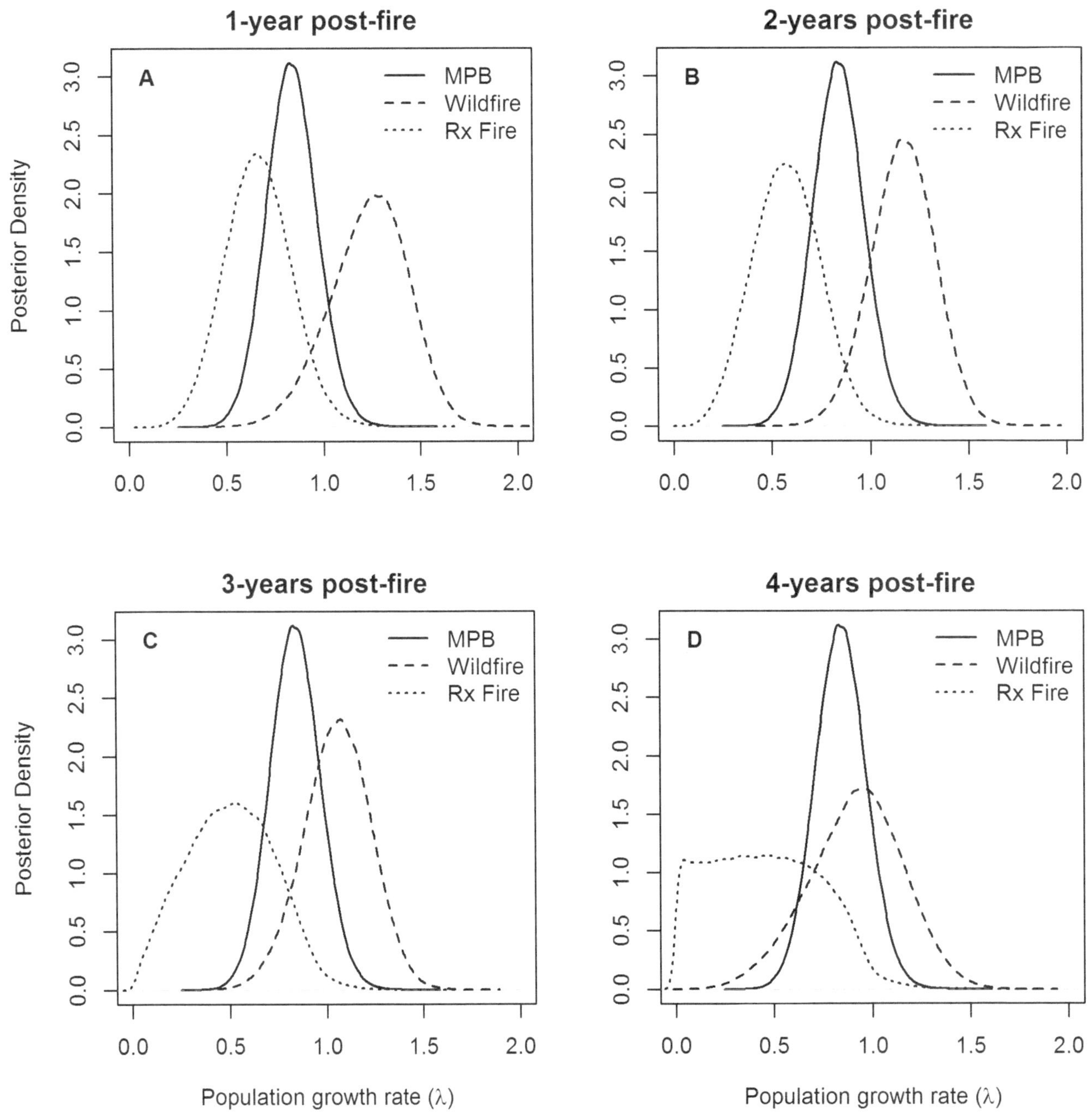

Figure 4. Posterior densities of habitat-specific population growth rates of black-backed woodpeckers in the Black Hills, SD. Plots represent (a) 1-year, (b) 2-years, (c) 3-years, and (d) 4-years post-fire in habitat created by wildfire and prescribed fire (Rx Fire). Note that the posterior density of population growth rates (λ) in mountain pine beetle (MPB) infestations remains constant across panels, since λ in MPB infestations was not estimated as a function of time post-disturbance.

obtained via radio-telemetry and dashed lines represent 95% probability contours of black-backed woodpecker home ranges estimated using kernel density methods [53]. The 95% home range contours were calculated from the points of the same color. Telemetry locations represented in this figure were collected between April 1 2009 and March 31 2010. The white space outside the shaded polygons represent forests not affected by 0–2 year-old MPB infestations as of autumn 2009.

Figure S3 Results from life stage simulation analysis evaluating sensitivity of black-backed woodpecker asymptotic population growth rates. Each point represents an estimate of asymptotic population growth rates (λ) corresponding to a random realization from the posterior distribution of A) annual adult survival probability, B) the probability a juvenile survives to an adult, C) the probability a nest successfully fledges at least 1 young, and D) the expected number of females fledged per successful nest. The coefficient of determiniation (r^2) is calculated by regressing λ against the corresponding demographic rate [48]. This analysis is based on 300,000 random draws from the posterior distribution of each demographic rate.

Table S1 Summary of posterior distributions of parameters included in the adult survival model.

Table S2 Summary of posterior distributions of parameters included in the juvenile survival.

Table S3 Summary of posterior distributions of parameters included in the nest survival model.

Table S4 Summary of posterior distributions of parameters included in the number fledged model.

Appendix S1 Statistical methods for modeling demographic rates of black-backed woodpeckers.

Acknowledgments

Disclaimer: The use of trade and company names is for the benefit of the reader; such use does not constitute an official endorsement or approval of any service or product by the US Department of Agriculture to the exclusion of others that may be suitable.

We thank Eileen Dowd-Stukel, Shelly Deisch, and Casey Mehls for support through their programs at the South Dakota Department of Game, Fish, and Parks. We thank M. Clawson, J. Dacey, B. Dickerson, M. Dolan, B. Furfey, K. Hansen, S. Hardaswick, M. Immel, W. Johnson, T. Juntti, S. Kolbe, R. Kummer-Bolie, N. Magliocco, E. Margenau, H. Mason, R. Mowry, B. Prochazka, B. Rota, H. Scott, L. Schulte-Welle, E. Seckinger, and C. Smithers for field assistance. We thank B. Muenchau and D. Roddy for assistance establishing Wind Cave National Park study sites. We thank W. Beatty, T. Bonnot, M. Ryan F. Thompson III, and C. Wikle for helpful insights on previous versions of this manuscript.

Author Contributions

Conceived and designed the experiments: CR JM CL MR DK. Performed the experiments: CR CL MR DK. Analyzed the data: CR. Contributed reagents/materials/analysis tools: JM MR DK. Wrote the paper: CR JM CL MR.

References

1. Connell JH (1978) Diversity in tropical rain forests and coral reefs. Science 199: 1302–1310.
2. National Interagency Fire Council (2013) Total wildland fires and acres (1960–2009). Boise: National Interagency Fire Council. Available: http://www.nifc.gov/fireInfo/fireInfo_stats_totalFires.html. Accessed 2013 Jan.
3. Saint-Germain M, Drapeau P, Hebert C (2004) Landscape-scale habitat selection patterns of Monochamus scutellatus (Coleoptera: Cerambycidae) in a recently burned black spruce forest. Environ Entomol 33: 1703–1710.
4. Keay JA, Peek JM (1980) Relationships between fires and winter habitat of deer in Idaho. J Wildl Manage 44: 372–380.
5. Zimmerman TJ, Jenks JA, Leslie DM (2006) Gastrointestinal morphology of female white-tailed and mule deer: Effects of fire, reproduction, and feeding type. J Mammal 87: 598–605.
6. Allen KK, Long DF, Cross FJ (2001) Evaluation of mountain pine beetle activity on the Black Hills National Forest, Biological Evaluation R2-002-02. Rapid City: U.S. Forest Service. 29 p.
7. Raffa KF, Aukema BH, Bentz BJ, Carroll AL, Hicke JA, et al. (2008) Cross-scale drivers of natural disturbances prone to anthropogenic amplification: The dynamics of bark beetle eruptions. Bioscience 58: 501–517.
8. Drever MC, Goheen JR, Martin K (2009) Species-energy theory, pulsed resources, and regulation of avian richness during a mountain pine beetle outbreak. Ecology 90: 1095–1105.
9. Edworthy AB, Drever MC, Martin K (2011) Woodpeckers increase in abundance but maintain fecundity in response to an outbreak of mountain pine bark beetles. For Ecol Manage 261: 203–210.
10. Hutto RL (1995) Composition of bird communities following stand-replacement fires in Northern Rocky Mountain (USA) conifer forests. Conserv Biol 9: 1041–1058.
11. Murphy EC, Lehnhausen WA (1998) Density and foraging ecology of woodpeckers following a stand-replacement fire. J Wildl Manage 62: 1359–1372.
12. Dudley JG, Saab VA (2007) Home range size of black-backed woodpeckers in burned forests of southwestern Idaho. West N Am Nat 67: 593–600.
13. Nappi A, Drapeau P (2009) Reproductive success of the black-backed woodpecker (*Picoides arcticus*) in burned boreal forests: Are burns source habitats? Biol Conserv 142: 1381–1391.
14. Dudley JG, Saab VA, Hollenbeck JP (2012) Foraging-habitat selection of Black-backed Woodpeckers in forest burns of southwestern Idaho. Condor 114: 348–357.
15. Russell RE, Royle JA, Saab VA, Lehmkuhl JF, Block WM, et al. (2009) Modeling the effects of environmental disturbance on wildlife communities: Avian responses to prescribed fire. Ecol Appl 19: 1253–1263.
16. Goggans R, Dixon RD, Seminara CL (1989) Habitat use by three-toed and black-backed woodpeckers, Technical report #87-3-02. Oregon: Oregon Department of Fish and Wildlife & Deschutes National Forest. 43 p.
17. Bonnot TW, Rumble MA, Millspaugh JJ (2008) Nest success of black-backed woodpeckers in forests with mountain pine beetle outbreaks in the Black Hills, South Dakota. Condor 110: 450–457.
18. Bonnot TW, Millspaugh JJ, Rumble MA (2009) Multi-scale nest-site selection by black-backed woodpeckers in outbreaks of mountain pine beetles. For Ecol Manage 259: 220–228.
19. Tremblay JA, Ibarzabal J, Dussault C, Savard JPL (2009) Habitat requirements of breeding black-backed woodpeckers (Picoides arcticus) in managed, unburned boreal forest. Avian Econservation and Ecology 1: 2. Available http://www.ace-eco.org/vol4/iss1/art2/. Accessed 2010 Feb 3.
20. Mohren SR (2002) Habitat evaluation and density estimates for the black-backed woodpecker (Picoides arcticus) and northern three-toed woodpecker (Picoides tridactylus) in the Black Hills National Forest. MS thesis. Laramie, University of Wyoming. 110 p.
21. Setterington MA, Thompson ID, Montevecchi WA (2000) Woodpecker abundance and habitat use in mature balsam fir forests in Newfoundland. J Wildl Manage 64: 335–345.
22. Hoyt JS, Hannon SJ (2002) Habitat associations of black-backed and three-toed woodpeckers in the boreal forest of Alberta. Can J For Res 32: 1881–1888.
23. Hutto RL, Gallo SM (2006) The effects of postfire salvage logging on cavity-nesting birds. Condor 108: 817–831.
24. Saab VA, Russell RE, Dudley JG (2007) Nest densities of cavity-nesting birds in relation to postfire salvage logging and time since wildfire. Condor 109: 97–108.
25. South Dakota Department of Game, Fish, and Parks (2006) South Dakota comprehensive wildlife conservation plan, Report 2006–09. Pierre: South Dakota Department of Game, Fish and Parks.
26. Federal Register (2013) Endangered and threatened wildlife and plants; 90-Day finding on a petition to list two populations of Black-Backed Woodpecker as endangered or threatened. Fed Regist 68: 21086–21097.
27. Dixon RD, Saab VA (2000) Black-backed Wooodpecker (*Picoides arcticus*). In Poole A, editor. The Birds of North America Online. Ithaca: Cornell Lab of Ornithology. Available: http://bna.birds.cornell.edu/bna/species/509. Accessed 2013 May.
28. Nappi A, Drapeau P (2011) Pre-fire forest conditions and fire severity as determinants of the quality of burned forests for deadwood-dependent species: the case of the black-backed woodpecker. Can J For Res 41: 994–1003.
29. Hanson CT, North MP (2008) Postfire woodpecker foraging in salvage-logged and unlogged forests of the Sierra Nevada. Condor 110: 777–782.
30. Hutto RL (2008) The ecological importance of severe wildfires: some like it hot. Ecol Appl 18: 1827–1834.
31. Graham RT, McCaffrey S, Jain TB (2004) Science Basis for Changing Forest Structure to Modify Wildfire Behavior and Severity, RMRS-GTR-120. Ogden: USDA Forest Service. 46 p.
32. Knapp EE, Estes BL, Skinner CN (2009) Ecological effects of prescribed fire season: A literature review and synthesis for managers, PSW-GTR-224. Redding: USDA Forest Service. 81 p.
33. Brown PM, Hull Sieg C (1996) Fire history in interior ponderosa pine communities of the Black Hills, South Dakota, USA. Int J Wildland Fire 6: 97–105.
34. Ferrenberg SM, Schwilk DW, Knapp EE, Groth E, Keeley JE (2006) Fire decreases arthropod abundance but increases diversity: early and late season prescribed fire effects in a Sierra Nevada mixed-conifer forest. Fire Ecol 2: 79–102.
35. Johnson SD, Horn KC, Savage AM, Windhager S, Simmons MT, et al. (2008) Timing of prescribed burns affects abundance and composition of arthropods in the Texas hill country. Southwest Nat 53: 137–145.
36. Cilimburg A, Smucker K, Hutto D (2006) Northern Region Landbird Monitoring Program 2006 final report: Black-backed Woodpeckers and the bird community in beetle outbreak areas. Missoula: Avian Science Center. Available: http://avianscience.dbs.umt.edu/projects/documents/beetle_report_final.pdf. Accessed 2013 May.

37. Hoffman GR, Alexander RR (1987) Forest vegetation of the Black Hills National Forest of South Dakota and Wyoming: a habitat type classification, Research Paper RM-RP-276. Fort Collins: USDA Forest Service. 48 p.
38. Lehman CP, Kesler DC, Rota CT, Rumble MA, Seckinger EM, et al. (2011) Netguns: a technique for capturing Black-backed Woodpeckers. J Field Ornithol 82: 430–435.
39. Rappole JH, Tipton AR (1991) New harness design for attachment of radio transmitters to small passerines. J Field Ornithol 62: 335–337.
40. Fair JM, Paul E, Jones J (2010) Guidelines to the use of wild birds in research. Washington: Ornithological Council. 210 p.
41. Ibarzabal J, Tremblay JA (2006) The hole saw method for accessing woodpecker nestlings during developmental studies. Ann Zool Fennici 43: 235–238.
42. Proudfoot GA (1996) Miniature video-board camera used to inspect natural and artificial nest cavities. Wildl Soc Bull 24: 528–530.
43. Huebner DP, Hurteau SR (2007) An economical wireless cavity-nest viewer. J Field Ornithol 78: 87–92.
44. Kéry M, Schaub M (2012) Bayesian population analysis using WinBUGS: A hierarchical perspective. Waltham: Academic Press. 535 p.
45. Dinsmore SJ, White GC, Knopf FL (2002) Advanced techniques for modeling avian nest survival. Ecology 83: 3476–3488.
46. Stanley T (2004) When should Mayfield model data be discarded? Wilson Bull 116: 267–269.
47. Caswell H (2001) Matrix population models: Construction, analysis, and interpretation. Sunderland: Sinaur Associates. 722 p.
48. Wisdom MJ, Mills LS, Doak DF (2000) Life stage simulation analysis: Estimating vital-rate effects on population growth for conservation. Ecology 81: 628–641.
49. Schütz S, Weissbecker B, Hummel HE, Apel K, Schmitz H, et al. (1999) Insect antenna as a smoke detector. Nature 398: 298–299.
50. Smucker KM, Hutto RL, Steele BM (2005) Changes in bird abundance after wildfire: Importance of fire severity and time since fire. Ecol Appl 15: 1535–1549.
51. Rota CT (2013) Not all forests are disturbed equally: Population dynamics and resource selection of Black-backed Woodpeckers in the Black Hills, South Dakota. PhD dissertation. Columbia: University of Missouri. 146 p.
52. Squires JR, Reynolds RT (1997) Northern Goshawk (*Accipiter gentilis*). In Poole A, editor. The Birds of North America Online. Ithaca: Cornell Lab of Ornithology. Available: http://bna.birds.cornell.edu/bna/species/298. Accessed 2013 May.
53. Rota CT, Rumble MA, Millspaugh JJ, Lehman CP, Kesler DK (2014) Space-use and habitat associations of Black-backed Woodpeckers (Picoides arcticus) occupying recently disturbed forests in the Black Hills, South Dakota. For Ecol Manage 313: 161–168.
54. Shinneman D, Baker W (1997) Nonequilibrium dynamics between catastrophic disturbances and old-growth forests in Ponderosa Pine landscapes of the Black Hills. Conserv Biol 11: 1276–1288.
55. Baker WL, Veblen TT, Sherriff RL (2007) Fire, fuels and restoration of ponderosa pine-Douglas fir forests in the Rocky Mountains, USA. J Biogeogr 34: 251–269.
56. USDA Forest Service, Rocky Mountain Region, Forest Health Management (2010) Annual aerial detection overview survey. Available: http://www.fs.usda.gov/detail/r2/forest-grasslandhealth/?cid=fsbdev3_041629. Accessed 2013 Feb 13.
57. USDA Forest Service, Rocky Mountain Region, Forest Health Management (2009) Annual aerial detection overview survey. Available: http://www.fs.usda.gov/detail/r2/forest-grasslandhealth/?cid=fsbdev3_041629. Accessed 2013 Feb 13.

18

Modelling Fire Frequency in a Cerrado Savanna Protected Area

Alfredo C. Pereira Júnior[1]*, Sofia L. J. Oliveira[2], José M. C. Pereira[2], Maria Antónia Amaral Turkman[3]

1 Divisão de Sensoriamento Remoto, Instituto Nacional de Pesquisas Espaciais, São José dos Campos, Brazil, **2** Centro de Estudos Florestais, Instituto Superior de Agronomia, Universidade de Lisboa, Lisboa, Portugal, **3** Centro de Estatística e Aplicações, Faculdade de Ciências, Universidade de Lisboa, Lisboa, Portugal

Abstract

Covering almost a quarter of Brazil, the Cerrado is the world's most biologically rich tropical savanna. Fire is an integral part of the Cerrado but current land use and agricultural practices have been changing fire regimes, with undesirable consequences for the preservation of biodiversity. In this study, fire frequency and fire return intervals were modelled over a 12-year time series (1997–2008) for the Jalapão State Park, a protected area in the north of the Cerrado, based on burned area maps derived from Landsat imagery. Burned areas were classified using object based image analysis. Fire data were modelled with the discrete lognormal model and the estimated parameters were used to calculate fire interval, fire survival and hazard of burning distributions, for seven major land cover types. Over the study period, an area equivalent to four times the size of Jalapão State Park burned and the mean annual area burned was 34%. Median fire intervals were generally short, ranging from three to six years. Shrub savannas had the shortest fire intervals, and dense woodlands the longest. Because fires in the Cerrado are strongly responsive to fuel age in the first three to four years following a fire, early dry season patch mosaic burning may be used to reduce the extent of area burned and the severity of fire effects.

Editor: Ben Bond-Lamberty, DOE Pacific Northwest National Laboratory, United States of America

Funding: This research was funded by the FCT - Fundação para a Ciência e a Tecnologia, through scholarship SFRH/BPD/48090/2008. SLJO was funded by FCT through scholarship SFRH/BD/44846/2008. MAAT was supported by National Funds through FCT projects PEst-OE/MAT/UI0006/2011 and PTDC/MAT/118335/2010. The funders had no role in study design, data collection and analysis, decision to publish, or preparation of the manuscript.

Competing Interests: The authors have declared that no competing interests exist.

* Email: alfredo@dsr.inpe.br

Introduction

The Cerrado, the Brazilian tropical savanna, is a complex vegetation formation characterized by a mosaic of physiognomies ranging from pure grasslands through woodland to closed forest [1–3] (Figure 1). It is a biodiversity hotspot that covers ≈ two million km^2 of central Brazil (about a quarter of the country area) and the second largest biome in South America, after the Amazon [4,5]. The Cerrado is a fire-prone biome [1,2,6,7] and anthropogenic wildfires have been playing an important role since at least the mid Holocene, from 6000 BP [8,9]. Indigenous people used fire to manage their resources and the decision to burn was based on several ecological indicators, denoting a clear social control [10]. Yet, as at many other places in the world, current land use and agricultural practices have considerably changed fire regimes in the Cerrado, and at least 49% of the vegetation has been converted to extensive monocultures and pastures [11]. Furthermore, 80% of the biome has been disturbed and most of the natural remnants are now located in the northern limit of its distribution [4,5]. This loss is twice the area lost in the Amazon forest and only 7.4% of the biome is inside protected areas [11].

Farmers use areas of the Cerrado for intensive farming, mostly soybean, but also corn, rice, cotton and sugarcane, and grass formations as natural pasture [2,5]. In the dry season, this grassy herbaceous biomass becomes especially dry and very flammable and it is burned to promote fresh growth in natural pastures. Crop residues are also burned to clear harvest in preparation for planting at the beginning of the wet season [1,2,7,12,13]. The current fire regime is characterised by extensive and frequent anthropogenic fires, that occur most often during the middle and latter months of the 6-month dry season (April to September), under severe fire weather conditions [1,2,6,7,12–14]. Fire return intervals are usually very short, from one to four years [1,2,7,12,13]. Such a fire regime favours herbaceous species and encourages the maintenance of the open Cerrado physiognomies, leading to soil degradation, overall biodiversity loss and an increased greenhouse gas emissions [1,2,7,12,13].

However, because the Cerrado is a fire-prone biome, periodical fires are important for the creation and maintenance of landscape structure, composition, function and ecological integrity [15–17]. Fire is an important land management tool in tropical savannas of Australia [18,19] and Africa [20,21], but its use to protect and promote biodiversity is very limited in the Cerrado, with the exception of Emas National Park [14]. Analysis of tropical savanna fire regimes in the presence of active prescribed burning programs [22,23] indicates a reduction in intensity, size, and wildfire hazard with favorable ecological and greenhouse gas emission implications. Progress has been made in recent years to understand fire regimes and their effects in the Cerrado [1,12–14,24–25]. Satellite remote sensing makes it possible to survey recent burned area history over large ecosystems, providing useful data to analyse features of the savanna fire regime, such as fire frequency. Mapping burned area from satellite imagery has been undertaken since the 1980s in the Brazilian Amazon [26–30] and in the Cerrado [31–34].

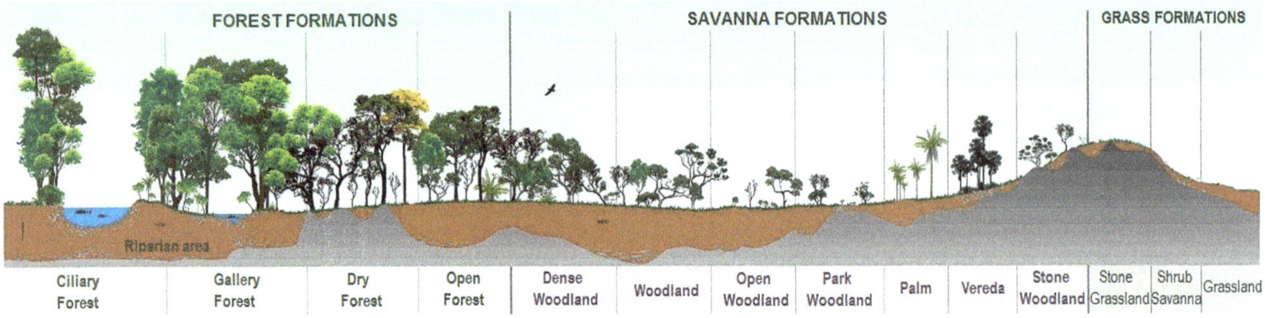

Figure 1. Physiognomies of the Cerrado. (adapted from [3]).

Remote sensing combined with statistical modelling is a powerful tool to quantify the uncertainty associated with fire regime descriptors, as shown in northern Australia [35,36] and South Africa [37]. According to [38], a statistical model is "a probability distribution constructed to enable inferences to be drawn or decisions made from data". As such, a statistical model here is a probability distribution assumed for the random variable of interest, namely the fire intervals over a 12-year time series. In the Cerrado there are few studies of fire frequency [14,39], and none addressing statistical modelling of fire return intervals. The choice of a particular model based on fitting statistical distributions is of great utility when working with fire interval data, because it provides rigorous quantitative understanding of the properties of fire regimes.

In this study, we analysed fire frequency over a 12-year time series (1997–2008) for Jalapão State Park, a protected area in the north of the Cerrado, based on burned area maps derived from Landsat imagery. Using a discrete lognormal model, we estimated fire interval, survival and hazard of burning distributions for different physiognomies of the Cerrado. This analysis is very useful for fire management and biodiversity conservation, and for estimating greenhouse gas emissions in protected areas.

Methods

Ethics Statement

Jalapão State Park is a protected area managed by the Instituto Natureza do Tocantins (NATURATINS). Fieldwork to develop the land cover assessment was undertaken in collaboration with NATURATINS, under the scope of a scientific-technical cooperation convention (registration no. 01.01.008.0/2006), and therefore no additional approval or permit was required. No protected species were sampled.

Study area

Jalapão State Park (JSP) was created in January 2001, with an area of 159,225 ha. It is located in Mateiros municipality, in the eastern part of Tocantins state, Brazil, between latitudes 10°08′53″S and 10°36′32″S and longitudes 46°24′24″S and 46°56′06″W (Figure 2). The park is located in the Jalapão region, which contains the largest officially recognized continuous preserved area of Cerrado in Brazil [40].

The climate at JSP is markedly seasonal (Aw in Köppen's classification: Group A - tropical/megathermal climates, w - tropical wet and dry, or savanna climate) with high rainfall in the wet season (around 1500 mm), from October to March, and an extended dry season from April through September/October (only 10% of the annual precipitation). Year round temperatures are mild, ranging from 22° to 27°C [5]. Population density is approximately 0.6 inhabitants/km^2, and the local economy is based on shifting cultivation, extensive cattle raising and, more recently, tourism and golden-grass (*Syngonanthus nitens, Eriocaulaceae)*) handcrafting [41,42]. Fire is used in cattle ranching as a management tool to stimulate grass regrowth in the dry season, when forage is in short supply and to stimulate golden-grass flowering, for use in local handicrafts. Ranchers and handcrafters use biennial fires, but the absence of firebreaks allows fires to spread over larger areas than required by those land use practices [42].

Jalapão State Park is an agricultural frontier and an area attractive for tourism, two economic activities that generate pressures for land use change. The legal status of JSP as a protected area severely restricts land use, only allowing activities related to conservation, and forbids the use of fire. However, over 150 farmer and extractivist families inhabited the area prior to the creation of JSP, own their lands, and remain in situ. They engage in land use and resource management practices, namely vegetation burning, which often conflict with the nominal, feebly enforced JSP nature protection status. The conflict is deemed irreconcilable, unless the state purchases all private land within the park boundaries, relocates the resident populations, and protects the area from inappropriate use by non-residents [41,42]. Only then, would it be possible to properly acknowledge the role of fire in Cerrado ecosystems and, rather than attempting full exclusion [15–17], implement fire management programs aimed at conserving and promoting biodiversity. The legal obligation to attempt fire exclusion has brought many problems to protected areas, such as disturbance of ecological processes, changes in species abundances, and biological invasions, which place the preservation of the biodiversity at risk, in the medium and long-term [15–19].

There are three main landscapes in JSP. First, the dryland on arenosol, where dominant physiognomies are Campo Sujo (Shrub Savanna - SS) and Cerrado Sensu Stricto (Woodland). The latter is divided into Cerrado Ralo (Open Woodland - OW), Cerrado Típico (Woodland - WD) and Cerrado Denso (Dense Woodland - DW) [3,41]. Second, the wetland on arenosol along the watercourses, where physiognomies are Vereda (VR), Gallery forest (GF) and Cilliary Forest (CF). Third, the plateau, where Open Woodland (OW) occurs on flat surfaces of ferralsol, and Dense Woodland (DW) occurs on hills of leptosols [3,41].

The Shrub Savanna (SS) is an herbaceous-shrubbery physiognomy with sparse shrubs over grassland. Woodland, divided into Open Woodland (OW), Woodland (WD) and Dense Woodland (DW), is a woody physiognomy with sparse trees and shrubs over grassland, where the woody layer is denser than in SS. Tree

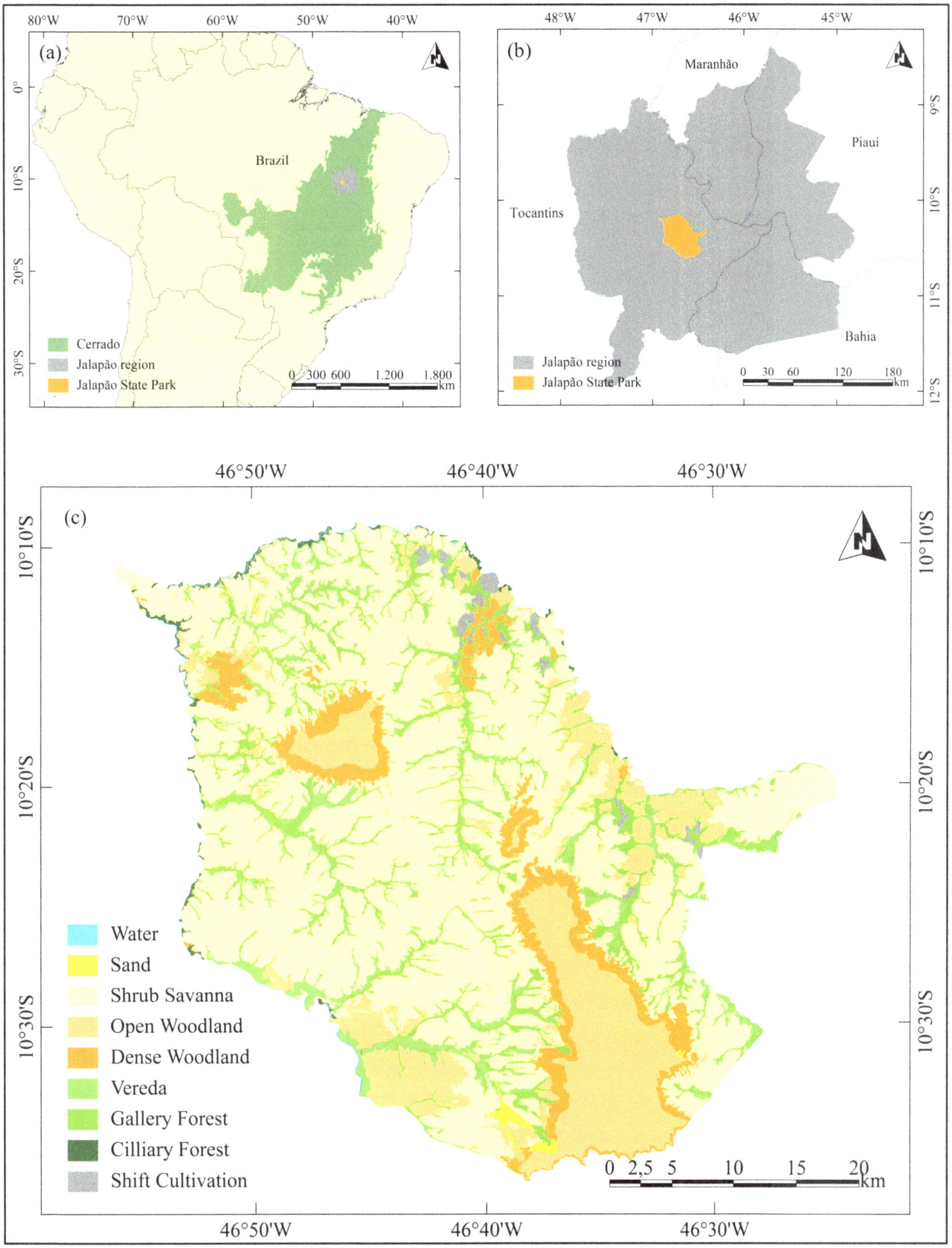

Figure 2. Location and land cover map of Jalapão State Park. (a) Cerrado biome (green) and Jalapão region (grey) in Brazil, (b) location of Jalapão State Park (orange) within Jalapão region (green), (c) land cover map adapted from the source map of [44] and modified based on fieldwork.

canopy cover in OW ranges from 5% to 20% and has an herbaceous-shrubbery layer denser than in WD and DW. Canopy cover in WD ranges from 20% to 50% and in DW ranges from 50% to 70% and both have a higher and more compact woody layer than OW [3]. Gallery Forest (GF) is a riparian physiognomy along narrow rivers, where the trees form a gallery over the watercourse with crowns overlapping and canopy cover from 70 to 95%. Belts of non-forest physiognomies surround both borders in an abrupt transition. Cilliary Forest (CF) is also a riparian physiognomy, present along medium and large rivers, where tree crowns do not form a gallery over the watercourse. Trees are predominantly deciduous and canopy cover ranges from 50 to 90% throughout the year. *Vereda* (VR) is a swampy area dominated by *buriti* palm (*Mauritia flexuosa,* Arecaceae) along narrow rivers (it can be associated with GF) and surrounded by belts of humid grassland [3], featuring golden-grass. Shifting cultivation (SC) is the agricultural system that occurs in JSP. It is defined as an "agricultural system that involves an alternation between cropping for a few years on selected and cleared plots and a lengthy period when the soil is rested" [43].

A land cover map of JSP was created by [44] using TM/Landsat-5 data acquired in 2007. The physiognomies WD and DW were not separable in this data and they were joined in one only class. This map was modified using field data collected in December 2005 and May 2008, that corrected errors existing in the source map, including land cover changes caused by fire (Figure 2c).

Burned area mapping

Burned area maps for JSP were produced with data from the Thematic Mapper (TM) sensor on board the Landsat-5 satellite and the Enhanced Thematic Mapper plus (ETM+) on board Landsat-7, for the period 1997 to 2008. We used 47 Landsat images (path/row 221/67) that were downloaded from the U.S. Geographical Survey Global Visualization Viewer (glovis.usgs.gov) and from National Institute for Space Research (INPE) Remote Sensing Datacenter (dgi.inpe.br/CDSR). At least three images were acquired for each dry season: at the end of the early dry season (April to May); at end of the middle dry season (June to July); and at the end of the late dry season (August to September). This bi-monthly sampling regime reduced omission errors in the classification, because burn scars in tropical savannas ecosystems can disappear within a few weeks [14].

The Landsat based burned area classification was performed using Object Based Image Analysis (OBIA) with the software eCognition Developer 8.0 [45]. OBIA has previously been used to map burned areas using Landsat images with promising results [46,47]. OBIA works in two steps: segmentation and classification. The segmentation splits an image into unclassified objects that are groups of pixels similar to one another based on a measure of spectral properties (i.e., color), size, shape, and texture, and pixel topology, controlled through parameters set by the user [48]. After segmentation, the image was classified by assigning each object to a class based on features and criteria set by the user. Segmentation and classification burned area outputs were considered final when they visually matched the original burned areas from the Landsat image. A minor amount of manual on-screen editing complemented the classification.

Fire rotation period

The annual burned area maps from 1997 to 2008, obtained from OBIA classification, were overlain with the land cover map for JSP, using the software TerraView 3.3.1 [49]. Overall fire frequency at JSP can be quantified by the fire rotation period (FRP), which is the time needed to burn an area of the same extent as the study area [50]:

$$\mathrm{FRP} = \frac{NS}{\sum A},$$

where N is the number of years in the study period; S is the area susceptible to burn in the study area; A is the annual area burned; and $\sum A$ is the sum of A within the period N. The inverse of the FRP, the annual percentage of area burned (APAB), is the mean percentage of the study area burned each year. Fire frequency was calculated for all of JSP and by land cover type.

Fire frequency analysis

Fire frequency analyses usually rely on the estimation of a time since fire ('survival') distribution and a fire interval ('mortality') distribution. These two distributions are linked by the hazard of burning function (force of mortality), which is the ratio of the mortality and survival distributions [51]. In tropical savanna ecosystems, the terms 'survival' and 'mortality' refer to a point or area having been unaffected or affected by fire. The most widely used statistical model in fire interval studies is the Weibull model [52,53]. However, [36] recently showed that the discrete lognormal model was more appropriate for fire interval analysis in Australian tropical savannas. The discrete lognormal was chosen for this study to model fire intervals in the Brazilian Cerrado because it has fire frequencies and vegetation types similar to those of the Australian savannas.

A 500 m regular point grid was used to sample intervals between successive fires, by land cover class, using the time-series of burned area maps (1997–2008). Fire survival analysis deals with analysis of time duration to until one or more fires happen. We sampled complete fire intervals, when both the starting and ending date of the fire-free period were known, and single-censored intervals. Single-censoring is a form of missing data that occurs often in survival analysis when either the starting or the ending date of the fire-free period is unknown. Right censoring occurs for all vegetation patches whose date of origin is known, but that remain undisturbed when the study ends. If the vegetation patch age at the start of the study is unknown, its lifetime is said to be left-censored. Ignoring single-censored data tends to lead to incorrect fire frequency estimates [36,53–55].

Fire interval data extracted with the point grid were used as input for the discrete lognormal model, to estimate fire interval, survival and hazard of burning distributions (e.g., [28,52,56]) using the survival package from software R [57]. The discrete lognormal model is a discrete version of a lognormal distribution, which is the distribution of a continuous random variable whose logarithm is normally distributed [58]. The fire interval distribution corresponding to the lognormal model, that is the probability that the time between consecutive fires is equal to t, is given by:

$$P(T=t|\mu,\sigma)=f(t|\mu,\sigma) = \begin{cases} \Phi\left(\frac{\ln(t)-\mu}{\sigma}\right)-\Phi\left(\frac{\ln(t-1)-\mu}{\sigma}\right), & t=1,2,\ldots, \\ 0, \text{otherwise,} & \end{cases}$$

where σ and μ are parameters of the model, and $\Phi(t)$ is the distribution function of the standard normal distribution. The survival distribution is the probability of any given vegetation patch having survived fire longer than t;

$$P(T>t|\mu,\sigma)=A(t|\mu,\sigma)$$
$$=\begin{cases} 1-\Phi\left(\dfrac{\ln(k)-\mu}{\sigma}\right), & k\leq t<k+1, \quad k=1,2,..., \\ 0, \text{ otherwise;} \end{cases}$$

The hazard of burning distribution is a conditional probability of burning during time interval t, given survival up to that point:

$$P(T>t|\mu,\sigma)=\lambda(t|\mu,\sigma)=\begin{cases} 1-\dfrac{\Phi\left(\dfrac{\mu-\ln(t)}{\sigma}\right)}{\Phi\left(\dfrac{\mu-\ln(t-1)}{\sigma}\right)}, & t=1,2,..., \\ 0, \text{ otherwise.} \end{cases}$$

In general, this hazard distribution initially increases, then decreases eventually approaching zero. Hence, vegetation patches following a discrete lognormal model have higher chance of burning as they age for some period of time, but after survival to a specific age (turning point), the probability of burning decreases as time increases. This turning point depends simultaneously on the ratio μ/σ and on the value of σ, in a nonlinear fashion [36]. If the turning point is one, the hazard function is always decreasing. The discrete lognormal parameters (σ and μ) were estimated with a maximum likelihood approach. The median fire interval is the smallest integer that is bigger than or equal to exp (μ) and the mode of the distribution is the smallest integer that is bigger than exp ($\mu - \sigma^2$).

A goodness-of-fit test for censored data developed by [59] was used to test the fit of the discrete lognormal distribution to the data. The test statistic, conveniently standardized, follows under the null hypothesis a standard normal distribution. Hence rejection of the proposed model is for values of the standardized test statistic above or below the adequate quantile of the standard normal distribution.

Results

Burned area mapping

The total area burned in JSP over the period 1997–2008 was 585,122 ha, which is equivalent to almost four times the size of the study area. The vast majority of the area affected by fire (77.6%) burned between three and six times (Figure 3). Only 6.4% of the area burned more than seven times and 16.0% burned once or twice. Unburned areas correspond to sand dunes and water bodies, and some areas of GF, CF and DW.

Fire rotation period

The overall estimated FRP for JSP was 3.0 years and the APAB was 34%. However, values varied with land cover type (Table 1). The lowest FRP occurred in SS and OW (2.72 years and 2.75 years, respectively). SS and OW occupy 77.6% of the study area (59.1% and 18.5%, respectively) and accounted for 84.2% of the total area burned during the study period (64.3% and 19.9% respectively). The highest FRP, 5.49 years, was observed in DW, that occupies 6.9% of the study area and accounted for 3.7% of the total area burned). Intermediate values were found in swampy (VR) and riparian physiognomies (GF and CF) and in SC (Table 1).

Fire frequency analysis

Estimated parameters for the lognormal model are shown in Table 1. The goodness-of-fit test never rejects the lognormal model, as shown by the test statistic and the corresponding p-values in Table 1. The turning point is larger than one for all land cover types, meaning that the hazard function follows the general pattern, increasing initially and then decreasing. Median fire intervals, can be aggregated in three groups of land cover: SS, OW, VR and SC with the lowest values (three years); GF and CF with intermediate values (four and five years); and WD with the highest median fire interval (six years).

The mode corresponds to the peak in the fire intervals distributions, which occurs at year two, except for DW, which peaked at year three (Figure 4). The probability of having a two-year interval between consecutive fires was highest in SS (0.32), followed by OW (0.28), VR (0.27), SC (0.26), GF (0.20) and CF (0.16). For DW the probability of having a three-year interval between consecutive fires was 0.11 (lowest peak value). Fire interval distributions can be aggregated in three groups according to the shape of the curves: SS, OW, VR, and SC, with probability higher than 0.26 at age two, falling abruptly to low values at older ages (e.g. $0.08\leq f(5)\leq 0.09$ and $0.02\leq f(8)\leq 0.03$); GF and CF with probability of fire between 0.16 and 0.20 at age two, and a less sharp decline of the curve (e.g. $0.09\leq f(5)\leq 0.10$ and $0.04\leq f(8)\leq 0.05$); DW, with a smooth decline ($f(3)=0.11$, peak value; $f(5)=0.09$; and $f(8)=0.05$).

Survival distributions (Figure 5) have high values during the first year for all land cover types ($A(1)\geq 0.83$), decreasing more or less abruptly, depending on land cover type: SS, OW, VR and SC, with curves falling abruptly (e.g. $0.11\leq A(5)\leq 0.20$ and $0.01\leq A(10)\leq 0.05$); GF and CF, with curves showing a less sharp decay (e.g. $0.32\leq A(5)\leq 0.41$ and $0.11\leq A(10)\leq 0.16$); DW, with the lowest decay rate of all curves (e.g. $A(5)=0.57$ and $A(10)=0.30$). The survival distributions indicate, for instance, that less than 10% of the landscape survives unburned up to year seven in SS, OW, VR and SC, up to year 10 in GF, up to year 13 in CF, and up to year 20 in DW.

Hazard of burning functions initially increase, reaching a peak at years three or four since the last fire (turning point), and then gradually decrease (Figure 6). The same three groups of land cover type can be identified, although there are differences in the turnings points within the same group: SS, OW, VR and SC with the highest probabilities of burning, given survival up to that point (hazard peak values of 0.42 and 0.37 in year four, and 0.33 and 0.31 in year three, respectively); GF and CF, with intermediate probabilities of burning (0.24 peaking at year three and 0.20 peaking at year four, respectively); DW, with the lowest probability of burning (hazard peak value of 0.13 at year four).

There was a negative relationship between hazard of burning and FRP. Sorting hazard of burning functions by ascending order we obtained, for every t, the following: $\lambda_{DW}(t)<\lambda_{CF}(t)<\lambda_{GF}(t)<\lambda_{SC}(t)<\lambda_{VR}(t)<\lambda_{OW}(t)<\lambda_{SS}(t)$ which is exactly the same order when sorting FRP values by descending order. Land cover types with the highest FRP are the ones with the lowest conditional probabilities of burning as a function of time since last fire.

Discussion

Burned area mapping and fire rotation period

The highest fire incidence in JSP was observed in SS and OW, physiognomies supporting large fuel loads forming a continuous fuel bed, where the grass and litter cure rapidly at the beginning of the dry season due to the low canopy cover. In these areas, cattle ranchers use fire every couple of years to stimulate grass regrowth

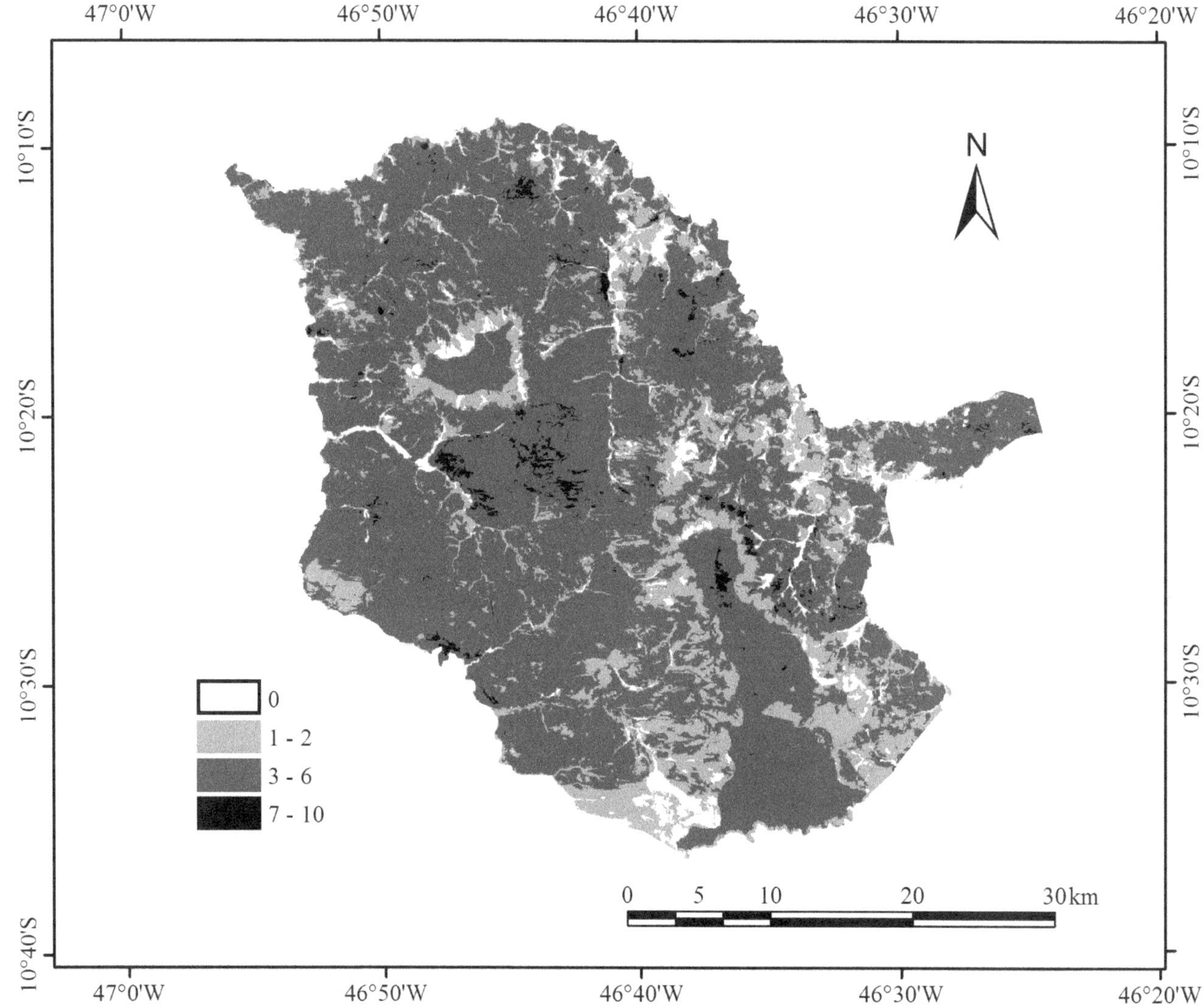

Figure 3. Number of times burned in Jalapão State Park, between 1997 and 2008.

for pastures. The lowest incidence of fire occurred in DW, where grassy fuels are highly discontinuous and fuel moisture tends to be higher, due to denser canopy cover. Anthropogenic fire is scarce in large areas of rocky substrate with rugged relief, which occur in 83% of DW in JSP. Intermediate values of fire incidence for VR and SC are likely due to burning practices to extract products directly from managed areas (use of fire to stimulate golden-grass flowering in VR, or to burn for clearing purposes in SC). Fire incidence was low in CF and GF because they are closed forest formations in humid habitats along rivers, and the fuel remains very moist throughout the year. In GF fire incidence was higher than CF because it is usually found in areas adjacent to VR, where the absence of fire breaks leads to fire spreading.

Fire return period of three years for JSP (APAB = 33%) is similar to the values observed in western Arnhem Land (WAL), a savanna region in northern Australia, where FRP varied between 2.7 and 2.8 years (APAB 36%–37%) [24,36]. [60] found higher APAB in Litchfield National Park (56%) and Nitmiluk National Park (41%), in northern Australia. On the other hand, [61] found lower APAB in Kakadu National Park, an area adjacent to WAL, with values ranging from 21% to 27% in sandstone habitats. Both Litchfield National Park and Kakadu National Park are protected areas and tourist attractions, like JSP. Controlled burning is practiced at the national parks in the coolest months to prevent severe and extensive fires in the hottest months. Western Arnhem Land is under aboriginal tenure, and traditional owners have been using fire as a land management tool for thousands of years, for 'cleaning up' the country and facilitate walking, hunting, and to encourage growth of plant and animal foods. In general, there were also similarities between fire frequency in the different physiognomies of JSP and those from northern Australia, reported by [36]. In both studies the fire incidence was higher in woodland and herbaceous-shrubbery physiognomies (SS and OW for JSP and woodlands and open forests in WAL). SS and OW had lower APAB values (37% and 36%, respectively) compared to woodlands and open forests (45% and 42%, respectively), but GF and CF (23% and 22%, respectively) had very similar values to closed forest (20%), in Australia. This highlights the strong similarities between these two tropical savannas and justifies analysis by land cover type.

Fire frequency analysis

Fire interval distributions reveal that a large fraction of the landscape burns every two to three years, and fire survival

Table 1. Fire frequency, fire interval descriptors, and goodness-of-fit tests, by land cover type for Jalapão State Park.

Land cover class	Area		FRP	APAB	μ	σ	Median	Mode	turning point	HP	p-values
	ha	%									
Shrub Savanna (SS)	93,680	58.9	2.72	37	0.82	0.66	3	2	4	−1.24	0.21
Open Woodland (OW)	29,310	18,4	2.75	36	0.93	0.69	3	2	4	−1.07	0.29
Dense Woodland (DW)	11,008	6.9	5.49	18	1.78	0.99	6	3	4	−0.90	0.37
Vereda (VR)	8,794	5.5	3.02	33	0.90	0.80	3	2	3	−1.03	0.31
Gallery Forest (GF)	13,459	8.5	4.44	23	1.19	0.91	4	2	3	−0.85	0.40
Cilliary Forest (CF)	908	0.6	4.57	22	1.39	0.92	5	2	4	−0.95	0.35
Shift Cultivation (SC)	1,233	0.8	3.71	27	0.86	0.89	3	2	3	−0.92	0.36

Other land cover classes not listed are sand (511 ha –0.3%) and water (250 ha –0.2%). Fire rotation period (FRP), annual percentage of area burned (APAB), estimated parameters μ and σ for the discrete lognormal model, median, mode and turning point, that is, the value where the hazard function attains its maximum. Hollander and Prochan (HP) test statistics provides a measure of the goodness of fit of the statistical model. P-values for the HP goodness of fit test are also shown. Since the p-values are above 0.05, there is statistical evidence to support the lognormal model.

distributions reveal that the extent of vegetation surviving this period and remaining available for burning at a later age decreases abruptly. Hazard of burning functions reach a peak three to five years after a fire and then decrease, meaning that because a large fraction of the landscape burns two to three years after the previous fire, the extent of landscape surviving this period and remaining available for burning at an older age decreases sharply. Thus, the probability of fire occurring in an older fuel age class is lower than that of it occurring in a younger class, mainly because long unburnt patches are quite rare in these landscapes. Although the decrease in the hazard of burning function does not imply a reduction in fuel loading, there is evidence that fuel loads in tropical savannas attain maximum values three to six years after a fire and then decrease [62,63]. Hazard of burning is also correlated with FRP since land cover types with the highest probabilities of burning are also the ones that burn more frequently, and need less time to burn an area equivalent to the study area.

Fire interval analysis indicated also clear patterns between physiognomies with low and high canopy cover, and riparian physiognomies. A short fire return interval (three years) was associated with physiognomies with lower density of trees (SS, OW and VR), where continuous fuels and geomorphologic features easily allow fire spread, and SC, where fire is used for crop residue burning and land clearing. Slightly longer fire return intervals (four or five years) were observed in riparian physiognomies (CF and GF) whereas longer fire return intervals were observed in DW, a high tree density physiognomy. These values also showed similarities with the median fire intervals estimated at WAL, Australia [36]; SS, OW, VR and SC had the same intervals as sandstone woodlands in WAL, and GF and CF approximately the same intervals as closed forests in WAL.

Fire ecology and management

According to our results, fires with a median fire interval of three to six years have been occurring since 1997 at JSP, a protected area of Cerrado. Those fires require large efforts and resources to be controlled, and demonstrate the ineffectiveness and inadequacy of the total fire suppression policy imposed by the Brazilian government agencies. Although the Cerrado is a fire-prone ecosystem and plants and animals show adaptations to frequent fires [17], frequently burned Cerrado areas tend to become more open and grassy [2,2464,]. There is also strong evidence that fire-sensitive vegetation (e.g. *Blepharocalix salicifolius, Sclerolobium paniculatum*) are in decline under the present-day Cerrado fire regime [6,17,65], in agreement to what is happening in other tropical savanna regions [66]. One exception is the golden-grass, in VR. Based on empirical knowledge of indigenous people, [42,67] verified that bi- or tri-annual fires promote the optimum population growth of golden-grass yet longer intervals do not diminish its population. With this in mind, [67,68] suggested joining both traditional ecological knowledge and conservation-oriented management, to extend current fire intervals. A similar strategy is being implemented in Australian tropical savannas, where indigenous aboriginal knowledge is incorporated into environmentally sustainable fire management [66].

Future fire management scenarios for JSP, including those that assume the absence of resident populations, will be able to use results of this study as a baseline against which to identify the need for modification of current fire regimes, and assess the effectiveness of subsequent interventions. Since our analysis was broken down by major vegetation type, it can support the development of specific management strategies, geared towards enhancing the

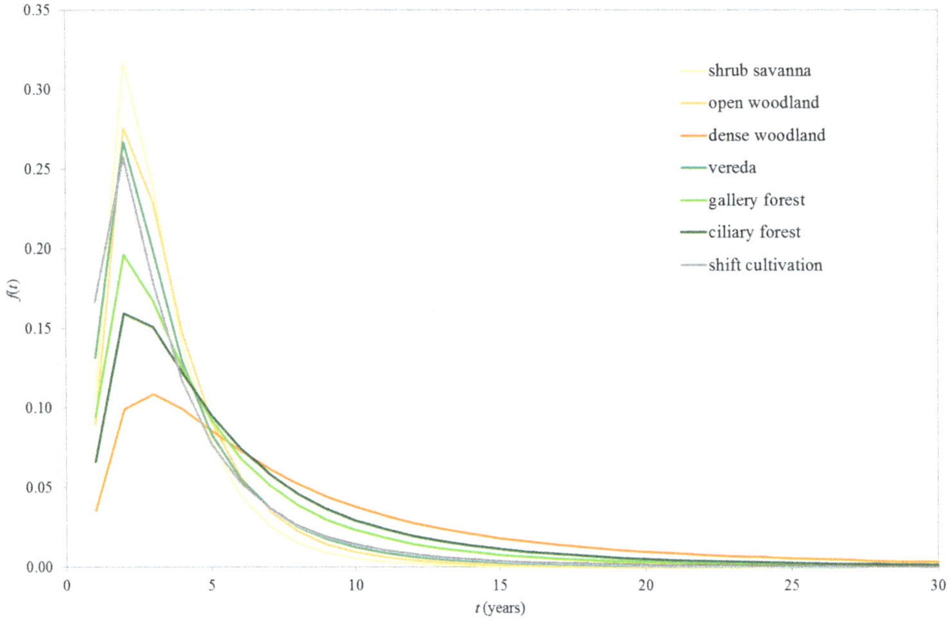

Figure 4. Fire interval distributions. Fire interval distributions (*f*) as a function of time since last fire (*t*) by land cover type for Jalapão State Park. Fire interval distribution is the probability that the time between consecutive fires is equal to *t*.

ecological values that justify the protected status of this area. For example, landscape formations with short fire return intervals, such as SS, OW, and VR (which occupy 83% of the JSP burnable area), more than a change in overall fire frequency, may benefit from a shift in the timing of burning towards earlier in the season, to reduce the severity of fire effects, including the amount of pyrogenic emissions of greenhouse gases [69]. On the other hand, landscapes with longer return intervals and vegetation that is more fire-sensitive, such as GF, CF and DW [68], probably ought to be the focus of efforts to contain escaped anthropogenic fires.

Conclusions

Fire frequency in JSP is very high, but not uncommon for tropical savannas, with median fire return intervals of three to six years, depending on the vegetation type. This high fire incidence is a function of the markedly seasonal climate of the Cerrado, fuel biomass accumulation, and traditional use of fire to stimulate the growth of palatable grasses for cattle, or to stimulate golden-grass flowering, used in local handicrafts. The probability of burning as a function of time since last fire reaches an early peak between

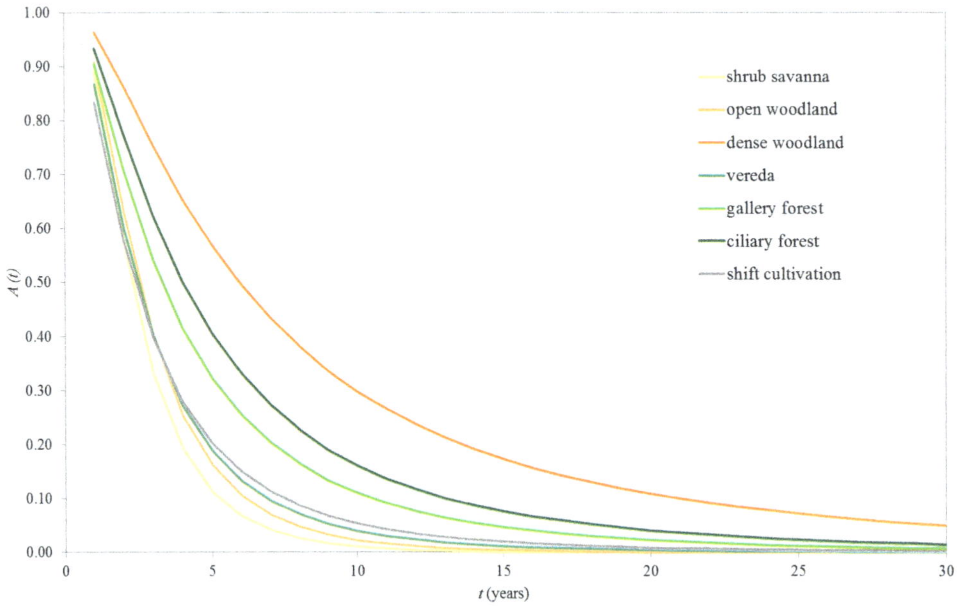

Figure 5. Survival distributions. Survival distributions (*A*) as a function of time since last fire (*t*) by land cover type for Jalapão State Park. The survival distribution is the probability of any given vegetation patch having survived fire for longer than *t*.

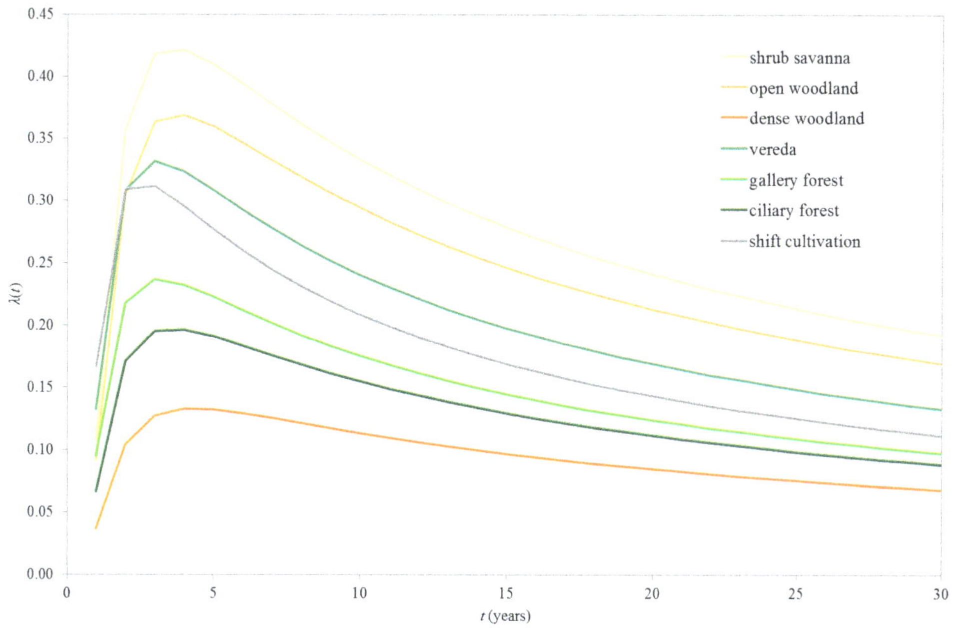

Figure 6. Hazard of burning distributions. Hazard of burning distributions (λ) as a function of time since last fire (*t*) by land cover type for Jalapão State Park. The hazard of burning distribution is a conditional probability of burning during time interval *t*, given survival up to that point.

three to four years, and subsequently declines due to fuel dynamics, reflecting a strong fuel age-dependency only in young vegetation patches. Open savanna physiognomies, that occupy most of the area of JSP, and shifting cultivation areas display the shortest fire intervals (median value of three years), and dense woodland formations the longest (median value of six years). Riparian physiognomies have intermediate median fire intervals of four to five years. The discrete lognormal model is appropriate for fire frequency analysis in tropical savannas, because it fits well regimes with short fire intervals, producing essential data for fire management, biodiversity conservation and greenhouse gas emissions estimates.

Sustainable management in protected areas of Cerrado should be a continuous process, incorporating scientific knowledge and traditional ecological knowledge of indigenous people and local communities [17,70]. Examples from countries such as Mali and Madagascar show that attempts to impose regulations and legal restrictions rarely contribute towards solving the savanna fire management problem. On the contrary, such rules often undermine traditional practices and weaken local land use management, including the capability for pro-actively managing a fire regime [71]. Ecologically sustainable management of fire regimes in Cerrado protected areas requires resources, time, planning capacity, and knowledge, and ought to be developed on a solid scientific foundation, and aim at enhancing the environmental values they are meant to protect.

Acknowledgments

The authors thank to Instituto Natureza do Tocantins (NATURATINS) for all the support during fieldwork.

Author Contributions

Conceived and designed the experiments: ACPJ JMCP. Performed the experiments: ACPJ. Analyzed the data: ACPJ SLJO MAAT JMCP. Contributed reagents/materials/analysis tools: ACPJ SLJO. Wrote the paper: SLJO ACPJ JMCP MAAT.

References

1. Coutinho LM (1982) Ecological Effects of Fire in Brazilian Cerrado. In: Huntley BJ, Walker BH, editors. Ecology of Tropical Savannas. Berlin Heidelberg: Springer. Ecological Studies, vol. 42, pp. 273–291.
2. Coutinho LM (1990) Fire in the ecology of the Brazilian Cerrado. In: Goldammer JG, editor. Fire in the Tropical Biota. Berlin Heidelberg: Springer-Verlag. Ecological Studies, vol. 84, pp. 82–105.
3. Ribeiro JF, Walter BMT (2008) As Principais Fitofisionomias do Bioma Cerrado. In: Sano SM, Almeida SPD, Ribeiro JF, editors. Cerrado ecologia e flora. Brasília - DF: Embrapa Informação Tecnológica, Vol. 1. pp. 151–212.
4. Myers N, Mittermeier RA, Mittermeier CG, da Fonseca GAB, Kent J (2000) Biodiversity hotspots for conservation priorities. Nature 403: 853–858. doi:10.1038/35002501.
5. Klink CA, Machado RB (2005) Conservation of the Brazilian Cerrado. Conserv Biol 19: 707–713. doi:10.1111/j.1523-1739.2005.00702.x.
6. Miranda HS, Bustamante MMC, Miranda AC (2002) The fire factor. In: Oliveira PS, Marquis RJ, editors. The Cerrados of Brazil: Ecology and Natural History of a Neotropical Savanna. New York: Columbia University Press. pp. 51–68.
7. Miranda HS, Sato MN, Neto WN, Aires FS (2009) Fires in the cerrado, the Brazilian savanna. In: Cochrane MA, editor. Tropical Fire Ecology: Climate Change, Land Use and Ecosystem Dynamics. Berlin Heidelberg: Springer Praxis. pp. 427–450.
8. Salgado-Labouriau ML, Casseti V, Ferraz-Vicentini KR, Martin L, Soubiès F, et al. (1997) Late Quaternary vegetational and climatic changes in cerrado and palm swamp from Central Brazil. Palaeogeogr Palaeoclim Palaeoecol 128: 215–226. doi:10.1016/S0031-0182(96)00018-1.
9. Salgado-Labouriau ML, Barberi M, Ferraz-Vicentini KR, Parizzi MG (1998) A dry climatic event during the late Quaternary of tropical Brazil. Rev Palaeobot Palynol 99: 115–129. doi:10.1016/S0034-6667(97)00045-6.
10. Posey DA (1985) Indigenous management of tropical forest ecosystems: the case of the Kayapó indians of the Brazilian Amazon. Agrofor Syst 3: 139–158. doi:10.1007/BF00122640.
11. Ministério do Meio Ambiente (MMA) (2011) Monitoramento do bioma Cerrado 2009–2010. Brasília - DF: Ministério do Meio Ambiente (MMA).
12. Mistry J (1998) Decision-making for fire use among farmers in savannas: an exploratory study in the Distrito Federal, central Brazil. J Environ Manage 54: 321–334. doi:10.1006/jema.1998.0239.
13. Mistry J (1998) Fire in the cerrado (savannas) of Brazil: an ecological review. Prog Phys Geogr 22: 425–448. doi:10.1177/030913339802200401.

14. França H, Ramos Neto MB, Setzer AW (2007) O Fogo No Parque Nacional Das Emas. Brasília - DF: Ministério do Meio Ambiente (MMA). 140 p.
15. Pivello VR, Coutinho LM (1996) A qualitative successional model to assist in the management of Brazilian cerrados. For Ecol Manag 87: 127–138. doi:10.1016/S0378-1127(96)03829-7.
16. Fidelis A, Pivello VR (2011) Deve-se usar o fogo como instrumento de manejo no Cerrado e Campos Sulinos? Biodiversidade Bras 2011: 12–25.
17. Pivello VR (2011) The Use of Fire in the Cerrado and Amazonian Rainforests of Brazil: Past and Present. Fire Ecol 7: 24–39. doi:10.4996/fireecology.0701024.
18. Andersen AN, Cook GD, Corbett LK, Douglas MM, Eager RW, et al. (2005) Fire frequency and biodiversity conservation in Australian tropical savannas: implications from the Kapalga fire experiment. Austral Ecol 30: 155–167. doi:10.1111/j.1442-9993.2005.01441.x.
19. Russell-Smith J, Whitehead P, Cooke P (2009) Culture, Ecology and Economy of Fire Management in North Australian Savannas: Rekindling the Wurrk Tradition. Collingwood: CSIRO Publishing.
20. Brockett BH, Biggs HC, van Wilgen BW (2001) A patch mosaic burning system for conservation areas in southern African savannas. Int J Wildland Fire 10: 169–183. doi:10.1071/WF01024.
21. Van Wilgen BW, Govender N, Biggs HC (2007) The contribution of fire research to fire management: a critical review of a long-term experiment in the Kruger National Park, South Africa. Int J Wildland Fire 16: 519–530. doi:10.1071/WF06115.
22. Edwards AC, Russell-Smith J (2009) Ecological thresholds and the status of fire-sensitive vegetation in western Arnhem Land, northern Australia: implications for management. Int J Wildland Fire 18: 127–146. doi:10.1071/WF08008.
23. Price OF, Russell-Smith J, Watt F (2012) The influence of prescribed fire on the extent of wildfire in savanna landscapes of western Arnhem Land, Australia. Int J Wildland Fire 21: 297–305. doi:10.1071/WF10079.
24. Pivello VR, Norton GA (1996) FIRETOOL: An Expert System for the use of Prescribed Fires in Brazilian Savannas. J Appl Ecol 33: 348–356. doi:10.2307/2404756.
25. Kauffman JB, Cummings DL, Ward DE (1994) Relationships of Fire, Biomass and Nutrient Dynamics along a Vegetation Gradient in the Brazilian Cerrado. J Ecol 82: 519–531. doi:10.2307/2261261.
26. Santos JR dos (1985) Análise automática de dados do sensor MSS/Landsat para a avaliação de áreas queimadas na instalação e conservação de projeto agropecuários. In: Memoria final. IV Reunion Plenaria SELPER. Santiago: Sociedad de Especialistas Latinoamericana en Percepcion Remota (SELPER). pp. 95–102.
27. Setzer AW, Pereira MC (1991) Amazonia biomass burnings in 1987 and an estimate of their tropospheric emissions. Ambio 20: 19–22.
28. Pereira MC, Setzer AW (1993) Spectral characteristics of deforestation fires in NOAA/AVHRR images. Int J Remote Sens 14: 583–597. doi:10.1080/01431169308904358.
29. Pereira MC, Setzer AW (1993) Spectral characteristics of fire scars in Landsat-5 TM images of Amazonia. Int J Remote Sens 14: 2061–2078. doi:10.1080/01431169308954022.
30. Setzer AW, Pereira MC, Pereira AC (1994) Satellite studies of biomass burning in Amazonia - some practical aspects. Remote Sens Rev 10: 91–103. doi:10.1080/02757259409532238.
31. Adámoli J, Fukuhara M, Silva JA da (1982) Aplicação de técnicas de sensoriamento remoto no estudo de queimas em pastagens nativas da região dos cerrados. In: Parada NJ, editor. Anais, II Simpósio Brasileiro de Sensoriamento Remoto. São José dos Campos: INPE. pp. 811–820.
32. Pereira Jr. A da C, Santos JR dos, Pereira M da C (1989) Monitoramento de queimadas nas savanas brasileiras: uma abordagem multisensor. In: Memorias, IV Simposio Latinoamericano de Percepcion Remota. São José dos Campos: INPE. pp. 678–686.
33. Pereira Jr. A da C (1992) Monitoramento de queimadas na região dos Cerrados com a utilização de dados AVHRR/NOAA corrigidos com dados TM/Landsat. MSc Dissertation. São José dos Campos: INPE.
34. Pereira AC, Setzer AW (1996) Comparison of fire detection in savannas using AVHRR's channel 3 and TM - images. Int J Remote Sens 17: 1925. doi:10.1080/01431169608948748.
35. Gill AM, Ryan PG, Moore PHR, Gibson M (2000) Fire regimes of World Heritage Kakadu National Park., Australia. Austral Ecol 25: 616–625. doi:10.1111/j.1442-9993.2000.tb00067.x.
36. Oliveira SLJ, Turkman MAA, Pereira JMC (2013) An analysis of fire frequency in tropical savannas of northern Australia, using a satellite-based fire atlas. Int J Wildland Fire 22: 479–492. doi:10.1071/WF12021.
37. Van Wilgen BW, Biggs HC, Mare N, O'Regan SP (2000) A fire history of the savanna ecosystems in the Kruger National Park, South Africa, between 1941 and 1996. South Afr J Sci 96: 167–178.
38. Davison AC (2003) Statistical Models. Cambridge: Cambridge University Press. 726 p.
39. Rivera-Lombardi RJ (2003) Estudo de recorrência de queimadas e permanência de cicatrizes de fogo em áreas selecionadas do cerrado brasileiro, utilizando imagens TM/LANDSAT. MSc Dissertation. São José dos Campos: INPE.
40. Silva JMC da, Bates JM (2002) Biogeographic Patterns and Conservation in the South American Cerrado: A Tropical Savanna Hotspot. BioScience 52: 225. doi:10.1641/0006-3568(2002)052[0225:BPACIT]2.0.CO;2.
41. Secretaria de Planejamento e Meio Ambiente (SEPLAN) (2003) Plano de Manejo do Parque Estadual do Jalapão. Palmas - TO: Secretaria de Planejamento e Meio Ambiente (SEPLAN). 132 p.
42. Schmidt IB, Figueiredo IB, Scariot A (2007) Ethnobotany and Effects of Harvesting on the Population Ecology of Syngonanthus nitens (Bong.) Ruhland (Eriocaulaceae), a NTFP from Jalapão Region, Central Brazil. Econ Bot 61: 73–85. doi:10.1663/0013-0001(2007)61[73:EAEOHO]2.0.CO;2.
43. Ruthenberg H (1971) Farming systems in the tropics. Oxford: Clarendon Press. 313 p.
44. Santiago AR, Pereira Jr. A da C (2010) Mapeamento da cobertura da terra dos Parques Estaduais do Jalapão (PEJ), Cantão (PEC) e município de Itaguatins (Tocantins). Ambiência 6: 109–124.
45. Definiens (2009) Definiens eCognition Developer 8– User Guide. München: Definiens AG. 250 p.
46. Mitri G, Gitas I (2004) A performance evaluation of a burned area object-based classification model when applied to topographically and non-topographically corrected TM imagery. Int J Remote Sens 25: 2863–2870. doi:10.1080/01431160410001688321.
47. Polychronaki A, Gitas IZ (2010) The development of an operational procedure for burned-area mapping using object-based classification and ASTER imagery. Int J Remote Sens 31: 1113–1120. doi:10.1080/01431160903334497.
48. Benz UC, Hofmann P, Willhauck G, Lingenfelder I, Heynen M (2004) Multi-resolution, object-oriented fuzzy analysis of remote sensing data for GIS-ready information. ISPRS J Photogramm Remote Sens 58: 239–258. doi:10.1016/j.isprsjprs.2003.10.002.
49. Projeto TerraView (2010) TerraView 3.3.1. São José dos Campos: INPE. http://www.dpi.inpe.br/terraview. Acessed 13 June 2014.
50. Heinselman ML (1973) Fire in the virgin forests of the Boundary Waters Canoe Area, Minnesota. Quat Res 3: 329–382. doi:10.1016/0033-5894(73)90003-3.
51. Johnson EA, Gutsell SL (1994) Fire Frequency Models, Methods and Interpretations. In: M Begon and A.H Fitter, editor. Advances in Ecological Research. London: Academic Press, Vol. 25. pp. 239–287. doi:10.1016/S0065-2504(08)60216-0.
52. Johnson EA (1992) Fire and Vegetation Dynamics: Studies from the North American Boreal Forest. Cambridge: Cambridge University Press. 146 p.
53. Polakow DA, Dunne TT (1999) Modelling fire-return interval T: stochasticity and censoring in the two-parameter Weibull model. Ecol Model 121: 79–102. doi:10.1016/S0304-3800(99)00074-5.
54. Moritz MA, Moody TJ, Miles LJ, Smith MM, Valpine P de (2009) The fire frequency analysis branch of the pyrostatistics tree: sampling decisions and censoring in fire interval data. Environ Ecol Stat 16: 271–289. doi:10.1007/s10651-007-0088-y.
55. Oliveira SLJ, Pereira JMC, Carreiras JMB (2012) Fire frequency analysis in Portugal (1975–2005), using Landsat-based burnt area maps. Int J Wildland Fire 21: 48–60. doi: 10.1071/WF10131.
56. Johnson EA, van Wagner CE (1985) The theory and use of two fire history models. Can J Res 15: 214–220. doi:10.1139/x85-039.
57. R Core Team (2013) R: A Language and Environment for Statistical Computing. Vienna: R Foundation for Statistical Computing. Available: http://www.R-project.org. Acessed 2014 Jun 13.
58. Crow EL, Shimizu K (1988) Lognormal distributions: theory and applications. New York: Marcel Dekker.
59. Hollander M, Proschan F (1979) Testing to Determine the Underlying Distribution Using Randomly Censored Data. Biometrics 35: 393–401. doi:10.2307/2530342.
60. Edwards A, Hauser P, Anderson M, McCartney J, Armstrong M, et al. (2001) A tale of two parks: contemporary fire regimes of Litchfield and Nitmiluk National Parks, monsoonal northern Australia. Int J Wildland Fire 10: 79–89. doi:10.1071/WF01002.
61. Edwards A, Kennett R, Price O, Russell-Smith J, Spiers G, et al. (2003) Monitoring the impacts of fire regimes on vegetation in northern Australia: an example from Kakadu National Park. Int J Wildland Fire 12: 427–440. doi:10.1071/WF03031.
62. Gill AM, Hoare JRL, Cheney NP (1990) Fires and Their Effects in the Wet-Dry Tropics of Australia. In: Goldammer DJG, editor. Fire in the Tropical Biota. Berlin Heidelberg: Springer-Verlag, Ecological Studies vol. 84, pp. 159–178.
63. Cook GD (2003) Fuel Dynamics, Nutrients, and Atmospheric Chemistry. In: Andersen AN, Cook GD, Williams RJ, editors. Fire in Tropical Savannas: The Kapalga Experiment. New York: Springer. Ecological Studies, vol. 169, pp. 47–58.
64. Pivello VR, Coutinho LM (1992) Transfer of macro-nutrients to the atmosphere during experimental burnings in an open cerrado (Brazilian savanna). J Trop Ecol 8: 487–497. doi:10.1017/S0266467400006829.
65. Ratter JA, Ribeiro JF, Bridgewater S (1997) The Brazilian Cerrado Vegetation and Threats to its Biodiversity. Ann Bot 80: 223–230. doi:10.1006/anbo.1997.0469.
66. Russell-Smith J, Edwards AC, Price OF (2012) Simplifying the savanna: the trajectory of fire-sensitive vegetation mosaics in northern Australia. J Biogeogr 39: 1303–1317. doi:10.1111/j.1365-2699.2012.02679.x.
67. Schmidt IB (2011) Effects of local ecological knowledge, harvest and fire on golden-grass (Syngonanthus nitens, Eriocaulaceae), a non-timber forest product (NTFP) species from the Brazilian savanna. PhD thesis, University of Hawai'i.

68. Schmidt IB, Sampaio MB, Figueiredo IB, Ticktin T (2011) Fogo e artesanato de capim-dourado no Jalapão – usos tradicionais e consequências ecológicas. Biodiversidade Bras 2011: 67–85.
69. Price OF, Borah R, Maier SW (2014) Role of weather and fuel in stopping fire spread in tropical savannas. Austral Ecol 39; 135–144. doi:10.1111/aec.12021.
70. Whitehead PJ, Bowman DMJS, Preece N, Fraser F, Cooke P (2003) Customary use of fire by indigenous peoples in northern Australia: its contemporary role in savanna management. Int J Wildland Fire 12: 415–425. doi:10.1071/WF03027.
71. Kull CA, Laris P (2009) Fire ecology and fire politics in Mali and Madagascar. In: Cochrane MA, editor. Tropical Fire Ecology: Climate Change, Land Use, and Ecosystem Dynamics. Berlin Heidelberg: Springer-Praxis, pp. 171–226.

19

The Phenology of Ticks and the Effects of Long-Term Prescribed Burning on Tick Population Dynamics in Southwestern Georgia and Northwestern Florida

Elizabeth R. Gleim[1,2,3]*, L. Mike Conner[3], Roy D. Berghaus[4], Michael L. Levin[5], Galina E. Zemtsova[5], Michael J. Yabsley[1,2]

1 Warnell School of Forestry and Natural Resources, University of Georgia, Athens, Georgia, United States of America, **2** Southeastern Cooperative Wildlife Disease Study, College of Veterinary Medicine, University of Georgia, Athens, Georgia, United States of America, **3** Wildlife Laboratory, Joseph W. Jones Ecological Research Center at Ichauway, Newton, Georgia, United States of America, **4** Department of Population Health, College of Veterinary Medicine, University of Georgia, Athens, Georgia, United States of America, **5** Centers for Disease Control and Prevention, Rickettsial Zoonoses Branch, Atlanta, Georgia, United States of America

Abstract

Some tick populations have increased dramatically in the past several decades leading to an increase in the incidence and emergence of tick-borne diseases. Management strategies that can effectively reduce tick populations while better understanding regional tick phenology is needed. One promising management strategy is prescribed burning. However, the efficacy of prescribed burning as a mechanism for tick control is unclear because past studies have provided conflicting data, likely due to a failure of some studies to simulate operational management scenarios and/or account for other predictors of tick abundance. Therefore, our study was conducted to increase knowledge of tick population dynamics relative to long-term prescribed fire management. Furthermore, we targeted a region, southwestern Georgia and northwestern Florida (USA), in which little is known regarding tick dynamics so that basic phenology could be determined. Twenty-one plots with varying burn regimes (burned surrounded by burned [BB], burned surrounded by unburned [BUB], unburned surrounded by burned [UBB], and unburned surrounded by unburned [UBUB]) were sampled monthly for two years while simultaneously collecting data on variables that can affect tick abundance (e.g., host abundance, vegetation structure, and micro- and macro-climatic conditions). In total, 47,185 ticks were collected, of which, 99% were *Amblyomma americanum*, 0.7% were *Ixodes scapularis,* and fewer numbers of *Amblyomma maculatum*, *Ixodes brunneus*, and *Dermacentor variabilis.* Monthly seasonality trends were similar between 2010 and 2011. Long-term prescribed burning consistently and significantly reduced tick counts (overall and specifically for *A. americanum* and *I. scapularis*) regardless of the burn regimes and variables evaluated. Tick species composition varied according to burn regime with *A. americanum* dominating at UBUB, *A. maculatum* at BB, *I. scapularis* at UBB, and a more even composition at BUB. These data indicate that regular prescribed burning is an effective tool for reducing tick populations and ultimately may reduce risk of tick-borne disease.

Editor: Kelly A. Brayton, Washington State University, United States of America

Funding: This work was funded by the Centers for Disease Control and Prevention/University of Georgia (UGA) collaborative grant, (http://www.cdc.gov/od/science/technology/innovation/seeds.htm) (8212, Ecosystem Health and Human Health: Understanding the Ecological Effects of Prescribed Fire Regimes on the Distribution and Population Dynamics of Tick-borne Zoonoses, MJY, ERG, LMC, and MLL). Additional funding was provided by the Joseph W. Jones Ecological Research Center at Ichauway, Warnell School of Forestry and Natural Resources (UGA), and the Southeastern Cooperative Wildlife Disease Study (SCWDS) (UGA) through the Federal Aid to Wildlife Restoration Act (50 Stat. 917) and through SCWDS sponsorship from fish and wildlife agencies of member states. The funders had no role in study design, data collection and analysis, decision to publish, or preparation of the manuscript.

Competing Interests: The authors have declared that no competing interests exist.

* Email: egleim@gmail.com

Introduction

In the past several decades, numerous novel tick-borne diseases have emerged and the incidence of other tick-borne diseases has increased [1–8]. While this increase is likely due to a number of factors, such as increases in reporting, diagnosis and host abundance, one of the primary drivers is thought to be human land modification and management practices [3,8,9]. This hypothesis has underscored the importance of understanding how human land management practices affect tick population dynamics as well as identifying methods to control tick populations and reduce human disease risk.

Perhaps one of the most promising methods for controlling tick populations is prescribed burning because it can be applied on a landscape level and is relatively time and cost efficient. Additionally, prescribed burning is a well-accepted form of ecosystem management and wildfire prevention [10–12]. In the southeastern United States, the value of prescribed burning is particularly apparent where forests are dominated by fire-adapted or fire-dependent tree species, such as longleaf pines (*Pinus palustris*). Specifically, burns reduce fuel load and cause regeneration and diversification of the understory of forests [11,13–16]. In addition, the immediate regeneration of understory vegetation after a burn and the long-term density and diversity of understory vegetation within regularly burned habitat provides resources for many wildlife species including the endangered or threatened gopher tortoise (*Gopherus polyphemus*), Eastern indigo snake (*Drymarchon couperi*), and red-cockaded woodpecker (*Picoides borealis*) [11,15,17–18].

Several past studies have been conducted on the impacts of prescribed burning on tick populations. Although there have been conflicting results, the majority of studies concluded that tick populations are reduced immediately after a burn event, but recover to pre-burn abundance within one year [19–27]. However, other studies have failed to observe a significant reduction in ticks and some even observed an increase [28–30]. These studies often did not simulate real-world management conditions in that burns in many studies were performed only once, over a small area of land, and/or were performed in areas that were previously unburned [19–21,23,24,26,31,32]. This is in direct contrast to the typical regimen for applying prescribed burns, which are conducted over hundreds to thousands of acres on a regular basis. These discrepancies have clear implications for tick re-establishment after a burn and should be accounted for in an analysis on the impacts of burning on tick populations. Furthermore, previous studies typically did not investigate other factors known to affect tick populations such as host abundance, microclimate and vegetation structure [21,24]. Failure to account for such factors may explain some of the conflicting results of past studies.

To address impacts of long-term prescribed burning on tick population dynamics and evaluate how prescribed burns might interact with other factors to affect tick abundance, we determined tick abundance, species composition, and seasonality at multiple plots with variable burn histories with >10 years of operational burn management. Additionally, other factors known to potentially affect tick abundance (e.g., host abundance, vegetation structure and micro- and macroclimate) were evaluated at each plot. Collectively, these data provided insight in to the efficacy of long-term prescribed burning for tick control and also revealed phenologies for numerous tick species in southwestern Georgia and northwestern Florida (USA), an area for which there is currently little to no data on ticks or tick-borne pathogens.

Methods

Study Area

This study was conducted in southwestern Georgia and northwestern Florida, which is dominated by pine and mixed pine forests and agriculture. Prescribed burning is commonly used throughout the region to maintain various pine ecosystems including, longleaf pine ecosystems. We selected 21 sites (two privately owned, eight state-owned, and 11 owned by the J.W. Jones Ecological Research Center [JERC]) based on the presence or absence of a long-term history of prescribed burning, with burned sites having had at least 10 years of regular prescribed burns and unburned sites having no history of prescribed burning for over 10 years (Table 1). Study sites (not including our control sites) ranged in size from 7.2–78.9 ha. To ensure that the same locations within the study sites, including our control sites, were being sampled on a month-to-month basis, specific plots were established which ranged from 0.35–3.22 ha (note that flagging effort was standardized by time). No permits were required for this work. Permission to work on public lands was given by the Georgia Department of Natural Resources. Permission to work on private lands was given by the respective land owners and future permissions should be addressed to Mike Conner.

To account for differences in land cover and management both within and immediately surrounding the site, each site was further categorized as being 1) burned surrounded by burned (BB), 2) burned surrounded by unburned (BUB), 3) unburned surrounded by burned (UBB) and 4) unburned surrounded by unburned (UBUB) (i.e., a control). When classifying prescribed fire status around a sampled site, surrounding land was defined based on burn status of land immediately surrounding the site. Because of the rarity of BUB sites, we classified these sites based upon the majority of the land immediately surrounding the plot being unburned. Selection of BUB and UBB was made possible by using or working adjacent to sites that were either part of a long-term fire exclusion site or using sites that were not burned due to logistical difficulties associated with smoke dispersion (e.g., stands bordered highways creating risk of automobile accident), or selecting sites associated with forest types that are not typically burned (e.g., hardwood forest). Common use of prescribed fire on the study landscape facilitated selection of BB sites. Lack of prescribed fire use within large tracts of land facilitated selection of UBUB sites (with tracts ranging from 165–385 ha).

Tick Collections

Each plot was sampled monthly for ticks for 24 months (January 2010-December 2011) using 1 m×1 m flags made of flannel cloth. To standardize conditions, flagging was conducted only when the temperature was above 7.2°C and after 10AM when vegetation was dry (no dew or moisture from precipitation). Additionally, flagging was not conducted during inclement weather (rain, snow, or excessive wind). Each plot was flagged for a minimum of one hour per plot per month. If more than 5 ticks were collected during a one hour effort, sampling would be continued for up to 30 minutes or until the entire plot had been covered. Flags were checked every 10 minutes and all nymphs and adults were collected and preserved in 70% ethanol. All larvae were removed en masse with masking tape with each clutch being kept separate. Timers were paused during flag checks and removal of ticks.

Tick Identifications

All adult and nymphal *Amblyomma* spp., adult *Dermacentor variabilis*, and adult *Ixodes* spp. were identified morphologically utilizing a microscope and published keys [33,34]. Identification of larvae, *Ixodes* spp. nymphs, and a subset of adult *Ixodes* spp. was conducted through polymerase chain reaction (PCR) and sequence analysis. DNA was extracted from ticks using a Qiagen DNeasy blood and tissue kit (Germantown, MD) utilizing the manufacturer's protocol. For *Amblyomma* larvae collected in 2010, a multiplex real-time PCR was used to differentiate between *A. maculatum* and *A. americanum* as described in Zemtsova et al [35]. For *Amblyomma* larvae collected in 2011 and all *Ixodes*, a PCR targeting the 16S rRNA gene was conducted using primers 16S-1 (5′-CCGGTCTGAACTCAGATCAAGT) and 16S+2 (5′-TTGGGCAAGAAGACCCTATGAA) as described in Black and Piesman [36]. DNA extraction and PCR reactions were performed in hoods dedicated strictly for their respective tasks and waters were included as negative controls during both DNA extraction and PCR in order to detect contamination. All amplicons from the 16S protocol were purified using a Qiagen gel extraction kit (Germantown, MD) and then bi-directionally sequenced at the Georgia Genomics Facility (Athens, GA).

Host Monitoring

Host occurrence, specifically mesomammals and deer, was determined through quarterly passive trail camera (Cuddeback Capture, Green Bay, WI) surveys. No permits or Institutional

Table 1. Locations and burn data for the 21 plots sampled during this study.

Location	GPS Coordinates	County†	Burn Treatment	Year Burned (during study)
JERC, BU 1*	N 31° 18.967, W 84° 26.495	Baker	BB	2010
JERC, BU 5	N 31° 18.422, W 84° 27.177	Baker	BB	2011
JERC, BU 7 East	N 31° 17.272, W 84° 27.877	Baker	BB	2011
JERC, BU 7 West	N 31° 17.770, W 84° 28.396	Baker	BB	2011
JERC, BU 64	N 31° 25.042, W 84° 48.285	Baker	BB	2010
JERC, BU 136	N 31° 11.627, W 84° 27.978	Baker	BB	2010
JERC, BU 137	N 31° 11.594, W 84° 27.976	Baker	BB	2011
JERC, BU 176	N 31° 14.484, W 84° 22.162	Baker	BB	2010
Chickasawhatchee WMA	N 31° 29.055, W 84° 26.744	Calhoun	BUB	2010
Hannahatchee WMA - plot 1	N 32° 8.850, W 84° 45.241	Stewart	BUB	2011
Hannahatchee WMA - plot 2	N 32° 06.559, W 84° 44.600	Stewart	BUB	2011
River Creek WMA	N 30° 51.023, W 84° 04.347	Thomas	BUB	2011
Silver Lake WMA	N 30° 47.270, W 84° 45.390	Decatur	BUB	2010
Flint River WMA	N 32° 08.431, W 84° 00.358	Dooly	UBB	n/a
JERC, BU 107	N 31° 16.367, W 84° 29.090	Baker	UBB	n/a
JERC, BU 140	N 31° 10.219, W 84° 28.017	Baker	UBB	n/a
JERC, BU 15	N 31° 29.066, W 84° 52.242	Baker	UBB	n/a
Lake Seminole WMA - Little Horseshoe Bend Tract	N 30° 50.649, W 84° 39.186	Decatur	UBB	n/a
Montezuma Bluff Natural Area	N 32° 20.235, W 84° 01.743	Macon	UBUB	n/a
Private Property 1	N 30° 49.578, W 84° 36.951	Decatur	UBUB	n/a
Private Property 2	N 30° 41.277, W 84° 49.023	Gladsden, Florida	UBUB	n/a

Prescribed burns took place every 2–4 years during the dormant season on pre-determined schedules.
JERC = Joseph W. Jones Ecological Research Center; BU = burn unit; WMA = wildlife management area; n/a = not applicable; BB = burned, surrounded by burned area; BUB = burned, surrounded by unburned area; UBB = unburned, surrounded by burned area; UBUB = unburned, surrounded by unburned area.
*JERC plots were distributed throughout this 12,000 ha area.
†Unless noted, all counties are in Georgia.

Animal Care and Use Committee approval is required for passive trail camera surveys and no endangered or protected species were involved. For each survey, one camera was placed at each plot on an unbaited, secondary dirt road or trail bordering or within 100 m of the plot. Each plot was surveyed for 76 hours with all plots being surveyed within an 11-day period (with the exception of the winter 2011 survey in which all plots were surveyed within 30 days due to logistical constraints). The species, date, and time of all animals captured were recorded.

Vegetation Surveys

To account for potential differences in vegetation among the plots, vegetation surveys were performed at all plots. To determine tree density, a singular point-centered quarter survey utilizing the equivalent of 20 points per hectare was performed with a quarter being considered empty if there were no trees within 10 m from the center point. These points were marked for understory vegetation surveys.

Understory vegetation and canopy closure were measured using quarterly surveys. To perform the survey, a 1 m×1 m frame was placed with the previously marked point falling at the center of the frame and the sides of the frame running directly north to south and east to west. Percent cover and composition of understory vegetation and ground litter was estimated [37]. Litter depth was measured at two random points within the frame. Canopy cover was measured from the center point using a spherical densiometer [38].

Microclimate and Weather

At least three microclimatic measurements (humidity, temperature, and median wind speed) were taken at each plot immediately after each monthly flagging using a Kestrel 3000 Weather Meter (Birmingham, MI). Measurements were taken at ground level and 1 m above ground. If a particular plot had areas with and without ticks, measurements were taken where ticks were collected and two additional measurements were taken at randomly assigned points where ticks were not detected. If no ticks were collected at a plot, measurements were taken at randomly selected points.

Weather data including maximum temperature and precipitation the current sampling day, maximum and minimum temperatures the previous day, and precipitation accumulation the previous 3 days were collected for all plots, each sample date from the JERC Weather Station (all JERC locations) and the National Weather Service (all other plots).

Statistical Analyses

A generalized estimating equation (GEE) negative binomial regression model [39] was used to evaluate the impacts of long-term prescribed burning, host abundance, vegetation composition, microclimate, and weather on tick abundance. This modeling approach is well-suited for repeated measures, accounts for clustering by plot, and is appropriate for use when there are large numbers of zeros in the data. Models were created for 1) total tick counts, 2) *A. americanum* counts, and 3) *I. scapularis* counts. We chose to include all lifestages captured by flagging because this method of collection acts as a proxy for questing behavior and ultimately the ability to locate a human host which has direct implications for pathogen transmission. When applicable, each larval cluster was counted as a single tick because 1) this would reduce skew in the data, and 2) larvae do not transmit pathogens to humans (except for *Rickettsia* spp.). However, we chose to include nymphs in our analyses because they are viewed as the primary lifestage responsible for transmitting pathogens to humans [40,41]. No other tick species were sufficiently abundant for further modeling effort.

Because host and vegetation surveys were only conducted quarterly, all data were evaluated on a quarterly basis (spring, summer, fall, and winter). Thus, for tick counts, microclimate, and weather data, we used data that were collected closest to the date of the vegetation surveys for each quarter. The relationship between continuous predictors and the outcome was graphically assessed by categorizing the predictors into quartiles and plotting the estimated coefficients for the resulting indicator variables against the category midpoints. Predictors having a non-linear relationship with the outcome were categorized for the analysis.

Multivariable model selection proceeded from a maximum model that contained all variables having an association ($P<0.2$) with tick counts in the univariate analysis. Non-significant predictors were removed from the maximum model using a manual stepwise procedure until only those having $P<0.05$ remained. Upon reaching a preliminary main-effects model, all previously excluded variables were added back to the final model one at a time to re-assess their significance, and all possible two-way interactions with burning were evaluated. Year and season were retained regardless of their significance because of their theoretical importance as predictors of tick abundance. Season was dichotomized into warm (spring and summer) and cool (fall and winter) periods because the individual seasons within these categories were homogeneous with respect to tick counts.

Results

In total, 47,184 ticks were collected with *A. americanum* being the most common tick detected followed by *I. scapularis, A. maculatum, D. variabilis*, and *I. brunneus* (Table 2). To account for varying effort among plots and collection dates, we standardized by evaluating ticks collected/hr. Seasonality trends were determined using all 21 plots for both 2010 and 2011 (Figure 1). In 2010, *A. americanum* activity peaked in May for adults, May and October for nymphs, and July and September for larvae. In 2011, periods of peak *A. americanum* activity were similar with adults peaking in April through June, nymphs in May and September, and larvae in July and October. Adult *A. maculatum* activity peaked June through August 2010 and in August 2011 (Figure 1). Single *A. maculatum* nymphs were collected in April 2010 and in March 2011 and a single *A. maculatum* larval clutch was collected in September 2011. *I. scapularis* adults peaked in November and March of 2010 and in February 2011. *D. variabilis* adult activity peaked in July 2010 and was not caught in enough abundance in 2011 to determine seasonality.

Within burn treatments, 43 ticks were collected in BB plots (one *A. americanum*, 37 *A. maculatum,* and five *I. scapularis*), 1,756 in BUB plots (1,687 *A. americanum,* 32 *A. maculatum*, 26 *I. scapularis,* and 11 *D. variabilis*), 3,719 in UBB plots (3,568 *A. americanum*, six *A. maculatum*, 135 *I. scapularis,* one *I. brunneus*, and nine *D, variabilis*), and 41,706 in UBUB plots (a minimum of 41,460 *A. americanum*, a minimum of four *A. maculatum*, 186 *I. scapularis*, and 18 *D. variabilis*) (Table 2). Importantly, species-specific numbers associated with *A. americanum* and *A. maculatum* in UBUB plots cannot be determined because a larvae cluster was captured that was assumed to consist of a single species; later PCR testing revealed that it contained both species. Our modeling efforts suggested that total tick counts were affected by long-term burning, season, litter cover, and trees density (Table 3). There was a significant interaction between burning and season. For plots in which burning occurred, tick counts did not change significantly by season [RR = 1.1 (95% CI: 0.57, 2.12); P = 0.774], confirming the trend that ticks were significantly reduced in burned plots (Figure 2). However, at plots in which there was no burning, tick counts were 10 times greater in the warm season than in the cold [Relative Risk (RR) = 10.7 (95% CI: 4.20, 27.18); $P<0.001$]. Having higher than 95% litter cover was positively associated with an approximately two-fold increase in tick counts and density of trees (>183 per ha) was associated with an approximately six-fold increase in tick counts.

Interestingly, dominant tick species differed by burn treatment, with *A. americanum* being the most prevalent tick in UBUB plots, *A. maculatum* being most common in BB plots, *I. scapularis* dominating UBB plots, and a more even distribution of the different tick species at BUB plots (Figure 3). To further investigate impacts of the measured variables on individual tick species, we constructed models specifically for *A. americanum* (Table 4) and *I. scapularis* (Table 5). *A. americanum* counts were best predicted by burning, season, bobcat presence/absence, and wind at ground level on day of tick counts. A significant interaction occurred between burning and season, with *A. americanum* not being significantly affected by season in burned plots [RR = 2.4 (95% CI: 0.3, 19.0); P = 0.409]; in other words, no seasonality was observed because there were so few ticks present in burned plots. However, at plots in which there was no burning, *A. americanum* counts were over 30 times greater in the warm season than in the cold [RR = 33 (95% CI: 14, 81); $P<0.001$], indicating what would be an expected seasonality pattern for *A. americanum*. Together, these findings suggest that prescribed burning reduced *A. americanum* abundance. The presence of bobcats (*Lynx rufus*) was positively associated with *A. americanum* counts, while an increase in wind by 1 kilometer per hour resulted in a 69% decrease in *A. americanum* detectability. Finally, *I. scapularis* counts were best predicted by burning, season, year, tree density, and precipitation. Long-term prescribed burning reduced *I. scapularis* counts by 78%. Furthermore, *I. scapularis* counts were reduced by approximately 85% in the warm season relative to the cold season and counts were 3.7 times greater in 2011 than in 2010. Lastly, *I. scapularis* counts were 18 times greater in plots with high tree density (>183 trees per ha) relative to plots with low tree density and were twice as likely to be collected if precipitation occurred during the 3 days prior to sampling.

Discussion

To our knowledge, this is one of the most comprehensive studies examining the effects of long-term operational prescribed burning while also incorporating other variables that are known to affect tick abundance. Furthermore, this is the first long-term study of tick phenology in southwestern Georgia, and provides valuable insight into seasonality and species composition within common habitats of the region.

Gaining a better understanding of tick species composition and seasonality in this rarely studied region provides important information for public health education and research. Our findings regarding seasonality of ticks were generally in agreement

Table 2. Number of ticks collected from different burn regimes during 2010 and 2011.

	Treatment	adult		nymph		larvae (clutches)	
		2010	2011	2010	2011	2010	2011
A. americanum	BB	0	1	0	0	0	0
	BUB	4	7	2	21	1,362 (1)	291 (2)
	UBB	5	4	24	4	2,354 (2)	1,177 (5)
	UBUB	272	170	1,171	981	26,432 (147)	12,433 (72)*
A. maculatum	BB	30	6	1	0	0	0
	BUB	25	6	0	1	0	0
	UBB	3	2	0	1	0	0
	UBUB	0	3	0	0	0	unknown*
I. scapularis	BB	2	3	0	0	0	0
	BUB	8	18	0	0	0	0
	UBB	40	94	1	0	0	0
	UBUB	47	128	5	1	5 (1)	0
D. variabilis	BB	0	0	0	0	0	0
	BUB	9	2	0	0	0	0
	UBB	7	2	0	0	0	0
	UBUB	13	5	0	0	0	0

BB = Burned, surrounded by burned, BUB = Burned, surrounded by unburned, UBB = Unburned, surrounded by unburned, UBUB = Unburned, surrounded by unburned.
*1 group of 40 larvae (not included in the numbers in this table) thought to be a single clutch was found to contain both *A. americanum* and *A. maculatum* upon PCR testing.

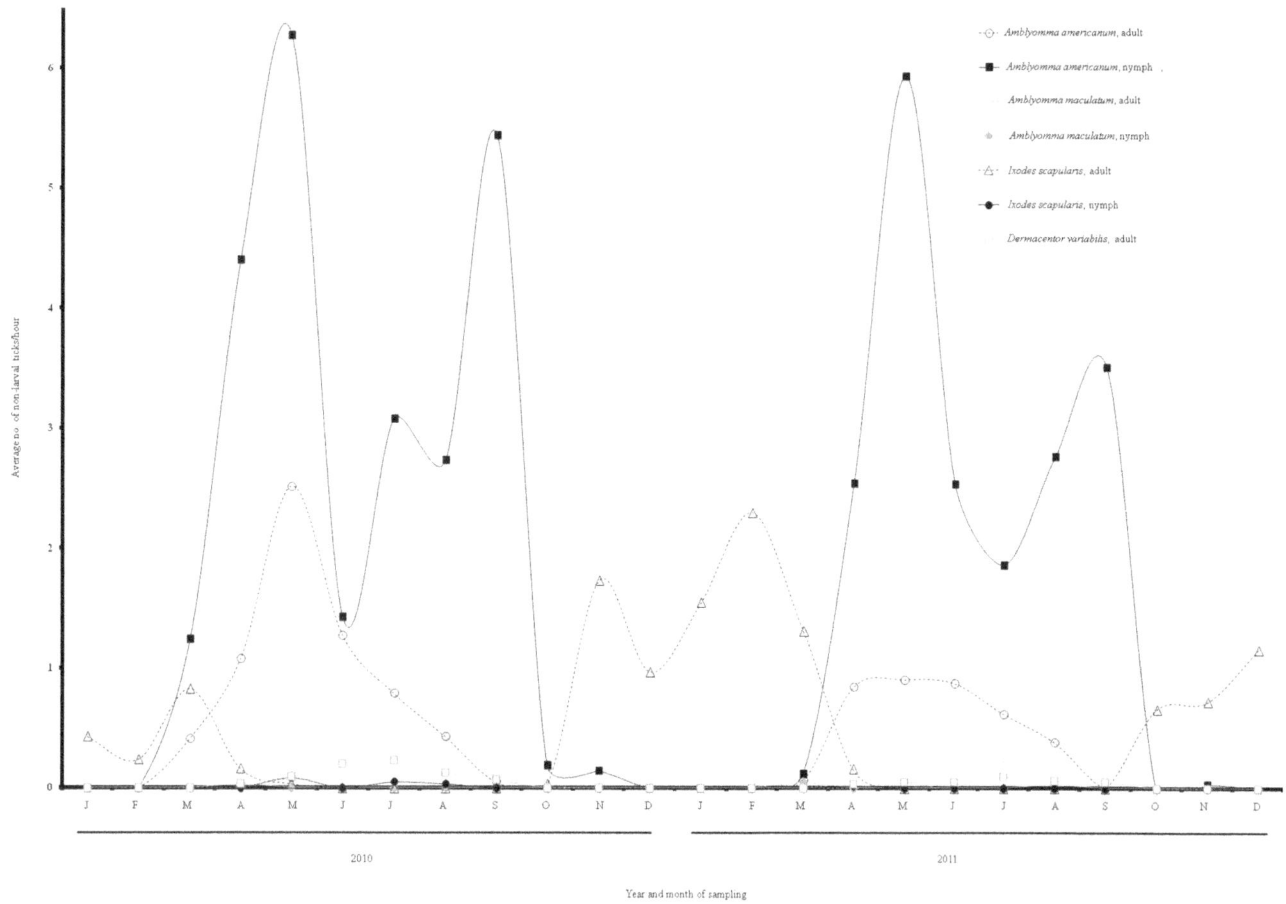

Figure 1. Seasonality of each tick species and life stage across all study plots for 2010 and 2011.

with studies done in other regions of Georgia and northern Florida [24,27]. One exception was that we detected a bi-modal peak for *A. americanum* nymphal activity similar to work conducted by Cilek and Olson [27] in northern Florida, whereas only a single peak was noted by Davidson et al. [24] in the Piedmont region of Georgia (i.e., farther north). This may indicate that nymphal *A. americanum* populations south of the fall line in Georgia have bi-modal peaks, whereas those north of the fall line have a single peak. Also, to our knowledge, this marks only the second longitudinal study on *A. maculatum* seasonality [42], providing important information on this tick which has gained interest in recent years due to expanding populations, transmission of numerous pathogens, and having been identified as a vector of the causative agent of *Rickettsia parkeri* rickettsiosis [43].

Past studies suggested that climate (e.g. temperature and rainfall) affects tick abundance and distribution [24,44] and we suspect that the differences observed in tick phenology from 2010 to 2011 were a result of climate differences. In 2011 there were significantly more *I. scapularis*. Furthermore, the bi-modal peak of *I. scapularis* occurred later in 2011 as compared to 2010. These differences may have been due to an exceptionally dry summer followed by a wet fall and mild winter (*I. scapularis* adult peak activity) in 2011 as compared to 2010.

Long-term prescribed burning significantly reduced tick counts. Although some previous studies have found either increases in tick populations or a recovery of tick populations following a burn [19–28], our results suggest that prior conclusions may be the result of failure to account for other variables and failure to simulate real-world management scenarios in which large-scale burns are regularly applied. Because operational burns typically occur on larger areas, re-colonization by ticks in these larger areas may take longer than observed in studies using smaller plots. Furthermore, repeated burn events occurring over time may deplete source populations of ticks.

Interestingly, we observed variation in the tick community composition relative to burn treatment which, to our knowledge, has never been reported. The most striking difference was related to greater abundances of *A. maculatum* in BB plots whereas *A. americanum* was most abundant in UBUB plots. In general, *A. americanum* are more abundant in hardwood forests [45], and some have suggested that *A. americanum* outcompetes other tick species, such as *D. variabilis* [46]. Thus, the greater abundances of *A. americanum* in UBUB plots which were hardwood dominated would arguably be expected. The greater question deals with the absence of *A. americanum* and dominance of *A. maculatum* in BB plots. Burning at BB plots directly altered the physical vegetation structure and maintained dominance of pines in the overstory. Thus, BB plots were generally open-canopied pine forests with a diverse understory and sparse midstory. These vegetation conditions result in a hotter, dryer environment than UBUB plots which are closed-canopy hardwood forests with dense mid-stories and sparse, litter-covered understories. We suspect that this plays a key role in both the long-term reduction of ticks and shift in tick species composition observed at these plots. Our models support

Table 3. Generalized estimating equation negative binomial regression model for the prediction of total tick counts at all 21 plots in 2010 and 2011.

Variable	Coefficient (SE)	RR (95% CI)	P
Any Burning (vs. No Burning*)	−1.5 (0.43)	ND	0.001
2011 (vs. 2010*	0.29 (0.24)	1.3 (0.84, 2.2)	0.22
Hot (Spring/Summer) (vs. Cool [Fall/Winter]*)	2.4 (0.48)	ND	<0.001
>95% Litter Cover (vs. <95%*)	0.81 (0.33)	2.2 (1.2, 4.3)	0.014
>183 Trees per Ha(vs. <183*)	2.3 (0.66)	6.4 (2.6, 35)	0.001
Any Burning X Season	−2.3 (0.59)	ND	<0.001
Constant	−2.0 (0.84)	NA	0.016
ln(Hours flagged)	1		

SE = Standard error. RR = Relative rate. ND = Not determined; RR is not given because it depends on the interacting variable. NA = Not applicable.
*Indicates the reference category.

this as vegetation structure (tree density and litter cover specifically) and microclimatic variables which would affect tick moisture retention were most important in predicting tick counts. Willis et al. [30] had similar findings in that *A. americanum* was found to be associated with litter cover and negatively associated with pine sapling density. Also in support of this theory, Gleim et al. [47] reported that *A. americanum* had significantly lower survival in experimental enclosures in burned habitats vs. unburned habitats compared with *A. maculatum*. Collectively, these data suggest that different physiologic and/or behavioral traits of different tick species increase survival success in certain environments [47,48].

Although host occurrence (specifically mesomammals and deer) was considered in all models of tick counts, only bobcat presence was an important predictor of *A. americanum* counts, suggesting that those host populations monitored did not generally impact tick counts. It is important to note, however, that our methodology (trail camera surveys) did not account for small mammal and bird

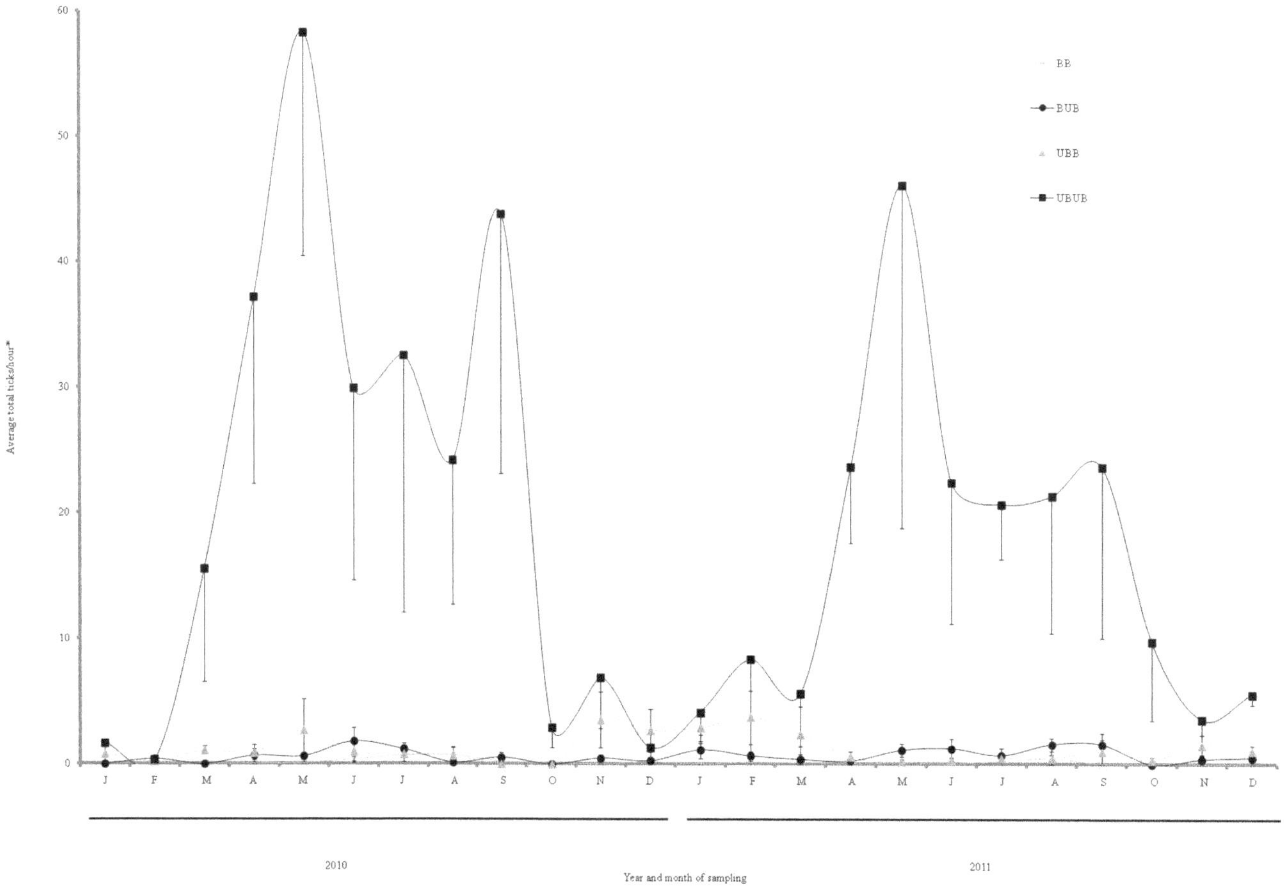

Figure 2. Effects of long-term prescribed burning on tick abundance at all study plots. *One clutch of larvae was counted as one tick.
doi:10.1371/journal.pone.0112174.g002

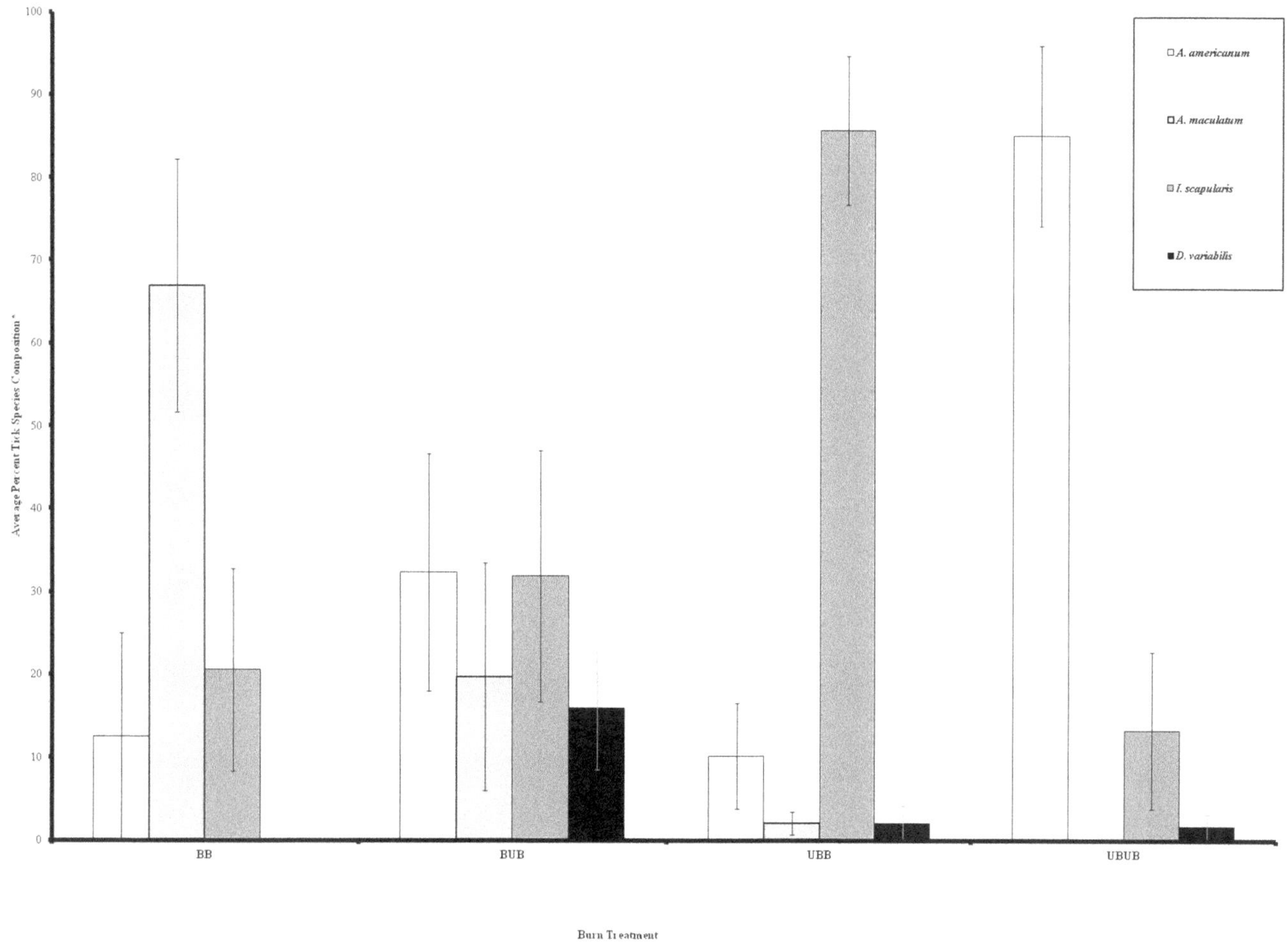

Figure 3. Ticks species composition at all study plots by burn treatment for 2010–2011. *One clutch of larvae was counted as one tick.

populations which are important hosts for both *A. americanum* and *A. maculatum* larvae and nymphs [49–51]. However, previous studies have found that small mammals and ground nesting birds tend to be found in greater abundance in burned areas where there is dense ground cover as compared to unburned areas in which ground cover is sparse to non-existent [52,53]. These data would therefore refute the assertion that the high tick abundances observed in UBUB were due to larger populations of small mammals and birds in these areas.

The failure to identify a relationship between host occurrence and tick counts questions acceptance of the dogma that host abundance drives tick abundance [54–56]. However, manipulative and observational studies have been inconclusive regarding a correlation between host and tick abundance [57,58]. Although

Table 4. Generalized estimating equation negative binomial regression model for the prediction of *A. americanum* counts at all study 21 plots.

Variable	Coefficient (SE)	RR (95% CI)	P
Any Burning (vs. No Burning*)	−2.5 (1.1)	ND	0.026
2011 (vs. 2010*)	0.07 (0.34)	1.1 (0.55, 2.1)	0.837
Hot (Spring/Summer) (vs. Cold [Fall/Winter]*)	3.5 (0.45)	ND	<0.001
Bobcats Present (vs. Absent)	1.2 (0.31)	3.2 (1.7, 5.8)	<0.001
Wind at Ground (mph)	−1.2 (0.34)	0.31 (0.15, 0.59)	<0.001
Any Burning X Season	−2.6 (1.1)	ND	0.019
Constant	1.3 (0.84)	NA	0.717
ln(Hours flagged)	1	(Effort)	

SE = Standard error. RR = Relative rate. ND = Not determined; RR is not given because it depends on the interacting variable. NA = Not applicable.
*Indicates the reference category.

Table 5. Generalized estimating equation negative binomial regression model for the prediction of *I. scapularis* counts at all 21 plots in 2010 and 2011.

Variable	Coefficient (SE)	RR (95% CI)	P
Any Burning (vs. No Burning*)	−1.2 (0.40)	0.29 (0.13, 0.63)	0.002
2011 (vs. 2010*)	1.3 (0.42)	3.8 (1.6, 8.7)	0.002
Hot (Spring/Summer) (vs. Cold [Fall/Winter]*)	−1.9 (0.47)	0.15 (0.06, 0.37)	<0.001
>183 Trees per Ha (vs.<183*)	2.9 (0.98)	18 (2.6, 122)	0.003
>0 cm Precipitation Previous 3 Days (vs. 0 cm*)	0.74 (1.1)	2.1 (1.2, 3.8)	0.014
Constant	−3.6 (1.1)	NA	0.001
ln(Hours flagged)	1	(Effort)	

SE= Standard error. RR= Relative rate. NA= Not applicable.
*Indicates the reference category.

bobcat presence was an informative predictor of *A. americanum* counts, we suggest that it is more likely bobcats were associated with other factors impacting *A. americanum* counts rather than actually acting as a causal mechanism affecting *A. americanum* counts. We suspect that while host abundance may play a minor role in tick dynamics, other factors such as burning and microclimate play larger roles in driving tick dynamics in the systems we studied.

In conclusion, operational prescribed burning in pine and mixed pine ecosystems significantly reduces tick populations which supports several previous studies [20,26]. A number of previous studies showed that although burning causes tick mortality, they recolonized the area within a few years [24,25,30]. We suggest that the repeated burnings associated with operational prescribed fires and the resulting vegetation and microclimate are ultimately responsible for the long-term reductions in tick populations observed in our study. These findings illustrate another benefit of prescribed fire and have important implications for public health as it is a time efficient, cost effective method for reducing tick populations and likely reduces the risk of human tick-borne diseases.

Future studies should include pathogen testing to gain a more complete understanding of tick-borne disease risk and dynamics in areas that utilize prescribed burns. Furthermore, studies examining the efficacy of prescribed burning in different management scenarios as well as determining the length of time required to provide overall reductions in tick populations within previously unburned areas is warranted.

Acknowledgments

We thank the many staff members at the J.W. Jones Ecological Research Center (JERC) for field support and assistance, particularly Jessica Rutledge, Gail Morris, and the many field technicians of the JERC Wildlife Laboratory. We also thank the Georgia Department of Natural Resources, Wildlife Resources Division, Region 5, particularly Brandon Rutledge, Brian Vickery, and the many wildlife technicians for their cooperation and field support. Thanks also to Murry and Marilyn Campbell, Perry Clements, Gordon Miller, and Chuck Wester for use of their land and field support. Finally, we thank Barbara Shock, Whitney Kistler, and Merrill Montgomery for laboratory assistance.

Author Contributions

Conceived and designed the experiments: ERG LMC MLL MJY. Performed the experiments: ERG GEZ. Analyzed the data: ERG RDB. Contributed reagents/materials/analysis tools: LMC MLL MJY. Wrote the paper: ERG LMC RDB MLL GEZ MJY.

References

1. Maeda K, Markowitz N, Hawley RC, Ristic M, Cox D, et al. (1987) Human infection with *Ehrlichia canis*, a leukocytic rickettsia. N Engl J Med. 316: 853–856.
2. Parola P, Paddock CD, Raoult D (2005) Tick-borne rickettsioses around the world: emerging diseases challenging old concepts. Clin Microbiol Rev. 18: 719–756.
3. Goodin DG, Koch DE, Owen RD, Chu Y, Hutchinson JMS, et al. (2006) Land cover associated with hantavirus presence in Paraguay. Glob Ecol Biogeogr. 15: 519–527.
4. Apperson CS, Engber B, Nicholson WL, Mead DG, Engel J, et al. (2008) Tick-borne diseases in North Carolina: is *Rickettsia amblyommii* a possible cause of rickettsiosis reported as Rocky Mountain spotted fever? Vector Borne Zoonotic Dis. 8: 597–606.
5. Bacon RM, Kugeler KJ, Mead PS (2008) Surveillance for Lyme disease—United States, 1992–2006. Morb Mort Weekly Rep. 57: 1–9.
6. Ogden NH, Lindsay M, Morshed M, Sockett PN, Artsob H (2009) The emergence of Lyme disease in Canada. Can Med Assoc J. 180: 1221–1224.
7. Thomas RJ, Dumler JS, Carolyn JA (2009) Current management of human granulocytic anaplasmosis, human monocytic ehrlichiosis, and *Ehrlichia ewingii* ehrlichiosis. Expert Rev Anti Infect Ther. 7: 709–722.
8. Allan B, Keesing F, Ostfeld R (2003) Effect of Forest Fragmentation on Lyme disease risk. Conserv Biol. 17: 267–272.
9. Patz JA, Campbell-Lendrum D, Holloway T, Foley JA (2005) Impact of regional climate change on human health. Nature. 438: 310–317.
10. Haines TK, Busby RL, Cleaves DL (2001) Prescribed burning in the South: Trends, purposes, and barriers. Southern J. Appl. Forestry. 25: 149–153.
11. Kirkman LK, Mitchell RJ, Helton RC, Drew MB (2001) Productivity and species richness across an environmental gradient in a fire-dependent ecosystem. Am J Bot. 88: 2119–2128.
12. Fernandes PM, Botehlo HS (2003) A review of prescribed burning effectiveness in fire hazard reduction. Int J Wildland Fire. 12: 117–128.
13. Townes G, Owensby C (1984) Long-term effects of annual burns at different dates in ungrazed Kansas tallgrass prairie. J Range Manage. 37: 392–397.
14. Rego FC, Bunting SC, da Silva JM (1991) Changes in understory vegetation following prescribed fire in maritime pine forests. Forest Ecol Manag. 41: 21–31.
15. Sparks JC, Masters RE, Engle DM, Palmer MW, Michael W, et al. (1998) Effects of late growing-season and late dormant-season prescribed fire on herbaceous vegetation in restored pine-grassland communities. J Veg Sci. 9: 133–142.
16. Knapp EE, Keeley JE, Ballenger EA, Brennan TJ (2005) Fuel reduction and coarse woody debris dynamic with early season and late season prescribed fire in a Sierra Nevada mixed conifer forest. Forest Ecol Manag. 208: 383–397.
17. Hobbs NT, Stowart RA (1984) Effects of prescribed fire on nutrition of mountain sheep and mule deer during winter and spring. J Wildl Manag. 48: 551–560.
18. Allan B (2009) Influence of prescribed burns on abundance of *Amblyomma americanum* (Acari: Ixodidae) in the Missouri Ozarks. J Med Entomol. 46: 1030–1036.
19. Hoch A, Semtner P, Barker R, Hair J (1972) Preliminary observations on controlled burning for lone star tick (Acarina: Ixodidae) control in woodlots. J Med Entomol. 9: 446–451.

20. Jacobson H, Hurst G (1979) Prevalence of parasitism by *Amblyomma americanum* on wild turkey poults as influenced by prescribed burning. J Wildl Dis. 15: 43–47.
21. Drew M, Samuel W, Lukiwski G, Willman J (1985) An evaluation of burning for control of winter ticks, *Dermacentor albipictus*, in central Alberta. J Wildl Dis. 21: 313–315.
22. Wilson M (1986) Reduced abundance of adult *Ixodes dammini* (Acari: Ixodidae) following destruction of vegetation. Journal of Economic Entomology. 79: 693–696.
23. Mather T, Duffy D, Campbell S (1993) An unexpected result from burning vegetation to reduce Lyme disease transmission risks. J Med Entomol. 30: 642–645.
24. Davidson WR, Siefkan DA, Creekmore LH (1994) Influence of annual and biennial prescribed burning during March on the abundance of *Amblyomma americanum* (Acari: Ixodidae) in central Georgia. J Med Entomol. 31: 72–81.
25. Stafford III KC, Ward JS, Magnarelli LA (1998) Impact of controlled burns on the abundance of *Ixodes scapularis* (Acari: Ixodidae). J Med Entomol. 35: 510–513.
26. Cully J (1999) Lone star tick abundance, fire, and bison grazing in tallgrass prairie. J Range Manage. 52: 139–144.
27. Cilek J, Olson M (2000) Seasonal distribution and abundance of ticks (Acari: Ixodidae) in northwestern Florida. J Med Entomol. 37: 439–444.
28. Padgett KA, Casher LE, Stephens SL, Lane RS (2009) Effect of prescribed fire for tick control in California chaparral. J Med Entomol. 46: 1138–1145.
29. Polito VJ (2012) Effects of patch mosaic burning on tick burden of cattle, tick survival, and tick abundance. M.S. Thesis. Oklahoma State University, Oklahoma.
30. Willis D, Carter R, Murdock C, Blair B (2012) Relationship between habitat type, fire frequency, and *Amblyomma americanum*, populations in east-central Alabama. J Vector Ecol. 37: 373–381.
31. Bailey J, Whitham T (2002) Interactions among fire, aspen and elk affect insect diversity: reversal of a community response. Ecology. 83: 1701–1712.
32. Frater HE (2011) Impact of prescribed burning for oak regeneration on forest vegetation, white-footed mouse populations, and Lyme disease. M.S. Thesis. Iowa State University, Iowa.
33. Kierans JE, Litwak TR (1989) Pictorial key to the adults of hard ticks, family Ixodidae (Ixodidae: Ixodoidea), east of the Mississippi River. J Med Entomol. 26: 435–448.
34. Kierans JE, Durden LA (1998) Illustrated key to nymphs of the tick genus *Amblyomma* (Acari: Ixodidae) found in the United States. J Med Entomol. 35: 489–495
35. Zemtsova GE, Watkins NE, Levin ML (2014) Multiplex qPCR assay for identification and differentiation of A. americanum, A. cajennense and A. maculatum tick species in the Eastern US. J Med Entomol. In press.
36. Black WC, Piesman J (1994) Phylogeny of hard- and soft-tick taxa (Acari: Ixodidae) based on mitochondrial 16S rDNA sequences. Proc Natl Acad Sci. 91: 10034–10038.
37. Daubenmire RF (1959) A canopy-cover method of vegetational analysis. Northwest Sci 33: 43–46.
38. Lemmon PE (1956) A spherical densiometer for estimating forest overstory density. Forest Sci 2: 314–320.
39. StataCorp (2009) Stata Statistical Software: Release 11. College Station, TX: StataCorp LP.
40. Piesman J, Mather TN, Dammin GJ, Telford SR III, Lastavica CC, et al. (1987) Seasonal variation of transmission risk of Lyme disease and human babesiosis. Am J Epidemiol. 126: 1187–1189.
41. Barbour AG, Fish D (1993) The biological and social phenomenon of Lyme disease. Science. 260: 1610–1616.
42. Goddard J, Paddock CD (2005) Observations on distribution and seasonal activity of the Gulf Coast tick in Mississippi. J Med Entomol. 42: 176–179.
43. Paddock CD, Sumner JW, Comer JA, Zaki SR, Goldsmith CS, et al. (2004) *Rickettsia parkeri:* A newly recognized cause of spotted fever rickettsiosis in the United States. Clin Infect Dis. 38: 805–811.
44. Ogden NH, Maarouf A, Barker IK, Bigras-Poulin M, Lindsay LR, at al. (2006) Climate change and the potential for range expansion of the Lyme disease vector *Ixodes scapularis* in Canada. Int J Parasitol. 36: 63–70.
45. Semtner PJ, Howell DE, Hair JA (1971) The ecology and behavior of the lone star tick (Acarina: Ixodidae): The relationship between vegetative habitat type and tick abundance and distribution in Cherokee, Co., Oklahoma. J Med Entomol. 8: 329–335.
46. Stromdahl EY, Hickling GJ (2012) Beyond Lyme: etiology of tick-borne human diseases with emphasis on the South-Eastern United States. Zoonoses Public Hlth. 59: 48–64.
47. Gleim ER, Conner LM, Yabsley MJ (2013) The effects of *Solenopsis invicta* (Hymenoptera: Formicidae) and burned habitat on the survival of *Amblyomma americanum* (Acari: Ixodidae) and *Amblyomma maculatum* (Acari: Ixodidae). J Med Entomol. 50: 270–276.
48. Cooley RA, Kohls GM (1944) The genus *Amblyomma* in the United States. J. Parasitol. 30: 77–111.
49. Teel PD, Hopkins SW, Donahue WA, Strey OF (1998) Population dynamics of immature *Amblyomma maculatum* and other ectoparasites on meadowlarks and northern bobwhite quail resident to the Coastal Prairie of Texas. J Med Entomol. 35: 483–485.
50. Barker RW, Kocan AA, Ewing SA, Wettemann RP, Payton ME (2004) Occurrence of the Gulf Coast tick on wild and domestic animals in North-Central Oklahoma. J Med Entomol. 41: 170–178.
51. Landers JL, Crawford RL (1995) NB66: A study of habitat relations of nongame birds and mammals in series of oldfield pinewoods following fire exclusion. Florida Nongame Wildlife Program Project Report GFC-84-004. Tallahassee, FL: Florida Game and Fresh Water Fish Commission, Nongame Wildlife Program.
52. Masters RE (2007) The importance of shortleaf pine for wildlife and diversity in mixed oak–pine forests and in pine–grassland woodlands. Shortleaf pine restoration and ecology in the Ozarks. Gen Tech Rep NRS-P-15, USDA Forest Service Northern Research Station, Newtown Square: 35–46.
53. Needham GR (1991) Off-host physiological ecology of ixodid ticks. Annu Rev Entomol. 36: 659–681.
54. Paddock CD, Yabsley MJ (2007) Ecological havoc, the rise of white-tailed deer, and the emergence of *Amblyomma americanum*-associated zoonoses in the United States. Curr Top Microbiol. 315: 289–324.
55. Brunner JL, Ostfeld RS (2008) Multiple causes of variable tick burdens on small-mammal hosts. Ecology. 89: 2259–2272.
56. Oorebeek M, Kleindorfer S (2008) Climate or host availability: what determines the seasonal abundance of ticks? Parasitol Res. 103: 871–875.
57. Daniels TJ, Fish D (1995) Effect of deer exclusion on the abundance on immature *Ixodes scapularis* (Acari: Ixodidae) parasitizing small and medium-sized mammals. J Med Entomol. 32: 5–11.
58. Schulze TL, Jordan RA, Hung RW (2001) Potential effects of animal activity on the spatial distribution of *Ixodes scapularis* and *Amblyomma americanum* (Acari: Ixodidae). Environ Entomol. 30: 568–577.

20

Using Unplanned Fires to Help Suppressing Future Large Fires in Mediterranean Forests

Adrián Regos[1,2]*, Núria Aquilué[1,2], Javier Retana[2,3], Miquel De Cáceres[1,2], Lluís Brotons[1,2]

1 CEMFOR-CTFC (Centre Tecnològic Forestal de Catalunya), Solsona, Spain, **2** CREAF (Centre de Recerca Ecològica i Aplicacions Forestals), Bellaterra, Spain, **3** Universitat Autònoma Barcelona, Bellaterra, Spain

Abstract

Despite the huge resources invested in fire suppression, the impact of wildfires has considerably increased across the Mediterranean region since the second half of the 20th century. Modulating fire suppression efforts in mild weather conditions is an appealing but hotly-debated strategy to use unplanned fires and associated fuel reduction to create opportunities for suppression of large fires in future adverse weather conditions. Using a spatially-explicit fire–succession model developed for Catalonia (Spain), we assessed this opportunistic policy by using two fire suppression strategies that reproduce how firefighters in extreme weather conditions exploit previous fire scars as firefighting opportunities. We designed scenarios by combining different levels of fire suppression efficiency and climatic severity for a 50-year period (2000–2050). An opportunistic fire suppression policy induced large-scale changes in fire regimes and decreased the area burnt under extreme climate conditions, but only accounted for up to 18–22% of the area to be burnt in reference scenarios. The area suppressed in adverse years tended to increase in scenarios with increasing amounts of area burnt during years dominated by mild weather. Climate change had counterintuitive effects on opportunistic fire suppression strategies. Climate warming increased the incidence of large fires under uncontrolled conditions but also indirectly increased opportunities for enhanced fire suppression. Therefore, to shift fire suppression opportunities from adverse to mild years, we would require a disproportionately large amount of area burnt in mild years. We conclude that the strategic planning of fire suppression resources has the potential to become an important cost-effective fuel-reduction strategy at large spatial scale. We do however suggest that this strategy should probably be accompanied by other fuel-reduction treatments applied at broad scales if large-scale changes in fire regimes are to be achieved, especially in the wider context of climate change.

Editor: Juan A. Añel, University of Oxford, United Kingdom

Funding: This study was supported by the research projects BIONOVEL (CGL2011-29539/BOS) and MONTES (CSD2008-00040) funded by the Spanish Ministry of Education and Science. The funders had no role in study design, data collection and analysis, decision to publish, or preparation of the manuscript.

Competing Interests: The authors declare that no competing interests exists.

* E-mail: adrian.regos@ctfc.es

Introduction

Wildland fires are a major component of disturbance regimes in many regions [1]. While climate and vegetation characteristics have been described as major determinants of fire regimes, in the Mediterranean Basin and similar regions where human influence is widespread, fire regimes emerge as a complex process in which landscape planning, economic activities and fire management can override the influence of natural factors [2,3,4].

Despite the huge amount of resources invested in fire prevention and suppression, the impact of wildfires has considerably increased since the second half of 20th century across different Mediterranean regions [1,3,5]. Fire suppression efforts have been stepped up in recent years, but while they appear to successfully deal with wildfires in mild weather conditions, there are doubts over the efficiency of these policies in climatically-adverse conditions [6,7,8]. Recent wildfires tend to be larger and more severe as a consequence of an increase in fuel accumulation and continuity (induced mainly by the abandonment of agriculture and livestock, and active afforestation policies) [3,9,10] coupled with drier and warmer climatic conditions [5]. In addition, the expansion of populations and the wildland–urban interface has also contributed to more fire ignition events [11].

In mesic regions of the Mediterranean basin (typically the Eastern Iberian Peninsula), fuel is now less limiting, and fire regimes appear to be mainly driven by the occurrence of climatically adverse conditions [12]. Different strategies can be envisaged to reduce the growing impact of wildfires in the Mediterranean region. Fire suppression policies have traditionally focussed on the preventive early detection of ignitions, but measures have recently shifted towards strategic planning and anticipation of fire spread to make optimal use of firefighting resources [13]. Other issues have been recognized as key, such as the management of fuel to reduce fire intensity and the extent and impact of large fires, but they pose difficulties in terms of effective implementation at large spatial scales [14,15,16,17]. Fuel reduction may increase the chances of suppressing large fires in adverse climate conditions [8,18]. Forest management (including grazing) has been proposed as a measure for reducing the accumulation of forest fuel. However, land abandonment is widespread in most regions affected by fire, and all available evidence suggests that forests are expanding in most of the Mediterranean [19,20]. Prescribed fires are increasingly used to reduce fuel, but many countries face strong public opposition to this preventive action, making prescribed fires more difficult to apply at large scales for efficient fuel reduction than in other regions with Mediterranean-type climate, such as Australia [16,21,22].

Given that wildfires are currently seen as one of the main drivers of forest landscape changes in many Mediterranean regions, wildfires occurring in mild weather conditions could become a tool to regulate the impact of undesired, destructive, large fires taking place in climatically adverse conditions [6,23]. Modulating fire suppression efforts in less adverse climatic conditions could allow a strategy to use unplanned fire events and the associated fuel reduction to create opportunities for efficient suppression of large fires in future adverse conditions [6,24]. While such a strategy has the potential to control and reduce fuel using the current pattern of ignitions, any attempt to change the basic firefighting principle of tackling "all fires" as soon as possible would obviously meet with controversy, especially since very little information is available on the potential effectiveness of such a fire management strategy. Here, we address these questions using landscape simulations under different scenarios combining fire suppression strategies and climatic severity. Specifically, we investigate the potential of this opportunistic strategy for reducing the impact of large fires in climatically adverse conditions. We also assess what amount of area would need to be burnt in mild weather conditions to obtain reductions in the area burnt by large fires under a future climatic warming scenario. Finally, we discuss the spatial scale at which this strategy may be implemented, and its possible future socio-economic implications.

Material and Methods

Study area

The study area was Catalonia, a region located in northeastern of Spain with a land area of 32,115 km^2 and an altitude that ranges from sea level to 3102 m. Catalonia has a typical Mediterranean climate with low winter precipitation and hot and dry summers. Moreover, its complex topography induces major variability in climatic and fire weather conditions across the territory. The vegetation is mostly comprised of forest and shrubland. Evergreen species occur in 60% of the total forest area, 73% of which is occupied by conifers (mainly *Pinus sylvestris*, *Pinus halepensis* and *Pinus nigra*), with sclerophyllous and deciduous species (*Quercus ilex*, *Quercus faginea* and *Quercus suber*) covering the remaining 40% [25]. According to the CORINE land cover map [26], shrubland of diverse species (and mainly evergreen) composition covers 37% of the total wildland area (Fig 1). Forest and shrubland were the most affected by fire during the 1975–1998 period [27].

During the same period, wildfires burnt about 13% of the wildland area (around 250,000 ha), and both the frequency of fire events and area burnt have increased since the pre-1970 period [12,27]. This shift in fire regime was mainly driven by the increase in fuel amount and continuity following rural land abandonment across Catalonia during the pre-1970 period [10,28]. As a consequence, resources allocated to fire suppression increased in the early 80s (*Focverd I* fire suppression program) and 90s (*Focverd II*) [29]. After 1999, fire-fighting capacities were improved through the creation of specific technical fire brigades (GRAF) whose mission is to understand fire behaviour and anticipate changes in fire propagation [29]. At the beginning of the 21st century, prescribed burning programs were also implemented among other efforts to reduce fire risk, but not at sufficiently broad scales to achieve effective reductions [13]. Despite all these increased fire suppression and prevention efforts, wildfires continue to burn thousands of hectares in Catalonia every year. In addition, the aridity trends observed over recent decades point to an increase in the number of dry days per summer [5].

MEDFIRE simulation model

The MEDFIRE model [4,30] is a spatially-explicit landscape model that is able to mimic changes in landscape composition derived from vegetation dynamics and fire disturbances.

We present here a **short overview** of the model. The complete description, calibration, and validation processes for the study area can be obtained from previous published work [4]. The model was implemented using the version 3.5 of SELES modelling platform [31] (http://www.seles.info/). The current version of MEDFIRE (version 2) has two main sub-models [4]: (i) after-fire succession and maturation of vegetation (vegetation dynamics sub-model), and (ii) wildfire disturbance (fire sub-model) (more details in https://sites.google.com/site/medfireproject/). The main purpose of the model is to examine the spatial interaction between wildfires, vegetation dynamics and fire suppression strategies. It was designed to assess how different drivers affect fire regime at short- and medium-term timescales through a quantitative evaluation of their effects on the distribution of the annual area burnt, fire size distribution, and landscape composition. Validation exercises carried out for different time windows with different climate and fire suppression data showed that the model was able to reproduce the basic descriptors of fire regime in our study area [4].

The **state variables** that MEDFIRE uses to describe landscape context and conditions are spatially explicit variables in raster format at 100 m resolution. Land cover type (LCT) and time since last fire (TSF) are dynamic variables, while the static variables are: ignition probability, bioclimatic region, fire spread type (relief- or wind-driven), elevation, aspect and main wind direction.

The **fire sub-model** uses a top-down approach: for each time-step (a year), fires are simulated until the potential annual area to be burnt is reached. Potential annual area refers to the area that is expected to burn according to the historical fire data (1975–99 period). According to previous research [5], climatically adverse years are characterized by a high number of weather risk days (hereafter referred to as "adverse years"), as opposed to years dominated by mild weather conditions (hereafter called "mild years"). Thus, potential burnt area and fire size distributions depend on the climatic severity of the summer. For each simulated fire, the model chooses an ignition location used to establish the fire spread type [32]. The spread rate is a function of TSF (as a proxy of fuel accumulation), LCT flammability (of burnable land covers), aspect and wind direction (in wind-driven fires) or topography (in relief-driven fires). Fire spread rate was parameterized in a calibration exercise comparing model outputs with historical fire data [4]. In the absence of fire suppression, all the pixels that could be reached within the timespan of the fire are recorded as burnt (i.e. post-fire transitions may occur, and the TSF is set to 0). Otherwise, if fire suppression occurs, the pixels recorded as burnt include only a subset of pixels that were not affected by fire suppression, and therefore the final fire size will be smaller than was potential. We assume that long-term droughts or long periods with high temperatures or strong winds can be predicted and, therefore, high-risk fire conditions can be anticipated before an ignition takes places thus permitting firefighters to distinguish fire conditions. Thus, although the model assumes that climate is the main driver of fire regime, key elements of fire regime such as fire size can be modulated by fire suppression.

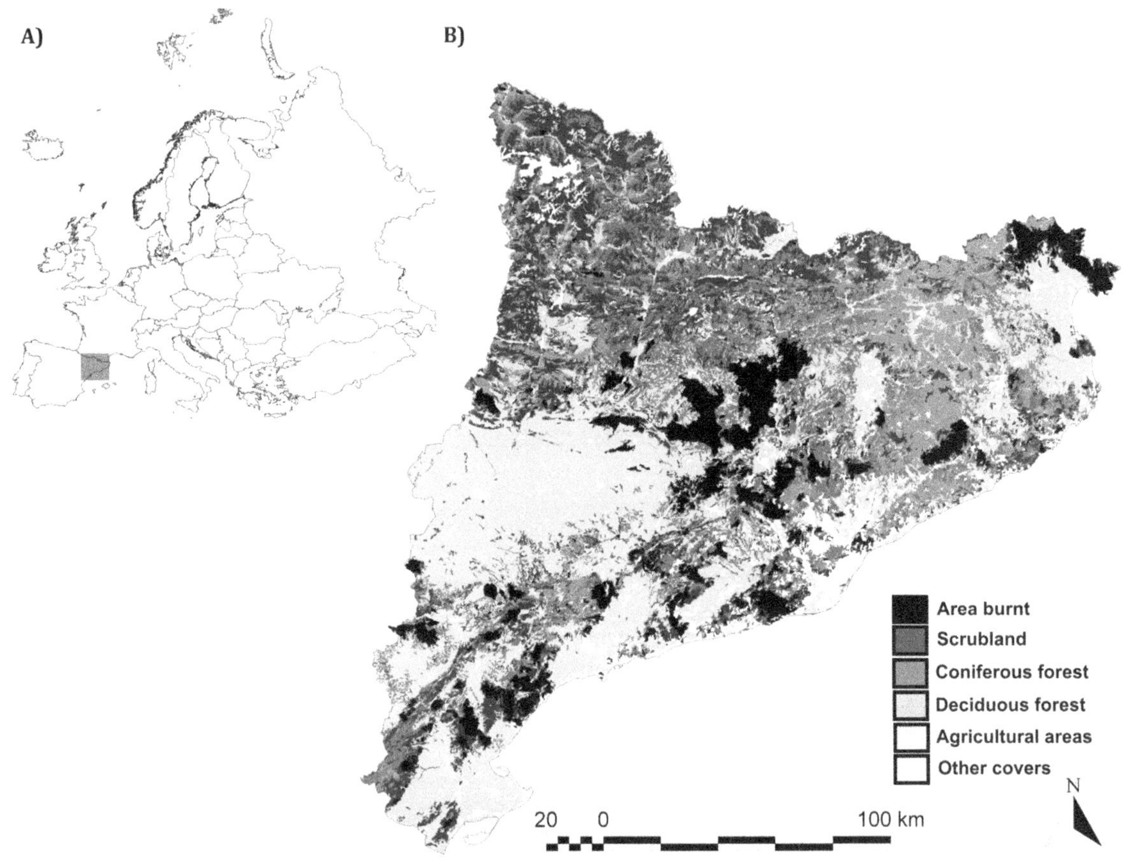

Figure 1. Location of the study area. Geographic location of the study area in south-western Europe (A). Land covers affected by fires between 1980 and 2000 in Catalonia (B).

The **vegetation dynamics sub-model** assumes that forest cover types are relatively stable, so a type-conversion can only occur after burning. Succession without burning can occur only from shrubland to forest. This land cover change takes place depending on the availability of mature forest in neighbouring cells and the TSF of shrubland that will potentially change. Once a cell is burnt, this sub-model updates the land cover according to two post-fire regeneration approaches:

- Applying non-spatial stochastic transitions, using a multinomial distribution with transition probabilities previously published by others [33] that depend on pre-fire cover class as well as on other factors such as aspect, bioclimatic region and TSF.
- By neighbourhood species contagion, considering the neighbours that were also burnt in the current year and shared the same pre-fire cover class. It is important to note that the model cannot handle the complexity derived from possible within-stand heterogeneity since each cell can only be described by a single dominant tree species.

Fire suppression strategies in MEDFIRE

Two fire suppression strategies are implemented in the MEDFIRE model:

- **The opportunistic strategy** mimics firefighting actions based on the ability to take advantage of opportunities derived from old fire scars. These fires provide firefighting opportunities since they are easy to detect by fire brigades and they strongly decrease fuel load and therefore fire intensity. The implementation of the opportunistic strategy in MEDFIRE suppresses burning in a cell whenever its TSF is below a pre-specified threshold (expressed in years) (Fig. 2).
- **The active fire suppression strategy** mimics the overall efficiency of firefighters to anticipate fire behaviour and reduce realized area burnt under determinate fire propagation conditions. This strategy was implemented through two different processes. First, we induced increases in the potential annual area to let burn so as to reproduce the effects of not suppressing small fires and therefore increase the total number of fires in a given scenario. By default, the MEDFIRE model simulates fires until the potential annual area to be burnt is reached, and so increases in potential area burnt result in a larger number of fires per year. Second, we introduced the concept of opportunities tied to fire-specific thresholds in spread rate. In areas in which spread rate is below a pre-specific threshold, firefighters are able to stop the fire spreading, which leads to decreases in the final area burnt [4].

Scenario definition

We assessed the effectiveness of an opportunistic fire suppression policy based on whether or not to allow unplanned fires to burn in mild weather conditions through scenarios characterized by a progressively decreasing active fire suppression policy (Fig. 3). Specifically, we designed eighteen future scenarios by combining

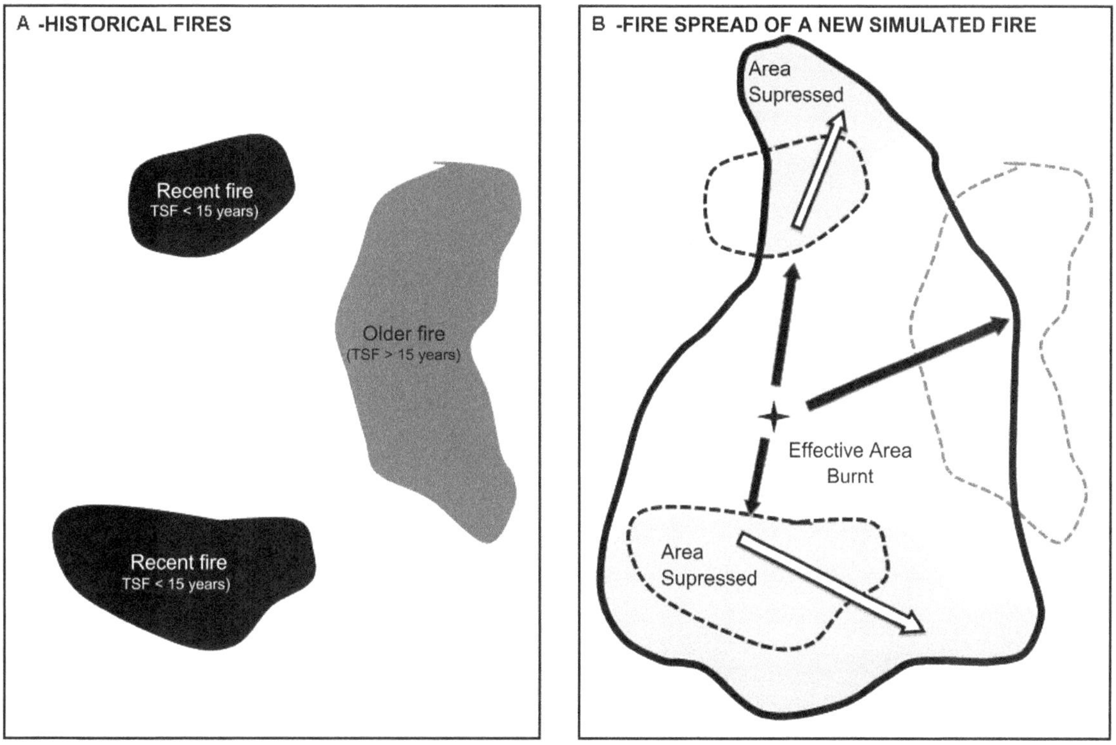

Figure 2. The figure, modified from previous work [4], illustrates the effects of opportunistic fire suppression on realized area burnt. (A) Historical fires in a region, where black patches are recent fire scars with time since last fire less than 15 years and grey patches correspond to older fire scars. (B) Fire spread of a new simulated fire in the area. Potential area (thick black line) is larger than the final area burnt due to opportunistic fire suppression generated by recent fires in (A). Suppressed areas are shown in grey and main spread axes are arrowed. Spread occurring within final area burnt (black arrows) and potentially within the suppressed area (white arrows) is shown.

different fire suppression strategies and levels of climatic severity (Tables 1 and 2). We considered **nine different treatments of active fire suppression** (acting only in years with mild climatic conditions) corresponding to situations of variability in general fire suppression efforts. The nine treatments were defined by combining three levels of potential area to be burnt with three levels of active suppression using spread rate thresholds: (a) the first three levels were simulated through variability in potential area to burn. Baseline annual distributions of area burnt were derived from 1975-99 wildfire statistics (using a lognormal distribution fitted for the available data). To reproduce the effects of not suppressing small fires (leading to increases in the total area burnt), we modified the mean of the lognormal distribution of area burnt in mild years according to the decreasing efforts of firefighters to suppress small fires: (1) *High* 7.74 (~6,500 ha/year); (2) *Average* 9.14 (~26,000 ha/year); and (3) *Low* 9.81 (~52,000 ha/year) (Tables 1 and 2). (b) The second three levels were defined using specific spread rate thresholds to reduce the final area burnt: (1) strong active suppression of opportunities corresponding to spread rate of heading fires, or descending fronts in pine forests (fire spread threshold 90); (2) medium active suppression of opportunities corresponding to spread rate in agricultural cover and sclerophyllous forest (fire spread threshold 40); (3) no active suppression of opportunities (fire spread threshold 0) (Tables 1 and 2). **Opportunistic fire suppression** was only allowed in climatically adverse years and was characterized by the number of years since the last fire in which the fire scars can be used as fire suppression opportunities by firefighters. In all simulation scenarios, the opportunistic fire suppression strategy was limited to fire scars generated in the last 15 years. The ongoing climatic trends show an increase in the number of years with very high fire-risk days [5]. Finally, we used **two climatic treatments** describing whether the percentage of adverse years (*i.e.* with dry and hot summers) will remain stable in the future (35% is the percentage of adverse years in the period 1980–1999) or is set to increase (up to 70%), following recent research [4] (Tables 1 and 2). Five hundred replicates of each scenario were simulated for a 50-year period (2000–2050).

Evaluation of simulation results

To evaluate the effect of opportunistic fire suppression strategies under the different scenarios, we calculated area suppressed as the difference between potential area to be burnt in a year and the final area burnt. We also tracked the area suppressed in adverse years according to the origin of the fire scar that created the firefighting opportunity: (1) scars of fires simulated in climatically adverse years, (2) scars of fires simulated in climatically mild years, and (3) scars of historical fires (those affecting the region before the simulation started, i.e. the 1975–99 period). The means and standard deviations of all these variables were used to describe the fire regime obtained under each simulated scenario. All statistical analyses were performed using R software, version 3.0.2 [34].

Results

Effects of decreasing fire suppression in mild weather on later undesired large fires

Scenarios with no and medium active fire suppression in mild years (see scenarios charted with white and light-grey box-plots in Fig. 4, A2) showed an increase in fire suppression opportunities in

Figure 3. Decision-making process for an opportunistic fire suppression policy based on whether or not unplanned fires should be let to burn in mild weather conditions.

adverse years compared to scenarios characterized by strong fire suppression (see dark grey box-plots in Fig. 4, A2). Specifically, the area suppressed in adverse years by opportunistic strategies increased from 18% to 29% in scenarios with strong active fire suppression (dark-grey box-plots in Fig. 4, A2), and from 22% to 50% with no active firefighting in mild years (white box-plots in Fig. 4, A2). Moreover, the opportunities derived from historic fire scars were the same in all scenarios (Fig. 5C), showing that the increased efficiency of the opportunistic strategy came from simulated fires.

Looking at the amount of area that would need to be burnt per year to reduce the impact of large fires in extreme fire weather, results show that the reference scenarios built on historical wildfire statistics (scenarios labelled *high* from 1 to 3, burning approxi-

Table 1. Description of the variables and levels used in the scenarios characterization.

Variables	Levels	Description
Active fire suppression	90	Strong active suppression corresponding to spread rate of heading fires or descending fronts in pine forests
	40	Medium active suppression corresponding to spread rate in agricultural cover and sclerophyllous forest
	0	No active suppression
Potential area to burn	6,500 (high)	Hectares/year to burn in climatically mild years according to historical fire statistics (1975–1999). Period characterized by strong efforts of firefighters to suppress small fires (*Focverd I* and *II* fire suppression programs)
	26,000 (average)	Hectares/year to burn in mild years considering an average efforts of firefighters to suppress of small fires
	52,000 (low)	Hectares/year to burn in mild years considering relatively little effort of firefighters to suppress small fires
Opportunistic fire suppression	15	Number of years since the last fire after which fire scars can be used as fire suppression opportunities by fire brigades
Climatic severity	35	Percentage of adverse years for the simulation period (2000–2050) according to the trends recorded in the period 1980–1999
	70	Percentage of adverse years for the simulation period (2000–2050) factoring in climatic warming

Table 2. List of simulation scenarios describing parameters used to reproduce active firefighting strategies and climate severity.

ID	Factors		
	Active Fire Suppression		**Climatic Severity**
	Potential annual area to burn (ha/year)	**Spread rate threshold**	**Adverse years (%)**
	Mild years	*Mild years*	
1	6,500	90	35
2	6,500	40	35
3	6,500	0	35
4	6,500	90	70
5	6,500	40	70
6	6,500	0	70
7	26,000	90	35
8	26,000	40	35
9	26,000	0	35
10	26,000	90	70
11	26,000	40	70
12	26,000	0	70
13	52,000	90	35
14	52,000	40	35
15	52,000	0	35
16	52,000	90	70
17	52,000	40	70
18	52,000	0	70

Potential annual area to burn follows a lognormal distribution (with mean 7.74 and standard deviation 1.43 in mild years fitted from 1975–99 wildfire statistics). The mean of that distribution is used as a scenario parameter with three possible values: (1) **High**-7.74 (~6,500 ha/year), (2) **Average**-9.14 (~26,000 ha/year), and (3) **Low**-9.81 (~52,000 ha/year).

mately 6,500 ha/year in mild years) allowed opportunistic suppression of 18–22% of the target annual area in adverse years (Fig. 4, A2). If mean area burnt in mild years was increased to 26,000 ha/year by decreasing efforts to suppress small fires (scenarios labelled *average* from 7 to 9), the area suppressed in adverse years would increase to 26–37% (Fig. 4, A2). Finally, when mean area burnt in mild years was increased to 52,000 ha/year due to a fire suppression policy aimed at engineering a shift in fire regime towards a higher number of small fires (scenarios labelled *low* from 13 to 15), the percentage of area suppressed in adverse years also increased to figures of up to 29–50% (Fig. 4, A2). However, while the opportunities derived from fires simulated in mild years (see scenarios 1–3 compared to scenarios 7–9 and 13–15 in Fig. 5, B) increased as the final annual area burnt in these years grew, the opportunities generated in earlier adverse years decreased (see scenarios 1–3 compared to scenarios 7–9 and 13–15 in Fig. 5, A). This increase in fire suppression opportunities was therefore partially modulated by the decrease in opportunities from adverse years.

Interaction between fire suppression and climatic severity

To evaluate the interaction between fire regime, fire suppression and climate warming, we compared scenarios with the same set of parameters while varying the percentage of adverse years from 35% up to 70% (see Table 1). When the number of years with adverse conditions increased, counterintuitive effects appeared on the potential opportunistic fire suppression strategies. Thus, the area suppressed was higher in the simulated scenarios under climate warming (scenarios 4–6 in Fig. 4, A2) than under the current climatic regime (scenarios 1–3 in Fig. 4, A2). Climate warming induced a greater area suppressed by opportunities derived from fires simulated in adverse years (compare scenarios under climate warming, boxed in a thick line, and scenarios without climate warming, boxed in a thin line, in Fig. 5, A). However, when we considered a more severe fire regime in mild years (~52,000 ha/year, scenarios with label *low*), the area suppressed was lower in scenarios with climate warming (scenarios 10–12 and 16–18 in Fig. 4, A2) than scenarios without climate warming (scenarios 7–9 and 13–15 in Fig. 4, A2). This increase in area burnt in mild years again led to a greater area suppressed by opportunities derived from fires simulated under mild weather conditions (compare scenarios labelled *high* and *low* in Fig. 5, B). However, it is also remarkable that climate change also led to fewer years with mild fire weather conditions, thus reducing the window of opportunity for the creation of opportunities under these conditions (compare scenarios with climate change, boxed in a thick line, against scenarios without climate warming, boxed in a thin line, in Fig. 5, B).

Discussion

We have shown that relaxing fire suppression efforts under relatively controlled conditions (opportunistic fire suppression policy) has the potential to substantially reshape fire regimes and decrease the amount of area burnt under undesired, extreme climate conditions. However, the potential of this strategy is

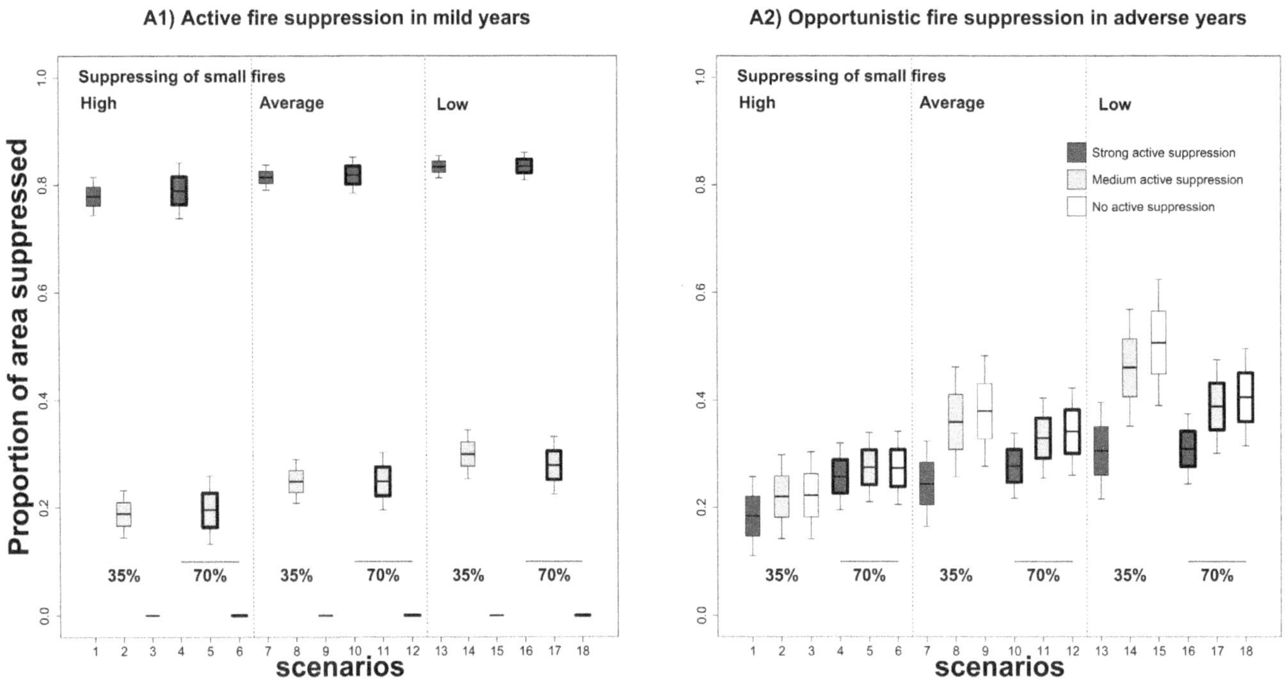

Figure 4. Proportion of area suppressed in relation to potential area to burn. Area suppressed by active fire suppression strategies in mild years (A1) and opportunistic fire suppression strategy in adverse years (A2). Potential annual area in mild years follows a lognormal distribution with means of 7.74 (**High** ~6,500 ha/year), 9.14 (**Average** ~26,000 ha/year), and 9.81 (**Low** ~52,000 ha/year), to reproduce a decreasing effectiveness of firefighter efforts to suppress small fires. Simulation scenarios characterized by thresholds of 90% in active fire suppression of mild years are represented in dark-grey box-plots, in light grey by thresholds of 40%, and in white of 0%. Box-plot elements are as follows _ lower and upper whiskers: approximately 68% of all data values (**mean** ± **SD**, standard deviation); lower and upper midhinges: the **mean** ± ½ **SD**; central black line: the **mean**. Box outline width represents the percentage of adverse years used in the simulations, i.e. 35% (thin width) and 70% (thick width).

somewhat limited due to the complexity of the interactions between fuel availability, fire impact, and fire suppression strategies.

Potential of opportunistic strategy for reducing the impact of large fires

Given the growing impact of wildfires, the effect of fire exclusion has been extensively studied and debated in many different

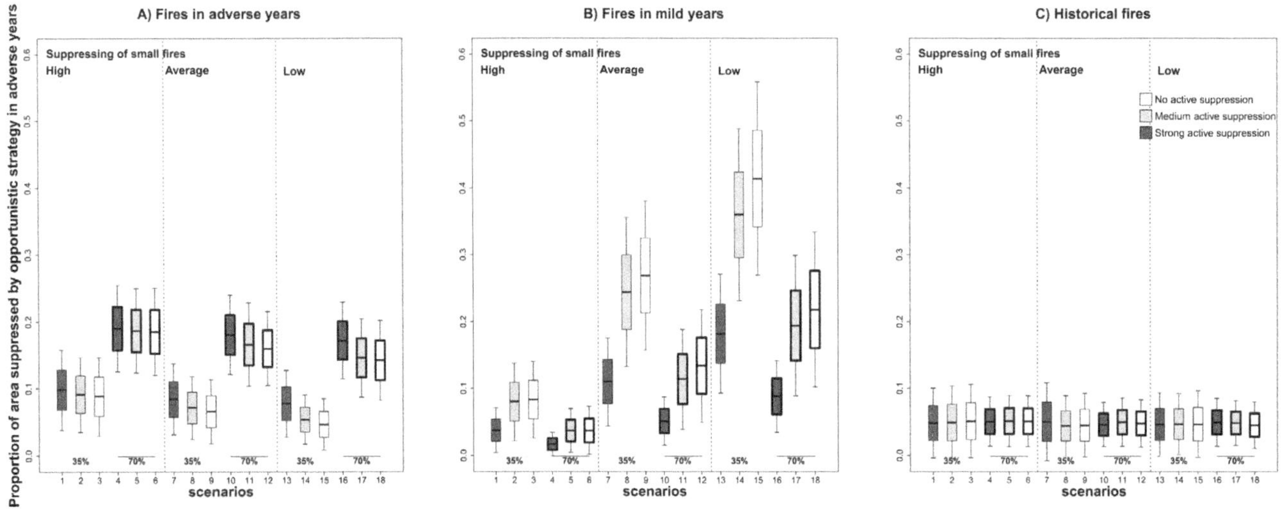

Figure 5. Proportion of area suppressed (in relation to potential annual area) by opportunistic fire suppression strategy in adverse years. Area suppressed by firefighting opportunities derived from: (A) fires in adverse years, (B) fires in mild years, (C) historical fires. Simulation scenarios characterised by thresholds of 90% in active fire suppression of mild years are represented in dark-grey box-plots, in light-grey by thresholds of 40%, and in white of 0%. The labels *high*, *average* and *low* refer to the distribution of potential annual area in mild years (see Fig 4 for more details). Box-plot elements and box-plot outline width are as in Fig. 4.

Mediterranean ecosystems for decades now [2,35,36]. Many authors claim that the systematic extinction of "all fires" leaves an accumulation of fuel that will be consumed in future large fires in years with extreme fire weather conditions (fire paradox) [35,37,38]. Recent studies carried out in the Mediterranean basin using fire-succession models have demonstrated that high fire suppression efforts may lead to a slightly higher proportion of large and more intense fires [2,39]. Our results concur with this hypothesis, since scenarios with strong active fire suppression in mild weather did lead to increases in the area burnt by large fires in adverse conditions due to feedbacks in the dynamics of fuel accumulation (Fig. 4). In addition, we demonstrated that relaxing the fire suppression efforts in mild years, in which benign fire weather conditions allow firefighters to tackle any fire event efficiently, provided additional fire scars associated to fuel reduction. These new burnt-area patches generated potential firefighting opportunities for later adverse years (Fig. 5, B). In our particular case, the area suppressed in adverse years by opportunities derived from fires simulated in mild years increased considerably, as the final annual area burnt in these years was higher (Fig. 5, B). In the reference scenarios derived from statistics on wildfires between 1975 and 1999 (which burnt 6,500 ha/year in mild years), the area suppressed by opportunities derived from previous fire scars only accounted for 22% of the potential annual area in adverse years (Fig. 4). This suggests that the current high-efficiency fire suppression policy may be decreasing the opportunities that arise from past fires. Our results showed that a progressive increase of area burnt in mild years by relaxing fire suppression efforts led to a reduction in area burnt in adverse years: a decrease of an additional 15% (up to 37%) would require a four-fold increase in the final annual area burnt in mild years (i.e. to 26,000 ha/year), whereas a decrease of an additional 23% (up to 50%) would require an eight-fold increase (i.e. to 52,000 ha/year). Therefore, to effectively reduce large fires under adverse weather conditions, we need to allow burning across larger areas in mild years. Bear in mind that reductions in the total area burnt in adverse weather conditions do not come for free _ here, the reductions involved a loss of future opportunities for firefighting (Fig. 5, A) [4]. Therefore, the amount of area treated in mild conditions has to be very high to reduce the amount of area burnt in adverse climate conditions and compensate for the loss of opportunities derived from fires avoided in such adverse conditions.

Fire management based on climate-adapted modulation of fire suppression shares some overlap with the management policy of prescribed burnings implemented in other Mediterranean ecosystems [15,21,22]. Both prescribed burning and the use of unplanned fires resulting from decreasing suppression efforts are tactics that use fire as a tool to fight larger wildfires and that aim to increase the effectiveness of fire suppression through fuel reduction [8,18]. Recent studies [2,39] used a simulation model to investigate whether large fires in the Mediterranean region are consequence of large fire suppression programs or, conversely, are driven by extreme fire weather conditions. Their results suggest that, although the total area burnt is much the same regardless of whether or not fire suppression or prescribed fire policies are used, prescribed burning does reduce fire intensity. Here, we show that like prescribed burning, unplanned fire events could be used to reduce fuel accumulation and fire intensity to create opportunities for effective fire suppression of large fires in future adverse conditions. We therefore suggest that designing treatments to minimize adverse fire effects may be a more effective strategy than designing treatments that attempt to extinguish "all fires".

Moreover, in our study and in the current fire regime context, unplanned fires increase landscape heterogeneity but do not seem enough to offset the decade-long general trend towards homogenization due to land abandonment and the coalescence of natural vegetation patches [28,40]. This landscape homogenization process is driving an increase in fuel continuity [41] and consequently fire spread and intensity [10,12,42]. To offset this ongoing trend and create new fire suppression opportunities, we envisaged a fuel-reduction strategy based on relaxing fire suppression efforts in mild years to create a novel fire regime with a large number of smaller fires. Our results have demonstrated that this strategy is associated with effective reductions in the area burnt by fires in adverse years. Thus, decreasing fire suppression in mild weather conditions may create landscapes in which wildfires occur with less devastating consequences. Fire may itself play a key role in maintaining these novel landscapes.

Opportunistic fire suppression under climatic warming scenarios

The interaction between area burnt, fire suppression and climate warming had counterintuitive effects on the potential for opportunistic fire suppression strategies to reduce the amount of area burnt in climatically adverse years. The area burnt in adverse years in scenarios with climate warming is consistent with climate change bringing warmer and drier summers and increased fire weather risk [43,44]. Overall, our results are in agreement with the trends reported for recent decades in the Mediterranean basin [5,45]. However, the area burnt in the simulated scenarios under climate warming was lower than expected due to increases in the area suppressed by opportunities derived from simulated fires in adverse years (Fig. 5, A). At the same time, climate warming also implies a lower number of mild years and therefore fewer windows for creating the opportunities targeted. In this context, exponentially larger areas need to be burnt in mild years to create additional fire suppression opportunities (Fig. 5, B). On the other hand, recent studies carried out in regions with Mediterranean climatic conditions highlight the role of landscape structure in shaping current and future fire–climate relationships at regional scale, and suggest that the future changes in fire regime under global warming may be different from what it is predicted by climate alone [46]. We argue that opportunistic fire suppression policies have the potential to substantially effect changes in this fire–climate relationship through the novel landscapes created by relaxing active fire suppression efforts in mild weather conditions. We also suggest that, given the large annual area burnt required to prevent large fires in adverse fire weather years, the spatial allocation of firefighting efforts and fire suppression resources will be a keystone for the optimization of this fuel-reduction strategy and its successful implementation in future firefighting programs forced to deal with climate change.

Implications for the future: some considerations

The effectiveness of this opportunistic fire suppression strategy is still relatively low compared to other fuel reduction strategies applied in different Mediterranean-type ecosystems [15,18,47]. In Australia, where prescribed burning is used as a cost-effective fuel-reduction treatment, previous works [22] found that three units of prescribed fires were required to reduce one unit of unplanned fire area. This negative relationship was stronger in the tropical savannas of northern Australia, where prescribed early-dry-season burning was able to substantially reduce late-dry-season fire area by direct one-to-one replacement [16]. In Australian eucalypt forest, other authors [21] found that each unit area reduction in unplanned fire required about four units of prescribed fire. We argue that strategic placement of these fuel treatments (*i.e.* prescribed burning or thinning followed by prescribed burning) is the likely key to effective implementation. In fact, recent research [48] found that strategic placement of fuel treatments reduced the predicted growth rates of simulated fires under adverse weather conditions more effectively than random placement in three study areas of western USA. Random placement of fuel treatments required about twice the treatment rate of optimally-placed fuel treatments to yield the same reduction in predicted fire growth rates [48]. However, opportunistic fire suppression is based on unplanned fire occurrence. Unplanned fires tend to be determined by the spatial arrangement of ignition factors. In Mediterranean countries where natural ignitions are scarce, fire regime is strongly linked to human activities, with the result that fire scars are not randomly distributed in the space but follow the auto-correlated pattern of human activities [1,49,50]. Therefore, we suggest that the identification of spatio-temporal patterns of fire occurrence at regional scales in these systems may optimize the opportunities created by unplanned fires, and thus mitigate the heavy impact of the undesired large fires in extreme fire weather.

From a purely economic standpoint, opportunistic 'let-burn' fire suppression strategies have a further benefit tied to the fact that they curb the economic losses caused by large fires while also saving on fire suppression resources. In fact, a recent study suggests that the potential savings associated to opportunistic strategies could be substantial [23]. The authors simulated unplanned fires at landscape scale over a 100-year period using existing models of fire behaviour, vegetation and fuel development and fire suppression effectiveness to estimate suppression costs using a suppression cost model. They found that estimated future suppression cost savings were positively correlated with fire size. Others authors [51] studied different spatial factors influencing large wildland fire suppression expenditures, and they also found that fire size and private land had a strong effect on expenditures. Note that just a tiny fraction of fires (around 1%) accounts for 85% of suppression expenditure in the western USA [6]. Given that the land in our study area is mainly privately-owned, we suggest that an opportunistic strategy aimed at mitigating large wildfires by relaxing efforts to suppress small fires could enable huge suppression savings. However, this strategy also implies large areas burnt in mild weather conditions. The decision on whether a particular fire should be let to burn must be made by weighing up the potential benefits in terms of the landowners' management objectives and the potential cost of damage from unsuppressed fire [52]. It is the net benefit of allowing a fire to burn that is the relevant criterion [23,53]. Therefore, factoring in the potential socio-economic and ecological costs (such as soil loss, destruction of wildlife habitat, loss of timber value, infrastructure and human life) and benefits discussed here is essential for identifying candidate areas suitable for using unplanned fires as a management tool.

To conclude, we suggest that to achieve the stand structure and fuel reduction goals required to minimize large fires in extreme fire weather, this strategy could be accompanied by other fuel-reduction treatments such as large-scale forest thinning or biomass extraction. Further studies are needed to assess the impact of novel fuel-reduction treatments on fire regime. Moreover, the possible impacts of these fire management options on biodiversity and a variety of ecosystem services should be carefully evaluated before cost–benefit analyses can be developed. These potential fuel-reduction treatments should therefore also be evaluated in terms of social, economic and environmental cost–benefit trade-offs.

Author Contributions

Conceived and designed the experiments: LB AR. Performed the experiments: AR NA LB. Analyzed the data: AR NA JR MC LB. Contributed reagents/materials/analysis tools: AR NA MC JR LB. Wrote the paper: AR NA JR MC LB.

References

1. Keeley JE, Bond WJ, Bradstock RA, Pausas JG, Rundel PW (2012) Fire in Mediterranean Ecosystems. New York: Cambridge University Press. 515 p.
2. Piñol J, Beven K, Viegas DX (2005) Modelling the effect of fire-exclusion and prescribed fire on wildfire size in Mediterranean ecosystems. Ecol Model 183: 397–409.
3. Moreira F, Viedma O, Arianoutsou M, Curt T, Koutsias N, et al. (2011) Landscape–wildfire interactions in southern Europe: Implications for landscape management. J Environ Manage 92: 2389–2402.
4. Brotons L, Aquilué N, De Cáceres M, Fortin MJ, Fall A (2013) How fire history, fire suppression practices and climate change affect wildfire regimes in Mediterranean landscapes. PLoS ONE 8(5): e62392.
5. Piñol J, Terradas J, Lloret F (1998) Climate warming, wildfire hazard and wildfire occurrence in coastal eastern Spain. Clim Change 38: 345–357.
6. Reinhardt ED, Keane RE, Calkin DE, Cohen JD (2008) Objectives and considerations for wildland fuel treatment in forested ecosystems of the interior western United States. For Ecol Manage 256: 1997–2006.
7. San-Miguel-Ayanz J, Moreno JM, Camia A (2013) Analysis of large fires in European Mediterranean landscapes: Lessons learned and perspectives. For Ecol Manage 294 11–22.
8. Williams J (2013) Exploring the onset of high-impact mega-fires through a forest land management prism. For Ecol Manage 294: 4–10.
9. Lloret F, Piñol J, Castellnou M (2009) Wildfires. In: Woodward JC, editor. The Physical Geography of the Mediterranean Basin. Oxford University Press, Oxford, pp. 541–558.
10. Loepfe L, Martinez-Vilalta J, Oliveres J, Piñol J, Lloret F (2010) Feedbacks between fuel reduction and landscape homogenisation determine fire regimes in three Mediterranean areas. For Ecol Manage 259: 2366–2374.
11. Keeley JE, Fotheringham CJ, Morais M (1999) Reexamining fire suppression impacts on brushland fire regimes. Science 284: 1829–1832.
12. Pausas J, Fernández-Muñoz S (2011) Fire regime changes in the Western Mediterranean Basin: from fuel-limited to drought-driven fire regime. Clim. Change 110: 215–216.
13. Costa P, Castellnou M, Larrañaga A, Miralles M, Kraus D (2011) La Prevenció dels grans Incendis Forestals adaptada a l'Incendi Tipus. Unitat Tècnica del GRAF. Divisió de Grups Operatius Especials. Direcció General de Prevenció, Extinció d'Incendis i Salvaments. Departament d'Interior. Generalitat de Catalunya. 87 p.
14. McIver J, Erickson K, Youngblood A (2012) *Principal Short-Term Findings of the National Fire and Fire Surrogates Study*. Gen. Tech. Rep. PNW-GTR-860. Portland, OR: U.S. Department of Agriculture, Forest Service, Pacific Northwest Research Station. 210 p.
15. Price OF, Bradstock RA, Keeley JE, Syphard AD (2012) The impact of antecedent fire area on burned area in southern California coastal ecosystems. J Environ Manage 113: 301–307.
16. Price OF, Russell-Smith J, Watt F (2012) The influence of prescribed fire on the extent of unplanned fire in savanna landscapes of western Arnhem Land, Australia. Int J Wildland Fire 21: 297–302.
17. Stephens SL, James D, Boerner J, Fettig J, Joseph B, et al. (2012) The effects of forest fuel reduction treatments in the United States. BioScience 62, 549–560.
18. McCaw WL (2013) Managing forest fuels using prescribed fire – A perspective from southern Australia. For Ecol Manage 294: 217–224.
19. Navarro LM, Pereira HM (2012) Rewilding abandoned landscapes in Europe. Ecosystems 15: 900–912.
20. Stellmes M, Röder A, Udelhoven T, Hill J (2013) Mapping syndromes of land change in Spain with remote sensing time series, demographic and climatic data. Land Use Policy 30: 685–702.
21. Boer MM, Sadler RJ, Wittkuhn R, McCaw L, Grierson PF (2009) Long-term impacts of prescribed burning on regional extent and incidence of wildfires—Evidence from 50 years of active fire management in SW Australian forests. For Ecol Manage 259: 132–142.
22. Price OF, Bradstock R (2011) Quantifying the influence of fuel management and weather on the annual extent of unplanned fires in the Sydney region of Australia. Int J Wildland Fire 20: 142–151.
23. Houtman RM, Montgomery CA, Gagnon AR, Calkin DE, Dietterich TG, et al. (2013) Allowing a wildfire to burn: estimating the effect on future fire suppression costs. Int J Wildland Fire. Available: http://dx.doi.org/10.1071/WF12157.
24. Adams MA (2013) Mega-fires, tipping points and ecosystem services: Managing forests and woodlands in an uncertain future. For Ecol Manage 294: 250–261.

25. Gracia C, Burriel C, Mata J, Ibanez T, Vayreda J (2000) Inventari Ecològic i Forestal de Catalunya. Centre de Recerca Ecològica i Aplicacions Forestals. Bellaterra: Centre de Recerca Ecològica i Aplicacions Forestals (CREAF).
26. CORINE (2006) Land-use land-cover database 1:250,000. European Environment Agency, Copenhagen, Denmark.
27. Díaz-Delgado R, Lloret F, Pons X (2004) Spatial patterns of fire occurrence in Catalonia, NE Spain. Landscape Ecol 19: 731–745.
28. Debussche M, Lepart J, Dervieux A (1999) Mediterranean landscape changes: evidence from old postcards. Global Ecol Biogeogr 8: 3–15.
29. DARP (1999) Foc Verd II. Programa de gestió del risc d'incendi forestall. Generalitat de Catalunya, Barcelona, Spain.
30. De Cáceres M, Brotons L, Aquilué N, Fortin MJ (2013) The combined effects of land use legacies and novel fire regimes on bird distributions in the Mediterranean. J Biogeogr 40: 1535–1547.
31. Fall A, Fall J (2001) A domain-specific language for models of landscape dynamics. Ecol Model 141: 1–18.
32. Castellnou M, Pagés J, Miralles M, Pique M (2009) Tipificación de los incendios forestales de Cataluña. Elaboración del mapa de incendios de diseño como herramienta para la gestión forestal, In: 5°Congreso Forestal. Ávila. pp. 1–15.
33. Rodrigo A, Retana J, Picó FX (2004) Direct regeneration is not the only response of Mediterranean forests to large fires. Ecology 85: 716–729.
34. R Core Team, 2013. R: A language and environment for statistical computing. R Foundation for Statistical Computing, Vienna, Austria. ISBN 3-900051-07-0. Available: http://www.R-project.org/.
35. Minnich R, Chou Y (1997) Wildland fire patch dynamics in the chaparral of Southern California and Northern Baja California. Int J Wildland Fire 7: 221–248.
36. Moritz MA, Keeley JE, Johnson EA, Schaffner AA (2004) Testing a basic assumption of shrubland fire management: how important is fuel age? Front Ecol Environ 2: 67–72.
37. Minnich RA (1983) Fire mosaics in Southern California and Northern Baja California. Science 219: 1287–1294.
38. Minnich RA (2001) An integrated model of two fire regimes. Cons Biol 15: 1549–1553.
39. Piñol J, Castellnou M, Beven KJ (2007) Conditioning uncertainty in ecological models: assessing the impact of fire management strategies. Ecol Model 207: 34–44.
40. Lloret F, Calvo E, Pons X, Díaz-Delgado R (2002) Wildfires and landscape patterns in the Eastern Iberian Peninsula. Landscape Ecol 17: 745–759.
41. Bielsa I, Pons X, Bunce B (2005) Agricultural abandonment in the north Eastern Iberian Peninsula: the use of basic landscape metrics to support planning. J Environ Planning Manage 48: 85–102.
42. Vega-García C, Chuvieco E (2006) Applying local measures of spatial heterogeneity to Landsat-TM images for predicting wildfire occurrence in Mediterranean landscapes. Landscape Ecol 21: 595–605.
43. Alcamo J, Moreno JM, Nováky B, Bindi M, Corobov R, et al. (2007) Chapter 12: Europe. In: Parry ML, Canziani OF, Palutikof JP, Van der Linden PJ, Hanson C.E., editors. Climate Change 2007: impacts, adaptation and vulnerability. Contribution of Working Group II to the Fourth Assessment Report of the Intergovernmental Panel on Climate Change. Cambridge University Press. Cambridge. pp. 543–566.
44. Liu Y, Stanturf J, Goodrick S (2010) Trends in global wildfire potential in a changing climate. For Ecol Manage 259: 685–697.
45. Loepfe L, Martinez-Vilalta J, Piñol J (2012) Management alternatives to offset climate change effects on Mediterranean fire regimes in NE Spain. Clim Change 115: 693–707.
46. Pausas JG, Paula S (2012) Fuel shapes the fire–climate relationship: evidence from Mediterranean ecosystems. Global Ecol Biogeogr 21: 1074–1082.
47. Vilen T, Fernandes PM (2011) Forest fires in Mediterranean countries: CO_2 emissions and mitigation possibilities through prescribed burning. Environ Manage 48: 558–567.
48. Finney MA, Seli RC, McHugh CW, Ager AA, Bahro B, et al. (2007) Simulation of long-term landscape-level fuel treatment effects on large wildfires. Int J Wildland Fire 16: 712–727.
49. Terradas J, Piñol J (1996) Els grans incendis: condicions meteorològiques i de vegetació per al seu desenvolupament, In: Terradas J., editor. Ecologia del Foc. Proa. Barcelona. España. pp. 63–75.
50. Gonzalez-Olabarria JR, Brotons L, Gritten D, Tudela A, Teres JA (2012) Identifying location and causality of fire ignition hotspots in a Mediterranean region. Int J Wildland Fire 21: 905–914.
51. Liang J, Calkin DE, Gebert KM, Venn TJ, Silverstein RP (2008) Factors influencing large wildland fire suppression expenditures. Int J Wildland Fire 17: 650–659.
52. Román MV, Azqueta D, Rodrígues M. (2013) Methodological approach to assess the socio-economic vulnerability to wildfires in Spain. For Ecol Manage 294: 158–165.
53. Houtman RM (2011) Letting wildfires burn: modeling the change in future suppression costs as the result of a suppress versus a let-burn management choice. MSc thesis. Oregon State University. Corvallis, OR.

21

Influence of Fuels, Weather and the Built Environment on the Exposure of Property to Wildfire

Trent D. Penman[1*¤], Luke Collins[2], Alexandra D. Syphard[3], Jon E. Keeley[4,5], Ross A. Bradstock[1]

1 Centre for Environmental Risk Management of Bushfires, Institute of Conservation Biology and Environmental Management, University of Wollongong, NSW, Australia, **2** Hawkesbury Institute for the Environment, University of Western Sydney, Penrith, NSW, Australia, **3** Conservation Biology Institute, La Mesa, California, United States of America, **4** U.S. Geological Survey, Western Ecological Research Center, Sequoia-Kings Canyon Field Station, Three Rivers, California, United States of America, **5** Department of Ecology and Evolutionary Biology, University of California Los Angeles, Los Angeles, California, United States of America

Abstract

Wildfires can pose a significant risk to people and property. Billions of dollars are spent investing in fire management actions in an attempt to reduce the risk of loss. One of the key areas where money is spent is through fuel treatment – either fuel reduction (prescribed fire) or fuel removal (fuel breaks). Individual treatments can influence fire size and the maximum distance travelled from the ignition and presumably risk, but few studies have examined the landscape level effectiveness of these treatments. Here we use a Bayesian Network model to examine the relative influence of the built and natural environment, weather, fuel and fuel treatments in determining the risk posed from wildfire to the wildland-urban interface. Fire size and distance travelled was influenced most strongly by weather, with exposure to fires most sensitive to changes in the built environment and fire parameters. Natural environment variables and fuel load all had minor influences on fire size, distance travelled and exposure of assets. These results suggest that management of fuels provided minimal reductions in risk to assets and adequate planning of the changes in the built environment to cope with the expansion of human populations is going to be vital for managing risk from fire under future climates.

Editor: Ritaban Dutta, CSIRO, Australia

Funding: Funding from the U.S. Geological Survey was provided to University of Wollongong to cover salary of LC for the study. The funders had no role in study design, data collection and analysis, decision to publish, or preparation of the manuscript.

Competing Interests: The authors have declared that no competing interests exist.

* Email: trent.penman@unimelb.edu.au

¤ Current address: Department of Forest and Ecosystem Science, University of Melbourne, Victoria, Australia.

Introduction

Wildfires can pose a significant risk to people and property [1–4]. Large losses of property and life have been recorded from individual fires or fire complexes in fire prone regions throughout the globe [5–7]. Such events can impact individuals and communities for many years [8–11]. As a result, fire management agencies invest significant budgets in reducing the risk of loss from wildfires, primarily through investment in fuel management and active fire suppression [12,13].

Fuel management is a commonly used risk management tool in fire prone landscapes [14,15]. The justification for the approach derives from the fundamentals of fire behaviour with a reduction in fuel loads expected to result in a subsequent lowering of the fire intensity and rate of spread [16–20]. In turn, these changes to fire behaviour are expected to increase the probability of successfully containing the fire with active fire suppression [21,22].

Fuel breaks are a commonly applied type of fuel management treatment in a variety of ecosystems. These are mechanical reductions or removal of fuel, typically as linear features along ridgetops, to enable safe access for fire suppression crews to manage fires [23]. Empirical analysis has found fuel breaks are more effective when readily accessible and well-maintained and when used for *backfire* operations [23,24]. Under these conditions the intensity and rate of spread are lower and containment of the fires through active suppression is more likely to be successful [21,22,24]. Simulation studies have found individual fuel breaks have the potential to reduce the size and intensity of wildfires [25,26]. While studies examining the impact of fuel breaks on the behaviour of individual fires are valuable, to quantify the extent to which fuel breaks reduce risk to lives and property, we need to examine the role of fuel breaks at the landscape scale [27]. Although simulation studies have shown management of fuels can alter fire regimes in forested ecosystems, there is a need to quantify how fuel breaks, in particular, can reduce the risk of exposure of assets. Here we examine the performance of fuel breaks in mitigating risk in Mediterranean shrubland (chaparral) landscapes.

Fuel breaks are the main fuel treatment carried out in the chaparral shrublands of southern California, with a long history of extensive deployment [23,24]. Thus case studies of their effectiveness in mitigating risk can provide valuable insight into mitigation of risk in a region with some of the highest exposure of fire-prone urban and peri-urban developments in the world. Case studies in this context may also be valuable for assessing fuel treatment options in other fire-prone, temperate environments which share similar elements of the problem [28].

Bayesian Networks (BN) provide a suitable methodology for the analysis of risk management problems [29–31]. They are depicted as directed acyclic graphs with nodes representing the variables and arrows representing the directional relationships between nodes [32]. There is a conditional probability table for each node that contains the joint probability distributions for the variable [33]. Root nodes occur at the top of the model and are not influenced by other variables in the model. These nodes have a conditional probability table containing a single probability for each state in that node. Child nodes are variables that are influenced by one or more variables (parent nodes). These nodes have a conditional probability table that represents the probability of a given state in the child node given the state(s) in the parent node(s). Uncertainty is propagated throughout the model, providing probability distribution for all output nodes. Results are likelihoods that form the basis for risk management calculations [29]. Bayesian Networks have been found to be a valuable method for examining fire risk management problems at the landscape scale [31,34].

Here we develop a Bayesian Network model to examine the role of fuel breaks in reducing the risk of assets being exposed to wildfire using San Diego county as a case study area. San Diego county has a history of major fire losses (circa 5000 houses destroyed between 2000 and the present), which reflects extensive and rapidly growing developments, exposure to regular episodes of severe fire weather and terrain and vegetation conducive to the rapid spread of intense fires [35]. Thus the county comprises encapsulates key elements and a significant portion of the wildfire risk problem in southern California. As a case study it therefore provides the potential for key insights into fire management that are regionally, nationally and globally significant. The BN model combines data from a fire simulation model (FARSITE) and environmental data. We specifically seek to determine the extent to which risk posed by wildfires to properties at the wildland urban interface is influenced by the environment (weather, fuel moisture, natural environment), developmental patterns (built environment) and fuel management (fuel load, fuel breaks).

Methods

The study area was San Diego County, California, USA, which supports a population of approximately 3.2 million people in an area of approximately 11 000 km^2 (US Census, http://quickfacts.census.gov/qfd/states/06/06073.html, Accessed December 2013). In the county, there is a long and complex wildland urban interface [36], along which thousands of homes have been destroyed in major fire events in the last decade [37]. Native vegetation of the area is dominated by chaparral, coastal sage scrub, and oak woodland. The county experiences a Mediterranean climate with hot dry summers and winter rainfall with moderate temperatures. Fires in the county occur most frequently in summer months, but most area burned occurs in the autumn when annual fuel moisture is lowest and Santa Ana winds are most frequent [38]. San Diego County was selected as it is dominated by highly flammable shrubland vegetation and falls within known Santa Ana wind corridors, making the region prone to recurrent large fire events [39,40]. The simulation landscape was defined as a 60 km×60 km area east of San Diego (Figure 1).

A Bayesian Network model was used to examine the relative influence of weather, the built and natural environment and fuel breaks on the risk of exposure to wildfire. Here we broadly follow the methods for developing Bayesian Networks recommended by Marcot *et al.* [41] and Chen and Pollino [42]. The primary steps used were to construct a conceptual model of the problem, develop influence diagrams to depict the relationships of the conceptual model and finally populate all the conditional probability tables within the model.

A basic conceptual model for the study was derived from previous fire management research [31,34]. This model assumed an ignition had occurred and predicted the subsequent spread and potential impact of the wildfire upon property, which was dependent upon environmental conditions and management decisions.

Influence diagrams encapsulated the conceptual framework and the relevant environmental factors (Figure 2). These were developed for the study area by the authors through a series of workshops held by the United States Geological Survey (December 2010, September 2011, May 2012) involving twelve researchers with expertise in fire management, Bayesian Network analysis and landscape ecology. Iterations of the influence diagrams were presented to the group, discussed and then further refined until a consensus was reached. In the final set of influence diagrams, fire size and distance travelled were assumed to be influenced by the key variables considered in the simulation modelling – weather, fuel moisture, landscape fuel load and the occurrence of fuel breaks within the National Forest. Elements of the natural environment at the ignition location, specifically fuel type, fuel load and elevation, were considered to have an influence on fire size and distance travelled. The built environment also influences fire spread and exposure of property. Exposure to fire was taken as simple function considering the distance the fire could potentially travel and the distance from the ignition point to property (Figure 2).

Data for the conditional probability tables in the analysis (Table 1) were derived from either a simulation study or empirical data for the study region. We undertook a comprehensive simulation of fires in the area using the Fire Area Simulator (FARSITE) using random ignition locations. FARSITE is a two dimensional spatially explicit model that models fire spread using Huygens' principle [43]. The simulations examined all combinations of fire weather (low, high and Santa Ana), live fuel moisture (LMF 60% and 90%), fuel loading (low and high) and the presence or absence of maintained fuel breaks. A total of 11,944 fires were simulated a FARSITE framework. Complete details of the approach are presented in Table S1 in Material S1. From the simulation data, we derived values for the nodes fire size and distance travelled. Weather, fuel moisture, landscape fuel load and the occurrence of fuel breaks were implemented as decision nodes to explore the relative influence of each factor. Environmental variables (i.e. elevation, ignition fuel load, ignition fuel type and distance to the interface) were derived from GIS data from the region (www.landfire.gov Accessed October 2012). All nodes and methods of discretisation are described in more detail in Table 1. The Bayesian Network model is available from the ABNMS data repository (www.abnms.org).

Two methods were used to examine the relative influence of variables. Firstly, the relative influence of each of the modelled factors was assessed using values of the terminal node – "exposure to fire". This node reflects the risk of property being exposed to a wildfire under the given conditions. We considered all 24 combinations of the key predictor variables – weather (3 levels), fuel moisture (2 levels), landscape fuel load (2 levels) and fuel breaks (2 levels) (Table 1). Secondly, the sensitivity of nodes was assessed using the sensitivity to findings function in Netica (http://www.norsys.com/netica.html, Accessed December 2013). This function examines the extent to which changes in one variable affects the variable of interest. We examined the sensitivity of

Figure 1. Location of the study area, San Diego County, USA.

findings for "exposure to fire" as the terminal node and "distance travelled".

Results

The size of fires after a 12 hour simulation ranged from 0.1 ha to 28,480 ha, with a mean of 2896 ha (±40 S.E.) and a median 882 ha. The distance travelled ranged from 18 m to 47,300 m with a mean of 6171 m (±57 S.E.). Responses of fire size and distance travelled to the predictor variables were consistent. Fire size and distance travelled increased with the severity of fire weather and the landscape fuel load, and decreased with increasing fuel moisture. The presence of fuel breaks had little influence on either individual fire size or distance travelled (Figure 3).

Risk of exposure to fires was influenced most strongly by weather (Figure 4). The risk of exposure was >99% for all fuel

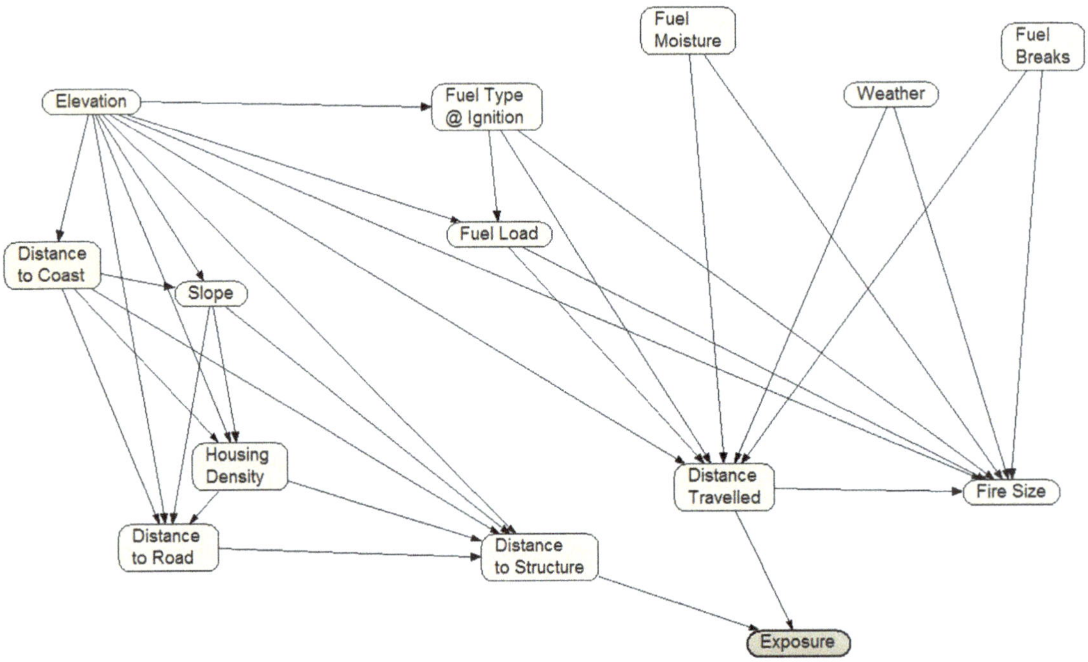

Figure 2. Influence diagrams for the Bayesian Network Model. See Table 1 for node definitions and states.

Table 1. Nodes, definitions and states used in the Bayesian Network model.

Node	Description	Levels
Distance Travelled (O)	The maximum distance the fire travelled.	0 to 0.5 km; 0.5 to 1.5 km; 1.5 to 4 km; 4 to 10 km; 10 to 15 km; >15 km
Distance to the coast (GIS)	Distance from the ignition point to the coast.	0 to 10 km; 10 to 25 km; 25 to 42 km; 42 to 68 km; 68 to 80 km; > 80 km
Distance to road (GIS)	The distance from the ignition point to the nearest mapped road.	0 to 50 m; 50 to 100 m; 100 to 500 m; 500 to 1000 m; 1000 to 2000 m; >2000 m
Distance to structure (GIS)	The distance from the ignition point to the nearest mapped house.	0 to 50 m; 50 to 100 m; 100 to 500 m; 500 to 1000 m; 1000 to 2000 m; >2000 m
Elevation (GIS)	Elevation of the ignition point	0 to 300 m; 300 to 600 m; 6000 to 1000 m; 1000 to 4000 m; >4000 m
Exposure (C)	Are houses exposed by the fire	Yes; No
Fire size (O)	The final size of the simulated fire	0 to 20 ha: 20 to 150 ha: 150 to 1000 ha: 1000 to 5000 ha: 5000 to 10000 ha: >10000 ha
Fuel breaks (S)	Whether fuel breaks are present	Yes; No
Fuel load (S)	Landscape fuel load	High (2001); Low (2008)
Fuel moisture(S)	Live fuel moisture level at the start of the simulation	60%; 90%
Fuel type at ignition (GIS)	Broad classification of the type of vegetation at the point of ignition.	Grass; Shrub; Tree
Housing density (GIS)	Number of houses per hectare	0 to 26; 26 to 33.5; 33.5 to 117; 117 to 205; 205 to 300; >300
Slope (GIS)	Slope under the ignition point	0 degrees; 0 to 5 degrees; 5 to 15 degrees; 15 to 30 degrees; >30 degrees
Weather (S)	Predominant conditions during the simulated fire	Low; High; Santa Ana

(C) = calculated variable; (GIS) = GIS derived variable; (O) = output of the simulation model; (S) = variable set in the simulation model.

scenarios considered under Santa Ana conditions. Under low and high weather conditions, fuel load and fuel moisture had a strong influence of risk of exposure, whereby risk was greater when landscape fuels were high compared to low and when fuel moisture was 60% compared to 90% (Figure 4). These effects were lost under Santa Ana conditions, where fuel load and fuel moisture had very little effect on the risk of exposure. Fuel breaks had very little influence on the risk of exposure. Risk varied significantly across elevation with fires starting in low elevation sites having significantly higher risk than fires starting at higher elevation sites (Figure 4).

Exposure to fires was most sensitive to changes in the built environment, as well as fire parameters, i.e. fire size and distance travelled (Figure 5a). Distance to structures from the ignition had the strongest influence, followed by the distance travelled by a fire and the fire size. This is expected as parent nodes are likely to have the strongest influence on a node. Of the variables a greater distance from the terminal node, variables depicting the built environment (housing density and distance to road) were the next most influential variables, followed by weather. Variables describing the natural environment had only a modest influence on exposure to fires, with the fuel variables having no influence (Figure 5a). Distance travelled by a fire was primarily influenced by the weather on the day of the fire (Figure 5b). Natural environment variables and fuel load all had minor influence (< 2.1%), with all built environment variables having low influence (<1%) (Figure 5b).

Discussion

Consistent with previous research, fire size and distance travelled is most sensitive to changes in weather [18,44–48] and the risk of exposure is most strongly influenced by attributes of the fire (size and distance travelled) and the nature of the built environment [34,49]. Measured attributes of fuels had only a minor influence on fire parameters and risk of exposure. Fuel breaks in the National Forest did not affect fire size, distance travelled or the risk of exposure at the interface.

Risk of exposure

Weather had the strongest influence on fire size and distance travelled (Figure 5b) and indirectly, the risk of exposure. It has been well documented that fire weather strongly influences fire size, the rate of spread, spotting distance, fire intensity and severity [16,18,40,50–52]. As a result, wildfires burning under extreme conditions account for the majority of area burnt for many regions [44,53–55]. Furthermore, it is under extreme fire weather conditions wildfires pose the greatest threat to people and property [50,56–59].

Exposure was also influenced by the built environment, namely distance to road and housing density (Figure 5). Higher densities of properties occur at lower elevations (less than 300 m) and these have a higher risk of exposure compared with higher elevation sites independent of weather (Figure 4). Fires starting close to the interface are more likely to impact upon assets under any weather conditions [60,61], whereas fires starting considerable distances from property are only likely to impact on property under extreme weather conditions conducive to fire spread [34,57]. These results suggest that adequate planning of the changes in the built environment to cope with the expansion of human populations is going to be important for managing risk from fire [49,62,63]. We do note that the model only considers exposure to fire and we have not considered the size of the fire or the extent of the interface exposed to the fire. Fires that start away from populations will be significantly larger when they do impact on the interface compared with those that start nearby [60]. Larger fires would

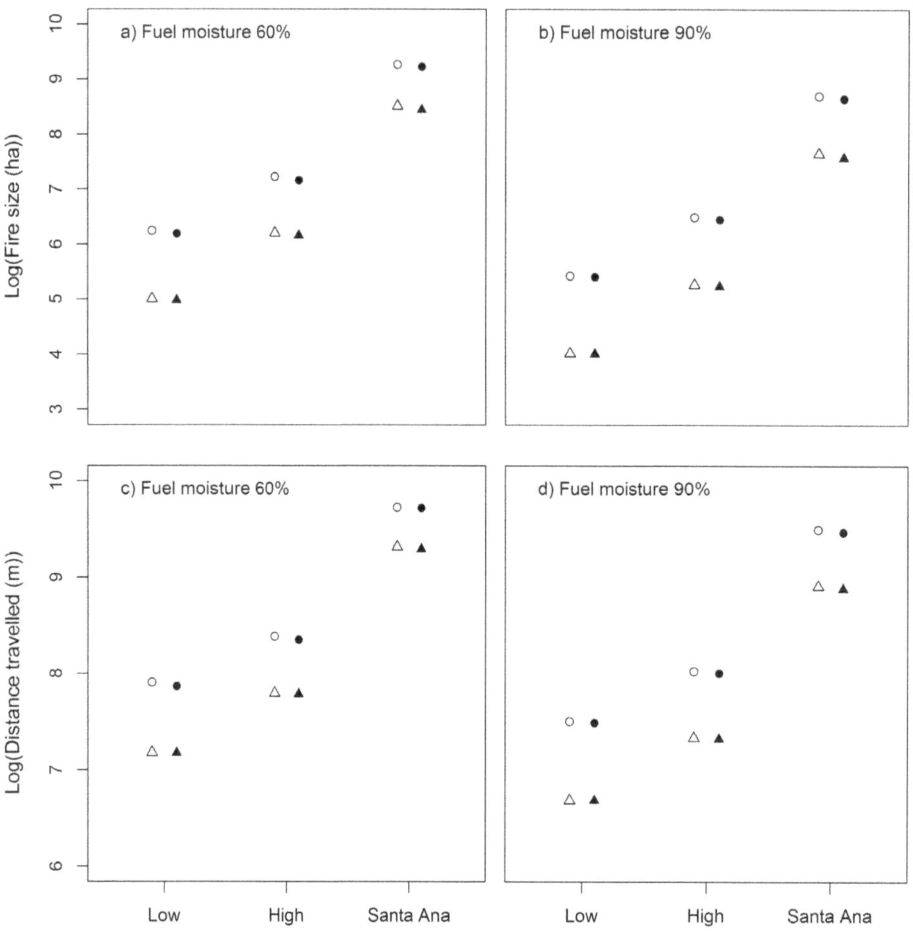

Figure 3. Relationships from the FARSITE simulation data between weather and fire size with fuel moisture of a) 60% and b) 90%, and distance travelled with fuel moisture of c) 60% and d) 90%. Open symbols are for simulations with no fuel breaks, closed symbols for simulations with fuel breaks. Circles represent a high landscape fuel load scenario and triangles represent a low landscape fuel load scenario. NB 95% confidence intervals were too small to depict in the graphics.

be expected to impact on a greater number of assets than smaller fires and these relationships require further investigation.

Of the factors relating to fuels, landscape fuel load had the strongest influence on fire size, distance travelled and risk of exposure (Figure 3; Figure 4). However, our model is more sensitive to the effects of weather and the built environment (Figure 5). Price et al. [64] found no effect of antecedent area burnt on the annual area burnt by wildfire in southern California coastal systems. The authors argue that the low effect of past fire is related to the low level of wildfire in the system (<2% per annum) and the rapid development of fuels (1–2 years). A large proportion of the study area (~22%) burnt in wildfires during 2007 and was consequently in the early stage of fuel development (i.e. 1 year old) in the 2008 fuel layer, which may explain the strong influence of fuel load under low and high fire weather. However, in San Diego County, the majority of annual area burnt occurs under extreme Santa Ana fire weather [55] where our model found no effect of fuel load. These results support the finding of Price et al. [64] that landscape fuel treatments in these systems are unlikely to reduce the risk of fire to assets. (Figure 3; Figure 4). Scenarios with 90% fuel moisture had significantly lower risk than those with 60% under low and high conditions, but not under Santa Ana (Figure 4). Live fuel moisture is related to fire activity in southern California, with large fires generally being associated with low levels (~60–80%) [65,66]. Early in the fire season live fuel moisture is generally greater than 90% (Keeley et al. 2009), the resulting area burnt in southern California typically is relatively small [65,66]. Our selection of 60% and 90% may not have truly captured the variable effect of live fuel moisture, particularly when these values exceed 100% and fire activity is expected to be low. However, the greatest risk to assets comes during Santa Ana weather conditions where there is no distinguishable effect of live fuel moisture, providing further support for our existing results. Regardless, it is typically lowest at the end of summer drought when Santa Ana winds and hence large fires are most likely [67].

Fuel breaks were ineffective at altering risk of exposure of property under any weather scenario in our study. Here we modelled fires assuming that all mapped fuel breaks in San Diego County were maintained (see Material S1; Figure 1), which exceeds current practice. Fuel breaks have been found to affect individual fire size and distance travelled [25,26]. The network of fuel breaks in San Diego County is highly clustered (Figure 1) presumably to protect particular assets from future wildfires, although fuel breaks continue to be constructed. Clustering of the fuel breaks will result in low encounter rates with wildfires that will result in a low efficacy of this management technique [24] when considering the landscape level risk. Here we assumed a random ignition model, however ignitions do not occur randomly across

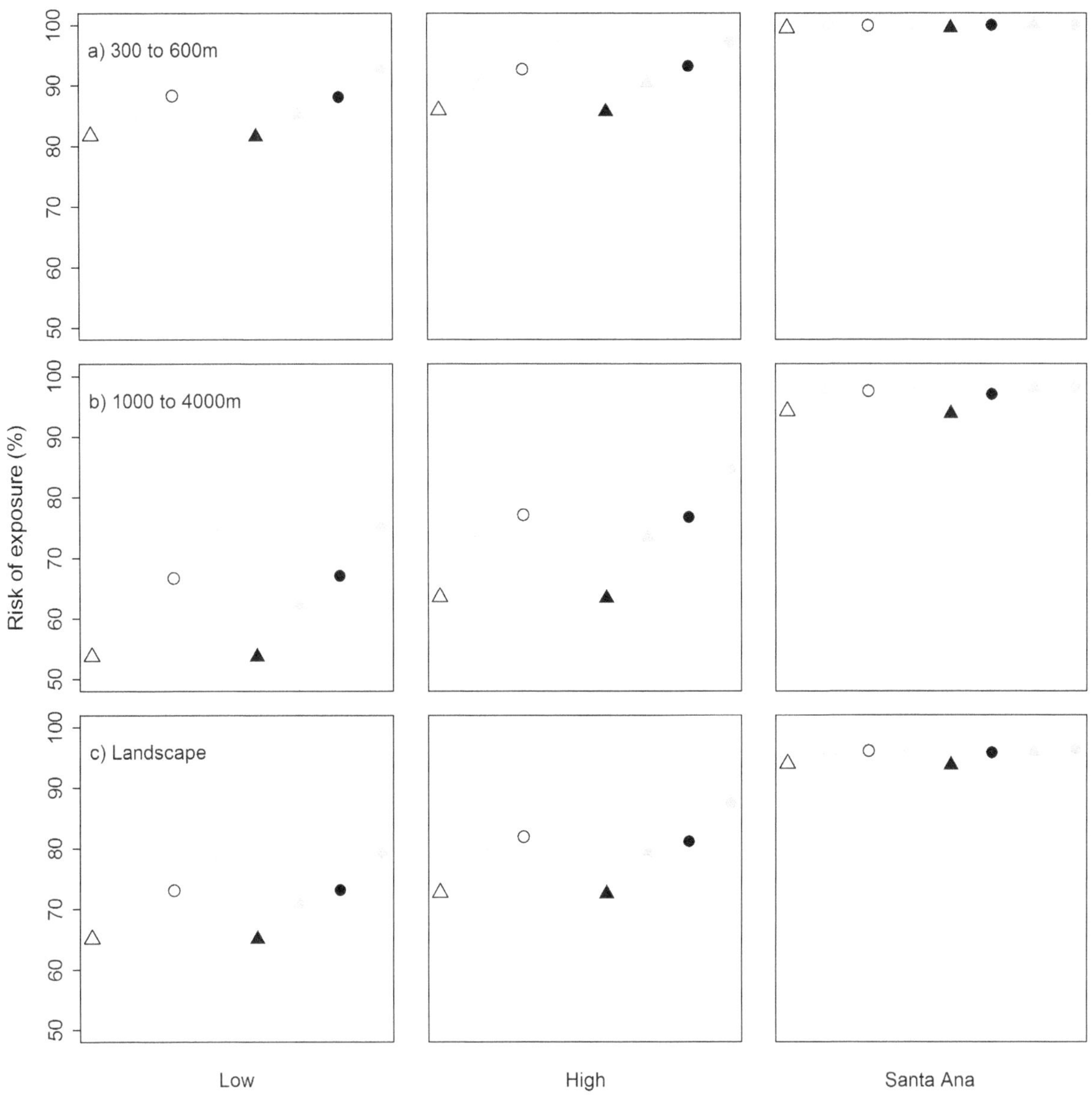

Figure 4. Risk of exposure for the 24 scenarios modelled for a) fires igniting at elevations of 300 to 600 m; b) fires igniting at elevations of 1000 to 4000 m; c) all locations across the landscape. Open symbols = fuel break scenarios; closed symbols = no fuel breaks; Grey symbols = fuel moisture of 60%; Black symbols = fuel moisture of 90%; Circles = high landscape fuel loads; Triangle = low landscape fuel loads.

landscapes [60,68–70] and tend to occur close to roads and development. This is important with regards to the result that fires that start closer to homes are most likely to reach those homes. Simulations have revealed that fire size and burn probability are sensitive to the use of random against non-random ignition locations, though these biases are minimised under extreme weather conditions [71] when the greatest risk of exposure occurs (Figure 4).

The effectiveness of fuel breaks is contingent on suppression resources and access [24]. In our study, the fuel breaks were constructed in FARSITE in a manner that simulated suppression along the fuel breaks. Fuel breaks tend to be constructed to allow for control of the fire flanks and not the head fire, i.e. the point of the greatest forward rate of spread. As a result, fuel breaks are unlikely to affect the distance travelled by a fire and have negligible impacts on total fire size. Our model did not model the interaction of suppression through direct attack of the fire front or indirect attack from other breaks in the landscape, e.g. roads and rivers. Inclusion of the suppression at these points may have altered the efficacy of fuel breaks when estimating risk. Similarly, we did not consider the impact on fires of igniting backburns from fuel breaks. However, as the severity of the fire weather increases the effectiveness of suppression actions are severely diminished [22,48,72]. Therefore, we would only consider fuel breaks in conjunction with suppression as having potential to further reduce risk under low fire weather and not under moderate fire weather or Santa Ana conditions [24,39].

Fire management

Management agencies seek to reduce risk to assets acknowledging that there are no practical means to remove the risk. Weather is the primary determinate of risk to assets from fire [34,55,57].

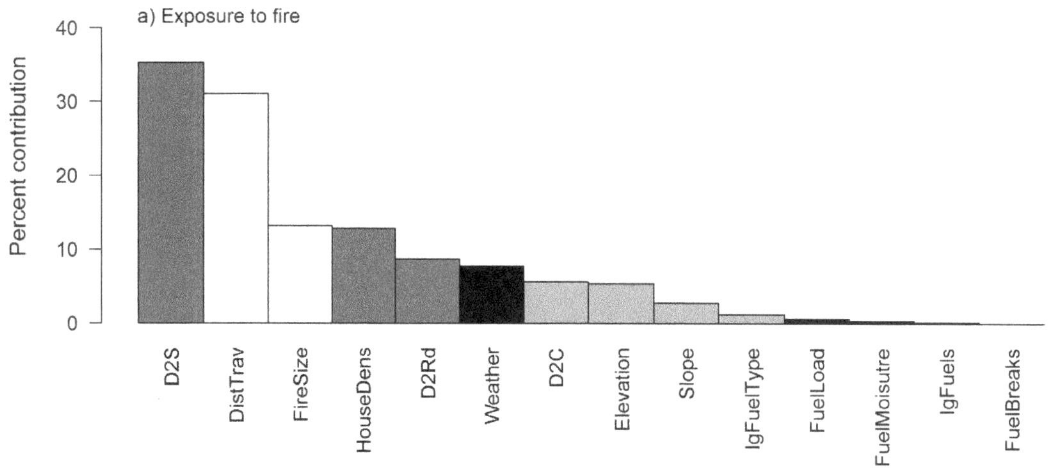

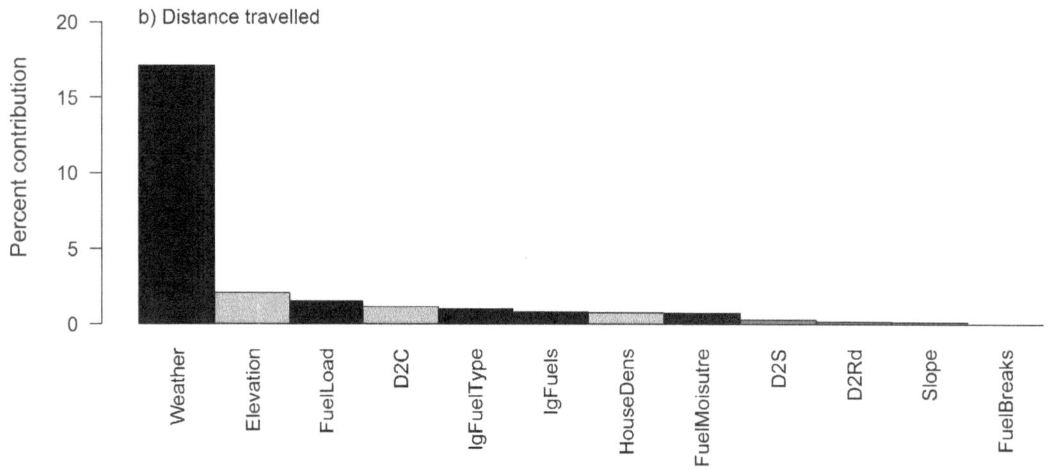

Figure 5. Sensitivity to findings for nodes a) Exposure to fire and b) Distance travelled. White bars = fire variables, dark grey bars = built environment variables; light grey bars = natural environment variables; Black bars = simulation model variables. D2S = distance to structure; DistTrav = distance travelled by the fire; HouseDens = housing density; D2Rd = Distance to road; D2C = distance to the coast; IgFuelType = fuel type at the point of ignition; IgFuels = fuel load at the point of ignition; FuelBreaks = presence of fuel breaks.

While management actions can be effective under relatively benign fire weather, understanding the effectiveness of management under extreme fire weather is fundamental to determining the extent to which management can reduce risk to people and property [48]. In our model, we considered the role of fuel treatments both fuel breaks for suppression and fuel treatments through examining the role of fuel loads. Neither of these was effective under extreme fire weather despite our model considering extreme levels of fuel treatment (>20% in 1 year old fuel) and fuel breaks (all mapped breaks in the county).

A range of other fire management approaches are available that were not tested here. Three broad management areas have the potential to significantly reduce risk to assets. Firstly, ignition management to reduce the occurrence of ignitions and subsequent fires will reduce the risk to assets [44]. Included in ignition management would be rapid response or initial attack [47,73], whereby resources are used to aggressively attempt to suppress fires before they become established fires. Secondly, improved urban planning policies to better develop the built environment to reduce the extent of the exposure [49,62,74]. This would include building in low risk areas outside Santa Ana wind corridors [35] and incorporating adequate offsets between vegetation and structures [74,75]. In southern California, the best urban planning practices would be to focus on infill-type development, as low to intermediate housing density, and isolated clusters of development are the strongest risk factors for a house being destroyed in a fire [63]. Finally, reduce the vulnerability of residents and properties at the urban interface. Residents can be educated to reduce the vulnerability of their property through adequate preparation [76,77]. Furthermore, properties can be built or retrofitted to appropriate construction standards to be more resilient to the impact of fire [78,79]. While each of these is likely to reduce risk, only through an expanded analysis of these approaches across all weather scenarios will it be possible to identify an optimal management strategy.

Conclusion

Weather determines the risk of exposure for assets in the landscape. Under extreme weather, where the risk of fire is greatest, landscape fuel treatments are unlikely to have a significant influence on risk. These results suggest that managing the occurrence of fire and the spatial distribution of the built environment across the landscape is likely to be the best way to alter the risk profile. Further research is needed to examine the cost trade-offs of each of these approaches.

Supporting Information

Material S1 Supplementary text outlines the modelling process in farsite. Table S1, Fuel moisture conditions used in the simulations. Dead fuel moisture values are from Scott and Burgin (2005). See text for description of LFM categories.

Acknowledgments

Data for the study was provided by U.S. Geological Survey. The paper was made possible due to the workshop participants - Teresa Brennan, Will Forney, CJ Fotheringham, Rick Halsey, Bill Labiosa, Ben Landis, Owen Price, Robert Taylor and Marti Winter. Stuart Brittain and Mark Finney kindly provided a command line version of FARSITE and assistance with its use. Robert Taylor provided advice on manipulating input layers for FARSITE to simulate spotting in shrubland fuels.

Author Contributions

Conceived and designed the experiments: TP LC AS JK RB. Performed the experiments: LC TP. Analyzed the data: TP LC AS. Contributed to the writing of the manuscript: TP LC AS JK RB. Designed the specifics of the computer simulation study: TP LC AS.

References

1. Cohen JD (2000) Preventing Disaster: Home Ignitability in the Wildland-Urban Interface. Journal of Forestry 98: 15–21.
2. Lampin-Maillet C, Jappiot M, Long M, Bouillon C, Morge D, et al. (2009) Mapping wildland-urban interfaces at large scales integrating housing density and vegetation aggregation for fire prevention in the South of France. Journal of Environmental Management 91: 732–741.
3. Gill AM, Stephens SL (2009) Scientific and social challenges for the management of fire-prone wildland–urban interfaces. Environmental Research Letters 4: 034014.
4. Gill AM, Stephens SL, Cary GJ (2013) The worldwide "wildfire" problem. Ecological Applications 23: 438–454.
5. McCaffrey S, Rhodes A (2009) Public Response to Wildfire: Is the Australian"Stay and Defend or Leave Early" Approach an Option for Wildfire Management in the United States? Journal of Forestry 107: 9–15.
6. Leonard J (2009) Report to the 2009 Victorian Bushfires Royal Commission Building performacnce in Bushfires. Highett, Victoria, Australia: CSIRO Sustainable Ecosystems. 74 p.
7. Boustras G, Boukas N, Katsaros E, Ziliaskopoulos A (2012) Wildland fire preparedness is Greece and Cyprus: Lessons learned from the catastrophic fires of 2007 and beyond. In: Paton D, Tedim F, editors. Wildfire and community: Facilitating preparedness and resilience. Springfield, Illinois, USA: Charles C Thomas Publisher Ltd. pp. 151–168.
8. Ganewatta G (2008) The economics of bushfire management. In: Handmer J, Haynes K, editors. Community Bushfire Safety. Collingwood: CSIRO Publishing. pp. 151–159.
9. McFarlane AC, Clayer JR, Bookless CL (1997) Psychiatric morbidity following a natural disaster: An Australian bushfire. Social Psychiatry and Psychiatric Epidemiology 32: 261–268.
10. Langley A, Jones R (2005) Coping Efforts and Efficacy, Acculturation, and Post-Traumatic Symptomatology in Adolescents Following Wildfire. Fire Technology 41: 125–143.
11. Papadatou D, Giannopoulou I, Bitsakou P, Bellali T, Talias MA, et al. (2012) Adolescents' reactions after a wildfire disaster in Greece. Journal of Traumatic Stress 25: 57–63.
12. Schoennagel T, Nelson CR, Theobald DM, Carnwath GC, Chapman TB (2009) Implementation of National Fire Plan treatments near the wildland-urban interface in the western United States. Proceedings of the National Academy of Sciences of the United States of America 106: 10706–10711.
13. Stephens SL, Ruth LW (2005) Federal forest-fire policy in the United States. Ecological Applications 15: 532–542.
14. Fernandes PM, Botelho HS (2003) A review of prescribed burning effectiveness in fire hazard reduction. International Journal of Wildland Fire 12: 117–128.
15. Penman TD, Christie FJ, Andersen AN, Bradstock RA, Cary GJ, et al. (2011) Prescribed burning: How can it work to conserve the things we value? International Journal of Wildland Fire 20: 721–733.
16. Noble IR, Bary GAV, Gill AM (1980) Mcarthurs Fire Danger Meters Expressed as Equations. Australian Journal of Ecology 5: 201–204.
17. Price OF, Bradstock RA (2010) The effect of fuel age on the spread of fire in sclerophyll forest in the Sydney region of Australia. International Journal of Wildland Fire 19: 35–45.
18. Bradstock R, Hammill K, Collins L, Price O (2010) Effects of weather, fuel and terrain on fire severity in topographically diverse landscapes of south-eastern Australia. Landscape Ecology 25: 607–619.
19. Fulé PZ, Crouse JE, Roccaforte JP, Kalies EL (2012) Do thinning and/or burning treatments in western USA ponderosa or Jeffrey pine-dominated forests help restore natural fire behavior? Forest Ecology and Management 269: 68–81.
20. Collins L, Bradstock RA, Penman TD (2013) Can precipitation influence landscape controls on wildfire severity? A case study within temperate eucalypt forests of south-eastern Australia. International Journal of Wildland Fire Accepted April 3rd 2013.
21. Hirsch KG, Podur JJ, Janser RF, McAlpine RS, Martell DL (2004) Productivity of Ontario initial-attack fire crews: results of an expert-judgement elicitation study. Canadian Journal of Forest Research 34: 705–715.
22. Plucinski MP (2012) Factors Affecting Containment Area and Time of Australian Forest Fires Featuring Aerial Suppression. Forest Science 58: 390–398.
23. Syphard AD, Keeley JE, Brennan TJ (2011) Comparing the role of fuel breaks across southern California national forests. Forest Ecology and Management 261: 2038–2048.
24. Syphard AD, Keeley JE, Brennan TJ (2011) Factors affecting fuel break effectiveness in the control of large fires on the Los Padres National Forest, California. International Journal of Wildland Fire 20: 764–775.
25. Agee JK, Bahro B, Finney MA, Omi PN, Sapsis DB, et al. (2000) The use of shaded fuelbreaks in landscape fire management. Forest Ecology and Management 127: 55–66.
26. Finney MA (2001) Design of Regular Landscape Fuel Treatment Patterns for Modifying Fire Growth and Behavior. Forest Science 47: 219–228.
27. Syphard AD, Scheller RM, Ward BC, Spencer WD, Strittholt JR (2011) Simulating landscape-scale effects of fuels treatments in the Sierra Nevada, California, USA. International Journal of Wildland Fire 20: 364–383.
28. Keeley JE, Bond WJ, Bradstock RA, Pausas, editors (2012) Fire in Mediterranean Ecosystems: Ecology, Evolution and Management. Cambridge: Cambridge University Press.
29. Marcot BG, Holthausen RS, Raphael MG, Rowland MM, Wisdom MJ (2001) Using Bayesian belief networks to evaluate fish and wildlife population viability under land management alternatives from an environmental impact statement. Forest Ecology and Management 153: 29–42.
30. Johnson S, Mengersen K, de Waal A, Marnewick K, Cilliers D, et al. (2010) Modelling cheetah relocation success in southern Africa using an Iterative Bayesian Network Development Cycle. Ecological Modelling 221: 641–651.
31. Penman TD, Price O, Bradstock RA (2011) Bayes Nets as a method for analysing the influence of management actions in fire planning. International Journal of Wildland Fire 20: 909–920.
32. Pearl J (1986) Fusion, propagation, and structuring in belief networks. Artificial Intelligence 29: 241–288.
33. Korb KB, Nicholson AE (2011) Bayesian Artificial Intelligence. Computer Science and Data Analysis. Second Edition. Boca Raton: CRC/Chapman Hall.
34. Penman TD, Bradstock RA, Price OF (2014) Reducing wildfire risk to urban developments: Simulation of cost-effective fuel treatment solutions in south eastern Australia. Environmental Modelling & Software 52: 166–175.
35. Syphard AD, Keeley JE, Massada AB, Brennan TJ, Radeloff VC (2012) Housing Arrangement and Location Determine the Likelihood of Housing Loss Due to Wildfire. PLoS ONE 7: e33954.
36. Hammer RB, Radeloff VC, Fried JS, Stewart SI (2007) Wildland-urban interface housing growth during the 1990 s in California, Oregon, and Washington. International Journal of Wildland Fire 16: 255–265.
37. Keeley JE, Syphard AD, Fotheringham CJ (2013) The 2003 and 2007 Wildfires in Southern California. In: Boulter S, Palutikof J, Karoly D, Guitart D, editors. Natural Disasters and Adaptation to Climate Change. Oxford: Cambridge University Press. pp. 42–52.
38. Keeley JE, Fotheringham CJ (2003) Impact of past, present, and future fire regimes on North American mediterranean shrublands. In: Veblen TT, Baker WL, Montenegro G, Swetnam TW, editors. Fire and climatic change in temperate ecosystems of the Western Americas. New York: Springer. pp. 218–262.
39. Keeley JE, Safford H, Fotheringham CJ, Franklin J, Moritz M (2009) The 2007 Southern California Wildfires: Lessons in Complexity. Journal of Forestry 107: 287–296.
40. Moritz MA, Moody TJ, Krawchuk MA, Hughes M, Hall A (2010) Spatial variation in extreme winds predicts large wildfire locations in chaparral ecosystems. Geophysical Research Letters 37: L04801.

41. Marcot BG, Steventon JD, Sutherland GD, McCann RK (2006) Guidelines for developing and updating Bayesian belief networs applied to ecological modeling and conservation. Canadian Journal of Forest Research 36: 3063–3074.
42. Chen SH, Pollino CA (2012) Good practice in Bayesian network modelling. Environmental Modelling & Software 37: 134–145.
43. Finney MA (1998) FARSITE: Fire Area Simulator-Model Development and Evaluation. United States Department of Agriculture Forest Service.
44. Cary GJ, Flannigan MD, Keane RE, Bradstock RA, Davies ID, et al. (2009) Relative importance of fuel management, ignition management and weather for area burned: Evidence from five landscape-fire-succession models. International Journal of Wildland Fire 18: 147–156.
45. Archibald S, Roy DP, Van Wilgen BW, Scholes RJ (2009) What limits fire? An examination of drivers of burnt area in Southern Africa. Global Change Biology 15: 613–630.
46. Bessie WC, Johnson EA (1995) The Relative Importance of Fuels and Weather on Fire Behavior in Subalpine Forests. Ecology 76: 747–762.
47. Arienti MC, Cumming SG, Boutin S (2006) Empirical models of forest fire initial attack success probabilities: the effects of fuels, anthropogenic linear features, fire weather, and management. Canadian Journal of Forest Research 36: 3155–3166.
48. Penman TD, Collins L, Price OF, Bradstock RA, Metcalf S, et al. (2013) Examining the relative effects of fire weather, suppression and fuel treatment on fire behaviour – A simulation study. Journal of Environmental Management 131: 325–333.
49. Syphard AD, Keeley JE, Bar Massada A, Brennan TJ, Radeloff VC (2012) Housing Arrangement and Location Determine the Likelihood of Housing Loss Due to Wildfire. PLoS ONE 7.
50. Price OF, Bradstock RA (2012) The efficacy of fuel treatment in mitigating property loss during wildfires: Insights from analysis of the severity of the catastrophic fires in 2009 in Victoria, Australia. Journal of Environmental Management 113: 146–157.
51. Collins BM, Kelly M, van Wagtendonk JW, Stephens SL (2007) Spatial patterns of large natural fires in Sierra Nevada wilderness areas. Landscape Ecology 22: 545–557.
52. Thompson JR, Spies TA (2009) Vegetation and weather explain variation in crown damage within a large mixed-severity wildfire. Forest Ecology and Management 258: 1684–1694.
53. Bradstock RA, Cohn JS, Gill AM, Bedward M, Lucas C (2009) Prediction of the probability of large fires in the Sydney region of south-eastern Australia using fire weather. International Journal of Wildland Fire 18: 932–943.
54. Minnich R, Chou Y (1997) Wildland Fire Patch Dynamics in the Chaparral of Southern California and Northern Baja California. International Journal of Wildland Fire 7: 221–248.
55. Keeley JE, Fotheringham CJ (2001) Historic Fire Regime in Southern California Shrublands. Conservation Biology 15: 1536–1548.
56. Blanchi R, Lucas C, Leonard J, Finkele K (2010) Meteorological conditions and wildfire-related houseloss in Australia. International Journal of Wildland Fire 19: 914–926.
57. Bradstock RA, Gill AM, Kenny BJ, Scott J (1998) Bushfire risk at the urban interface estimated from historical weather records: Consequences for the use of prescribed fire in the Sydney region of south-eastern Australia. Journal of Environmental Management 52: 259–271.
58. Keeley JE, Fotheringham CJ, Moritz MA (2004) Lessons from the October 2003 Wildfires in Southern California. Journal of Forestry Research (Harbin) 102: 26–31.
59. Bar Massada A, Radeloff VC, Stewart SI, Hawbaker TJ (2009) Wildfire risk in the wildland-urban interface: A simulation study in northwestern Wisconsin. Forest Ecology and Management 258: 1990–1999.
60. Syphard AD, Radeloff VC, Keuler NS, Taylor RS, Hawbaker TJ, et al. (2008) Predicting spatial patterns of fire on a southern California landscape. International Journal of Wildland Fire 17: 602–613.
61. Price OF, Bradstock RA (2013) The spatial domain of wildfire risk and response in the Wildland Urban Interface in Sydney, Australia. Natural Hazards and Environmental Management 1: 4539–4564.
62. Syphard AD, Radeloff VC, Keeley JE, Hawbaker TJ, Clayton MK, et al. (2007) Human influence on California fire regimes. Ecological Applications 17: 1388–1402.
63. Syphard AD, Bar Massada A, Butsic V, Keeley JE (2013) Land Use Planning and Wildfire: Development Policies Influence Future Probability of Housing Loss. PLoS ONE 8: e71708.
64. Price OF, Bradstock RA, Keeley JE, Syphard AD (2012) The impact of antecedent fire area on burned area in southern California coastal ecosystems. Journal of Environmental Management 113: 301–307.
65. Dennison PE, Moritz MA (2009) Critical live fuel moisture in chaparral ecosystems: a threshold for fire activity and its relationship to antecedent precipitation. International Journal of Wildland Fire 18: 1021–1027.
66. Dennison PE, Moritz MA, Taylor RS (2008) Evaluating predictive models of critical live fuel mois-ture in the Santa Monica Mountains, California. International Journal of Wildland Fire 17: 18–27.
67. Keeley JE, Zedler PH (2009) Large, high intensity fire events in southern California shrublands: debunking the fine-grained age-patch model. Ecological Applications 19: 69–94.
68. Penman TD, Bradstock RA, Price O (2013) Modelling the determinants of ignition in the Sydney Basin, Australia: implications for future management. International Journal of Wildland Fire 22: 469–478.
69. Hering A, Bell C, Genton M (2009) Modeling spatio-temporal wildfire ignition point patterns. Environmental and Ecological Statistics 16: 225–250.
70. Diaz-Avalos C, Peterson DL, Alvarado E, Ferguson SA, Besag JE (2001) Space-time modelling of lightning-caused ignitions in the Blue Mountains, Oregon. Canadian Journal of Forest Research-Revue Canadienne De Recherche Forestiere 31: 1579–1593.
71. Bar Massada A, Syphard AD, Hawbaker TJ, Stewart SI, Radeloff VC (2011) Effects of ignition location models on the burn patterns of simulated wildfires. Environmental Modelling & Software 26: 583–592.
72. Hirsch KG, Corey PN, Martell DL (1998) Using expert judgement to model initial attack fire crew effectiveness. Forest Science 44: 539–549.
73. Plucinski MP, McCarthy GJ, Hollis JJ, Gould JS (2012) The effect of aerial suppression on the containment time of Australian wildfires estimated by fire management personnel. International Journal of Wildland Fire 21: 219–229.
74. Gibbons P, van Bommel L, Gill AM, Cary GJ, Driscoll DA, et al. (2012) Land Management Practices Associated with House Loss in Wildfires. PLoS ONE 7: e29212.
75. Cohen JD (2004) Relating flame radiation to home ignition using modeling and experimental crown fires. Canadian Journal of Forest Research 34: 1616–1626.
76. Penman TD, Eriksen C, Blanchi R, Chladil M, Gill AM, et al. (2013) Defining adequate means of residents to prepare property for protection from wildfire. International Journal of Disaster Risk Reduction 6: 67–77.
77. McLennan J, Elliott G, Omodei M, Whittaker J (2013) Householders' safety-related decisions, plans, actions and outcomes during the 7 February 2009 Victorian (Australia) wildfires. Fire Safety Journal 61: 175–184.
78. Standards Australia (2009) A.S.3959–2009: Construction of buildings in bushfire prone areas. Sydney, Australia: SAI Global Limited. pp. 129.
79. CBC (2007) California Build Code Materials and construction methods for exterior wildfire exposure Title 24, part 2, volume 1 of 2. Sacramento, CA.

PERMISSIONS

All chapters in this book were first published in PLOS ONE, by The Public Library of Science; hereby published with permission under the Creative Commons Attribution License or equivalent. Every chapter published in this book has been scrutinized by our experts. Their significance has been extensively debated. The topics covered herein carry significant findings which will fuel the growth of the discipline. They may even be implemented as practical applications or may be referred to as a beginning point for another development.

The contributors of this book come from diverse backgrounds, making this book a truly international effort. This book will bring forth new frontiers with its revolutionizing research information and detailed analysis of the nascent developments around the world.

We would like to thank all the contributing authors for lending their expertise to make the book truly unique. They have played a crucial role in the development of this book. Without their invaluable contributions this book wouldn't have been possible. They have made vital efforts to compile up to date information on the varied aspects of this subject to make this book a valuable addition to the collection of many professionals and students.

This book was conceptualized with the vision of imparting up-to-date information and advanced data in this field. To ensure the same, a matchless editorial board was set up. Every individual on the board went through rigorous rounds of assessment to prove their worth. After which they invested a large part of their time researching and compiling the most relevant data for our readers.

The editorial board has been involved in producing this book since its inception. They have spent rigorous hours researching and exploring the diverse topics which have resulted in the successful publishing of this book. They have passed on their knowledge of decades through this book. To expedite this challenging task, the publisher supported the team at every step. A small team of assistant editors was also appointed to further simplify the editing procedure and attain best results for the readers.

Apart from the editorial board, the designing team has also invested a significant amount of their time in understanding the subject and creating the most relevant covers. They scrutinized every image to scout for the most suitable representation of the subject and create an appropriate cover for the book.

The publishing team has been an ardent support to the editorial, designing and production team. Their endless efforts to recruit the best for this project, has resulted in the accomplishment of this book. They are a veteran in the field of academics and their pool of knowledge is as vast as their experience in printing. Their expertise and guidance has proved useful at every step. Their uncompromising quality standards have made this book an exceptional effort. Their encouragement from time to time has been an inspiration for everyone.

The publisher and the editorial board hope that this book will prove to be a valuable piece of knowledge for researchers, students, practitioners and scholars across the globe.

LIST OF CONTRIBUTORS

Lisa A. Eby and Olga Helmy
University of Montana, Wildlife Biology Program and Department of Ecosystem and Conservation Sciences, College of Forestry and Conservation, Missoula, Montana, United States of America

Lisa M. Holsinger
U.S. Department of Agriculture Forest Service, Rocky Mountain Research Station, Fire Sciences Laboratory, Missoula, Montana, United States of America

Michael K. Young
U.S. Department of Agriculture Forest Service, Rocky Mountain Research Station, Forestry Sciences Laboratory, Missoula, Montana, United States of America

David L. A. Gaveau, Elis Molidena, Husna Yaen and Robert Nasi
Center for International Forestry Research, Bogor, Indonesia

Sean Sloan
Centre for Tropical Environmental and Sustainability Science, School of Marine and Tropical Biology, James Cook University, Cairns, QLD, Australia

Doug Sheil
Center for International Forestry Research, Bogor, Indonesia

Department of Ecology and Natural Resource Management (INA), Norwegian University of Life Science (NMBU), Ås Norway

Nicola K. Abram
Durrell Institute of Conservation and Ecology, University of Kent, Canterbury, Kent, United Kingdom
Borneo Futures project, People and Nature Consulting International, Ciputat, Jakarta, Indonesia

Marc Ancrenaz
Borneo Futures project, People and Nature Consulting International, Ciputat, Jakarta, Indonesia
Sabah Wildlife Department, Kota Kinabalu, Sabah, Malaysia

HUTAN, Kinabatangan Orang-utan Conservation Programme, Kota Kinabalu, Sabah, Malaysia
North England Zoological Society, Chester Zoo, Chester, United Kingdom

Marcela Quinones and Niels Wielaard
SarVision, Wageningen, The Netherlands

David B. McWethy
Department of Earth Sciences, Montana State University, Bozeman, Montana, United States of America

Erik Meijaard
Center for International Forestry Research, Bogor, Indonesia
Borneo Futures project, People and Nature Consulting International, Ciputat, Jakarta, Indonesia
School of Biological Sciences, University of Queensland, Brisbane, QLD, Australia

Abbey Rosso and Peter Neitlich
National Park Service, Winthrop, Washington, United States of America

Robert J. Smith
Department of Botany and Plant Pathology, Oregon State University, Corvallis, Oregon, United States of America

David D. Gustine and Layne G. Adams
U. S. Geological Survey, Alaska Science Center, Anchorage, Alaska, United States of America

Todd J. Brinkman, Michael A. Lindgren, Jennifer I. Schmidt and T. Scott Rupp
Scenarios Network for Alaska and Arctic Planning, University of Alaska Fairbanks, Fairbanks, Alaska, United States of America

Heather Keith, David B. Lindenmayer, David Blair, Lauren Carter, Lachlan McBurney and Sachiko Okada
The Fenner School of Environment and Society, Australian National University, Building 48, Canberra, ACT, Australia

Brendan G. Mackey
Griffith Climate Change Response Program, Griffith University, Queensland, Australia

Tomoko Konishi-Nagano
Fujitsu Laboratories Ltd., Kawasaki, Japan

Jennifer L. Schafer and Michael G. Just
Department of Plant and Microbial Biology, North Carolina State University, Raleigh, North Carolina, United States of America

António M. Lopes
Institute of Mechanical Engineering, Faculty of Engineering, University of Porto, Porto, Portugal

J. A. Tenreiro Machado
Institute of Engineering, Polytechnic of Porto, Porto, Portugal

Jamie R. Wood and Matt S. McGlone
Landcare Research, Lincoln, New Zealand

John M. Mola
Department of Biological Sciences, Humboldt State University, Arcata, California, United States of America
Graduate Group in Ecology, University of California Davis, Davis, California, United States of America

J. Morgan Varner
Department of Forestry, Forest & Wildlife Research Center, Mississippi State University, Mississippi State, Mississippi, United States of America

Erik S. Jules
Department of Biological Sciences, Humboldt State University, Arcata, California, United States of America

Tova Spector
Florida Park Service, Department of Environmental Protection, Panama City, Florida, United States of America

Kingsley O. Odigie and A. Russell Flegal
WIGS Laboratory, Environmental Toxicology, University of California Santa Cruz, Santa Cruz, California, United States of America

Rosemary L. Sherriff
Department of Geography, Humboldt State University, Arcata, California, United States of America

Rutherford V. Platt
Department of Environmental Studies, Gettysburg College, Gettysburg, Pennsylvania, United States of America

Thomas T. Veblen, Tania L. Schoennagel and Meredith H. Gartner
Department of Geography, University of Colorado, Boulder, Colorado, United States of America

Alec M. Kretchun, Robert M. Scheller and Melissa S. Lucash
Portland State University, Department of Environmental Science and Management, Portland, Oregon, United States of America

Kenneth L. Clark
Silas Little Experimental Forest, USDA Forest Service, New Lisbon, New Jersey, United States of America

John Hom and Steve Van Tuyl
United States Department of Agriculture Forest Service, Newtown Square, Pennsylvania, United States of America

Janet M. Wilmshurst
Landcare Research, Lincoln, New Zealand
School of Environment, University of Auckland, Auckland, New Zealand

Cathy Whitlock
Department of Earth Sciences, Montana State University, Bozeman, Montana, United States of America
Institute on Ecosystems, Montana State University, Bozeman, Montana, United States of America

Marcos D. Robles, Robert M. Marshall and Jeanmarie A. Haney
The Nature Conservancy Center for Science and Public Policy, Tucson, Arizona, United States of America

Frances O'Donnell
Northern Arizona University, Flagstaff, Arizona, United States of America

Edward B. Smith
The Nature Conservancy in California, Sacramento, California, United States of America

David F. Gori
The Nature Conservancy in New Mexico, Santa Fe, New Mexico, United States of America

Catherine J. Payne, Euan G. Ritchie and Dale G. Nimmo
Centre for Integrative Ecology, School of Life and Environmental Sciences, Deakin University, Melbourne, Victoria, Australia

Luke T. Kelly
Australian Research Council Centre of Excellence for Environmental Decisions, School of Botany, University of Melbourne, Melbourne, Victoria, Australia

Hugh W. McGregor
School of Biological Sciences, University of Tasmania, Hobart, Tasmania, Australia
Australian Wildlife Conservancy, Derby, Western Australia, Australia

Sarah Legge
Australian Wildlife Conservancy, Mornington Wildlife Sanctuary, Derby, Western Australia, Australia

Menna E. Jones and Christopher N. Johnson
School of Biological Sciences, University of Tasmania, Hobart, Tasmania, Australia

Sean A. Parks and Carol Miller
Aldo Leopold Wilderness Research Institute, Rocky Mountain Research Station, USDA Forest Service, Missoula, Montana, United States of America

Marc-André Parisien
Northern Forestry Centre, Canadian Forest Service, Natural Resources Canada, Edmonton, Alberta, Canada

Solomon Z. Dobrowski
Department of Forest Management, College of Forestry and Conservation, University of Montana, Missoula, Montana, United States of America

Christopher T. Rota, Joshua J. Millspaugh and Dylan C. Kesler
Department of Fisheries and Wildlife Sciences, University of Missouri, Columbia, Missouri, United States of America

Mark A. Rumble
Forest and Grassland Research Laboratory, United States Forest Service Rocky Mountain Research Station, Rapid City, South Dakota, United States of America

Chad P. Lehman
Custer State Park, South Dakota Department of Game, Fish, and Parks, Custer, South Dakota, United States of America

Alfredo C. Pereira Júnior
Divisão de Sensoriamento Remoto, Instituto Nacional de Pesquisas Espaciais, São José dos Campos, Brazil

Sofia L. J. Oliveira and José M. C. Pereira
Centro de Estudos Florestais, Instituto Superior de Agronomia, Universidade de Lisboa, Lisboa, Portugal

Maria Antónia Amaral Turkman
Centro de Estatística e Aplicações, Faculdade de Cieˆ ncias, Universidade de Lisboa, Lisboa, Portugal

Elizabeth R. Gleim
Warnell School of Forestry and Natural Resources, University of Georgia, Athens, Georgia, United States of America
Southeastern Cooperative Wildlife Disease Study, College of Veterinary Medicine, University of Georgia, Athens, Georgia, United States of America
Wildlife Laboratory, Joseph W. Jones Ecological Research Center at Ichauway, Newton, Georgia, United States of America

Luke Collins
Hawkesbury Institute for the Environment, University of Western Sydney, Penrith, NSW, Australia

L. Mike Conner
Wildlife Laboratory, Joseph W. Jones Ecological Research Center at Ichauway, Newton, Georgia, United States of America

Roy D. Berghaus
Department of Population Health, College of Veterinary Medicine, University of Georgia, Athens, Georgia, United States of America

Michael L. Levin and Galina E. Zemtsova
Centers for Disease Control and Prevention, Rickettsial Zoonoses Branch, Atlanta, Georgia, United States of America

Michael J. Yabsley
Warnell School of Forestry and Natural Resources, University of Georgia, Athens, Georgia, United States of America
Southeastern Cooperative Wildlife Disease Study, College of Veterinary Medicine, University of Georgia, Athens, Georgia, United States of America

Adrián Regos, Núria Aquilué, Miquel De Cáceres and Lluís Brotons
CEMFOR-CTFC (Centre Tecnològic Forestal de Catalunya), Solsona, Spain
CREAF (Centre de Recerca Ecològica i Aplicacions Forestals), Bellaterra, Spain

Javier Retana
CREAF (Centre de Recerca Ecològica i Aplicacions Forestals), Bellaterra, Spain
UniversitatAutònoma Barcelona, Bellaterra, Spain

Trent D. Penman and Ross A. Bradstock
Centre for Environmental Risk Management of Bushfires, Institute of Conservation Biology and Environmental Management, University of Wollongong, NSW, Australia

Alexandra D. Syphard
Conservation Biology Institute, La Mesa, California, United States of America

Jon E. Keeley
U.S. Geological Survey, Western Ecological Research Center, Sequoia-Kings Canyon Field Station, Three Rivers, California, United States of America
Department of Ecology and Evolutionary Biology, University of California Los Angeles, Los Angeles, California, United States of America

Index